THE ENCYCLOPÆDIA OF NEW AND REDISCOVERED ANIMALS

The okapi and the Vu Quang ox. (William M. Rebsamen)

Komodo dragons. (Pius Lee)

Irwin's turtle, Elseya irwini, described in 1997. (Ian Sutton)

THE ENCYCLOPÆDIA OF NEW AND REDISCOVERED ANIMALS

From the Lost Ark to the New Zoo—and Beyond

DR. KARL P.N. SHUKER

COACHWHIP PUBLICATIONS
LANDISVILLE, PENNSYLVANIA
2012

DEDICATION

To the Harry Johnstons, Richard Meinertzhagens, Hans Schomburgks, James Chapins, Marjorie Courtenay-Latimers, Geoffrey Orbells, Kitti Thonglongyas, Ralph Wetzels, Robert Ballards, Alain Delcourts, Salim Alis, Tim Flannerys, and Marc van Roosmalens of the future, and to all of those spectacular new and rediscovered animals that they will assuredly find; but most especially to Dr. Reinhardt Møbjerg Kristensen, to whom the entire zoological world owes an immense, unique debt of gratitude for bringing to attention at least two totally new *phyla* of animals, the loriciferans and the cycliophorans—as well as a third remarkable taxon that might well constitute a third new phylum!

THE ENCYCLOPAEDIA OF NEW AND REDISCOVERED ANIMALS
ISBN-13 978-1-61646-130-0
Coachwhip Publications (Landisville, Pennsylvania)
© 2012 Karl P.N. Shuker
All Rights Reserved.

Front and back cover images by William M. Rebsamen.

All images within this publication are public domain, used with permission, or licensed (including Creative Commons licensing, 2.0 or 3.0 with attribution).

QL706.8 S59 2012   591.68

CONTENTS

Foreword to the Third Edition: by Loren Coleman	7
Foreword to the Second Edition: by Dr. Lee Durrell	9
Foreword to the First Edition: by Gerald Durrell	11
Acknowledgments	13
Introduction: Arrivals and Revivals	15
Part 1: New and Rediscovered Animals of the 20th Century	21
The Mammals	23
The Birds	115
The Reptiles and Amphibians	161
The Fishes	191
The Invertebrates	221
Part 2: New and Rediscovered Animals of the 21st Century	271
The Future	311
A Last Word	313
Appendix: The Scientific Classification of Animals	315
Bibliography	319
Index	353
Stop Press	363
Author Biography	367

Pygmy hippopotamus. (Craig Dingle)

Emperor tamarins. (West Chester Dumonts)

FOREWORD TO THE THIRD EDITION

The remarkable life journey of Karl Shuker is nicely captioned within and parallels the recent history of cryptozoology, the study of hidden or unknown animal species, as yet unidentified by Western Science.

"Cryptozoology," as a word, issued from the tradition of animal collectors, field workers, researchers, zoologists, naturalists, and science writers who would detail, before the coining of this term, what was referred to as "romantic zoology."

The Scottish naturalist Ivan T. Sanderson first coined the word "cryptozoology" in the 1930s and 1940s. After moving to America in the late 1940s, Sanderson would write the earliest articles on cryptozoology in *Saturday Evening Post*, about tropical dinosaur-like animals and living aquatic monsters worldwide.

The Belgian science writer and zoologist Bernard Heuvelmans read those Sanderson-authored contributions. When Heuvelmans read the January 3, 1948, *Saturday Evening Post* article ("There Could be Dinosaurs"), in which biologist Ivan T. Sanderson sympathetically discussed the evidence for relict dinosaurs, Heuvelmans decided to pursue his vague, unfocussed interest in hidden animals in a systematic way. At the time, he was translating numerous scientific works, among them *The Secret World of the Animals* by Dr. Maurice Burton, which was republished afterward in seven volumes under the title *Encyclopedia of the Animal Kingdom*. Heuvelmans became personally inspired to research and pen "cryptozoology" articles.

Heuvelmans would be incorrectly credited for years as the inventor of the word "cryptozoology."

Heuvelmans, late in the 1940s, began to gather material about yet-to-be-discovered animals in what he would later refer to as his growing "dossiers" on them. From 1948 on, Heuvelmans exhaustively sought evidence in scientific and literary sources. Within five years he had amassed so much material that he was ready to write a large book. That book turned out to be the two volumes in French, *Sur la Piste des Bêtes Ignorées*, published in 1955, and better known in its English translation three years later as *On the Track of Unknown Animals* (London: Hart-Davis, 1958; NY: Hill and Wang, 1959).

Soon, Heuvelmans was engaged in massive correspondence as his library and other researches continued. In the course of letter writing, he often used and re-invented the word "cryptozoology." (It does not appear in *On the Track of Unknown Animals*.) That word saw print for the first time in 1959, which would be an important year in Shuker's life, when French wildlife official Lucien Blancou dedicated a book to Bernard Heuvelmans, as the "master of cryptozoology."

Years later, in 1965, Heuvelmans would acknowledge Sanderson's role in being the first to coin the word, "cryptozoology." Sanderson, in fact, was the first author to publish the word "cryptozoological" in English, in 1961, in his book, *Abominable Snowmen: Legend Come to Life —The Story of Sub-Humans on Five Continents From The Early Ice to Today* (Philadelphia: Chilton, 1961).

In the midst of all of this occurring, Karl Shuker would appear, apparently just at the right time. In 1959, new animal species were still being found, including, for example, the chunyi or dwarf brocket (*Mazama chunyi*), which had been discovered in Peru and Bolivia. Meanwhile in 1959, the "Howard Hughes of Cryptozoology," Texas millionaire Tom Slick, inspired by Bernard Heuvelmans' *On the Track of Unknown Animals*, was sponsoring the first expeditions in pursuit of Abominable Snowmen in Nepal, Bigfoot in California, and Sasquatch in British Columbia. It made for an interesting junction in eras, as the world opened to the new insights at the mid-century modern mark. In this same year and next, I would find myself, a budding cryptozoologist, actually beginning to read of Yetis and Sea Serpents, Loch Monsters and Giant Mystery Cats. These were exciting times.

It is into this environment that Karl Shuker was born in 1959. This was the first year after the publication of London's English translation and in the same year of America's first edition of Heuvelmans' now classic book, *On the Track of Unknown Animals*. This tome would later have a significant impact on Shuker's life. Indeed, Shuker's direct interest in cryptozoology began at the age of thirteen, in 1973, when he came across a copy of On the Track of Unknown Animals, while at a bookstore near his home in the West Midlands, England. Though unable to afford the book at the time, he soon received it as a birthday gift and then commenced to read it cover to cover many times.

Shuker was thoroughly hooked by the topic. He fell in love with cryptozoology and soon began to collect all the information on the subject that he could, including newspaper clippings, magazine articles, and books on a wide variety of cryptids (as he would learn the animals of cryptozoology would be coined in 1983, by John E. Wall of Manitoba, in an issue of the International Society of Cryptozoology's newsletter).

When it came time to attend college, Karl Shuker chose zoology as his major. However, during his first year as a zoology undergraduate in 1979, Shuker began to lose weight suddenly and was admitted to a local hospital. A week later it was determined that the young man was a diabetic and reemerged, at the start of a new phase of his life, as an insulin-dependent diabetic. In a sense, Shuker merely turned this sudden discovery to his advantage and decided to dedicate his life more deeply to his still growing passion for cryptozoology. The diagnosis did not slow him down. Instead, he went on to finish his studies in zoology at the University of Leeds and obtained a Ph.D. in zoology and comparative physiology at the University of Birmingham. But Shuker had to forego the rigors of stressful scientific inquiry.

It was at this time that Karl Shuker decided to turn his entire life-long interest in cryptozoology into a writing career. Soon Karl found himself writing articles for popular magazines and thoroughly researching several more scholarly book ideas.

Down through the years, Shuker has written regular cryptozoological columns for various English-language magazines, and became a contributing editor and cryptozoological columnist for other publications. Karl Shuker continues to appear regularly on television and radio programs, and is today widely recognized as one of the leading experts in cryptozoology because of his attention to scientific detail.

Through many years of hard work, Shuker was able to see many of his projects come to print, including the early versions of the book you are reading today. Further labor and updating of his studies would follow, and Shuker has toiled to revise his research continuously with the latest new data, discoveries, and discussions. His volumes have become some of the most important cornerstone works for the new generation of cryptozoologists in the 21st century. You will not be disappointed by the near decade of fine-tuning and massive revisions that Shuker has put into this book.

After decades of intensive research and revisions, I am happy to see so many of Karl Shuker's books remain in print, and especially this one. This book's impact has been enormous. Because Karl Shuker's research is based on rigorous dedication to scientific method and scholarship and due to his solid background in zoology, Shuker's works are respected throughout the scientific community *and* the popular cryptozoology fraternity, as they should be.

This volume will not disappoint you, and, mark my word, it will influence the field of cryptozoology for decades to come.

LOREN COLEMAN, DIRECTOR
INTERNATIONAL CRYPTOZOOLOGY MUSEUM
PORTLAND, MAINE
DECEMBER 6, 2011

FOREWORD TO THE SECOND EDITION

Rereading the words of my late husband, Gerald Durrell, in the original forward to *The Lost Ark*, I see that he was expounding on one of his favourite theories—if naturalists used their observational powers to the fullest, they would be rewarded with the most extraordinary surprises and insights about the natural world. Karl Shuker reminds us again in the revised edition, re-titled *The New Zoo*, that biodiversity still has not been fully explored or described even now at the dawn of this new century by we who pride ourselves on the scientific and technological advances made in the last.

Before becoming involved in conservation, I trained as a scientist, and Karl's last sentence strikes a deep chord within me: 'A wise scientist does not take pride in how much he knows, but rather takes heed of how little he knows.'

The Lost Ark has become *The New Zoo* in only eight years, and the wise scientists have been busy. Karl has provided updates and reported discoveries for several hundred animal species by name and possibly even thousands by implication with the recognition of two, maybe three, new phyla of invertebrate. Gerry's and Karl's theories about the zoological payoff from meticulous observational work have been borne out.

Even more important, I feel, is their equal and deep concern about the current biodiversity crisis.

Unfortunately, although we were alerted to the situation decades ago, the natural world is still taking a beating. Pollution of soil, sea and air, destruction of forest, marsh and reef, disruption of fragile plant and animal communities by the introduction of robust 'aliens', harvest of plants and animals in numbers beyond their capacity to recover—all these contribute to the decline and demise of species, i.e. to the biodiversity crisis. The situation seems to be worsening, and more and more often circumstances require a conservationist to take action even when knowledge about species and their ecological contexts is inadequate.

Enter the 'conservation biologist', an amalgam of the scientist and the conservationist, who gathers the best knowledge available to date, assesses the consequences of basing actions on such knowledge and then makes decisions. Conservation biology is still a young discipline, but as it matures and its practitioners gain experience, one has more and more confidence that the decisions will protect endangered species and their habitats rather than risk contributing to their decline. At the very least we know that they are the best decisions that can be made right now—who can do better?

As Karl points out, the discovery of new species is a good antidote to depressing news about the demise of others. Thus this new edition of his book is a joy to read. If it also strengthens its readers' already lively interest in the natural world and inspires a commitment to grappling with the biodiversity crisis, then it will have fulfilled its purpose.

DR. LEE DURRELL, JERSEY, JULY 2001.

Vampire squid. (William M. Rebsamen)

Hispaniolan solenodon, taxidermy specimen. (Frank Wouters)

FOREWORD TO THE FIRST EDITION

It was Cuvier, at the beginning of the 19th century, who made the rather pretentious and unwise statement that now all large creatures on the planet had been discovered and described. Since his day, of course, a host of creatures, ranging from pygmy hogs to white rhinos, Komodo dragons to coelacanths, have turned up to confound him. Usually, the local people were aware of the animal, but it was not known to European science. The pygmies, for example, knew all about the okapi long before Johnston 'discovered' it. In many cases, it is lack of observation that keeps a creature a secret from science for so long.

I remember being on a live animal collecting trip in Cameroon, and I had the fruits of my labours—some 250 specimens of 14 different species—housed in a large marquee on the river bank. The local District Officer asked if he might come and see them, so I welcomed him with warm beer and showed him round. He was absolutely dumbfounded. He told me that he had worked for 25 years in West Africa and he had never seen any of these creatures. Where, he asked, had I found them? I told him that I had caught a good number of them in the tiny strip of forest at the end of his garden. I think he thought I was making fun of him, for our relationship remained somewhat cool from that moment onwards.

However, I was rather neatly hoist with my own petard for, on that particular expedition, I quartered miles of forest searching for the nesting site of the rare Bald Rock Crow *Picathartes oreas* with no success. Returning a year later, I made my base camp in the same village and the hunters told me with pride that they had found the nest site of the rare bird I wanted so badly. To my embarrassment and chagrin, they led me to a rock face approximately 200 yards from where, the year previously, I had made my base camp. *Picathartes* builds swallow-like mud nests and I could see this nest site was an old one. So while I had been blundering about the forest looking for the 'elusive' bird, it had been happily nesting almost in my tent.

I cannot understand why scientists on the whole look scornfully at the idea of some large unknown animal lurking in a lake or in the sea or on land. They don't believe it until they have a specimen—preferably dead—in their hands. I believe everything is possible and those of you who read and relish this book as I have done will see why. With delighted anticipation, I await the discovery of a sea serpent or, better still, an Abominable Snowman, and, if he proves too close to us in appearance for comfort, what do we do with him? Put him in a cage or send him to university?

This fascinating, encouraging book should be part of every naturalist's library to give them hope that one day they themselves may make a wonderful discovery of a new species.

GERALD DURRELL, 3 SEPTEMBER 1992.

Rothschild's mynah. (Jonathan Kriz)

Neon tetra. (Oleg Korotkov)

ACKNOWLEDGMENTS

It would not have been possible to prepare a book of this scope without the willing interest and assistance of a great many persons, societies, and organisations. In particular, I wish to offer my most sincere and grateful thanks to the following:

United Kingdom:
W.F.H. Ansell, Dr. Nick Arnold, *Avicultural Magazine*, Colin Bath, Dr. Simon K. Bearder, Endymion Beer, Trevor Beer, BIOSIS, Birdland (Bourton-on-the-Water), *Birmingham Evening Mail*, Birmingham Public Libraries, Janet and Colin Bord, Lena and Paul Bottriell, Michael Bright, Bristol Zoo Garden, British Library, British Museum (Natural History), G.H.H. Bryan, Owen Burnham, Centre for Fortean Zoology, Chester Zoological Gardens, D.N. Clark-Lowes, Prof. John L. Cloudsley-Thompson, Dr. Nigel Collar, Dr. N. Mark Collins, Martin Cotterill, Jonathan Downes/*Animals and Men*, Drayton Manor Park & Zoo, Dudley Public Libraries, John Edwards, Nick G. Ellerton, Excalibur Books, Fauna and Flora Preservation Society, Reginald Fish, Fortean Picture Library, *Fortean Times*, Philippa Foster, Richard Freeman, Errol Fuller, Etienne Gilfillan, Glasgow Zoo, Geoffrey R. Greed, Gina Guarnieri, Harrap Publishing Group, Dr. Peter Henderson, David Heppell, R.R. Hepple, Dr. John Edwards Hill, Richard Hill, Amanda Hillier, the late Mary Harvey Horswell, Gordon Howes, Jean and Keith Howman, International Congress for Bird Preservation, Dr. Gareth Jones, Dr. Jonathon Kingdon, Michael Lyster, Dr. Desmond Morris, Richard Muirhead, Dr. Darren Naish, Nature Photographers Ltd, *New Scientist*, the late Richard J.P. O'Grady, Edward Orbell, Mark O'Shea, Paignton Zoological & Botanical Gardens, Paul Pearce-Kelly, Michael Playfair, Dave Powers, Alan Pringle, Prof. David Pye, Bob Rickard, the late Roy Robinson, Craig Robson, Sandwell Public Libraries, the late Lady Philippa Scott, Paul Screeton/*Folklore Frontiers*, Steven Shipp/Twilight Books, Mary D. Shuker, Paul Sieveking, Kenneth G.V. Smith, Dr. Eve Southward and Prof. Alan J. Southward, Dr. Paul Sterry, Michael K. Swales, Dr. Ian R. Swingland, Ann Sylph, Spencer Thrower, *The Times*, the late Gertrude and Ernest Timmins, University of Birmingham, University of Leeds, Walsall Public Libraries, *Wild About Animals*, the late Jan Williams, *Wolverhampton Express and Star*, Wolverhampton Public Libraries, the late Gerald L. Wood, World Conservation Monitoring Center, World Pheasant Association, Dr. Nathalie Yonow, World Wide Fund For Nature UK, Zoological Society of London.

Overseas:
Dr. Shane T. Ahyong (Australia), American Museum of Natural History (U.S.A.), Dr. Alan N. Baker (New Zealand), Dr. Aaron M. Bauer (U.S.A.), the late Mark K. Bayless (U.S.A.), Berlin Zoological Gardens (Germany), Dr. Ian Best (Bahrain), Matthew Bille/*Exotic Zoology* (U.S.A.), Dr. Wolfgang Böhme (Germany), Chris Brack (Germany), Bronx Zoo (U.S.A.), Geert Brovad (Denmark), Markus Bühler (Germany), the late Mark Chorvinsky/*Strange Magazine* (U.S.A.), Janice Clark (U.S.A.), Loren Coleman (U.S.A.), cz@onelist.com/cz@egroups.com (U.S.A.), Department of Conservation (New Zealand), Department of Conservation & Environment, Victoria (Australia), Dr. Chris R. Dickman (Australia), Dr. Dominique A. Didier (U.S.A.), Prof. John M. Edmond (U.S.A.), Elsevier Science Publishers B.V. (Netherlands), Dr. Don D. Farst (U.S.A.), *Fate* (U.S.A.), Dr. Richard Faust (Germany), Dr. Tim Flannery (Australia), Angel Morant Forés (Spain), Dr. Jacques Forest (France), Matthias Forst (Germany), Dr. John Forsythe (U.S.A.), Dr. Hans Frädrich (Germany), Frankfurt Zoological Gardens (Germany), Phyllis Galde (U.S.A.), Gladys Porter Zoo (U.S.A.), Dr. Dennis Gordon (New Zealand), the late J. Richard Greenwell (U.S.A.), Prof. Karl G. Grell (Germany), Dr. Colin P. Groves (Australia), Bob Hay (Australia), Craig Heinselman/*Crypto* (U.S.A.), Markus Hemmler (Germany), Dr. Robert R. Hessler (U.S.A.), the late Dr. Bernard Heuvelmans (France), Dr. Shoichi Hollie (Japan), *Honolulu Advertiser* (U.S.A.), *Honolulu Star-Bulletin* (U.S.A.), Prof.

G. Imadate (Japan), Prof. Yoshinori Imaizumi (Japan), International Society of Cryptozoology (U.S.A.), J.L.B. Smith Institute of Ichthyology (South Africa), Dr. C.M. King (New Zealand), Jürg Klages (Switzerland), Dr. Reinhardt Møbjerg Kristensen (Denmark), Dr. Anne LaBastille (U.S.A.), Marcel Lecoufle (France), Gerard van Leusden (Netherlands), Dr. Roy P. Mackal (U.S.A.), Dr. Bernhard Meier (Germany), James I. Menzies (Papua New Guinea), Dr. Christopher Mercer (Papua New Guinea), Dr. Adam C. Messer (Japan), Dr. Ralph E. Molnar (Australia), Tim Morris (Australia), National Museum of Natural History (U.S.A.), Nature Production (Japan), New York Zoological Society (U.S.A.), Michael Newton (U.S.A.), Prof. Jörgen G. Nielsen (Denmark), Dr. Mark Norman (Australia), the late Scott T. Norman (U.S.A.), Dr. John P. O'Neill (U.S.A.), Dr. Pierre Pfeffer (France), the late Dr. Jordi Sabater Pi (Spain), Dr. Chris Raxworthy (U.S.A.), Michel Raynal (France), William M. Rebsamen (U.S.A.), Dr. David C. Rentz (Australia), Dr. William Robichaud (Canada), Dr. Marc van Roosmalen (Brazil), Dr. Clyde Roper (U.S.A.), Royal Society of New Zealand (New Zealand), Lorraine Russell (New Zealand), Yohei Sakamoto (Japan), Ron Scarlett (New Zealand), Dr. José E. Serrão (Brazil), SIR Publishing (New Zealand), Malcolm Smith (Australia), Dr. David Stein (U.S.A.), Richard Svensson (Sweden), Dr. Thomas Teyke (Germany), Lars Thomas (Denmark), Dr. Ralph Tiedemann (Germany), Robert Timmins (Laos), Prof. Michael J. Tyler (Australia), University of Michigan Press (U.S.A.), Wanganui Regional Museum (New Zealand), Robert M. Warneke (Australia), Bob Warth (U.S.A.), Dr. Vern Weitzel (Vietnam), Albert G. Wells (Australia), the late Dr. Ralph Wetzel (U.S.A.), Rudy Wicker (Germany), Wildlife Photos W.A. (Australia), World Wide Fund for Nature International (Switzerland), Dr. Jill Yager (U.S.A.), Prof. Yin Wen-Ying (People's Republic of China), and Prof. Zhou Kaiya (People's Republic of China).

I would like to reiterate my especial thanks to the late Gerald Durrell for his delightful foreword to this book's first edition, and, through his own many wonderful books, for encouraging my childhood fascination with natural history; to Dr. Lee Durrell for her inspiring foreword to this book's second edition; to Loren Coleman for his compelling foreword for this third edition, and also to my publisher, Chad Arment, and to my agent, Mandy Little of Watson, Little Ltd, for their belief in this book, which gave me the opportunity to carry out my longstanding intention to prepare a third, fully-updated edition, spanning not only the entire 20th century but also the opening decade of the 21st century.

Picture Credits
The author has sought permission for the use of all illustrations known by him to be still in copyright. Any omission brought to his attention will be rectified in future editions of this book. All uncredited illustrations are, as far as he is aware, in the public domain.

Skeleton of Congo hero shrew—revealing its backbone's extraordinary degree of protective reinforcement, enabling it to withstand immense pressure (Neg. #37485/Photo: Kay C. Lenskjold/Courtesy Department Library Services, American Museum of Natural History).

INTRODUCTION

Arrivals and Revivals –
A New Zoo of Creatures From the Lost Ark

> Nothing is rich but the inexhaustible wealth of Nature.
> She shows us only surfaces, but she is a million fathoms deep.
> RALPH WALDO EMERSON—'RESOURCES', IN *LETTERS AND SOCIAL AIMS*

The phrase 'lost ark', when applied to animals, may well conjure up images of creatures that no longer share our world—extinct animals, ranging from dodos to dinosaurs, whose lifeless remains, whether recent or fossilised, are all that we have left to remind us of their former existence. There is, however, a second, much less familiar lost ark, from which the subjects of this current book have originated—a veritable new zoo of creatures that are very much alive, but which have only recently been discovered or rediscovered.

They are all visually spectacular, zoologically significant, or both. Yet, surprisingly, every one of them successfully eluded scientific discovery until the 20th century, or was confidently written off by science as extinct until rediscovered during the past eleven decades. In many cases, these animals were well known to our ancestors, to so-called 'primitive' native tribes, and even to western laymen, but had been dismissed by zoologists as fantasy or folklore. While other species sharing their habitats had become known and duly documented, these more elusive, secretive, or inaccessible species had instead experienced a bizarre exile from zoological reality—alive yet anonymous, real yet unrecognised, a very sizeable catalogue of animals effectively lost to scientific study for untold years.

This is not a new phenomenon. The western gorilla *Gorilla gorilla* was familiar to tribes sharing its African forests, but its existence was not accepted by zoologists until 1847. Baird's tapir *Tapirus bairdii*, Mesoamerica's largest mammal, was hunted by the local people, but was unknown to science until 1865. The golden snub-nosed monkey *Rhinopithecus roxellanae* (brought to zoological attention as a living species in 1870), the gerenuk or giraffe antelope *Litocranius walleri* (1878), and Grévy's zebra *Equus grevyi* (1882) had hitherto been known to science only from ancient, sometimes exaggerated depictions (and had thus been discounted as either extinct or non-existent). Hence there seemed no reason why this progression from scientific obscurity to rubber-stamped acknowledgement should not continue throughout (and beyond) the 20th century—and, in fact, that is precisely what happened. But very few people actually realise this. Why?

'*On a sondé ces régions voilées, les bornes du possible ont été reculées. Un mortel a pu voir, armé d'un oeil géant, osciller des lueurs aux confins du néant!*' [Man has plumbed these veiled realms, the boundaries of the possible have been extended. A mortal, armed with the eye of a giant, has been enabled to see gleams of light oscillating on the confines of empty space!]

So wrote J.J. Ampère more than 100 years ago, marvelling at the spectacular advances in astronomy made possible by the telescope's invention. How much more would he have marvelled, were he to have known that one day man would set foot on the moon, and receive detailed photographs of the Solar System's planets and satellites transmitted to Earth from sophisticated data-seeking probes voyaging ever onwards and outwards through Space?

How ironic, then, that while man's knowledge of other worlds is ever increasing, considerable expanses of his own, on land and under the water, are still virtually unexplored and unknown. One particularly sad outcome of this is that much of the wildlife inhabiting these regions remains mysterious and sparsely documented. Yet, very surprisingly, this tragic situation is largely ignored (or simply unrecognised) by most people—scientists and laymen alike—who maintain or presume that in this ultra-scientific age our world has been thoroughly charted and, accordingly, that few (if any) major new animals still await discovery. Instead, at least as far as they are concerned, the days when such creatures were added to the zoological catalogue ended decisively with the scientific unveiling of the okapi in 1901—and they consider the rediscovery of supposedly extinct animals to be as likely as the alchemic conversion of iron into gold.

Some, it is true, admit that the capture in 1938 of a living coelacanth—a peculiar lobe-finned fish belonging to an archaic group hitherto believed extinct for more than 60 million years—was indeed something of a surprise. However, they reassure themselves that this was doubtlessly a lone exception, a nonconformist novelty in an age when the mere concept of finding major new animals or encountering reputedly extinct ones is widely considered (especially in scientific circles) to be insufferably anachronistic—an antiquated, atavistic, incorrigibly romantic, and thoroughly unrealistic dream of eccentrics, young children, and others who ought to know better.

The perpetuation of this remarkably narrow-minded and naive attitude even until as recently as the scientifically-enlightened 1990s was due in no small way to a certain, quite extraordinary, gap in the vast spectrum of modern-day wildlife books. There are—and quite rightly so—a very great number of works that deal with the tragic plight of endangered, vanished, and extinct animal species. Conversely, until the publication of this book's first edition in 1993 there had not been a single volume devoted to the more optimistic, obverse side of this coin—the side concerned with the histories of those many zoologically significant and often visually striking new animal species that continue to be uncovered year after year, and of the equally impressive quantity of supposedly extinct species that defy all expectations by reappearing after many years (sometimes centuries) of 'official' non-existence. As a result, the number of people actually aware of this unique array of recently-unearthed zoological treasure trove was a pitifully small one.

As someone with a lifelong interest in new and rediscovered animals of the 20th century, and having amassed an extensive archive of material covering a vast range of such creatures—which include some of the most fascinating, unexpected, engaging, and bizarre species ever to come to scientific notice—I had grown increasingly surprised as the years passed by with no sign of such an outstanding omission in the literature ever being rectified. Consequently, I eventually decided to attempt this myself, and after five years of research and writing, 1993 finally saw the publication of what would be the first edition of this present book—and which was entitled *The Lost Ark*—as the first to be devoted solely to the 20th century's quota of newly-revealed and recently-revived multicellular animals (thus excluding only the protozoans, i.e. single-celled animals).

The reason why I chose to publish it then, rather than in 2000, was that the preparation of such a book was long overdue, and not just to fill a very surprising gap in the worldwide library of wildlife works, but also to emphasise that in these days of ever-decreasing natural wildernesses there is so much that could be lost before its very existence is even realised.

The wholesale destruction of tropical rainforests, together with the obscene pollution of the oceans and freshwater ecosystems, has alerted scientists and the public to frightening truths about the likely consequences of such desecration. Among these is the disturbing awareness that our world could lose unique, irreplaceable species of plant and animal that may yield vital additions to modern medicine's stock of life-saving drugs and antidotes. And there may be other equally irrevocable consequences that we are unable even to comprehend at present—with so incomplete a knowledge of global biodiversity, how can we?

In any event, the right of all wildlife species to an existence on Earth is as basic as our own. Moreover, their formal investigation would undeniably lead to a greater understanding of our own place and purpose in the scheme of Nature, and their perpetuation would enrich our lives and maintain the delicate ecological balance that ensures the survival and stability of our entire planet.

Thanks to the illogical, blinkered attitude to the subject of new and reputedly extinct animals that existed through much of the 20th century, however, many important species could well have already been allowed (albeit unknowingly) to slip into extinction without ever becoming known to science. And certain 'officially' extinct species that may in reality have persisted in small numbers within remote, rarely-visited localities could well have truly died out by now, through lack of serious attention concerning the possibility of their survival.

Why, therefore, I reasoned, should these deplorable trends be allowed to continue unchallenged until 2000? If my book could in the smallest way inspire interest in seeking formally unrecognised and technically demised animals, and in so doing offer an additional reason for the conservation of wildlife habitats of every type throughout the world, then the optimum date for its publication was today, not tomorrow, or next year, or in a decade's time. And indeed, by the end of 1993 my book had attracted considerable interest and favourable reviews throughout the world, alerting and informing not only the general public but also the international scientific community that the days of great zoological discoveries (and rediscoveries) was far from over.

So profound and encouraging, in fact, was the attention given to my book that I soon decided to prepare a second, wholly updated edition at the close of the 20th century, which would document all of the major zoological arrivals and revivals from the period 1900-1999. The result was *The New Zoo*, published in 2002.

Nine years have passed since then, but the stream of notable zoological discoveries and rediscoveries has continued in unabated fashion. So too has the proliferation to a previously-unimaginable extent of online information appertaining to all such creatures, from the present and the past. That has enabled me to update and amend earlier entries to a degree that I could never have even conceived, let alone achieved, in pre-internet days (when The *Lost Ark* appeared) or even back in 2002 (when *The New Zoo* appeared).

Consequently, this third incarnation of *The Lost Ark*, which you are now reading, not only contains entries on all of the most significant animals to have been found or refound since 2002, but also provides a meticulous revision of all previous entries from the earlier two editions, in which taxonomic updates (including as many as possible of the most notable name changes, splittings, lumpings, and total reclassifications of the species in question), changes in conservation status, additional information concerning the species' history of discovery and rediscovery, new illustrations, and much else besides have been incorporated here. The result is, quite simply, the single most extensive documentation of modern-day new and rediscovered animals ever published—the standard, definitive work, in fact, on this subject.

Moreover, one thing is certain. I am not alone in my views regarding the likelihood that there are still new creatures awaiting discovery and long-lost ones awaiting rediscovery (habitat destruction permitting). On the contrary, so great is the variety of those mystery beasts whose alleged existence is supported by reliable eyewitness accounts, photographs, and sometimes even a complete specimen or two, that in 1982 a scientific organisation was established for the specific purpose of reviewing and investigating such cases in a methodical, serious manner—far removed from the traditional flippant or dismissive approaches adopted by others in the past.

Known as the International Society of Cryptozoology (ISC), it is, sadly, no longer around, but its success during its decade of existence was such that it led the way to the establishment of several other notable societies in this field, and which are still very much around today—such as the Centre for Fortean Zoology (CFZ) in the UK, the British Columbia Scientific Cryptozoology Club in Canada (BCSCC), and the Pangea Institute in the USA, to mention just three.

Furthermore, of especial significance is that the ISC's directors, officers, and members included some of the world's foremost zoological authorities—experts who have come to recognise (unlike so many of their peers) that it is not the pursuit of such creatures which is antiquated, but rather the obstinate refusal of scientists to examine their credentials in an impartial, objective manner. Since the ISC's establishment, some of its members have been instrumental in revealing a number of remarkable new and rediscovered animals, as will be disclosed in this book. Clearly then, there is good reason for believing that cryptozoology—commonly translated as 'the study of hidden animals'—has much to recommend it.

Yet whatever types of creature still await detection, they will only be the latest in a long and immensely varied parade of zoological arrivals and revivals that have emerged out of obscurity from 1900 onwards, creatures that in the vast majority of cases were once mystery beasts themselves, known to their local human neighbours (i.e. ethnoknown) but dismissed as fantasy by science—until conclusive evidence of their reality was finally obtained. Of course, most new and rediscovered animals have been small, inconspicuous forms—insects, rodents, tiny lizards, fishes, bats, songbirds, etc, or ones barely distinct from others already recorded. To document every single new and rediscovered animal listed since 1899 would require a volume of literally voluminous proportions! Moreover, as so many of those contained inside it would be of little general interest (especially when even the greatest cryptozoological sceptics have never denied that many creatures in this category of undistinguished, unsurprising species await formal detection), no useful purpose would be served in preparing such a volume.

Instead, I have concentrated upon the 20th and 21st centuries' more spectacular discoveries and rediscoveries—unexpected creatures defying the well-established belief that there were few if any notable zoological novelties still to be revealed. Accordingly, the animals that I have selected for inclusion within this book fall into at least one of the following categories:

1) They are relatively large, visually impressive creatures—or allegedly extinct forms that had been the subject of extensive searches in the past—whose scientific concealment until the 20th century was therefore quite remarkable.
2) They are small in size, but are nonetheless of great scientific significance by constituting animals *dramatically* different from any previously documented.
3) They are not particularly important in a strictly scientific sense, but their histories or some

feature(s) of their appearance or lifestyle are so extraordinary that they thoroughly deserve inclusion within any book devoted to zoological finds of the 20th and 21st centuries.

Finally, all that remains to be discussed within this Introduction are some matters concerning the scientific documentation and naming of new species—intrinsic aspects of animal discovery and discrimination, which adhere to certain strict rules and conventions that require a little elucidation before progressing further (see also the Appendix, p. 315, for an explanatory account of animal classification).

DESCRIPTION AND NAMING OF NEW SPECIES

The species is the fundamental 'unit' of taxonomy (the scientific classification of living organisms); and although it has been defined in a number of different ways (so that there is still some dispute regarding its precise definition), a species is considered by most authorities to constitute one or more populations of organisms whose members can freely interbreed with one another to yield fit, fertile offspring (see also the Appendix).

The discovery of a creature that seems to represent a new species is followed by a standard series of procedures in science. First of all, at least one specimen of it is collected, preferably a few more (including one of each sex and a juvenile) if such collection does not endanger the species' survival. Nowadays, there is a very laudable tendency for researchers to seek already-dead specimens for this purpose, rather than killing living ones. Occasionally, a mere sample of blood extracted from a living specimen has proven sufficient.

These specimens are then closely compared with those already-known species to which they seem most closely related—a meticulous process involving detailed comparisons of external morphology, internal anatomy, and also (particularly in recent times) various genetic and biochemical characters if possible. If the researchers carrying out such comparisons are satisfied that the new creature's representative specimens do indeed differ in taxonomically important ways from all ostensibly similar, currently-known species, then its status as a new species is accepted by them.

Its existence must then be officially reported to the scientific world and its distinguishing features fully documented. This is achieved by preparing a detailed *description* of the new species for publication in a scientific journal. Whenever possible, this description is based principally upon just one of the specimens collected; the specimen selected for this purpose is normally one that appears to be a typical representative of that species, and is (preferably) complete and undamaged. This specimen is designated as the new species' *type specimen* or *holotype* (and the precise locality in which it was collected is referred to as that species' *type locality*). If there is need to refer to other specimens (e.g. to include mention of any differences between sexes, and between adults and juveniles, or if the species exhibits notable variation in external colouration, overall size, and so forth), these specimens are designated as this species' *paratypes*.

In view of the great scientific value of type specimens and paratypes, they are carefully preserved afterwards within the collection of a natural history museum or some other scientific institution. As a consequence, if at some stage in the future some specimens are collected of a creature that may belong to that species, but whose identity is currently undetermined, researchers can compare its specimens with the earlier-described species' type and paratypes in an attempt to ascertain their creature's identity.

When a new species is formally described, it must also be given, as part of that description, its own scientific name. This is a unique, two-part (binomial) name that distinguishes it from all other species, is generally of Latin or Greek origin (or both), and normally is *always* set entirely in italics, *not* in Roman typeface (unless the main text is set in italics), and *never* in a combination of italics and Roman. For example, the wolf's scientific name is *Canis lupus*; the name *lupus* is its specific or trivial name (which always begins with a lower case letter), and *Canis* is its generic name or genus (which always begins with a capital letter). A new species' scientific name can provide clues concerning that species' closest relatives, because it will often share their generic name (see the Appendix for further details)—but not always, because sometimes a species may be sufficiently different even from those to which it is evidently most closely related for a completely new genus to be required for it.

More rarely, a new species may be so distinct from all other species that it warrants its own taxonomic family, more rarely still its own taxonomic order, and even more rarely its own class. During the past 100 years, very many thousands of new animal species have been described (of which the vast majority have been insects), but only four have each been sufficiently different from all others to warrant the erection of an entirely new phylum—the taxonomic category at the very pinnacle of the hierarchy of animal classification (p. 315). (NB—all taxonomic categories above the level of genus are always set in Roman typeface, never in italics.)

Sometimes, later studies of a species will reveal that its original classification was incorrect; as a result, its scientific name may need to be changed. Equally, there are occasions when a new species is formally described by more than one researcher or group of researchers, wholly independently of each other; in this situation, the description that is published first has precedence over all of the others, so that the scientific name included within it for the new species is the one that becomes accepted.

Also worth noting is that for a variety of different reasons, the year in which a new species is discovered is not always the same as the one in which it is officially described—the latter sometimes does not occur for a few years after the species' discovery. For example, the megamouth shark (p. 210) was discovered in 1976 but was not described until 1983, on account of the immensely detailed study that was required to yield a definitive description of such a huge and radically separate species as this one. In the case of the ningaui marsupial mice (p. 75), although the first species was discovered in the 1950s its distinctiveness was not recognised for many years, and it was not formally described until 1975.

BIBLIOGRAPHICAL BENEFITS
As readers of this book may wish to seek out additional details regarding some or all of the animals featured within it, I have supplied an extensive bibliography containing all of the principal sources consulted by me during the book's preparation. Whenever possible, these have included, for each newly-discovered species, the full reference to the published scientific paper or book in which it was formally described and named following its discovery. These very important works are distinguished from other references by an asterisk (*) prefix, and have never been brought together before within a single publication—thereby rendering this book's bibliography unique, and an essential, invaluable addition to a work concerned with the finding of new species. Also included in the bibliography are the full references to papers, articles, and other publications that announced the rediscovery of those species documented here that were once thought to be extinct.

Despite the universal adoption of the metric system by modern-day science, during my extensive correspondence while preparing this book I was intrigued to learn that most people still visualise the size of animals more easily when they are expressed in imperial units. Also, this third edition is published directly in the U.S.A., unlike its two previous U.K.-published versions, in which the imperial system of measurement has not been supplanted by the metric one. For these reasons, therefore, I have retained the imperial system here wherever possible.

Finally: among this book's illustrations are several rare archive items, comprising some of the earliest photographs in existence of certain species documented here, as well as several specially-commissioned full-colour paintings and artistic renditions not previously published (for which I give sincere thanks to wildlife artists Philippa Foster, Tim Morris, Markus Bühler, and William M. Rebsamen).

Preparing this book's three editions has given me a very great sense of pleasure and an unexpected degree of optimism. Researching the histories of so many new and rediscovered animals of the 20th and 21st centuries, so frequently filled with wonder and intrigue, seems to have acted as an uplifting antidote to the oppressively smug, disdainful attitude displayed for far too long by far too many whenever the subject of such creatures has arisen.

My sincerest hope is that this book's readers will share this experience of mine, and, as I do, look to the future in anticipation of the many equally exciting animals that are undoubtedly still waiting to step forth from the new zoo during future decades of the 21st century. As the American writer John Burroughs once said:

There is always a new page to be turned in natural history if one is sufficiently on the alert.

Baby mountain gorilla. (Dmitry Pichugin)

James's flamingos. (Pedro Szekely)

PART 1:
20TH-CENTURY DISCOVERIES AND REDISCOVERIES

Rothschild's giraffes. (Nigel Wedge)

Baby giant panda. (Kitch Bain)

THE MAMMALS

From Okapi to Onza—and Beyond

One can have no idea today of the romance surrounding the discovery of the Okapi, nor of the excitement caused in natural history circles, first by the vague reports of its presence, and later by its actual finding.
DR. MAURICE BURTON—*THE STORY OF ANIMAL LIFE*, VOL. II

This issue features what *may* be the most significant cryptozoological find since the Society was founded 4 years ago; that is, the acquisition of a complete specimen of an Onza, the legendary Mexican cat which may have been overlooked by zoology because of its resemblance in coloration and size to the puma.
J. RICHARD GREENWELL—*INTERNATIONAL SOCIETY OF CRYPTOZOOLOGY NEWSLETTER* (SPRING 1986)

Although the okapi is certainly the most famous mammalian discovery of the 20th century, it is joined by a host of other distinguished fur-bearing arrivals, and revivals—including a giant ape of the mountains, one of zoology's biggest pygmies, the world's favourite wild animal and the seas' most mysterious whale, a foxy-furred goat-antelope first made known to science as a carpet, a mélange of supposedly extinct marsupials, a porcine recluse resurrected from the Ice Ages, a minuscule bat no bigger than a bumblebee, a Mexican pseudo-cheetah called the onza, the Queen of Sheba's gazelle, the sensational Vu Quang ox, a contradictory ground-living tree kangaroo, and much more!

COTTON'S WHITE RHINOCEROS

Following the discovery of Grévy's zebra *Equus grevyi* in 1882 and a last flourish of new antelopes in the early 1890s, zoologists confidently asserted that no further large animals remained unknown to science in Africa. How wrong they were was demonstrated dramatically when, within the space of just a few years, the Dark Continent unfurled a series of spectacular new mammals, to astonish the scientific world.[1-3]

The first of these was truly extraordinary—a completely new subspecies of white rhinoceros, the world's fourth largest species of land mammal. Zoologists traditionally believed that the white rhino was restricted to areas of southern Africa located south of the Zambezi, including Botswana, Mozambique, and Zululand, but in 1900 Major Powell-Cotton bagged several specimens more than 2000 miles further north—on the Upper Nile, bordering the Sudan, Congo, and Uganda. Also in 1900, and from that same locality, Captain A.St.H. Gibbons brought back a white rhino skull.[1-3]

Not surprisingly, zoologists were amazed to learn that populations of such an enormous mammal were existing undocumented by science, but this was remedied by mammalogist Oldfield Thomas's *Nature* report of 18 October 1900. And in 1908, Dr. Richard Lydekker named the beast *Ceratotherium simum cottoni*, thereby differentiating it from its slightly hairier, southern counterpart.[1-3] (Intriguingly, unconfirmed encounters with Central African white rhinos were reported by Dr. John Gregory and Count Samuel Teleki in the 1880-90s.[1]) Tragically, fewer than a dozen specimens of Cotton's white rhinoceros now exist at most, all in captivity, and the IUCN (International Union for the Conservation of Nature) considers it to be extinct in the wild. Moreover, making this situation even more disturbing, recent research suggests that Cotton's white rhinoceros may well

Science was unaware of the white rhinoceros's existence in Central Africa until 1900. (Owen Burnham)

be sufficiently distinct from its southern counterpart to be classed as a separate species, *C. cottoni*, with distinct genetic and morphological differences indicating that the two white rhinoceroses have been separated for at least one million years.[3a]

OKAPI—THE CONGO'S INCONGRUOUS SHORT-NECKED GIRAFFE

Although, as this entire book demonstrates, it is inaccurate to refer to the okapi as the *last* major zoological discovery (as so many people still do), there is assuredly ample justification for calling it the 20th century's *greatest*—because in so many, very varied ways it is unique, and wholly unmistakable, embodying a bewildering plethora of morphological and historical paradoxes. With a richly-hued coat more comparable to the gleaming plumage of some sultry jungle bird than to the pelage of a mammal, adorned by vivid zebra-like stripes, and equipped with an inordinately long blue tongue that would seem more at home in the mouth of a chameleon, this exotic-looking creature has always been a living contradiction in terms—first believed to be a cloven-hoofed zebra, then proving to be a short-necked giraffe!

It all began in 1890 when, in his book *In Darkest Africa*, the famous explorer Sir Henry Morton Stanley briefly mentioned a strange, supposedly ass-like beast familiar to the Wambutti pygmies inhabiting the dense, little-known Ituri Forest—nowadays housed within the Democratic Congo (originally the Belgian Congo, then Zaire) but at that time contained within Uganda's borders. According to Stanley:[4]

> The Wambutti know a donkey and call it "atti". They say that they sometimes catch them in pits. What they find to eat is a wonder. They eat leaves.

If the Wambutti did indeed know of a forest donkey, then they knew more than any western zoologist, because no specimen or remains of such a creature existed in any museum or scientific institution during that period. The anomalous *atti* attracted considerable interest from one of Stanley's friends, Sir Harry Johnston, who spoke about it to Stanley in London before going out to Uganda as its new Governor. Johnston would have liked to follow up Stanley's information, but did not receive any opportunity to do so—until 1899. This was when he succeeded in rescuing a band of Wambutti pygmies, captured in the Ituri Forest by an unscrupulous German impresario, who had planned to display them as 'ape-men' at the Paris Exhibition after leaving Uganda with them.[3,5-8]

Naturally, the Wambutti were extremely grateful to Johnston, who entertained them at his home before sending them safely back to their forest, and were only too happy to answer his many questions regarding the mysterious *atti* or forest donkey. They did indeed know of such a creature, but disclosed that its name was not *atti*; instead, it was pronounced *o'api*, the apostrophe being pronounced like the gasp-like Arabic 'k', i.e. '*okapi*'.[3,5-8]

They described it as a shy creature, captured only by digging large camouflaged pits, into which it would fall as it walked across the forest floor. Its body was dark grey or dun-coloured, with striped legs, and in overall form it resembled a large donkey or mule. Accordingly, Johnston felt sure that the okapi was a type of forest zebra, and, as such, was a completely new species awaiting formal description.[3,5-8]

In 1900, he set out on its trail, travelling into what was then the Congo Free State, and reaching the Belgian fort at Mbeni, whose officers he questioned thoroughly regarding the okapi's existence and whereabouts. To his delight, he learnt not only that they knew of this elusive creature, but also that they had frequently seen the bodies of specimens, brought back to the fort by native militia, who found its meat very tasty. Most exciting of all, the officers felt sure that a complete skin of an okapi was still at the fort. Sadly, however, after searching in vain, they discovered that it had recently been cut into strips by the native soldiers to make waistbelts and head bands (bandoliers)—these latter being made from the striking striped portions of the skin.[5-7]

Although Johnston failed to obtain a complete skin, the officers were able to describe the okapi, and their description tallied with the account given the previous year by his pygmy friends. They also noted that its snout was markedly 'effilée' (long and drawn-out), and that its feet each bore more than one hoof. This latter feature was very significant, because whereas all modern-day horses, donkeys, and zebras only have one hoof on each foot, the extinct primitive horse *Hipparion* had three hooves per foot. Thus, Johnston revised his view concerning the okapi's identity, from a forest zebra to a surviving *Hipparion*. Suddenly the okapi's scientific standing had increased greatly—it now seemed likely that this reclusive beast was a living fossil![5-7]

The climax to Johnston's investigations at Mbeni came when its officers found and gave to him two native bandoliers made from the striped sections of the skin. At last Johnston had some physical proof of the okapi's existence and evidence of its distinctness, for the stripes' appearance and form were unlike those of any species known to science at that time.[5-7]

Johnston was clearly closing in on the okapi, but then missed a wonderful opportunity of meeting it face-to-face in its native homeland. On entering the Ituri Forest, he was shown some tracks that the natives insisted had been made by an okapi. To his disappointment, however, the tracks were of a cloven-hoofed mammal—one that had an even number of hooves on each foot rather than an odd number. Due to his preconceived notion that the okapi was a member of the horse family (and hence odd-toed), Johnston dismissed the natives' testimony, believing instead that the tracks had been made by some form of large forest antelope. And so the trail was never followed, and soon afterwards the expedition had to be abandoned, on account of the inhospitable conditions within the forest's humid, fever-infested depths.[3,5-8]

On 21 August 1900, Johnston wrote to Dr. Philip Sclater, Secretary of London's Zoological Society, informing him of his investigations to date and promising him the bandoliers. When they arrived, Sclater exhibited them at a meeting of the society on 18 December, and formally documented them in its *Proceedings* on 5 February 1901. Agreeing with Johnston's verdict that the okapi was an unknown species of horse, but preferring his original concept of a forest zebra to his later vision of a surviving *Hipparion*, Sclater named the new species *Equus? johnstoni*, cautiously adding the question mark after its genus, as he knew that it would be imprudent to adopt too dogmatic a stance regarding the identity of a creature known from as limited an amount of material as two slender strips of skin—a judgement that soon proved to be fully warranted.[7]

The okapi is probably the most famous new animal of the 20th century and is the giraffe's closest living relative. (Bristol Zoo Gardens)

At much the same time as Sclater's report appeared, Johnston finally received all the material that he needed to ascertain conclusively the okapi's true status. Karl Eriksson, a Swedish officer who had been the Commandant at Fort Mbeni, had succeeded in obtaining a complete okapi skin and two okapi skulls, all of which were sent on to Johnston, now back at Entebbe, Uganda. Studying these priceless specimens carefully, Johnston prepared a detailed water-colour painting, which he dispatched with the skin and skulls in June 1901 to British Museum (Natural History) director Prof. Edwin Ray Lankester—along with his surprise conclusion as to what the okapi had proven to be.[5,9]

Against all expectations, it was neither a zebra nor a present-day *Hipparion*. Indeed, it was not an equid at all, but—just as the natives had insisted—a cloven-hoofed creature, thereby related to the cattle, deer, camels, pigs, and hippopotami, plus one other notable beast, the giraffe. And it was the giraffe that held the key to the okapi's identity—because the skulls showed unquestionably that the okapi was a specialised, forest-dwelling giraffe. Millions of years ago, there had been a wide variety of giraffe types. Some were tall and long-necked like today's giraffe; some were sturdy deer-like species (sivatheres) with spreading antler-like horns; and some were smaller, short-necked forms very like the newly-discovered okapi. One of these last-mentioned forms, belonging to the genus *Helladotherium*, seemed especially similar.[5,9]

So, undeterred by his failure to resurrect *Hipparion*, Johnston saw a second chance to revive a fossil genus, by naming the okapi *Helladotherium tigrinum*, which he did in a letter to Lankester enclosed with the specimens and painting. The more knowledgeable Lankester, however, recognised that the okapi was more closely related to the extinct short-necked giraffe genera *Palaeotragus* and *Samotherium*. Furthermore, he could see that certain features of its skull distinguished it even from these. So when he formally described the okapi on 18 June 1901 Lankester created for it a new genus, *Okapia*, and according to the rules of nomenclatural precedence its scientific name thus became *Okapia johnstoni*.[9]

The okapi proved to be a singularly striking animal—resembling an unlikely hybrid of giraffe, mule, antelope, and zebra, standing just under 5.5 ft at the shoulder (and marginally over 6 ft in total height, thanks to its slightly prolonged neck). It had a long tuft-tipped tail, and was clothed in a short, dark violet coat of a glistening velvet texture, vividly emblazoned with black and white stripes upon its rump and the upper portions of its limbs (especially the hind pair). Added to this was a pale ghostly face with an elongate muzzle accommodating its lengthy, leaf-tearing blue tongue, a pair of long ass-like ears, and (in the male) a pair of short horns.

News of this extraordinary animal's discovery made newspaper headlines worldwide, which in turn inspired several, thankfully unsuccessful, attempts by Western hunters to add an okapi or two to their trophy walls and cabinets. Eventually, however, the desire to become famous by shooting okapis was superseded by a comparable aim to achieve immortality by being the first to observe and photograph specimens in their native habitat, and especially to bring a *living* okapi back to the west. Between 1909 and 1915, an expedition sent to the Ituri by the New York Zoological Society and including Dr. James Chapin (later to earn fame by discovering another major new African species—p. 123) uncovered a great deal about the okapi's docile, vegetarian lifestyle in its jungle demesne, and also captured two specimens. Sadly, however, both died shortly afterwards, but on 19 August 1919 a live okapi finally reached a western zoo, when a young specimen caught during the previous December arrived at Belgium's Antwerp Zoo.[10-11]

Yet only 50 days after its Antwerp debut this first Western okapi also died—a sad story that would be repeated with a succession of other captured okapis until Tele, a female specimen, arrived at Antwerp in 1928. Adjusting well to zoo life, she thrived there for the next 15 years, and even her death in 1943 was due only to lack of sufficient food caused by World War II. America's first okapi, a male called Congo who arrived at the Bronx Zoo in autumn 1937, also lived for 15 years. By the 1950s, there were several okapis worldwide, and the species had begun to breed in captivity too,[10-11] with Britain's Bristol Zoo in particular earning great renown for accomplishing this difficult feat.

Unlike so many discoveries that initially engendered international attention and interest but swiftly lost their wonder in the wake of newer finds, the okapi has never lost its romantic appeal as a truly unique animal totally unlike any other, and as living proof that the age of animal discovery did not end with the close of the 19th century. For wildlife enthusiasts and scientists everywhere, it is the epitome of the unexpected and the unknown that still await disclosure within every aspect of Nature. And in recognition of this, the okapi received a further accolade in 1982, when it was adopted as the official emblem of the International Society of Cryptozoology—a fitting tribute to the animal that remains the premier zoological discovery of the 20th century.

THE FIVE-HORNED GIRAFFE AND A GIANT GENET

Sir Harry Johnston is principally remembered for the discovery of the okapi. Less well known is that within weeks of this becoming major news throughout the world, his name resurfaced in connection with the finding of a *second* new type of giraffe!

*Rothschild's giraffe—
the male has five horns.*

The true giraffe *Giraffa camelopardalis* exists in a number of different subspecies—some with two horns and some with three. Shortly after unveiling the okapi, Johnston travelled into the Ngishu Plateau of what was then the Uganda Protectorate, just south-east of Mount Elgon, and saw some unusually-marked giraffes whose males were further distinguished from all previously recorded forms by having *five* horns. In addition to the single median horn on the forehead and the pair sited just behind it (as in other giraffes), the males bore a second pair, at the back of the skull. Greatly intrigued, Johnston obtained the skulls and head skins of two males and two females, which he sent to the British Museum.[12-13]

This latest new giraffe attracted the interest of Lord Walter Rothschild, who funded an expedition to bring back a complete specimen for the museum. This was eventually achieved, and the specimen was duly mounted and displayed, revealing the new giraffe form to be a particularly handsome one, whose large brown blotches on the neck, body, and upper limbs each contained an attractive star-shaped motif; and whose slim lower limbs were pure white, in contrast to the darker shading characterising most of the other giraffe races. In recognition of Lord Rothschild's efforts, the new subspecies was dubbed *G. c. rothschildi*—Rothschild's giraffe.[12-13] (In recent years, some researchers have suggested that this and certain other giraffe subspecies should be elevated to the rank of separate species.)

And to complete Sir Harry Johnston's run of zoological success in 1901, this year also saw the description of a magnificent new species of genet, based upon a skin that he had obtained a little earlier. Genets and civets belong to a family of cat-related carnivores known as viverrids, and, traditionally, there had never been any problem in distinguishing the various African genets from the African civet. Whereas the former were small, thin-bodied creatures with long, slender tails and readily-distinguished markings, the civet was much larger and bulkier, with a thicker, brush-like tail, and darker, less clearly-delineated markings. Johnston's new genet, however, made short work of disrupting this orderly classification, because although its anatomy undeniably allied it with the genets, outwardly it displayed a distinct resemblance to the civet—not just on account of its large, dense markings and very thick, rather bushy tail, but also by virtue of its enormous size. With a heavily-built body and a total length of 3.5 ft, it was almost twice the size of any other genet, thereby approaching the civet's own dimensions.[14]

Rothschild's giraffes—among the rarest of giraffe subspecies (or, as some now believe, species).

Unquestionably a new species, the giant genet was named *Genetta victoriae* by British Museum zoologist Oldfield Thomas, who believed that the type skin supplied by Johnston had been collected at Entebbe, on the shores of Lake Victoria. In fact, it had originated from the okapi's home, the Ituri Forest-Semliki River region (nowadays contained within the Democratic Congo's territory), but its species has since been recorded from western Uganda too, once again from dense lowland forests, though it remains little-known and rarely-spied.[14]

MOUNTAIN GORILLA—
GIANT APE OF THE VOLCANOES

Zoologists had scarcely recovered from the shock of the okapi's discovery when the Dark Continent unfurled another important new mammal. Prior to the 20th century, the gorilla (whose own existence was not officially recognised by science until the 1840s) was thought to be confined to the lowlands of west-central Africa. Yet as far back as 1860, the explorer John Speke had collected native reports of a huge monster, resembling a giant ape, which supposedly inhabited the lofty Virunga (Mufumbiro) Volcanoes range of mountains, constituting the borders of eastern Democratic Congo, Rwanda, and Uganda. And in 1898, long-distance walker Ewart Grogan, passing through the Virungas, encountered the skeleton of an enormous ape. Unfortunately, he left it where he found it, but sufficient interest had been aroused in scientific circles for others to take notice of further reports.[2-3,15]

By 1901, it seemed certain that an unknown form of gorilla did indeed exist in this mountain range—all that was needed for official verification was a specimen. And in October 1902, this vital requirement was supplied by Captain Robert 'Oscar' von Beringe, from the Belgian Army, and his companion, a medical officer called Dr. Engeland, who bagged a couple of gorillas on the Virungas' Mount Sabinio, and sent the skull and partial skeleton of one of them on to Europe.[2-3,15]

By comparison with the western lowland form, the mountain-dwelling Virunga gorilla was larger and sturdier in overall build, with a broader chest and body, a jet-black (rather than a brownish-grey) coat with longer fur, and longer jaws with larger teeth, but relatively shorter arms. In 1902, Berlin Museum's director, Dr. Paul Matschie, formally named the new ape *Gorilla beringei*, popularly known thereafter as the mountain gorilla (and later demoted to subspecific status as *Gorilla gorilla beringei*).[15]

Modern studies by researchers such as Dr. George Schaller and Dian Fossey have shown that far from being a savage man-eater, this giant ape is a gentle, intelligent vegetarian. It is also highly endangered; hunting and habitat erosion have reduced its numbers to around 700.

Gorillas occur in Congo's lowlands too, but differ from the mountain subspecies by way of their shorter fur and jaws, smaller teeth, and longer arms. They have traditionally been classed as a separate subspecies—*G. g. graueri*, the eastern lowland gorilla.[16]

However, during spring 2000, experts at a major primatological conference held in Orlando, Florida, proposed a radical revision of gorilla taxonomy, which has since become widely accepted. Based upon results obtained from molecular genetics research in conjunction

Early photos of mountain gorilla—dismissed as an ogre of native folklore until its scientific discovery in 1902.

with more traditional taxonomic methods, they recommended that this single species should be split into two distinct species. These are: the western gorilla *Gorilla gorilla* (containing two subspecies), and the eastern gorilla *G. beringei*, containing three. The two western subspecies are the western lowland gorilla *G. g. gorilla*, and Nigeria's Cross River State gorilla *G. g. diehli*. As for the eastern species, its three subspecies are the mountain gorilla *G. beringei beringei*, the eastern lowland gorilla *G. b. diehli* (formerly *G. g. graueri*), and the Bwindi gorilla (as yet unnamed) inhabiting Uganda's Impenetrable National Park.[16a]

Conversely, the so-called pygmy gorilla, dubbed *Pseudogorilla mayema* in 1913 by Dr. Daniel G. Elliot and known from several skins,[17-18] is not recognised today as a genuinely distinct taxonomic form.

SAGA OF THE SEA MINK—
THE HISTORY OF A TRAGEDY

The official description of the sea mink in 1903 was no cause for scientific celebration; instead, it brought to a sorry close the exceptionally tragic story of a remarkable species driven into extinction before science had even become aware of its existence.

As late as the 1850s, fur trappers working for the European fur industries and based along the coasts of New England and western Canada were familiar with an extremely large form of mink. It was not only half as long again as the longest individuals of the common American mink *Mustela vison* on record, but also much fatter, so that it yielded a pelt twice the size of any from American mink. It was further distinguished by the coarser texture and pale, reddish colouration of its pelage; its body odour was also said to be totally different. As for its habitat, whereas the American mink is an inland species the giant form was apparently a coastal dweller, confined exclusively to the shores and offshore islands within its recorded distribution range. Thus it was known to the fur trappers as the sea mink.[19-20]

Yet despite its distinctive appearance and specialised habitat, the sea mink attracted no interest from science; specimens were presumably dismissed as nothing more than freakishly large individuals of the ordinary American mink. The trappers, however, were only too keen to take notice of it, because its large pelt made it a highly sought-after fur-bearer (especially the males, which were about a fifth larger than the females).[19-20]

By the 1870s it had become much rarer, and in 1880 a specimen killed on one of the islands of the Maine township of Jonesport (and sold to a fur buyer) may have been the very last sea mink in existence. Even without its tail it measured over 26 in long (thereby exceeding the *total* length of the longest New England specimens of *M. vison*). However, there is also a record of a very large mink of pale colouration taken in 1894 or thereabouts on New Brunswick's Campobello Island.[19-20]

Even so, it had certainly died out by the turn of the century, when piles of its remains, found in Indian shell-heaps along the New England coasts, attracted the attention of American scientist Dr. Daniel Webster Prentiss. These convinced him that this giant mink was a distinct species, which he formally described in 1903, christening it *Mustela macrodon* (emphasising its large teeth).[19-21] The sea mink had finally gained scientific recognition—but over a decade (if not two) after it had become extinct. There seems little doubt that if it had been identified earlier, it could have been saved by captive breeding. Indeed, the great value of its oversized pelt would surely have been encouragement enough for the establishment of such an enterprise. Instead, it had died as it had lived, in zoological obscurity, and remains today one of the least-known of all modern-day mammalian carnivores, rarely accorded even the briefest of mentions in books.

Sea mink. (Tim Morris)

What has often been claimed to be the only known preserved specimen of a sea mink is a striking (albeit greatly faded) light reddish-tan taxiderm mink, mounted on a wooden base with the skull and limb bones still in place. Its present location is unknown to me, but it was originally owned by Clarence H. Clark—a businessman, politician, historian, and county commissioner of Maine's Washington County, who claimed in 1924 that it was taken at Campobello Island in 1894. According to a newspaper report in 1929, it had been sought by several museums and was "beyond price". By 1964, however, Clark's mink and his other zoological specimens had passed into the ownership of James C. Sullivan, living at that time in Dennysville, Maine, who did not wish to retain any of them, but deemed it likely that he could sell the mink for a large sum. He approached the U.S. National Museum, who offered to examine it before deciding whether or not to purchase it.[21]

This famous specimen duly arrived at the museum in May 1965, and was scrutinised there by no less than

20 different mammalogists before it was returned to Sullivan (living by then in East Winthrop, Maine). Deemed to be probably an adult male, it measured 72 cm (28.8 in) in total length, of which 21 cm (8.3 in) constituted its tail. However, although larger in total length (but this may have been due to stretching during the taxiderm process), its other measurements (including dental ones) all placed it within the size limits of *M. vison* and well below those of the sea mink *M. macrodon*.[21]

As a result, the U.S. National Museum announced that this specimen was not a sea mink, but was either an unusually large example of *M. vison*, or an intergrade (intermediate) specimen between *M. vison* and *M. macrodon*. In 1966, this latter possibility helped to persuade museum mammalogist Dr. Richard H. Manville, within a formal paper documenting the known history of *M. macrodon* (and including the museum's findings regarding the taxiderm mink), to reclassify the sea mink as a subspecies of *M. vison*, thereby renaming it *M. v. macrodon*. Most authorities, however, still choose to treat it as a distinct species.[21] (Recently, the American mink has been allocated its own genus, *Neovison*, thus becoming *Neovison vison*, and the sea mink has also been placed within it, becoming *Neovison macrodon*.)

DWARF SIAMANG—AN AMBIGUOUS APE

In 1821, the world's largest species of gibbon, the siamang *Symphalangus syndactylus* of Sumatra, was described by Sir Stamford Raffles. Almost a century later, zoologist C. Boden Kloss discovered what seemed to be a dwarf counterpart, living on South Pagi, one of the Mentawi Islands off Sumatra's western coast. Much smaller and rarer than the true siamang, with very soft, silky black fur, it was described in 1903 by Smithsonian Institution mammalogist Dr. Gerrit Miller,[22] but its precise affinities remained somewhat controversial. Certain authorities maintained that it was truly most closely related to the siamang and referred to it as *S. klossii*, the dwarf siamang. Various others installed it within a genus all to itself, *Brachytanites*; but nowadays most researchers prefer to ally it with the other gibbons, thereby naming it *Hylobates klossii*, the Mentawi gibbon.

PACARANA—COUNT BRANICKI'S 'TERRIBLE MOUSE'

1904 was a momentous year for mice, for it marked the rediscovery of a truly remarkable rodent known to science as 'the terrible mouse', due to the fact that it was as large as a fox terrier!

Needless to say, any mouse the size of a small dog is no ordinary mouse, and in truth this species is not a bona fide mouse at all. If anything, it more closely resembles a long-tailed, spineless porcupine in general shape, and sports a handsome grey-black pelage decorated with longitudinal rows of white spots, which compares well with that of the South American paca *Cuniculus paca*, a fairly large relative of the guinea pig. Indeed, in its native Andean homeland, the 'terrible mouse' is known locally as the pacarana ('false paca'). Yet it is neither paca nor porcupine either. Instead, it is sufficiently removed from all living rodents to require an entire taxonomic family to itself, Dinomyidae, thereby making it one of the most important mammalian discoveries of the past 150 years—not to mention one of the most elusive.[23]

Measuring up to 3.25 ft long, the pacarana is the world's third largest living rodent (exceeded only by the capybaras and beavers), and was discovered in 1873 by Prof. Constantin Jelski, curator of Poland's Cracow Museum. Financed by Polish nobleman Count Constantin Branicki, Jelski was engaged in zoological explorations in Peru when, one morning at daybreak, he observed an extremely large but wholly unfamiliar rodent. It had very long whiskers and a fairly long tail, and was wandering through an orchard in the garden of Amablo Mari's hacienda near Vitoc, in the eastern Peruvian Andes. He swiftly dispatched the poor creature, and sent its skin and most of its skeleton back to Warsaw, where it gained the attention of Prof. Wilhelm Peters, Berlin Zoo's director, who meticulously studied its anatomy. Recognising that this huge rodent represented a dramatically new species, by the end of 1873 he had published a scientific description of it, in which he named it *Dinomys branickii*—'Branicki's terrible mouse'. The pacarana had made its scientific debut.[23-4]

Pacarana—Count Branicki's 'terrible mouse'.
(Jürg Klages/Photo und Graphik, Zurich)

Peters's studies disclosed that its anatomy was a bewildering amalgamation of features drawn from several quite different rodent families. In terms of its pelage and limb structure, it compared well with the paca, but unlike the five-toed (pentadactyl) configuration of the latter's paws the pacarana's each possessed just four toes. Many of its cranial and skeletal features (not to mention its long, hairy tail) also set it well apart from the paca, especially the flattened shape of the front section of its sternum (breast bone), and the development of its clavicles (collar bones). Certain less conspicuous features of its anatomy were reminiscent of the capybara, but various others (including the shape of its molar teeth) corresponded most closely with those of the chinchillas. There were also some additional characteristics that seemed to ally it with the West Indian hutias (see p. 56). Little wonder then that Peters elected to create a completely separate taxonomic family for it![23-4]

Rare early photograph of a captive pacarana.

The pacarana was clearly a major find—yet no sooner had it been discovered than it vanished. For three decades nothing more was heard of this 'false paca', and zoologists worldwide feared that it was extinct. Then in May 1904, Dr. Emil Goeldi, director of Brazil's Para (now Belem) Museum, received a cage containing two living pacaranas (an adult female and a subadult male). These precious animals had been sent from the upper Rio Purus, Brazil, and proved to be extremely docile, inoffensive creatures, totally belying their 'terrible mouse' image. They were swiftly transferred to Brazil's Zoological Gardens, but tragically the adult female died shortly afterwards, following the birth of the first of two offspring that she was carrying.[23,25]

In 1919, a more unusual-than-normal pacarana was described by Alipio de Miranda Ribeiro. Instead of being greyish-black in colour, it was brown, and so Ribeiro designated it as the type specimen of a new species, christened *D. pacarana*. Three years earlier, the first pacarana recorded from Colombia had been collected (near La Candela, Huila); in 1921, this became the type of a third species, *D. gigas*. During the early 1920s, a series of pacaranas was procured by Edmund Heller from localities in Peru and also Brazil, so that by the 1930s a number of museum specimens existed, which were then examined carefully by Dr. Colin Sanborn in the most detailed pacarana study undertaken at that time. Publishing his findings in 1931, he revealed that *D. pacarana* and *D. gigas* were merely forms of *D. branickii*, which meant that only a single species existed after all.[26]

A rarely-glimpsed inhabitant of mountain forests, the pacarana feeds on leaves, fruit, and grass, and is hunted as a source of food by its Indian neighbours, but little else is known about its lifestyle in the wild. It is currently classed as a vulnerable species by the IUCN, yet as a result of its secretive habits and relatively inconspicuous habitat it may be more abundant than hitherto suspected (nowadays it is known to be fairly common, for instance, in Bolivia's Cotapata National Park).[23]

GIANT FOREST HOG—
THE WORLD'S LARGEST PIG

Those critics of cryptozoology who believed that the okapi would surely be the last large mammal to be discovered in Africa were effectively silenced less than three years after its debut, when in 1904 this ever-startling continent offered up another outstanding zoological surprise—a previously undescribed species of wild pig attaining quite colossal proportions.

As far back as 1668, in his *Naukeurige Beschryvinge der Afrikaensche gewesten van Egypten*, Dr. Olfert Dapper had referred to an extremely large black pig from Liberia, known to the natives as *couja quinta*. Two centuries later, during his Emin relief expedition of 1888-90, Sir Henry Stanley collected reports of a gigantic pig, reputedly reaching 6 ft in length, which allegedly inhabited the Ituri Forest. At much the same time, rumours of a comparable creature were also issuing from Kenya. Some of these eventually reached the ears of Lieutenant (later Captain) Richard Meinertzhagen of the British East-African Rifles, who happened to be stationed in Kenya in 1903, and as a keen hunter he became determined to track down this formidable animal.[2-3,27]

His first piece of good fortune came in February 1904, when he learnt that one of these creatures had been killed by native hunters at a small village on Mount Kenya. By the time that he reached the spot, however, the carcase had been badly damaged, but one of his companions obtained from it a couple of pieces of skin. Meinertzhagen readily recognised these as being from some form of pig, but none that was known to him. Not long afterwards, he was even more successful, discovering in a village on the mountain's southeastern slopes an almost complete, relatively fresh skin, plus a second, somewhat older skin. Parcelling all of his finds

Giant forest hog—an early photo highlighting this species' grotesque, fungus-like facial swellings.

together, he sent them to Oldfield Thomas at the British Museum, and resolved to seek out additional evidence of the giant pig's existence.[2-3,27]

Meinertzhagen's crowning triumph came in May, when he learnt that another individual had been killed, this time in the Nandi country near to Lake Victoria. He arrived on the scene in time to secure not only a further piece of skin, but also the entire skull—which measured 3 ft in length! He also obtained an incomplete skull from a somewhat older specimen. Well-pleased with his good fortune, and anxious to discover whether they were of any scientific significance, Meinertzhagen sent these latest specimens to London, and awaited Thomas's verdict—which proved to be far more exciting than anything that he had dared to anticipate.[2-3,27]

The giant forest hog, as it soon became known, was indeed a new species, but so unlike anything seen before that it also warranted the creation of a new genus, *Hylochoerus*. And in recognition of the valiant efforts made by its discoverer to bring it to the attention of western science, Thomas named it *H. meinertzhageni* in his official description of the species, published on 15 November 1904.[27]

The most striking features of the giant forest hog, other than its huge size—it can exceed 7 ft in length, and reach 3 ft at the shoulder—are its relatively long legs and massive head; its bristly, shiny black or dark brown coat; the pair of formidable tusks, curved and bulky, projecting from its upper jaw; a huge pair of fungus-like warts positioned beneath its eyes and stretching down as far as its cheeks; and a most peculiar but very noticeable indentation on top of its head, large enough to accommodate a man's fist![27]

A predominantly nocturnal denizen of dense jungles, the giant forest hog is now known to exist in three principal localities across Africa. One comprises those regions of Kenya and Tanzania bordering Lake Victoria, and stretches northeastwards into the Democratic Congo (formerly Zaire) and southern Ethiopia. The second overlaps Cameroon and the People's Republic of the Congo (where it was once believed to constitute a separate species, named *H. gigliolii*). And a third, deep in West Africa (where it was once believed to comprise another separate species, named *H. rimator*), extends over the border region between Ghana, Burkina Faso, and the Ivory Coast, westwards into Liberia. There are also unconfirmed reports that it exists even further west, in Guinea and in Guinea Bissau.[8,27-8]

Regrettably, however, it is rare in all of these areas, although this was not always the case. Judging from native testimony, prior to the late 1800s it was apparently much more common, but in 1891 domestic pigs throughout Africa were devastated by rinderpest, a disease to which the giant forest hog seems to be particularly susceptible, and which almost certainly had a profound effect upon its numbers thereafter. Fortunately, although uncommon its continuing survival does not appear to be threatened. Indeed, it has become a somewhat unexpected tourist attraction—thanks to its occasional forays into the forests overlooked by the famous Kenyan hotel Tree Tops.[8]

Goeldi's monkey. (Zoological Society of London)

GOELDI'S PERPLEXING LITTLE PRIMATE

Based upon a skull-less, incomplete skin, in 1904 Oldfield Thomas described a new species of monkey that was destined to disrupt very severely the traditional division of South American primates into two distinct taxonomic families—the marmosets and tamarins (Callithricidae) and the true monkeys (Cebidae). The skin had been obtained from Brazil's Para Museum, and Thomas named its species *Midas goeldii*, after the museum's director, pacarana rediscoverer Dr. Emil Goeldi.[29]

In December 1911, a small female monkey, formerly living in the Para Museum's adjoining zoological gardens, was described by Alipio de Miranda Ribeiro under the name *Callimico snethlageri*, as Ribeiro considered it to be sufficiently different from all other South American species to require its own genus.[29] When its remains were examined in 1913 by Thomas, he agreed with Ribeiro's views concerning its notable distinctness, but he also recognised that this specimen and the incomplete skin that he had personally described back in 1904 clearly belonged to the same species. Accordingly, the rules of nomenclatural precedence dictated that the species would thereafter be known as *Callimico goeldii*, Goeldi's monkey—a profoundly enigmatic creature that embodies a unique combination of features from both families of New World primate.[29-30]

Externally, it appears quite an ordinary-looking animal, superficially resembling a predominantly black or dark brown marmoset, with an average head-and-body length of 8 in, and an extra 10-13 in of tail, plus a mane of fur over its neck and shoulders. Examination of its cranial, dental, and skeletal features, however, exposes its bemusingly intermediate position between the two primate families. Whereas the claw-like nails and the overall skeletal structure of its feet, plus the basic configuration of its face, are all more comparable to those of genuine marmosets and tamarins, its dentition and cranial structure are more like those of South America's true monkeys.[30]

So how can it be classified? Even today, there is no firm agreement concerning Goeldi's monkey. Some authorities (like Oldfield Thomas) have chosen to house it within a subfamily of Cebidae (as a peculiar true New World monkey), others (like Reginald Pocock) within its own subfamily of Callithricidae (as an aberrant marmoset). A few (like Captain Guy Dollman) have preferred to place it within a family of its own—Callimiconidae. Goeldi's perplexing little primate succinctly demonstrates that however well-established a system of classification for a given group of species seems to be, there is always the possibility that a new species will come along and completely overturn it![30]

Hawaiian monks seals. (USFWS)

A SHOAL OF NEW SEALS

The Hawaiian monk seal *Monachus schauinslandi* is the world's most recently-described, accepted species of seal, named in 1905 by Dr. Paul Matschie after Prof. H.H. Schauinsland, a German zoologist who had discovered the first scientifically-known skull from this species on the Hawaiian island of Laysan.[31] Up to 9 ft in length, with grey-brown upperparts and silvery-hued underparts (sometimes acquiring a greenish sheen caused by growth of algae amongst its fur), this was once an abundant species. Relentless hunting by man, however, has reduced its total population to 1000-1500 individuals, so that today it is one of the world's rarest seals, and is officially classed by the IUCN as critically endangered.[32]

Speaking of monk seals: the last confirmed report of a Caribbean monk seal *M. tropicalis* was in 1952, and in 2008 the U.S. government officially deemed it extinct.

However, a survey conducted by I.L. Boyd and M.P. Stanfield among fishermen in Haiti and Jamaica during 1997 yielded some persuasive circumstantial evidence for believing that it may still survive.[32a]

In 1904, a new species of fur seal, *Arctocephalus galapagoensis*, the world's smallest, was described from the Galapagos Islands, but some authorities merely treat it as a subspecies of the South American fur seal *A. australis*. Similarly, the Australian fur seal *A. doriferus*, described in 1925 by F. Wood Jones, is usually demoted to a subspecies of the South African fur seal *A. pusillus*.[32] In 1963, a new species of seal from the Pacific's Kurile Islands was reported by a Russian scientist, but its separate specific status is rarely accepted nowadays.[33]

THE EMPEROR WHO SHOULD HAVE BEEN A MANDARIN

One of the most delightful and frequently illustrated species of South American monkey is the emperor tamarin of northwestern Brazil. Yet it was unknown to science until 1907, when it was officially described and named by Dr. Emil Goeldi. With a principally black head and body measuring up to 9 in long, and a rufous-coloured tail around 12 in long, the emperor tamarin is instantly recognised and distinguished from all other species by way of its enormous white moustache, which sweeps majestically downwards in faithful imitation of a venerable Chinese mandarin's.[34]

Its 'imperial' title arose through a mistake made by the taxidermist responsible for preserving the species' type specimen. As he had never seen a living example, he apparently assumed that in life its extraordinarily large moustache would curl *upwards*, just like that of the German emperor Wilhelm, and so that is precisely how he arranged it. Accordingly, this engaging little monkey became known as the emperor tamarin, and Goeldi duly christened it *Midas imperator* (later changed to *Saguinus imperator*). Of course, the procurement of more specimens swiftly exposed the true appearance of this species' moustache, but by then its imperial appellation had become too widely used to be dropped[10]—all of which explains why the emperor tamarin is *not* known, much more appropriately, as the mandarin tamarin!

SLOW LORISES SMALL, INTERMEDIATE—AND TAILED!

Those Asian, near-tailless relatives of lemurs, the slow lorises, were known only from two species—the Sunda slow loris *Nycticebus coucang* and the Bengal slow loris *N. bengalensis*—until 1907, when a third was described by J.L. Bonhote. Its type specimen had been collected by a Dr. Vassal on 13 November 1905 at Nha-trang, Annan (now part of Vietnam); and although an adult, measuring a mere 8 in long it was only half the total length of the other slow lorises. Its dentition was also very distinct, and so Bonhote dubbed it *N. pygmaeus*, the pygmy slow loris.[35]

Intriguingly, a single adult female *Nycticebus*-like loris was obtained many years later from Hoa Binh, northern Vietnam, that was midway between the two large slow lorises and *N. pygmaeus* in size. This inspired D.V. Tien in 1960 to designate it as the type of a new species, which he fittingly named *N. intermedius*,[35] but few zoologists accept this. (Having said that, however, some zoologists have lately advocated splitting the *Nycticebus* lorises into no less than six separate species, including *N. intermedius*, and have split the slender loris *Loris tardigradus* into six species too.)

Far more extraordinary were the white-coated, woolly-furred *Nycticebus*-like lorises captured and photographed near Fort Lungleh in Assam's Lushai Hills during December 1889 (but undocumented until 1908). With short but stout limbs, a large rounded head, flat face and small muzzle, short roundish ears, large eyes each encircled by a dark triangular patch, and a narrow black stripe running from its skull's occipital region along its back's entire length, this form differed dramatically from all other lorises in one very conspicuous way—it had a thick bushy tail![35] Unless it is a teratological, freak variety of *N. bengalensis*, a radically new

Emperor tamarin
(A. van den Nieuwenhuizen)

species of slow loris still awaits official recognition—for which, in this present book's first edition, *The Lost Ark* (1993), I proposed *Nycticebus caudatus* ('tailed slow loris') as a suitable name.

While speaking of pygmy primates, another controversial form has recently gained taxonomic respectability. In 1921, an unusually small specimen of tarsier collected four years earlier from upper montane rainforest in central Sulawesi (Celebes) became the type of a new species, dubbed *Tarsius pumilus*, the pygmy tarsier. However, its status as a valid species was not generally accepted—until 1987, when an extensive study of this mysterious mammal by Dr. Guy G. Musser and Marian Dagosto demonstrated unequivocally that it deserved separate specific status, thereby adding (after 70 years as a primatological *persona non grata*) a fourth tarsier to the zoological catalogue of formally recognised species.[36]

And just a year later, in 1988, another new species of tarsier was revealed, again in central Sulawesi, which in 1991 was christened Dian's tarsier *Tarsius dianae* (later renamed *T. dentatus*, as *dianae* was found to be a junior pseudonym of *dentatus*). Moreover, taxonomic splitting of existing species in recent times has yielded a total of nine currently-recognised tarsier species.[36]

ANDREWS' BEAKED WHALE— THE ONSET OF AN OCTET

Beaked whales are medium-sized cetaceans superficially dolphin-like in shape but characterised by a pair of longitudinal grooves that meet on the throat, and by long tapering jaws that usually house no more than one or two pairs of teeth (depending upon the genus concerned). Of the eight new species of beaked whale reported during the 20th century, six belong to the genus *Mesoplodon*, each species of which has only a single pair of teeth.

The first member of this *Mesoplodon* sextet to receive scientific attention was Andrews' beaked whale *M. bowdoini*. Distinguished by the splayed manner in which its lone pair of teeth projects from its jaws, it was described in 1908 by Dr. Roy Chapman Andrews from the American Museum of Natural History, who named it after George S. Bowdoin, a donor to the museum. The type specimen, a mounted adult skeleton, had been purchased by the museum four years earlier, and had originated from a stranded individual that had met its death ashore on Brighton Beach, near Canterbury, New Zealand.[32,37-8]

Since then, some 35 stranded specimens have been recorded, at least six of which have duly been obtained for study, all originally stranded on various shores of New Zealand, mainland Australia, or Tasmania, and with an average total length of 13 ft. Anatomically, this species seems in many ways to be a smaller, southern counterpart of Stejneger's beaked whale *M. stejnegeri*; but with no available records or studies relating to living examples, Andrews' beaked whale remains one of the least-known cetaceans.[32,37-8]

MOUNTAIN NYALA—BUXTON'S MAJOR SURPRISE FROM ETHIOPIA

Just when it seemed that Africa must surely have exhausted its supply of significant zoological surprises, yet another new mammal of magnitude was discovered here. In mid-December 1910, zoologist Dr. Richard Lydekker received the skull, spiralled horns, and grey-coated skin of a handsome kudu-like antelope, courtesy of Rowland Ward, proprietor of the famous Piccadilly firm of taxidermists. They had been sent to Ward by Major Ivor Buxton, who had shot the animal that summer in southern Ethiopia, and Ward believed that it may represent a new species. Upon examining the relics, Lydekker came to the same conclusion.[39]

Lydekker duly wrote a letter to *The Times*, which was published on 23 September. As a result, he was contacted shortly afterwards by Major Buxton, who

Tailed loris—mystifying loris still unrecognised by science.

Painting of the mountain nyala's type specimen—a subadult buck.

informed him that the antelope's precise provenance was a locality to the west of the Arusi plateau of Gallaland, on the open stony ground of Ethiopia's Sahatu Mountains, at an altitude of about 9000 ft. Equipped with this extra information, Lydekker prepared a formal description of the new species, which was published on 29 September in *Nature*, and in which he named it *Strepsiceros buxtoni* (thereby allying it with the kudus). However, in a later account, appearing in *The Field* on 22 October and based upon his examination of some additional material received from Buxton, Lydekker changed his mind. He now announced that the antelope was actually most closely related to the nyala *Tragelaphus angasi*, and suggested that it should therefore be referred to popularly as the mountain nyala and be known scientifically as *T. buxtoni*.[40]

Although there can be no doubt that these two species are indeed closely allied, there are various morphological differences that readily distinguish them externally. The mountain nyala is almost 1 ft taller than the common nyala, standing 52 in or so at the shoulder, and it lacks the latter species' very striking series of transverse creamy-white body stripes, although it does bear a longitudinal line of nine white spots on each haunch.

Subsequent investigations within its native habitat concluded that the mountain nyala was very rare, numbering less than 2000 individuals in total, but in 1966 this assessment was shown by a joint World Wildlife Fund/American Geographic Society survey to be an underestimate. Nevertheless, the mountain nyala is by no means abundant (it is currently believed to number around 2500 and is classed as endangered by the IUCN), and is still one of the least-studied of Africa's larger antelope species.[8,40]

QUARLES'S MOUNTAIN ANOA— A BANTAM-WEIGHT BUFFALO

With a shoulder height of little more than 3 ft, the lowland anoa *Bubalus* [formerly *Anoa*] *depressicornis* is a small but nonetheless belligerent species of dark-coated, white-legged, southeast Asian buffalo, native to the lowland forests of Sulawesi (Celebes). Smaller still, however, is its relative the mountain anoa *B. quarlesi*, which inhabits Sulawesi's montane forests, and has a shoulder height not exceeding 2.5 ft. This species was officially described in 1910 by Major P.A. Ouwens, director of the Botanical Gardens at Buitenzorg, Java, who had received two living specimens that year from the mountains of Sulawesi's central Toradja region. They were distinguished from the lowland anoa not only by their smaller size and shorter tails but also by their thicker, lighter-coloured coats, absence of white colouration on their legs, and by their horns' conical bases (those of the lowland anoa are triangular).[41]

There has been much confusion and controversy regarding anoa taxonomy over the years, and some authorities attest to the existence on Sulawesi of a *third* anoa species, even smaller than the mountain anoa, measuring no more than 27 in at the shoulder, and sharing the montane habitat of the latter species.[42] Most researchers, however, prefer to treat these two forms as one. Moreover, in some publications the mountain anoa is referred to as *A. fergusoni*; however, this is the name given by Dr. Richard Lydekker in 1905 to a zoo specimen of anoa long thought to have been a mountain anoa but now known to have been a lowland anoa—hence '*A. fergusoni*' is nowadays suppressed in favour of '*B. quarlesi*' for the mountain anoa.[41]

HERO SHREWS—STRONGMEN OF THE SMALL MAMMAL WORLD

The year 1910 saw the discovery of the first of two very small but thoroughly amazing species of mammal known as hero shrews. Their name derives from the belief held by some native tribes that if they consume (or wear as an amulet) any part of these animals, they will become indestructible. Dubbed *Scutisorex somereni*, this species was found in rainforests near Kampala, Uganda; and in 1913 a second species, later christened *S. congicus*, was revealed in the Democratic Congo's Ituri Forest.[43] (Today, many authorities class these two as a single species.)

Measuring no more than 9 in long (half of which comprises the tail) and weighing only a few ounces, these diminutive species outwardly look much like other shrews, but their skeleton possesses an incredible secret—which was first disclosed in 1916, with *S. congicus*, when a team of American mammal collectors (led by Herbert Lang) visited the terrain of the Ituri's Mangbetu tribe. Prior to this, science only had skulls and skins of hero shrews, and therefore was totally unaware of their unique capability—one that the Mangbetu people exposed in what seemed initially to be a singularly barbaric manner.[44]

A tribesman placed one of these shrews on the ground, and then, to the American collectors' horror, he stood directly on top of its back, balancing on one foot! Naturally, the Americans fully expected the tiny creature to be instantaneously squashed to pulp. Instead, and to their absolute astonishment, not only was the shrew completely undamaged, but when the man finally stepped off its back several minutes later, it merely shook itself and scampered back into the undergrowth, clearly unaffected by the experience![44]

This extraordinary episode's denouement came when the American team sent back some complete *S. congicus* skeletons for study at the American Museum of Natural History (*S. somereni* skeletons were also obtained in due

Congo hero shrew. (Neg. #36883/Photo: Kay C. Lenskjold/ Courtesy Department Library Services, American Museum of Natural History)

course). These revealed that the hero shrew's backbone incorporated an unparalleled degree of protective reinforcement. The vertebrae were much broader than those of other shrews, but in addition the surface of each one was effectively buttressed with a formidable array of bony knobs, ridges, and spines. These interlinked so closely and tightly with those of the vertebrae immediately in front and behind that the backbone constituted an invincible bridge resistant to immense pressure that would instantly crush the backbone of much larger animals, let alone other shrews (see photo, p. 14).[44]

Most mysterious of all, however, is the reason *why* the two species of hero shrew have evolved such an exceptional skeleton, because their insectivorous lifestyle does not seem to differ from that of normal shrews.[44] A century has now passed since the first hero shrew was discovered, but science has yet to provide a convincing answer to this baffling riddle.

PYGMY HIPPOPOTAMUS—RESURRECTION OF A VERY LARGE DWARF

The common hippopotamus *Hippopotamus amphibius* has been known to the western world for many centuries, but in 1849 Dr. Samuel G. Morton, vice-president of Philadelphia's Academy of Natural Sciences, described a second, much smaller species, based upon two skulls originating from Liberia. He dubbed it *H. minor*, but changed that to *H. liberiensis* when he found that *H. minor* was already the name of a fossil species. In 1852, the new hippo's name changed again, this time to *Choeropsis liberiensis* ('pig-like creature from Liberia')—given to it by anatomist and fossil-seeker Dr. Joseph Leidy, whose own examination of the two skulls convinced him that the presence of certain pig-like features rendered this species sufficiently distinct from the much larger, familiar hippopotamus to warrant its own genus.[45] (Since the late 1970s, incidentally, some researchers have reassigned it to a hitherto-fossil genus, *Hexaprotodon*, as *Hexaprotodon liberiensis*, but more recently others have chosen to reallocate it within *Choeropsis*.)

This marked the beginning of a long and heated dispute in scientific circles, with some authorities supporting Leidy's views, certain others preferring to retain it within the genus *Hippopotamus*, and some merely dismissing this so-called 'pygmy hippopotamus' as a freak, stunted variety (i.e. a dwarf) of the ordinary species, or even as an immature form of it.

Over the years, some other skulls and a few skeletons were brought back to the West from Liberia, but these could not stem the flow of conflict that ensued whenever the pygmy hippo attracted scientific discussion. In 1870, there was even a living specimen on show (albeit very briefly) in Europe, when an extremely small individual arrived at Dublin Zoo from Liberia. True to form, it was initially dismissed as a very young common hippo; but when it died a few weeks later, it was carefully examined by Irish zoologist A. MacAllister, who announced that it was a juvenile specimen of the controversial pygmy hippopotamus.[46]

Moreover, its untimely death provided MacAllister with a unique opportunity to study in detail this very mysterious species' external appearance, and as a result he recognised that its critics were totally wrong in assuming that it was merely a scaled-down version of the common hippopotamus (and that it did not therefore deserve separate specific status).[46]

On the contrary, MacAllister found that it was more pig-like than its larger relative, with a proportionately

Pygmy hippo—discounted as a freak, stunted variety of the common hippo until its rediscovery in 1911.

smaller head and longer limbs, and an arched back, and was much lighter in overall build. No longer were its porcine parallels confined to cranial characteristics. Nonetheless, its basic skeletal structure, as well as its large jaws and absence of a delineated snout, confirmed its closer affinity to the hippopotamus than to the pigs.[46]

Yet despite such findings as these, the 'stunted freak' identity still drew a surprising amount of support from scientists. And, to make matters even worse, little additional news reached the West concerning this perplexing pygmy, so zoologists finally wrote it off as extinct (almost as if to rid themselves of a troublesome enigma).[2]

Fortunately, the famous animal trader Carl Hagenbeck was not prepared to let this matter rest so readily and passively. So when he learnt from one of his collectors that Liberians still spoke of a strange beast called the *nigbwe* whose description sounded suspiciously similar to that of the supposedly demised pygmy hippo, he decided to sort out this unseemly taxonomic tangle once and for all—by sending to Liberia the renowned explorer-naturalist Hans Schomburgk. If anyone could find an answer to this anomaly, Schomburgk could.

And after diligent searches and long periods of time spent patiently questioning the natives, Schomburgk did succeed—on 13 June 1911, he found himself standing face to face with a living pygmy hippopotamus, deep in the Liberian forests. Sadly, however, he did not have the means with which to catch it, and he did not wish to kill it—his trip had been primarily a fact-finding mission, to determine whether or not the species still existed. Now that he knew it did, Schomburgk took scant notice of the disbelief and disdain that he received from scientists when he returned home to Europe empty-handed. Instead, he assembled the necessary equipment to catch some pygmy hippos, arrived back in Liberia at the end of 1912, and captured a living specimen on 1 March 1913. Bringing his search to a triumphant conclusion, he returned home in August accompanied by no less than five thriving pygmy hippos, and a wealth of information concerning this 'extinct' species.[47]

No longer could there be any doubt. The pygmy hippopotamus, less than 6 ft long and only 2.5 ft high, with a glossy black skin and a preference for terrestrial life in forests rather than an aquatic existence in lakes and rivers, was indisputably a valid species in its own right. Another Dark Continent denizen had finally acquired scientific recognition.[48]

Pygmy hippos.

What is now recognised to have been a distinct subspecies of pygmy hippo lived in small numbers in Nigeria until as recently as the 20th century. However, its separate taxonomic status was not confirmed until 1969, when four skulls sent by hunter I.R.P. Heslop to the British Museum (Natural History) during the early 1940s (making them the last known specimens of the Nigerian pygmy hippo) were finally evaluated. This extinct subspecies was formally dubbed *C. l. heslopi*.[48a]

SPECTACLED PORPOISE AND TRUE'S BEAKED WHALE

At the turn of the 19th century, an exceptionally handsome porpoise was captured off Argentina, and taken to the Buenos Aires Museum, where scientists were in little doubt that it represented a hitherto unknown species. Its jet black upperparts contrasted sharply against its snowy white flanks and underparts; but most distinctive of all were its eyes, each of which, housed within the white portion of its head, was encircled by a black ring.[38,49]

Tragically, however, before this unique specimen could be studied fully it was somehow lost, but in 1912 a second specimen appeared, stranded at Punta Colares on Argentina's Rio de la Plata. This time, a detailed examination was carried out, by the museum's whale specialist Fernando Lahille, who confirmed that it was a species new to science and, in recognition of its most striking external feature, named it *Phocaena* [now *Phocoena*] *dioptrica*, the spectacled porpoise.[49] It is a relatively small species, no more than 6.5 ft in total length. Less than 15 specimens have been obtained since 1912, and no living individual has ever been studied in detail. As many cetaceans are notoriously elusive, however, rarity of collected specimens is by no means synonymous with rarity of species.[38]

True's beaked whale also became known to science in 1912, when a specimen was stranded on Bird Island Shoal, outside North Carolina's Beaufort Harbour. The following year it was described fully by mammalogist Dr. Frederick True, who named it *Mesoplodon mirum* [now *mirus*].[50] Dark greyish-blue dorsally, with paler, sometimes yellowish underparts (often speckled with small spots), and averaging 17 ft or so in total length, True's beaked whale has since been recorded elsewhere in the U.S.A.'s New England area, as well as from beaches in western Britain, on the opposite side of the North Atlantic. Stranded specimens in South Africa and Australia support the existence of a discrete, southern hemisphere population in temperate waters in the Indian Ocean, which may constitute a separate subspecies.[38,50]

On 9 July 2001, the first living True's beaked whale ever confirmed breaching in European waters was sighted and photographed by Jon Stokes. Together with fellow passengers Audrey and Martyn Harrison (who videoed it), he spied it roughly 30 miles north of the Spanish coast while aboard the *Pride of Bilbao*, as the whale breached no fewer than 24 times.[50a]

Like those of certain other *Mesoplodon* species, the females do not have teeth emerging through the gums, but the males have a single, flattened pair of small, triangular teeth at the end of the lower jaw.[38,50]

TWO SURPRISING VIVERRIDS—
ONE INSECTIVOROUS(?), ONE AMPHIBIOUS(?)

Two of the world's most mystifying members of the viverrid family (housing the civets, genets, and mongooses) came to zoological attention within the space of just two years.

Owston's banded civet *Chrotogale owstoni* is an obscure Asian species, measuring up to 3.5 ft in total length. Named after Alan Owston, whose native collector procured its type specimen on 16 September 1911 at Yen-bay, on Tonkin's Song-koi River in southern China, it was officially described in 1912 by Oldfield Thomas, who designated it as the sole member of a new genus. The visually arresting pattern created by contrasting light and dark, transverse bands on its body and the basal portion of its tail closely resembles that of the banded palm civet *Hemigalus derbianus*, but the latter species lacks the dark spots visible on the neck, shoulders, flanks, and thighs of *Chrotogale*.[48,51]

Furthermore, anatomical comparisons uncovered distinct differences in cranial structure and dentition between the two species, differences sufficiently marked to warrant these civets' respective residence in separate genera. Most remarkable of these contrasts were the very slender muzzle of *Chrotogale*, and its incongruous incisors—these latter teeth are surprisingly broad and close-set, and arranged almost in a semi-circle, a condition more comparable to that of certain insectivorous marsupials than to any species of viverrid.[48,51]

Whether *Chrotogale* too is predominantly insectivorous, however, remains uncertain, as even today it is still a very mysterious animal, known from less than two dozen preserved specimens originating variously from northern Vietnam, Laos, and from Tonkin and Yunnan in China.[48,51] A live individual was captured in Vietnam in 1991, followed by others more recently. These latter include (in 1999) three males and seven females at Hanoi Zoo, six males and four females at Ho Chi Minh City (formerly Saigon) Zoo, four of each sex at Pittsboro Zoo in North Carolina, and one female at Frankfurt Zoo. More recently, an international conservation and breeding programme for them was established in co-operation with Vietnam's Cuc Phuong National Park working with various zoos including Newquay Zoo.[51]

Owston's banded civet. (Chris Brack)

Even more enigmatic, however, is the second member of this distinctive duo of new viverrids. Unrecorded by science until 1913, never studied alive by scientists, and virtually unknown even to the local natives (a rare event indeed!), the water civet *Osbornictis piscivora* is one of the world's most mystifying mammals.

Yet the water civet is an exceedingly handsome, strikingly-coloured creature, with a densely-furred chestnut head and body, a black bushy tail (comprising almost half of the animal's total length of 3 ft), and white facial markings. The type specimen of this secretive species was obtained on 1 December 1913 in a forest stream at Niapu, in northeastern Zaire (now the Democratic Congo), by Drs. James P. Chapin and Herbert Lang during the American Museum of Natural History's Congo Expedition. Six years later, the species was formally described by J.A. Allen from the museum; its generic name honours Prof. Henry Fairfield Osborn (who was greatly interested in the Congo Expedition), and records its fish-eating proclivity.[48,52]

Although its anatomy suggests that it is most closely related to the genets (despite its civet appellation), the water civet exhibits several features markedly at variance with typical genet morphology. Most obvious of these is its vulpine colouration, totally different from the black-and-white coat patterning of spots and bands synonymous with genets. In addition, the soles of its paws are unfurred, its teeth are much weaker and narrower than those of correspondingly-sized genets, its nose is somewhat smaller, its muzzle is shorter, and its overall size rivals that of the giant genet (see p. 27), the largest of all *Genetta* species.[52] (Having said that, in recent years some researchers have reassigned the water civet to the genet genus, *Genetta*, and refer to it as the aquatic genet.)

Most books state that the water civet was *totally unknown* to the natives prior to its scientific discovery in 1913; this is not true. Along with the holotype, Lang and Chapin also obtained an incomplete specimen (lacking skull, tail, and feet) from a native[52]; and in the local Kibila and Kipakombe languages, it has its own specific name—*esele*.[53] Nevertheless, for the most part it is truly as much a mystery to them as it is to science, with virtually no information available concerning its natural history, and very few museum specimens.

In 1996, however, a major new chapter was written in this species' sparse history, when veteran wildlife film-maker Alan Root announced that he had succeeded in filming a water civet in its native Congolese habitat, hunting for fishes by gently tapping the water with its paws and then trailing its long white whiskers on the surface to detect any movements. This unique footage formed part of a special one-hour film of Congolese wildlife by Root entitled *A Space in the Heart of Africa*, which was first screened on British television within the long-running ITV *Survival* series in July 1996.[53a]

BAIJI—CHINESE RIVER DOLPHIN, OR REINCARNATED PRINCESS?

The baiji or Chinese river dolphin *Lipotes vexillifer* is revered in China as the reincarnation of a drowned princess, and has therefore been featured extensively in generations of Chinese poetry, legends, and literature stretching back as far as 200 BC. All of which makes it very surprising that this distinctive cetacean's existence only became known to western science as recently as 1916 (*not* 1914, as many books allege). On 18 February 1916, a visiting American, Charles M. Hoy, killed a specimen in Lake Tung Ting, about 600 miles up the Yangtze River, and sold its skull and neck vertebrae to the United States National Museum, where they were examined by mammalogist Dr. Gerrit S. Miller. He perceived that they were from a completely new species, which he described in 1918, assigning it to a new genus.[54]

A very attractive fish-eating species measuring 7-8 ft in total length, the baiji is silver-grey dorsally and almost pure-white ventrally (*baiji* translates as 'white dolphin'), with a long and slender upturned beak, a triangular

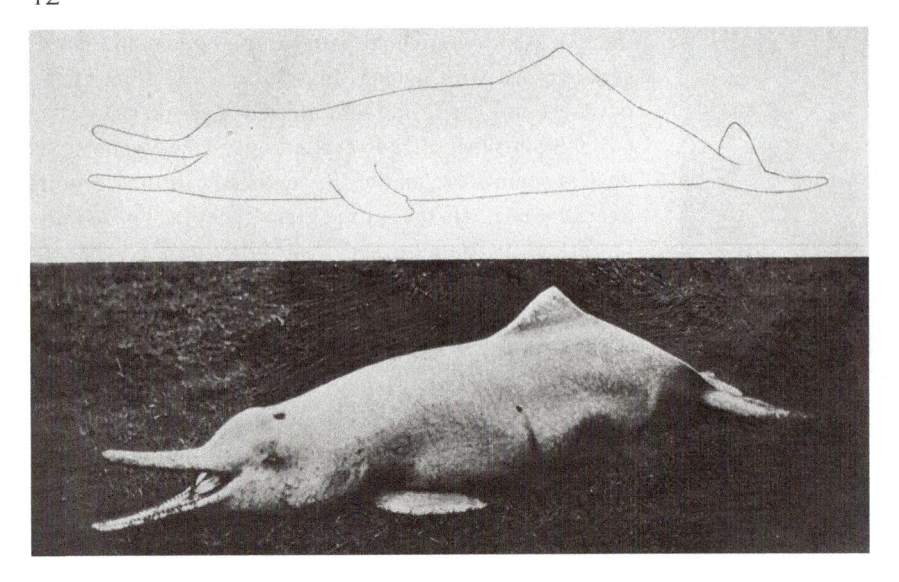

A baiji or Chinese river dolphin, photographed in 1916.

dorsal fin, squat flippers, but almost sightless eyes.[55] According to local reports, it frequently emits loud roaring sounds at night.[38,48] Worth noting is that similar noises have been reported from certain remote Chinese lakes that supposedly house mysterious water monsters. Could these monsters be forms of freshwater dolphin, perhaps even species still undescribed by science?

Few specimens have been taken since Hoy's, but many have been accidentally killed through entanglement with fishing equipment as well as via river pollution, and the species is gravely endangered—if, indeed, it still exists. In *The New Zoo* (2002), I noted that it was restricted to stretches of the Yangtze downstream from Yidu, and that there may be fewer than 100 individuals remaining.[36,48] I also noted that attempts at semi-captive breeding had been made at Tongling, where the baiji is the municipal mascot, and is honoured by a spectacular sculpture depicting five baijis leaping out of the water.[55]

In the years following my book's publication, however, the baiji's fortunes plummeted disastrously. The last captive baiji, Qi Qi, died in 2002. Moreover, during a six-week survey of the Yangtze River in 2006 by the Yangtze Freshwater Dolphin Expedition, no baiji sightings were recorded. And although video footage obtained by Zeng Yujiang of a very large white animal tentatively identified by Wang Kexiong (of the Institute of Hydrobiology of the Chinese Academy of Sciences) as a single baiji swimming in the Yangtze was publicly released in 2007, this iconic species is currently deemed extinct—the first unequivocal extinction of a cetacean species due to human action—though some researchers still hold out hope that it survives.[55a]

WYULDA—A SCALY-TAILED PUZZLE

Whereas most Australian possums are marsupial equivalents of squirrels, tree rats, dormice, or other rodents in basic appearance, the scaly-tailed possum more closely resembles the Madagascan lemurs. Unlike lemurs, however, and unlike other possums too, most of its tail is hairless. Only the basal portion is furred, the remainder being covered with thick, non-overlapping scales instead, from which it derives its common and scientific names.[56-7]

This unusual creature has short pale-grey fur, a dark stripe running along its back from shoulders to rump, a fairly wide head, and uses its unique tail for tree climbing. However, it was unknown to science until 1917, when a female specimen was captured alive at Violet Valley Station, near Turkey Creek in north Western Australia's Kimberley district. In 1919 it became the type specimen of a new species (and genus), named *Wyulda squamicaudata* by W.B. Alexander (*wyulda* is the scaly-tailed possum's Western Australian aboriginal name),[56] and was maintained thereafter in captivity at the South Perth Zoological Gardens.[57-8]

Scaly-tailed possum. (Wildlife Photos W.A.)

Following this, the scaly-tailed possum eluded scientists for 23 years, until in 1942 a second specimen (this time a male) was obtained, by the Reverend J.R.B. Love, close to his mission at Kunmunya—once again in the Kimberley district but on the opposite peninsula coast to the first specimen's provenance. In 1954, Specimen No. 3 was recorded—a female containing a baby in her pouch, collected at Wotjulum Mission, south of Kunmunya, by Kenneth Buller of the Western Australian Museum. No others were obtained until 1965—so that despite extensive searches, from 1917 until the mid-1960s only three adults and one infant had been documented, suggesting that this species was exceedingly rare.[42,48,57-8]

At the end of 1965 and during the beginning of 1966, however, a number of specimens were procured by Harold Butler at Kalumburu Mission, near to Western Australia's northernmost edge, thereby greatly increasing their species' known distribution range. Furthermore, Butler learned from one of the older aborigines there that this possum's numbers fluctuated quite considerably; mostly it was rare, but sometimes it became quite common, although the reason for such oscillations in numbers (if true) has yet to be ascertained. It is quite possible that the recorded 'rarity' of Western Australia's wily *Wyulda* is not so much an accurate assessment of its status as a reflection of the fact that it seldom seems to frequent areas inhabited or studied to any extent by western man.[48,57-8]

KING CHEETAH—A STRIPED ENIGMA

A bizarre beast known to Rhodesian natives as the *nsuifisi* ('leopard-hyaena') was assumed by zoologists to be merely a myth—until the skin of one of these creatures was brought to scientific attention. This was in 1926, when Major A.C. Cooper of Salisbury (now Harare) documented in *The Field* for 14 October the existence at Salisbury's Queen Victoria Memorial Library and Museum of an extraordinary skin obtained from an exceedingly strange felid now known to have been trapped at Macheke, about 62 miles southeast of Salisbury. Royally adorned with ornate blotches and curved stripes upon its flanks and upper limbs, a series of longitudinal stripes running along its back and shoulders, and distinctive rings encircling much of its tail, this exceptionally handsome skin was ultimately identified by Reginald Pocock of the British Museum (Natural History) as that of a cheetah—but one that was drastically different in coat patterning from the typical spotted version.[59-61]

Pocock was convinced that this wonderful striped cheetah constituted a hitherto unknown species, and in his scientific description of it, published in 1927, he christened it *Acinonyx rex*—the king cheetah, a name inspired not only by its regal coat but also by its prominent mane. By 1939, however, he had changed his mind—other 'king' skins had been obtained by then, and some seemed distinctly intermediate between the type specimen recorded by Cooper and normal, spotted skins. This suggested that the king cheetah was nothing more than a freak—a genetic mutant variety of the spotted cheetah *A. jubatus*. Accordingly, from that time onwards it was simply referred to as *A. jubatus* var. *rex*.[59-61]

Breeding experiments during the early 1980s at the de Wildt Cheetah Breeding and Research Centre of Pretoria's National Zoological Gardens confirmed this, revealing that the king cheetah's spectacular coat pattern was due to a recessive mutant allele (i.e. an often-hidden, abnormal version of a gene), probably homologous (equivalent) to the allele creating the blotched tabby pattern in the domestic cat. Nevertheless, the mystique of the king cheetah has not been completely dispelled.[60-1]

Modern-day interest in this most elegant and enigmatic of felids has been due largely to the extensive investigations carried out since the late 1970s by king cheetah experts Lena and Paul Bottriell. Not only have they tracked down and filmed over 20 preserved skins, they have also exposed some anomalous facets of the king cheetah's natural history that cannot be readily

King cheetah—discounted as a native myth until 1926. (Alan Pringle)

reconciled with the belief that it is nothing more than a simple mutant of the normal cheetah. For example, hair samples collected by them have revealed that the cuticular scale pattern of king cheetah guard hair more closely resembles that of the leopard than that of the spotted cheetah. Moreover, whereas the spotted cheetah is primarily a diurnal savannah-dweller, its striped counterpart appears to favour a nocturnal, forest-inhabiting lifestyle, one in which its richly-marked coat would be of particular benefit, providing it with effective camouflage.[60-1]

Hence in 1987 the Bottriells boldly postulated that the king cheetah may actually be demonstrating evolution in action—the divergence of a mutant form from the normal, wild-type, which, if separated reproductively from the latter for a sufficiently lengthy period of time via habitat and behavioural differences, might ultimately become a separate species in its own right. Perhaps Pocock's original view concerning the king cheetah's taxonomic status was not incorrect after all, but merely (in evolutionary terms) a little premature?[60-1]

Also of note in the felid world during the mid-1920s was the discovery in 1925 of a strange wildcat-like skull in China, for its finding marked the rediscovery of *Felis bieti*, the Chinese desert cat. One of the world's most obscure felids, it was previously known solely from two skins purchased in the Tibetan fur markets of Torgolo and Tatsienlu in 1889 by a collector of zoological specimens for Prince Henry d'Orleans; their species was formally described in 1892.[62]

LONGMAN'S BEAKED WHALE—
A MAJOR MARINE MYSTERY

In 1822, a very eroded whale skull, nearly 4 ft long and originating from a male individual, was discovered on a beach near Mackay, Queensland. It was presented by E.W. Rawson to the Queensland Museum, where it lingered in scientific obscurity for the next 104 years. It finally received formal attention in 1926, when it was included by H.A. Longman in a paper dealing with cetaceans of Queensland. Longman could see that the skull had clearly belonged to some form of beaked whale, but as it did not correspond with any of the species hitherto recorded, he designated it as the type specimen of a new one—which he named *Mesoplodon pacificus*, because the skull's structural characteristics most closely resembled those of the *Mesoplodon* species.[63]

Nothing more was heard of this belatedly-described species until 1955, when a second skull was found, albeit in a somewhat unexpected location—the floor of a fertiliser factory near Mogadiscio, Somalia. Fortunately, its discoverer, Italian scientist Dr. Ugo Funaioli, recognised its worth, and traced its origin by learning that it had come in only a few weeks earlier, having been collected on a beach near Danane by some local fishermen. This second skull, which proved to be from a female, was sent by Funaioli to the University of Florence, where in 1968 its identity as a Longman's beaked whale was confirmed by cetologist Maria Azzaroli.[64] That year also saw the creation for this species of its own genus, *Indopacetus*, by Dr. Joseph Curtis Moore, an expert on beaked whales, who cited the skulls' very long beak and shallow tooth sockets as features justifying separate generic status for their species. In or around 1968, a third skull (but lacking the mandible) was found on the Kenyan coast and was given to Nairobi's International School of Kenya, though it was not initially recognised as being from a Longman's beaked whale.[63-6]

In late July 2002 a unfamiliar-looking beaked whale, tan and grey in colour, measuring 21 ft long, had swum aground and died shortly afterwards on the coast in southern Japan's Kagoshima prefecture. Scientists initially buried it, but it was disinterred a week later for identification, and to everyone's amazement was found to be a female Longman's beaked whale *Indopacetus pacificus*—the first complete adult specimen of this ultra-rare species to be recognised by science (in 2000, a complete adult female with foetus had been obtained from Felidhu Atoll in the Maldives, but its taxonomic identity had not been immediately revealed). Prior to the Japanese female's appearance, two complete specimens had been washed ashore in South Africa, in 1976 and 1992 respectively, but these were both juveniles.[66]

Moreover, in August 2002 cetologist Dr. Vic Cockcroft of South Africa's Centre For Dolphin Studies announced that a juvenile Longman's beaked whale had been found washed up on a South African beach. A male, it measured over 15 ft long. However, this specimen was subsequently dismissed by some workers as merely a misidentified Cuvier's beaked whale *Ziphius cavirostris*.[66]

There has also been a tantalising fin-note to this saga. In 1987, Robert L. Pitman documented an unidentified form of beaked whale that has been regularly reported from the eastern tropical Pacific, and can be either uniformly grey-brown or black-white in colour (different sexes or ages?). One of the identities contemplated by Pitman for it was Longman's beaked whale, and following the discoveries of the various complete Longman's specimens, this has now been confirmed.[66a]

DENOUEMENT OF DASOGALE—
A MISIDENTIFIED MADAGASCAN

Indigenous to Madagascar, tenrecs are small, insectivorous mammals that can be readily divided into two major groups—spiny hedgehog-like species, and soft-furred shrew- or mole-like species. For every good rule, however, there is usually an exception to test it, and in the case of the tenrecs the exception was, until quite recently, *Dasogale fontoynonti*.

This species was discovered in 1928 in the forests of eastern Madagascar and described a year later by G. Grandidier (who named it in honour of Dr. Fontoynont, a former president of the Madagascan Academy). It is exceptional not only because it is known from just one specimen, but also because that specimen's skull and dental characteristics are perplexingly intermediate between those of the hedgehog tenrecs and those of the other group. Externally it is also transitional, with spines on its back and flanks, but not on its head or underparts.[48,67]

Inevitably, this anomalous little creature attracted notable zoological interest, and its extreme scarcity earned it a longstanding place in *The Guinness Book of Records* as the world's rarest mammal. In 1987, however, its celebrity status finally ended, when, after detailed comparative studies of tenrec anatomy, Dr. R.D.E. MacPhee of North Carolina's Duke University stated that its type (and only) specimen was actually an immature individual of the greater hedgehog tenrec *Setifer setosus*. Another memorable zoological mystery had been solved.[68]

PYGMY CHIMPANZEE—
A NEW APE, INCOGNITO

Until the mid-1920s, science only accepted the existence of one species of chimpanzee, the familiar common species *Pan troglodytes*, generally divided into three subspecies. In 1928, however, zoologist Dr. Ernst Schwarz was examining some specimens at the Congo Museum in Tervueren, Belgium, when he came upon a series of skeletons and skins of a chimpanzee type that appeared very different from any that he had seen before.[2-3,69]

Obtained by a M. Ghesquiere, they revealed that this strange form of chimpanzee was smaller in size than the common species, and was much more slender in build. Its head was smaller too, but its face was longer and narrower, its dense fur was also long and was uniformly black except for a small white patch on the rump, and it lacked the common chimpanzee's familiar white beard. Schwarz discovered that these odd-looking chimps had been collected in an area of tropical rainforest on the *south* bank of the Congo River—all previously recorded chimpanzees had been obtained from localities *north* of this river.[2-3,69]

Evidently, the southern chimps comprised an important find, and their distinctive morphology persuaded Schwarz that they represented a currently undescribed subspecies. Hence in 1929 he formally documented it within the museum's journal, naming this new ape *P. t. paniscus*—the pygmy chimpanzee. Five years later, it was elevated to the level of a full species—*P. paniscus*.[2-3,48,69]

A pygmy chimpanzee exhibited at Amsterdam Zoo, before its species' official scientific discovery. (John Edwards)

Once science became aware of this freshly-found primate, investigations uncovered that the Tervueren examples were not the only specimens of pygmy chimp to have been collected without recognition of their separate taxonomic status. The British Museum's collections had contained one in 1895, for instance, and a living example of what was almost certainly a pygmy chimp had been on show for a time in 1923 at New York's Bronx Zoo.[10] Similarly, John Edwards, a zoological historian and Fellow of London's Zoological Society, has kindly brought to my attention a picture postcard in his private collection depicting a chimp housed in Amsterdam Zoo, again during the early 1920s, which was quite obviously a pygmy chimpanzee.

Since the species' 'official' discovery, specimens (correctly identified) have been exhibited at several zoological gardens, particularly in European collections such as Antwerp Zoo and Vincennes Zoo. In 1962, Frankfurt Zoo succeeded in breeding pygmy chimps for the first time in captivity, and has repeated this feat on a number of occasions since then. Studies of captive individuals such as these have revealed that the pygmy chimpanzee is much more docile and even-tempered than its less placid, better-known relative.[10]

There has been a fair amount of controversy concerning the pygmy chimpanzee's precise taxonomic identity. Whereas some experts are so convinced of its distinct specific status that they have even placed it within its own genus, as *Bonobo paniscus* (*bonobo* is its native name), certain others still prefer to treat it merely as a subspecies of the common chimpanzee. Recent biochemical and genetic comparisons between common and pygmy chimps, however, indicate that the latter ape is certainly sufficiently distinct to justify classification as a separate species (though whether separate *generic* status for it is warranted is still a matter for conjecture).[70]

It seems surprising that a wholly new species of ape could remain undetected by science until as recently as the late 1920s, especially when specimens were actually preserved in scientific museums and exhibited alive in zoos. In fact, there is an even more ironic twist to this tale. Following extensive researches on the subject, Vernon Reynolds revealed in the 1960s that the individual designated by Linnaeus way back in 1758 as his type specimen for the common chimpanzee was actually a pygmy chimpanzee![42]

GIANT PANDA—RETURN OF THE WORLD'S FAVOURITE WILD ANIMAL

The giant panda is a leading contender for the title of the world's favourite wild animal. Its unique charisma has even charmed that most fickle of popularity promoters, the cuddly toy industry. And as a modern-day endangered species, its fight for survival has engendered international interest, concern, and assistance on a scale equalled by few (if any?) other forms of threatened wildlife. In view of this overwhelming evidence for the giant panda's immense popular appeal, it seems scarcely believable that there was once a time prior to 1929 when zoologists feared that it had been allowed to slip into extinction through lack of interest—neglected by science, and largely unknown to the world outside China.

Referred to by the Chinese as *bei-shung* and prized for its extremely handsome fur (used in earlier days for making ornamental and sleeping rugs), the giant panda was 'officially' discovered by western science in the second half of the 19th century, thanks to the enquiring mind of French missionary-naturalist Père Armand David. At that time, western scholars were aware that ancient Chinese paintings sometimes featured a strange bear-like creature that was predominantly white in colour, but these were generally assumed to be depictions of polar bears brought back to China by hunters—even though the creatures in the paintings usually had black legs (hardly a typical feature of polar bears!).[2,71]

On 11 March 1868, while residing with a landowner at his home in Szechwan, Père David noticed that his host possessed a skin of China's mysterious 'black-legged polar bear', which revealed categorically that the paintings had represented the animal accurately and that, whatever it was, it certainly was *not* a polar bear. His interest keenly awakened, Père David was anxious to see some complete specimens of this exotic-looking species—an inhabitant, he learnt, of the Hsifan Mountains' bamboo forests.[2,71]

Accordingly, the hunters employed by him for the procurement of zoological specimens captured a young panda, which, unhappily, they killed in order to transport it more conveniently. Père David received it from them on 23 March 1869, and on 1 April he received another dead specimen, this time an adult.

Early picture postcard of young captive giant panda. (John Edwards)

Suspecting that their species was a member of the bear family, and one that was new to science, Père David named it *Ursus melanoleucus* ('black-and-white bear'), but when his specimens reached Europe they were examined by mammalogist Prof. Alphonse Milne-Edwards, who came up with a very different identity. Although the creature was indeed new and indisputably bear-like in outward appearance, a number of its dental and skeletal features (including the structure of its feet) convinced Milne-Edwards that in reality it was not a bear, but rather a giant relative of the raccoons—and especially of a fiery-furred Chinese mammal now known as the lesser or red panda *Ailurus fulgens*, described in 1825 by Baron Cuvier and held to be the only living Old World member of the raccoon family.[72-3]

Consequently, when he published his views on this matter in 1870, Milne-Edwards changed the scientific name of Père David's 'black-and-white bear' to *Ailuropoda melanoleuca* ('black-and-white with panda's feet'), emphasising its apparent affinity to *Ailurus*.[72-3] Thus began an extraordinary saga of conflict and contradiction regarding the taxonomic relationship of *Ailuropoda*—one that persisted right up to modern times, as will be discussed later. With respect to this animal's common name, in 1901 its large size and supposed kinship with the lesser panda (which, until then, had simply been called the panda) led to its becoming known as the giant panda.

Its unique appearance and taxonomic interest made zoologists anxious to learn more about the giant panda, and to observe it alive in its native habitat—a desire that proved singularly difficult to satisfy. A few specimens were shot shortly after its official discovery, and found their way into various Western museums. There is even an unconfirmed report that Père David had sent some *living* specimens to Europe, where they were reputedly displayed for a time in Paris. However, most authorities believe that this story arose from nomenclatural confusion, with 'panda' reports based upon *lesser pandas* (i.e. *Ailurus*) being erroneously interpreted as accounts describing the giant panda.[71]

On the whole, therefore, this intriguing bicoloured beast remained intransigently elusive. Except for a single specimen, a young cub given to German zoologist Hugo Weigold as a pet by the local people during his expedition to China's Wassu Province in 1916 (sadly it died only a few days later, in spite of every effort made by Weigold to rear it), no living giant panda was seen by any Westerner from the 1880s to the late 1920s. Eventually, Western science began to fear that this mystifying *sui generis* of the mountains had died out.[2,71]

Not everyone, however, shared this pessimistic view, and among those who intended to spy a living panda were a number of notable big-game hunters from America—eager to add a near-legendary creature to their array of trophies. So it was that the morning of 13 April 1929 found Colonel Theodore Roosevelt and Kermit Roosevelt—sons of the famous U.S. president 'Teddy' Roosevelt—clambering through the snow near Yehli in the Hsifan Mountains, seeking a giant panda to shoot. And after following panda tracks for 2.5 hours, they found one, peacefully asleep in the hollow of a huge spruce tree. As they drew near, it arose, and emerged, still half asleep, looking slowly from side to side before ambling towards some bamboo nearby—whereupon the Roosevelt brothers opened fire at it simultaneously, the resulting hail of bullets killing the poor animal instantly.[2,71,74]

News of this 'success'—the shooting of a half-asleep, inoffensive panda adding a further page of triumph to the hallowed annals of 'sport'—was flashed around the world, enticing other hunters and inciting further killings over the coming years, in what must surely be a uniquely macabre manner of celebrating the rediscovery of a long-lost species.[2,71]

Nevertheless, although the Roosevelts' shooting of their panda is hardly likely to be commemorated by conservationists, in an ironic way it did have one effect that has acted very much in the giant panda's favour ever since, and which is now the major force behind wholehearted attempts sponsored worldwide to prevent the species from dying out. What it did was to bring the giant panda to public notice—introducing to the world at large a creature hitherto little-known beyond scientific circles but which exuded an unbridled degree of charm and appeal. And it did all of this extremely successfully. Very thankfully, by the mid-1930s the craze for panda shooting was past, and a new objective had emerged—to bring back a *living* giant panda to the west.

On 18 December 1936 this goal was accomplished, when the first giant panda known (officially) to have reached the western world alive was gently carried off the *President McKinley* liner at San Francisco, to be greeted with the kind of rapturous welcome normally reserved only for the Hollywood elite. Su-Lin had arrived, causing wholesale pandemonium (or should that be panda-monium?). Still only a cub, she had been captured by Ruth Harkness, widow of William Harvest Harkness, Jr., who had failed in his own, earlier attempts to achieve this zoological coup.[75] Tragically, however, less than two years after her arrival in America, Su-Lin choked on a branch at her home in Chicago's Brookfield Zoo and died; but in the years to come, other giant pandas would arrive in the west, most would thrive for many years, and all would be major mammalian mega-stars.[71,76]

Equally well-publicised have been the many vain attempts to breed giant pandas successfully in the west—the most celebrated, long-running episode involving the chaperoned visits to one another, back and forth between London and Moscow, by Britain's female panda Chi-Chi and Russia's male An-An. In contrast, giant pandas in Chinese zoos have been successfully bred and reared for many years.[71,76]

Needless to say, the adoration that giant pandas inspire in the west has not gone unnoticed by the Chinese, whose awareness of their country's most popular animal swiftly turned to alarm as they recognised that it has progressively become one of its rarest. Current estimates place the total number of wild pandas in China at less than 1000. For many years, therefore, the Chinese government has been at the forefront of continuing, concerted efforts to perpetuate and safeguard the species in its native habitat. Just to give a single example of this: as the giant panda feeds principally (though not—as many books erroneously claim—exclusively) upon bamboo shoots, large areas of bamboo forest have been set aside as protected zones, as with Sichuan's Wolong Panda Reserve, where the punishment for harming or killing a giant panda is often the death penalty.[77]

To end the giant panda story—an update on the much-debated matter of its taxonomic allegiances. Just what *is* the giant panda—an enormous raccoon, or a small bear? Drawing upon anatomical comparisons and considerations, both schools of thought acquired formidable champions down through the years—with Dr. Dwight Davis (Mammal Curator at Chicago's Field Museum of Natural History), for example, supporting the bear candidature; and Dr. R.F. Ewer (Ghana University carnivore expert) among the raccoon identity's adherents. There was even a middle-of-the-road contingent, promoted by (among others) Dr. George Schaller (Vice-President of the New York Zoological Society), recommending that the giant panda should be classed within a family of its own (or with the lesser panda as its only other member).[71]

In short, a century's worth of anatomy-based arguments failed to provide a conclusive answer to the problem. The year 1956, however, marked the beginning of a very different approach, when a paper was published by Kansas University researchers Dr. Charles Leone and Alvin Wiens. They had carried out a painstaking serological study of the giant panda, and their results indicated a bear identity for it.[78] This was followed by other, more detailed and specific biochemical research, culminating in 1983 by utilisation of the most accurate and advanced tests so far devised for investigating a species' taxonomy—DNA hybridisation.[79]

Carried out by Prof. Stephen O'Brien and colleagues at the U.S. National Cancer Institute, this basically involved making direct comparisons (via a complex means of chemical 'pairing') of the giant panda's genetic ma-

Giant panda—the world's favourite wild animal. (Dr. Karl P.N. Shuker)

terial (DNA) with that of bears and raccoons, to discover which of these displayed the greatest similarity to it—and which was therefore the giant panda's closest relative. Additional tests, of a more straightforward biochemical nature, were also run, to check for consistency. The results obtained were indeed consistent, and provided some notable surprises. They indicated that whereas the lesser panda was truly a raccoon, the giant panda represented a specialised side-branch of the bears. Everyone had been wrong! The two pandas were only distantly related, and the giant panda was indeed a bear after all—just as Père David had believed, more than a century earlier![79]

Moreover, just as one taxonomic controversy drew to a close, another one hit the headlines. Ever since its scientific discovery just over 140 years ago, zoologically-speaking this iconic species has always been looked upon as being very much one of a kind. Now, however, a major new study of this singular species by a team of Chinese scientists has disclosed that in reality it appears to be two of a kind instead. Led by Zhejiang University panda expert Prof. Fang Shengguo, the team studied and compared the genetics and morphology of giant pandas from Sichuan with those of some from Qinling, and discovered that the two geographically discrete populations exhibited sufficiently marked differences to warrant their classification as two distinct subspecies—the first time that the Qinling giant panda has been separated taxonomically as a panda subspecies in its own right.[80]

The team's study revealed that this subspecies has been isolated from the Sichuan giant panda for at least 12,000 years, and is closer to the common ancestor of both subspecies than is the Sichuan, whose evolution has been faster. In addition, the Qinling giant panda's head is rounder and more feline in appearance than the longer, more ursine head of the Sichuan panda. Even the colours of their chest and chest fur are different, with the Qinling giant panda's being browner than the more familiar black chest and white chest fur colouration of the Sichuan giant panda.[80]

Finally: those Western scholars who had assumed (prior to the giant panda's discovery by Père David) that the white bear-like beasts depicted in ancient Chinese manuscripts were polar bears may not have been completely wrong after all. Since 1964, a number of creamy-white bears have been captured in Shennongjia, China, and some are currently housed in various Chinese zoos. Previously unknown to western science, these bears have naturally attracted much interest—especially as some Chinese researchers have expressed the opinion that they do not merely constitute an albinistic morph of the brown bear *Ursus arctos*, but may instead represent a form quite separate from all others previously documented.[81-2]

In *The Giant Panda Book* (1981), Jenny Belson and James Gilheany offer a wonderfully succinct summary of the entire history of this, the world's most charismatic animal:[76]

> Once Giant Pandas were hunted down and killed for their coats. Now they are treated like royalty, their courtships, births, and deaths chronicled around the world with an almost obsessive curiosity. The Giant Panda is the ultimate Chinese puzzle: one of the few creatures left on Earth which continue to mystify zoologists.

Judging from the recent discoveries recorded above, it seems more than likely that the giant panda will succeed in its profound ability to mystify for a long time to come!

QUEMI—FORGOTTEN FOR 300 YEARS

Sadly, the tragic history of the sea mink *Neovison macrodon* (see p. 29), which had already become extinct by the time that science had officially discovered and described it, is not unique—as the quemi's own history can confirm.

Not so long ago, a close relative of the pacarana *Dinomys branickii* (see p. 30) existed. In his 16th-Century account of the West Indian island of Hispaniola (comprising Haiti and the Dominican Republic), explorer Ganzalo Fernández de Oviedo y Valdés mentioned a mysterious rodent that he called the quemi, which was said to be brown in colour like the island's hutias, but larger in size. It was apparently a traditional item of food for the Hispaniolan natives, but after Oviedo's report nothing further was heard of it.[19, 48]

Then in the 1920s, bones of a pacarana-like rodent were discovered in a cave near a plantation at St Michel, Haiti. After studying them, Dr. Gerrit Miller of the Smithsonian Institute identified their owner as a representative of Oviedo's obscure quemi, and in 1929, within his formal description of the bones, Miller named their species *Quemisia gravis*. It seems that the quemi died out soon after the arrival on Hispaniola of the Spaniards, and certainly no later than the 16th century's close.[19, 48]

GOLDEN HAMSTER—FROM EXTINCT ENIGMA TO POPULAR PET

Many pet owners will undoubtedly be surprised to learn that the world's most popular species of small pet mammal was virtually unknown to science until 1930, and that its rediscovery that year apparently snatched it from the very jaws of extinction. All of this, and more, feature in the remarkable history of the golden hamster.

It began in 1839, when the skin of a hamster-like creature, smaller than known hamsters and captured near Aleppo, northern Syria, was described by George R. Waterhouse at a meeting of London's Zoological Society. He named it *Cricetus auratus*—the golden hamster—after that particular specimen's unusually rich fur colour. Nothing more emerged regarding this newly-described species until 1879, when some live examples were brought back to England by James Skene, who had been working in the diplomatic service in Syria. These thrived and bred for 30 years (during which time their species was renamed *Mesocricetus auratus*); but when, in 1910, the newest generation of progeny descended from those original specimens died without issue, the golden hamster once again sank into obscurity.[82]

During the late 1920s, zoologist Prof. Israel Aharoni, from Jerusalem's Hebrew University, was reading through some ancient Aramaic and Hebrew documents when he came upon a passage telling of a special kind of Syrian mouse that had once existed in the district of Chaleb (nowadays the site of Aleppo), and which had been taken to Assyria, where specimens had been kept as docile pets in cages by the Assyrian children. Aharoni was very intrigued by this, as he knew of no living species of animal from the Aleppo area that fitted the description in the ancient document. In April 1930 he visited Aleppo, and there, in an 8-ft-deep earth burrow, an adult female golden hamster and eleven young were found, whose appearance tallied with that of the ancient

Golden hamster—amazingly, all of today's millions of pet golden hamsters are descended from three specimens captured in Syria in 1930.

text's mysterious 'Syrian mouse'. (Later that same month, three old females were also caught, and are now preserved at Berlin's Zoological Museum.)[82-3]

Aharoni brought back nine of the eleven young, which he reared and presented to his university's parasitology department in July 1930. Five escaped, and a sixth was killed, but the remaining three—two females and one male—survived, and bred successfully in captivity. To safeguard this rediscovered species' continuing existence, specimens were later sent to universities and zoos all over the world, thereby ensuring that if one establishment's stock was lost or somehow destroyed, there would still be many others elsewhere.[82-3]

The golden hamster soon became very popular as a laboratory species, particularly in the study of genetics as it bred so readily in captivity. Even more significantly, as a result of its docility it carved out a peerless niche for itself as a children's pet throughout the world, so that its numbers are today counted in the many millions! Most remarkable of all, however, is the fact that every single one of these is descended from the three specimens housed at the Hebrew University's parasitology department in the 1930s, because no others were obtained in the wild until Dr. Michael Murphy procured 12 at Aleppo in 1971 (these were afterwards maintained at the National Institutes of Health at Bethesda, Maryland). In 1978, two more were brought from Aleppo to the U.S.A., this time by Bill Duncan of the Southwestern Medical School in Dallas, Texas.[83]

During September 1997 and again in March 1999, hamster researchers visited northern Syria and southern Turkey to determine whether any wild specimens of golden hamster still existed. The team discovered and mapped 30 burrows, but none of those inhabited contained more than a single adult. The team captured 13 specimens, which included a pregnant female that shortly afterwards gave birth to six young. All 19, plus a further three wild-caught specimens obtained from the University of Aleppo, were then shipped to Germany to establish a new breeding stock.[83a]

DESERT RAT KANGAROO—
THE CASE OF THE MISSING MARSUPIAL

Closely paralleling the golden hamster's early history is the saga of the desert rat kangaroo *Caloprymnus campestris*. A small, sandy-furred species of hopping marsupial, it is characterised by its long ears, naked muzzle, and enormous hind feet. It was first made known to science in 1843, when three specimens were collected from an unrecorded locality in South Australia by Sir George Gray, who sent them to the British Museum (Natural History). Nothing more was heard of it until the early 1930s, when it was rediscovered by Adelaide University zoologist Dr. H.H. Finlayson, who recorded in September 1931 that he had sighted a number of specimens in a large expanse of plains country within Lake Eyre Basin, straddling the borders of South Australia and Queensland.[84]

Desert rat kangaroo.

Following a sighting at Ooroowilanie, east of Lake Eyre, South Australia, in 1935, however, no additional, conclusive observations have been reported, so that many fear that this scarcely-known marsupial, the only member of its genus, is extinct. Yet in view of its previous success at remaining hidden, and acknowledging that its terrain's harsh nature deters most people from visiting this region, there must still be hope that it survives.

Shepherd's beaked whale skeleton—discovered in 1933, this peculiar species has 90 teeth, far more than any other beaked whale. (Wanganui Regional Museum)

TASMACETUS—THE WORLD'S MOST PRIMITIVE BEAKED WHALE

Nine years after the type specimen of *Indopacetus pacificus* had been officially documented by Longman, another, even more unusual, species of beaked whale made its existence known to science. On 7 November 1933, the carcase of a dead beaked whale (with pale underside but dark back and flippers) was washed onto a beach at Ohawe, in the New Zealand province of Taranaki. Over the next month, various reports concerning its appearance ashore, and subsequent disappearance back into the sea, were published in a local newspaper, the *Hawera Star*.[85]

In early December, one of these attracted the attention of G. Shepherd, curator of the nearby Wanganui Alexander Museum, who alerted beach patrols to look out for its possible restranding. And sure enough, a few days later the decomposing remains were once again cast up onto the shore. News of this was duly passed to Shepherd, who came over the next morning with two assistants to undertake the unenviable task of dissecting out all existing skeletal and dental material—the stench arising from the putrefying carcase of a 16-ft-long, one-month-dead whale has to be experienced to be believed![85-6]

Wrapped profusely and soaked liberally in Lysol® (in a valiant attempt to counteract its overtly odiferous nature), the precious material was transported by Shepherd back to the museum for preservation and study. It soon became clear that the remains were from a quite extraordinary and totally new species, so Shepherd invited cetologist Dr. W.R.B. Oliver, director of the Dominion Museum at Wellington, to prepare a formal description. This was published in 1937, and in recognition of Shepherd's commendable actions in procuring and preserving its type specimen Oliver named the new species *Tasmacetus shepherdi*.[86]

Sole member of a new genus, Shepherd's beaked whale possesses a number of cranial features conspicuously different from those of other species. In fact, it appears to be the most primitive of all known present-day beaked whales, as readily indicated by its most noticeable characteristic—in stark contrast to the mere handful (at most) of functional teeth possessed by all other beaked whales, *Tasmacetus* has no less than 90. Also of interest is that its lower jaw is slightly longer than its upper one.[86]

By the early 1970s, four more *Tasmacetus* specimens had been identified, all from New Zealand (more recently, in April 1994, a specimen died after becoming stranded on a beach near Nelson), but it offered up a surprise in 1973 when a beached individual was recorded from Patagonia, followed later by a record from Chile—these examples thereby expanding its known distribution range very considerably. A beached example was subsequently reported from South Australia too; and by 2010, a total of 28 stranded specimens had been documented from around the world, as well as a handful of confirmed live sightings dating from the mid-1980s onwards. There are also records of possible sightings of living specimens from earlier years. In 1964, one such sighting was made by William Watkins from a cliff top near Christchurch, New Zealand; another was reported from the Seychelles.[38-38a]

Moreover, J. Vollewens's detailed account and sketch of a mysterious sea creature sighted by him in 1904, while he was serving as third officer on the steamer *Ambon* in the Straits of Bab-el-Mandeb (between the Red Sea and the Gulf of Aden),[87] readily recall to mind images of a Shepherd's beaked whale. If this was indeed the creature's identity, we thus have on file an eyewitness record of a seldom-seen species made almost 30 years *before* that species' discovery by science.

SOLENODONS—LIVING FOSSILS FROM THE CARIBBEAN

Looking like large rats with very long, attenuated snouts, solenodons may not seem very prepossessing in appearance, but they have great zoological significance, as they represent the last of an ancient line of insectivorous mammals stretching back 30 million years. Now wholly confined to the West Indies, there are just two surviving species—the Hispaniolan solenodon *Solenodon paradoxus*, and the Cuban *S. cubanus*. Both are extremely rare, and have been written off as extinct on several occasions in the past.

The black-and-red Hispaniolan solenodon, measuring almost 2 ft in length, was formally described in 1833, but was seldom reported thereafter, and following the introduction of the mongoose onto Hispaniola in 1870 it seemed to have totally vanished. In 1907, however, it

*Hispaniolan and Cuban solenodons.
(Fortean Picture Library)*

was rediscovered in the island's northeastern interior by A. Hyatt Verrill. Five specimens were captured alive, and were sent to Washington Zoo and New York's Bronx Zoo. A few others were also caught—after which their species disappeared once more, not turning up again until 1935, when it was tracked down by German animal collector Paul Thumb, assisted by his dog. Some live specimens were obtained and sent to various German zoos in the hope that they would breed and thrive in captivity, but although some lived for up to 11 years the greatly-desired establishment of a captive breeding population was never achieved.[19]

The last pair in any European zoo died in 1973 at Frankfurt Zoo, the female in August and the male on 12 December. The latter, named Soli, was a very friendly little creature, who enjoyed climbing into the laps of visitors and poking his generally damp and extremely sensitive snout into their ears! Soli's species is totally protected on its island home, but is still very much on the endangered animals list.[88]

The Cuban solenodon or almiqui was once represented by two readily differentiated subspecies—the very distinctive black-and-white *S. c. poeyana*, and the buff-headed *S. c. cubanus*. The species as a whole was first brought to attention in 1838, and scientifically distinguished from its slightly larger Hispaniolan relative in 1863. Intermittently reported and captured for the next two years, it vanished afterwards until the capture of a living specimen in Cuba's eastern mountains during

1909. Sadly, this individual died only a short time later, and its species promptly disappeared again, remaining undetected for over 30 years.

By the early 1940s, however, there were examples in Cuba's La Habana Zoo, and subsequently in the private zoo of a school in Santiago de Cuba as well, but these were all dead by the mid-1950s. No specimen of the buff-headed subspecies has been reported since 1944, but in 1974 a single male of the black-and-white *poeyana* version was caught, and small numbers, wholly protected, have been recorded in Cuba's eastern section. Even so, it remains extremely scarce.[19,88]

Having said that, after last being reported on the island during 1999 in the eastern province of Holguin, a single specimen of this exceedingly rare creature was captured alive and well in September 2003 by a farmer in the mountains of Holguin. Dubbed Alejandrito, the adult male almiqui was held in captivity for two days while it was studied and medically examined, then released back into the wild where it had been caught. The solenodons' main hope seems to be in successful captive breeding.[88a]

A smaller, third species of solenodon, Marcano's solenodon *S. marcanoi*, is known only from geologically-recent skeletal remains, found in the Dominican Republic on Hispaniola, whose species was formally described and named in 1962. Because they were found in association with remains of *Rattus* rats (introduced onto Hispaniola by European settlers), it is believed that this diminutive solenodon persisted beyond Hispaniola's initial European colonisation by Columbus during the late 1400s, but was wiped out soon afterwards by the rats once they arrived aboard Spanish vessels during the early 1500s.[88b]

Another family of unusual insectivores exclusive to the West Indies consisted of the nesophontids—shrew-like mammals of varying sizes (one species was as large as a chipmunk) that supposedly died out during the 17th century on the cluster of Caribbean islands constituting their homeland. In 1930, however, some nesophontid bones and tissues extracted from a mass of owl pellets discovered in the Dominican Republic, Hispaniola, were found to be so fresh that it seemed possible that the individual(s) from which they had derived had been killed only a short time before. This encouraged Smithsonian Institution zoologist Dr. Gerrit Miller to speculate that some nesophontids may still exist after all, but radiocarbon-dating of such fresh-seeming pellets has so far failed to substantiate any 20th-century survival.[19,89]

Nevertheless, certain researchers have suggested that some nesophontids may indeed have persisted until at least the early 1900s.[89a] And given the extremely elusive nature of their solenodon relatives on Hispaniola and Cuba, perhaps there may come a time when the nesophontid family will itself be resurrected.

KOUPREY—A SPECTACULAR NEW MAMMAL FROM CAMBODIA

A favourite subject for temple carvings and statues of the Khymer Culture of 400-800 years ago,[90] and one of Asia's largest mammals, yet undescribed by science until as recently as the 1930s and still of uncertain identity more than 70 years later, the wild grey ox of Cambodia is a pre-eminent paradox in the annals of zoology.

During the early 1930s, various scientists visiting northern Cambodia's hilly areas took note of native reports describing a large, dark-bodied, white-limbed type of wild ox, standing a little over 6 ft at the shoulder, bearing a very long, pendulous dewlap (flap of skin) extending from its throat to its mid-chest region, and armed with a pair of lengthy, slender, widely-spreading horns. This ox was known locally as the kouprey.[2-3,91]

Such information greatly interested Prof. Achille Urbain, the director of Vincennes Zoo near Paris. In 1936, while visiting northern Cambodia, he met resident vet Dr. R. Sauvel, and saw at his home a most impressive set of ox horns that Sauvel said were from a kouprey. Intrigued by these, Urbain implored Sauvel to do whatever he could to capture a living specimen for exhibition at Vincennes Zoo, a request that Sauvel succeeded in fulfilling not long afterwards, in the shape of a young bull—which became the first and only kouprey (to date) to be displayed at any zoo in the world. Sauvel also shot a kouprey, an adult bull, which he allowed Urbain to examine closely.[91]

Based upon these two specimens, in late 1937 Urbain officially described the kouprey, classing it as a new species and naming it *Bos sauveli*, thereby allying it with the other two species of large Asian wild ox—the comparably-sized banteng *B. banteng*, and the somewhat larger gaur *B. gaurus*.[91]

Nesophontes micrus. (Dr. Karl P.N. Shuker)

In addition to its grey-black body colour, white 'stockings', and prominent dewlap, the kouprey is distinguished from other Asian oxen by bearing only a very insignificant ridge along its back—differing markedly from the well-developed version exhibited by the other species. Furthermore, its horns are distinctive not just on account of their length, slender form, and wide span, but also by way of their fringes. Just behind the tip of each horn in adult male koupreys is a fringe of horny splinters—the remains of the juvenile horn sheath that has been pierced by the permanent, adult horn that grows underneath as the kouprey matures. In all other species of oxen, this fringe of juvenile horn splinters is swiftly removed when the animal rubs its horns on the ground, so that only a ridge is left behind. The complete curve of each of the adult male kouprey's horns, however, is such that this type of rubbing is not possible, so the fringe remains for a long time afterwards.[2-3,10,91-2]

The kouprey also has many internal distinguishing anatomical features. Indeed, following his meticulous examination of the kouprey's anatomy, in 1940 Harvard University mammalogist Dr. Harold Jefferson Coolidge was so convinced of the kouprey's distinctly separate scientific status that he proposed a new genus for it—*Novibos* ('new ox').[93]

All of this, therefore, makes all the more surprising (at least on first sight) the many claims since then that the kouprey is not a genuine species, and controversy regarding its precise scientific identity continues to the present day. The reason for this uncertainty is the kouprey's unexpected amalgamation of features from different species of oxen.

For example, some of its cranial and external characteristics are more comparable with those of the gaur and (especially) the banteng, but certain others appear closer to those of the domestic zebu *B. indicus*. Consequently, some researchers, such as F. Edmond-Blanc, have suggested that the kouprey may be a hybrid form, probably descended from interbreeding between the above species. Yet Urbain sighted entire herds of kouprey after its discovery—the existence of herds would not be expected if it were merely a hybrid.[10,92]

In contrast, Dr. Charles Wharton has opined that the kouprey may have been domesticated during the early Khymer Culture but became feral after the latter's decline, so that today's koupreys are modified versions of the original stock. Certain other workers have expanded upon this theory, suggesting that the kouprey may actually have *originated* as a domestic breed, produced deliberately by banteng-zebu crossbreeding but later running wild, and existing ever since in a feral state. Confounding matters even further, there are (or were) domestic cattle in the same locality as the kouprey that combine certain of its features with those of the zebu.

Kouprey. (Dr. Pierre Pfeffer/ World Wide Fund for Nature)

Such cattle are referred to as boeufs des Stiengs (named after the area's local people, the Stieng).[10,90,92]

Whether or not the kouprey is itself of domestic origin, however, is another matter entirely. Mammalogists Dr. H. Bohlken and Dr. Theodor Haltenorth feel that it is wisest to consider the kouprey as a genuine wild species as long as there is no conclusive evidence pointing to its derivation from a feral domestic form.[10,90,92]

Australian National University mammalogist Prof. Colin Groves considers that the kouprey may be a last surviving remnant of the wild species that gave rise to the zebu. A similar viewpoint has been proffered by mammalogist Dr. F.W. Braestrup, who regards it as the world's most primitive species of wild ox alive today. Finally, cattle expert Dr. Caroline Grigson has suggested that the banteng, gaur, kouprey, zebu, and boeuf des Stiengs are all members of an extensive species complex.[90,92]

When officially described in 1937, the kouprey was believed to number at least 1000 in Cambodia, with similar counts in Laos, Vietnam, and Thailand. By 1953, however, it had seemingly died out in Thailand, and disturbance by hunters dramatically decreased its numbers elsewhere. Despite being chosen in 1964 by Prince Sihanouk as Cambodia's national animal, the kouprey continued its downward gallop towards extinction. After 1970, this was hastened along by years of intense warfare within the major portion of its distribution range between the Khymer Rouge and the collective Vietnamese and Kampuchean forces. Indeed, for some time zoologists feared that the kouprey had totally died out, but sporadic reports during the 1970s of small numbers in various localities within southern Laos and northern Cambodia refuted this. During 1980, it was rediscovered in Thailand, in Si Sa Ket Province's Dongrak mountain range, on the border with Cambodia, prompting the king of Thailand to proclaim the setting aside of 79,000 acres in the province as a kouprey sanctuary.[90,94-5]

In 1988, the entire world total of koupreys was estimated by biologists working in Indochina at no more than 300 individuals, the majority of which existed in Cambodia, with 100 at most in Laos, plus a few specimens more recently sighted in Vietnam. The Thailand contingent appears to consist of individuals that wander back and forth across the border with Cambodia. It is indisputably one of the world's rarest wild animals, with 21st-century searches for it in the wild having been unsuccessful, leading the IUCN to categorise the kouprey as critically endangered, possibly extinct. There are no captive specimens.[90,96]

Kouprey (male). (Helmut Diller/ World Wide Fund for Nature)

Any attempts to safeguard the kouprey from demise in the wild are fraught with problems arising from its relatively remote habitat, and its vulnerability to local hunters, who see a 6-ft-tall ox as too good a source of fresh meat to be missed. Obviously, the best hope for the kouprey would be the establishment of a captive herd, thereby allowing the species to multiply in safety, and also enabling some specimens to be sent periodically to other zoos or parks worldwide in order to establish additional herds (mirroring the Duke of Bedford's successful efforts with Père David's deer *Elaphurus davidianus* earlier in the 20th century.

The need for a captive breeding herd has been recognised by Hanoi University zoologist Prof. Vo Quy. After two years of planning, and sponsored by many zoos worldwide as well as by the World Wide Fund for Nature (WWF) and the IUCN, in November 1988 he led a team of workers into a region of Vietnam where a kouprey sighting had been made in 1987, in the hope of capturing some specimens. This herd was to be maintained in one of Vietnam's wildlife parks, but some specimens would be sent to various of the funding zoos. Sadly, however, the search was not successful, yet during the late 1990s a herd of ten was sighted in Cambodia, near the border with Vietnam. If specimens are eventually located and captured one day, this would provide zoologists with the opportunity to carry out comprehensive DNA comparisons between koupreys and other forms of ox.[96] In short, captive breeding could not only free the kouprey from the imminent threat of extinction, but also reveal its true scientific identity, so that we will at last know just what this enigmatic beast really is.

Meanwhile, the controversy continues. A research team from Illinois's Northwestern University, led by biologist Dr. Gary J. Galbreath, compared mitochondrial DNA samples from the kouprey with those from Cambodian specimens of a related wild ox called the banteng, and announced in 2006 that they were so similar that the kouprey may in reality have been nothing more than a domestic strain of banteng x zebu hybrid that ran wild (i.e. a feral crossbreed ox). Not all scientists supported this possibility, however, and in 2007 the Galbreath team itself retracted its earlier claim, with other, subsequent genetic studies also discounting a hybrid identity for it. [96a]

And as if all of this were not complicated enough, in 1994 scientists in Cambodia began investigating local lore alleging the existence of a second, smaller species of kouprey![96b] Nothing further appears to have emerged since, however, regarding this startling claim of a pygmy kouprey, but with Cambodia's enigmatic wild ox, nothing should ever be ruled out!

SELEVIN'S DORMOUSE— AN INSECTIVOROUS RODENT

In 1938, zoologist W.A. Selevin discovered an extraordinary little rodent on the sandy plains of Kazakhstan, near the village of Betpak-Dala. Accordingly, when formally described the following year by Soviet scientists Drs. B.A. Belosludov and V.S. Bashanov, it was named *Selevinia betpakdalaensis*.[97] Although only mouse-sized, it looks rather plump, because of its relatively long hair, and it behaves in a most un-rodent-like manner. Instead of following a predominantly vegetarian diet, it is exclusively insectivorous, devouring grasshoppers and

Hutia—several of these species famously appeared or reappeared during the 20th century.

other insects. Generally called the desert or Selevin's dormouse, it does bear a superficial resemblance to dormice, but its collective anatomy is sufficiently distinct from this family of rodents (and, indeed, from all others) for it to have been housed for many years in an entire taxonomic family of its own (though in more recent times it has been reassigned to the dormouse family).[97]

Despite its relatively recent scientific debut, this unique little creature's nomadic human neighbours, the Kazakhstanis, have always known of its existence (even though it is mainly nocturnal), and refer to it as the *shalman-kulak* or *kalkan-kulak*, both names inspired by this species' large round ears.[10,97]

HUTIAS—HIDE-AND-SEEK RODENTS OF THE WEST INDIES

Hutias are large coypu-related rodents indigenous to the West Indies, currently numbering 20 species (half of which may be extinct), and famed for their long-running series of impromptu appearances and disappearances during the 20th century.

One of the most celebrated examples involved Cuvier's hutia *Plagiodontia aedium*. Described in 1836 by, and named after, the French naturalist Frederick Cuvier (brother of eminent zoologist Baron Georges Cuvier), and based upon a single specimen obtained in 1826 by Alexander Ricord from Hispaniola, no more specimens or reports of this 1.5-ft-long species were brought forth for over a century. So zoologists felt more than justified in classing it as extinct. Thus it was with much surprise that they learnt in 1947 that a living individual had been captured in a little-explored area of the island, and that according to local hunters it was quite common there.[19,98]

When first discovered in the 19th century, Cuvier's hutia was thought to be the only modern-day member of its genus, but in 1923 Hispaniola proved science wrong when a second living *Plagiodontia* species was collected in the island's northeastern section, within the Dominican Republic. Based upon 10 adults and three young obtained close to Samaná Bay that year by Dr. W.L. Abbott, the new species was christened *P. hylaeum*, the Dominican hutia. It is distinguished from Cuvier's by its narrower feet and sharper, longer claws, its proportionately larger body and shorter tail, and its darker fur, but some authorities, including the IUCN, have lately reclassified it as a subspecies of Cuvier's hutia.[99]

Moreover, recent surveys of this species in Haiti uncovered some unconfirmed reports a mystery rodent known locally as the comadreja that may indicate modern-day survival on Hispaniola of a third *Plagiodontia* form—the Samaná hutia *P. ipnaeum* (=*velozi*). Formally described in 1948, it is officially believed to have become extinct following the European colonisation of Hispaniola during the late 1400s and early 1500s. (Intriguingly, some researchers have speculated that this rodent, and not *Quemisia gravis*, may have been the true identity of the quemi—see p. 49.)[99a]

The Cuban dwarf hutia *Capromys nana* (subsequently renamed *Mesocapromys nanus*) first became known to science as a seemingly extinct species, officially described in 1917 from some apparently ancient jawbones found in cave deposits in the Sierra de Hato Nuevo and near Limones in Cuba's western-central region.[100] Independently, however, 1917 also saw the discovery of some living specimens, collected in the island's famous Zapata Swamp[19] (from where a new wren, the Zapata wren *Ferminia cerverai*, would be formally described in 1926, and a new rail, the Zapata rail *Cyanolimnas cerverai*, in 1927[101]). Sadly, this hutia species has not been recorded alive since 1930, but findings of tracks and droppings in the Zapata Swamp give hope that it still survives.

A couple or so centuries ago, there were many more species of hutia alive in the West Indies than are present today; the disappearances of those no longer living can be largely attributed to indirect interference or deliberate destruction by humans. Unhappily, such extinctions have not ceased—indeed, certain species of hutia have been described in recent times only to disappear completely thereafter. In 1967, for example, a new species, named Garrido's hutia *Capromys* [now *Mysateles*] *garridoi*, was described from a single individual collected on the tiny islet of Cayos Maja, off southern Cuba; only two others have been sighted (in 1989, on two nearby islets).[102-4]

The year 1970 saw the discovery of the little earth hutia, restricted to Juan Garcia Cay, also sited just off southern Cuba. Four specimens were collected, two of which were captured alive and maintained at Cuba's Institute of Biology by L.S. Varona, who named their species *C.* [now *Mesocapromys*] *sanfelipensis*. Sadly, however, since the collection of 43 other specimens on the island in 1978 (why were so many collected?!), no more have been recorded there.[103,105] Similarly, Cabrera's hutia *C.* [now *Mesocapromys*] *angelcabrerai*, confined to the Cayos de Ana Maria islets, once again situated just off southern Cuba, is known only from a small series of specimens collected there during the 1970s, plus a small coastal mainland population at Jucaro Este.[103,106]

There seems little doubt that some if not all of those latter hutias are extinct; their island homes are so small that it is unlikely that viable populations could remain hidden, especially when, as in the case of the little earth hutia in 1980, scientific expeditions arrive specifically to seek out all existing specimens and are unable to find any.

JENTINK'S DUIKER—RETURN OF THE WORLD'S RAREST ANTELOPE

Duikers are diminutive, short-limbed, primitive antelopes native to tropical Africa. They derive their name from their tendency to dive for cover into nearby foliage at the slightest indication of danger; *duiker* is Afrikaans for 'diver'.

With a shoulder height of no more than 31 in, Jentink's duiker *Cephalophus jentinki* is actually one of the taller species. Regrettably, however, for quite some time it also laid claim to the much less desirable title of the world's rarest antelope. It first came to attention in 1884, when a female was collected near the Liberian coast by Dutch zoologist Fredericus A. Jentink, working for F.X. Stampfli. Two more were obtained in 1887, and in 1892 the species was described and named by Oldfield Thomas of the British Museum (Natural History), which still has the type specimen. Reports of its existence in Sierra Leone also emerged at that time, but were not confirmed.[42]

Jentink's duiker exhibits a very unusual, diagnostic colour scheme. Except for its white muzzle, its head is jet black, as are its neck and shoulders, whereas the rest of its body is medium grey, but these two contrasting colours are separated by a wide collar of very light grey that encircles its body from the back to the lower chest. The outer side of each limb is the same shade of grey as its body, but the inner side is much paler. Its tail is also pale, but terminates in a darker tuft of hair. Its pointed horns are very small and almost straight.[42]

It is clearly not an antelope to be overlooked or confused with any other, which is why zoologists later began to fear that it was extinct. Following its scientific description, nothing more was heard of Jentink's duiker for half a century. Not even an unconfirmed sighting or two were reported; it was as if this distinctive antelope had never existed. Then in 1948, P.L. Dekeyser and A. Villiers successfully refuted that impression when they obtained the skull of an adult male in Liberia's Dyiglo region. Later searches revealed that it was present not only in Liberia, where it inhabits deep, secluded forests, but also in the Ivory Coast, where the natives call it *nienagbé*. It was once believed to number only in the low hundreds, but is now estimated to number around 3500, though it remains endangered and is formally categorised as such by the IUCN.[42,103]

Jentink's duiker. (Alan Pringle)

On 1 December 1971, the Gladys Porter Zoo in Brownsville, Texas, achieved a significant captive-breeding coup with the birth of Alpha, the first Jentink's duiker ever born in captivity. Happily, Alpha, a female, thrived, and by 1979 she not only had been joined by a mate, Beta, but also had given birth to two offspring.[107]

Most recently, in 1988, this reclusive species was rediscovered in Sierra Leone, ending decades of doubt as to whether it had ever existed here to begin with.[108]

ANDEAN WOLF—A MYSTERY MAMMAL FROM THE MOUNTAINS

The maned wolf *Chrysocyon brachyurus* is a highly distinctive species whose pointed muzzle, bright chestnut coat, and bushy tail make it look more like a fox than a wolf—except for its extremely long, slender legs, which set it totally apart from all true foxes, and enable it to peer over the tall grass of its pampas homelands, extending from Brazil to Argentina. Its unusual appearance is aptly described by its native name—*aguara guaza* ('fox on stilts').

Prior to the 1920s, the maned wolf was believed to have no close relatives; it appeared to be a species well-delineated from all other members of the dog family. Then in 1927, German animal dealer Lorenz Hagenbeck was visiting Buenos Aires when he saw a most unusual pelt for sale. It was reminiscent of the maned wolf's, but was much longer, thicker, and darker in colour, varying from black on its upperparts to dark brown on its neck and underparts. Like the maned wolf, it possessed a particularly dense, mane-like covering over its neck, but its ears were much smaller and rounder. Making enquiries, Hagenbeck learnt that the pelt was from a strange dog-like beast that inhabited the Andes.[2,23,109]

Hagenbeck bought the pelt, and submitted it for scientific examination upon his return to Germany. Its unusual appearance generated a great deal of discussion and bemusement, but the final verdict was that it seemed to represent a previously unrecorded mountain-dwelling counterpart of the maned wolf. Naturally, however, such an opinion was very provisional—a complete specimen, or at least a skull and/or other remains, would be needed to ascertain more accurately its status. The problem with a unique pelt (especially a unique canid pelt), however distinctive it may be, is that there is little way of determining whether it is from a genuinely discrete species or merely from some strange hybrid or mongrel.[2,23,109]

One of the experts who inclined towards the identity of a mountain maned wolf for the pelt was German mammalogist Dr. Ingo Krumbiegel—especially when he learnt from Hagenbeck in 1947 that during his Buenos Aires visit he had seen three other pelts of that same type. Canid hybrids are notoriously variable, hence the known existence of at least four such pelts favoured the reality of a distinct species (rather than crossbreeding between domestic dogs and/or wild species) as the most likely explanation for them.[2,23,109]

In addition, Krumbiegel recalled an odd-looking canid skull that he had examined in 1935, and which had been present within a collection of specimens originating from the Andes. At the time, he had dismissed it as an unusual maned wolf skull, but now he sought it out again, and meticulously re-examined it. Although it did indeed have close similarities to maned wolf skulls, with a length of 31 cm it was more than 5 cm (2 in) longer than the average recorded from a sample of 23 from this species. Moreover, the maned wolf is an exclusive plains-dweller, not known from mountainous areas. But what if it had a high-altitude counterpart? What would such a creature look like? Krumbiegel realised that it would certainly be a densely-furred, short-eared form, which would conserve heat more efficiently than the typical shorter-furred, large-eared maned wolf, and would therefore match the appearance of Hagenbeck's obscure pelt.[2,23,109]

Pelt of Andean wolf—the only known skin of this still-controversial canid. (Alan Pringle)

It all seemed to fit, and so in 1949, Krumbiegel published a cautious formal description of this putative mountain-modified maned wolf, which he named *Oreocyon hagenbecki* ('Hagenbeck's mountain wolf'), and which is commonly referred to nowadays as the Andean wolf. In his paper, he provided comparative sketches of the typical maned wolf, and the likely appearance of this newly-described Andean equivalent, which, in addition to the pelt characteristics and shorter ears already noted, would most probably have shorter limbs and more powerful claws. Not long afterwards, Krumbiegel discovered that the name *Oreocyon* had already been given by palaeontologist Othniel Charles

Marsh in 1872 to a primitive species of fossil carnivorous mammal called a creodont. Consequently, he had no option but to change the name of the Andean wolf, rechristening it *Dasycyon hagenbecki* in 1953, emphasising the thickness of its coat.[2,23,109]

Krumbiegel and the rest of the zoological community waited eagerly for fresh specimens with which to resolve the issue of the Andean wolf's identity, but nothing else has been obtained so far. In the meantime, Dr. Fritz Dieterlen has carried out some detailed hair analyses, using samples of the fur from its type (and only available) specimen's pelt and comparing their form and structure with those of fur from other types of canid. In 1954 he concluded that there were certain similarities between the Andean wolf's fur and that of the alsatian (German shepherd dog), but that they were not sufficiently strong to confirm a direct relationship (i.e. a derivation of the Andean wolf from some form of hybrid between alsatians and other dogs, etc). In contrast, three years later Dr. Angel Cabrera unequivocally classed the Andean wolf pelt as that of a domestic dog.[23,110]

In 2000, an attempt was made to analyse DNA samples from the pelt. Unfortunately, the outcome was unsatisfactory, because the samples were found to be contaminated somehow with dog, wolf, human, and even pig DNA, and to make matters worse still, the pelt had been chemically treated.[110]

Today, this enigmatic pelt resides at the Zoological State Museum in Munich, but more than 60 years after Krumbiegel's scientific description of *D. hagenbecki*, its mystery remains unsolved. Is it merely from some unusual crossbreed, or does a large and significant species of wild dog exist in the Andes mountains, still awaiting official acceptance by science? To be continued . . . ?

DON FELIPE'S AQUATIC WEASEL

The early 1950s were good years for mammalian discoveries in South America. The black-shouldered opossum *Caluromysiops irrupta*, a new genus and species of marsupial native to Peru and Brazil, was first revealed in 1950,[111] and just a year later saw the discovery of what would seem to be a South American counterpart of the water-loving mink.[112]

On 2 October 1951, mammalogist Dr. Philip Hershkovitz (affectionately known to the locals as Don Felipe) captured an unusual weasel at Santa Marta, near San Agustin in Huila, Colombia. An adult specimen with very dark upperparts and pale buff-orange underparts, what made it so unusual was the presence of distinct webbing between its toes, particularly extensive between the second, third, and fourth on each foot, indicating an aquatic lifestyle. On 29 September 1956, another specimen, an adult male, was collected—this time by Kjell von Sneidern at Popayán, Cauca, in Colombia.[112]

Both of these specimens were presented to Chicago's Field Museum of Natural History; yet surprisingly, the new species that they clearly represented remained undescribed and un-named until as recently as 1978. In that year, the necessary duty was undertaken by Drs. Robert J. Izor and Luis de la Torre, who named it *Mustela felipei*, after its discoverer.[112] A further skin was later procured, this time from Ecuador; and more recently a living specimen was captured in Colombia's Cueva de los Guacharos National Park.[113] Its species is variously referred to as the Colombian, tropical, or Don Felipe's weasel.

PYGMY KILLER WHALE—
A LITTLE-KNOWN LITTLE WHALE

Whereas the common killer whale *Orcinus orca* is one of the world's most familiar cetaceans, the pygmy killer *Feresa attenuata* is one of its least known, and at no more than 9 ft long is less than one third of the length of its mighty relative.

The first two specimens—just a couple of skulls—were obtained in the 19th century, and deposited in the British Museum. In 1874, their species was formally named and described, after which this scarcely-known sea-dweller was not recorded again for more than 70 years. In 1952 it was rediscovered, when a specimen was obtained by whale hunters near Taiji, on Honshu, Japan, providing scientists with the first opportunity of gaining a complete skeleton of this species.[114] Many others have since been procured, mostly from dead stranded individuals, frequently from Japanese waters, but also documented from the Caribbean, southern Africa, Hawaii, and the Gulf of Mexico. Thus, this dark-bodied, paler-bellied cetacean, sometimes with white lips and chin, is now known to be widely distributed, but its natural history remains virtually unknown.[38]

Incidentally, the pygmy killer whale should not be confused with the dwarf killer whale. The latter is something of a mystery species, formally described by a team of Soviet cetologists in 1981, who named it *Orcinus nanus*,[115] but its status as a valid species has not been accepted by everyone. In 1983 another allegedly new species of killer whale was described, this time by Drs. A.A. Berzin and V.L. Vladimirov, who named it *O. glacialis*, and recorded it from Prudes Bay in the high latitudes of the Antarctic's Indian Ocean sector. Again, however, its classification as a separate species has not received full acceptance.[116]

More recent studies of killer whales have spilt Antarctic specimens into three morphologically distinct forms, labelled Types A, B, and C (smaller in size than A and B), but whether any or all of these represent distinct species is as yet unresolved.[116a]

Golden langur—scientifically undescribed until 1955. (Don-iv79/Wikipedia)

OTTER SHREWS IN MINIATURE

Discovered in 1861 by the French-American explorer Paul du Chaillu, the giant otter shrew *Potamogale velox* of West Africa is an extraordinary creature. Although unrelated to them, its pursuit of an aquatic existence comparable with that of otters has engineered, via evolutionary convergence, a corresponding similarity in its outward appearance too. Thus it very closely resembles a miniature otter, measuring just over 2 ft in total length. For almost a century, *Potamogale* was believed to be the only species of otter shrew in existence, in a family all to itself, but in 1954 a hitherto unknown, second species was officially described.

Discovered at Ziéla, at the foot of the Nimba Mountains in the Guinea Republic, it was only 10 in long, looked more like the familiar water shrews (*Neomys*) than an otter shrew (or an otter), and had a rounded tail instead of the noticeably flattened version used so effectively by *Potamogale* for propulsion through the water. It didn't even have webbed feet. Nonetheless, internal anatomy clearly indicated a close relationship with the giant otter shrew; and so to underline this, while concomitantly stressing its much smaller size, the new species was named *Micropotamogale lamottei*—'Lamotte's pygmy otter shrew' (after French zoologist Dr. Maxime Lamotte, who collected animal specimens in West Africa).[117]

Only a year later, a second species of pygmy otter shrew was described. Originally named *P. ruwenzorii* but subsequently reclassified as a *Micropotamogale*, it was somewhat intermediate between *M. lamottei* and *Potamogale*. For although it was most similar to the former in overall appearance, it was longer, reaching up to 14 in, and, like *Potamogale*, had webbed feet. Its type specimen had been obtained in 1953, caught in a native's basket fish-trap set at Mutsora, on the River Talya—a River Lume tributary on the western slopes of the Ruwenzori Mountains in what is now the Democratic Congo.[117]

Close examination of the few pygmy otter shrew specimens on record indicate that they form a link with the Madagascan tenrecs (see p. 44), so nowadays tenrecs and all three species of otter shrew are usually classed together within a single taxonomic family.[48]

GOLDEN LANGUR—A NEW, BRIGHTLY-FURRED MONKEY FROM INDIA

One of Asia's most attractive primates eluded capture and classification by science until as recently as 1955. Yet as far back as 1907, when mentioned by E.O. Shebbeare, sightings of an unidentified form of monkey with light silvery-golden fur had been reported from the hills of northern Assam, close to the India-Bhutan border. No specimens were obtained, however, and the monkeys gained little scientific notice, as they were simply assumed to be golden snub-nosed monkeys *Rhinopithecus roxellanae* from Bhutan that had wandered southwards into India.[118]

Then in 1947, while spending some time at the Jamduar Forest Rest Home on the eastern bank of the Sankosh River in northern Assam's Goalpara District, sportsman-naturalist C.G. Baron made his own sighting of these unexamined, brightly-furred monkeys, and judged them to be an unidentified species of langur. Also known as leaf monkeys, langurs are slender, long-bodied and lengthy-tailed simians—their gracility contrasting with the more robust build of the golden snub-nosed monkey. There are many species of langur spread through much of Asia, but none matched the description of the mysterious examples from northern Assam.[118]

By now they had attracted the attention of E.P. Gee, a noted authority on Indian wildlife, and in November

1953 he visited the Jamduar Forest area to see them for himself. On the river's eastern bank at a locality close to Bhutan, he spotted two troops, collectively comprising more than 40 individuals, which he photographed, filmed, and observed closely for many days. They were certainly langurs, but their exquisite colouration distinguished them at once from all others known. During the next year, Gee communicated with various primatologists regarding them, and in January 1955 his films and photos had attracted the keen interest of Dr. S.L. Hora, director of the Zoological Survey of India, who sent a party in search of some specimens to enable the form to be officially classified.[118]

Six were duly collected during the first week of April 1955 from the forests around the Jamduar Rest Home, by the party's leader, zoologist Dr. H. Kharjuria. Upon examination of these, Kharjuria decided that although its cranial characteristics were similar to those of the capped langur *Presbytis* [now *Trachypithecus*] *pileatus*, its paler colouration and various other pelage differences were sufficient to necessitate the new monkey's delineation as a separate species—which, in honour of its leading investigator, he named *P.* [now *T.*] *geei*. In popular parlance, it is most commonly termed the golden langur.[119]

No more than two years had elapsed before another new species of langur was named—the white-headed langur *P.* [now *T.*] *leucocephalus*, discovered in China's Fusui County (Funan, southern Kwangsi). Initially known from a single skin obtained in 1953 during a scientific expedition led by T'an Pang-Chien, by 1957 ten more had been collected from the same area, convincing T'an Pang-Chien that this eyecatching monkey (known to the locals as *paiyuan*—'white ape') truly constituted a distinct species. Other authorities, conversely, prefer to regard it merely as a subspecies of *T. poliocephalus*, or even as a partial albinistic population of Francois's langur *T. francoisi*.[120]

FRASER'S DOLPHIN—THE WORLD'S MOST COMMON 'RARE' SEA MAMMAL

Fraser's dolphin is a species with an uncommonly eventful and unusual scientific history, which began as far back as 1895, when its type specimen, a skull, was collected by Charles E. Hose at the mouth of the Lutong River, Baram, in the portion of Borneo nowadays called Sarawak. It subsequently arrived at the British Museum (Natural History), in the company of several specimens of the Indopacific humpback dolphin *Sousa chinensis*—which led in 1901 to its own classification as a member of that species. It attracted no further attention for the next 54 years—not until 1955, when merely by chance its remains were noticed by cetologist Francis C. Fraser, who swiftly realised that it was something markedly different from any form of humpback dolphin. He proceeded to study it very carefully, and perceived that its anatomy combined certain features characteristic of the ploughshare dolphins (*Lagenorhynchus*) with others more typical of the common dolphin *Delphinus delphis*.[38,121]

It was without question a new species, and one worthy of a new genus too—so in 1956 Fraser created the genus *Lagenodelphis*, stressing its shared affinities with *Lagenorhynchus* and *Delphinus*. Its full scientific name is *Lagenodelphis hosei*, honouring its discoverer.[121]

Following its official recognition by science, cetologists avidly awaited fresh news regarding this newly-unmasked species, but their expectations were not fulfilled. At the beginning of the 1970s, *The Guinness Book of Records* referred to Fraser's dolphin as probably the world's rarest species of marine mammal, known only from the single Sarawak specimen—but what a difference a few months can make. For by the end of 1971, this cryptic cetacean had become one of the world's most widely distributed cetaceans![38]

The explanation for such a tremendous turnabout was that within the first five months of 1971, a number of stranded specimens had turned up in a variety of widely separated localities—from Cocos Island in the eastern Pacific, and New South Wales in Australia, to an area of shoreline close to Durban in South Africa. In later years, Fraser's dolphin was also recorded from the Caribbean, as well as from the coasts of Japan and Taiwan. Sometimes, entire schools have been sighted too, so that the species is now known to be relatively common—a far cry from its earlier status.[38]

A somewhat small cetacean, attaining a mere 8.5 ft in total length when fully grown, and characterised externally by surprisingly short fins and snout, Fraser's dolphin has medium-grey upperparts and pinkish-white underparts, with two distinctive lateral stripes—the upper one pale grey and stretching from just above its eye to its tail's base, the lower one a more striking, jet-black band spanning its eye and anus.[38]

JAPANESE BEAKED WHALE—TEETH LIKE THE LEAVES OF THE MAIDENHAIR TREE

The first specimen of the Japanese beaked whale *Mesoplodon ginkgodens* to come to scientific attention did so in a particularly tragic way. In September 1957, a large, uniformly blue-black whale swam close to the shore at Tokyo's Oiso Beach, attracted by the sight of some boys playing ball at the water's edge and no doubt inquisitive to learn more about their activity. In response, the boys waded into the water and promptly clubbed the poor animal to death with their baseball bats. The only positive aspect to emerge from this very unpleasant incident was that the whale's body was salvaged and carefully examined by cetologists Drs. M. Nishiwaki and T. Kamiya from the Japanese Whale Research Institute. They found that its relative body proportions and certain anatomical features set it apart from all other beaked whales, so in 1958 they designated it as the type specimen of a new species.[122]

Yet another member of the genus *Mesoplodon*, its specific name is derived from the most unusual, diagnostic shape of its single pair of teeth (sited midway along the lower jaw in males, but failing to erupt through the gums in females), because when viewed from the side these teeth bear a close resemblance to the lobe-edged leaves of the Chinese maidenhair tree *Ginkgo biloba*, one of the plant kingdom's most famous 'living fossils'.[38,122]

In 1963, a second specimen appeared, washed ashore at Ratmalana, Sri Lanka. This was initially described as another new species, and given the name *M. hotaula*, but then a third specimen turned up, originally cast up onto Delmar's public beach in southern California during 1954 (though not identified at that time). Comparisons of these three individuals by beaked whale expert Dr. J.C. Moore revealed that they were all of the same species, so '*M. hotaula*' thereafter became a synonym of *M. ginkgodens*.[38,123]

During his studies on the Japanese beaked whale, which appears to attain a total length of up to 17 ft when mature, Moore uncovered some other preserved remains previously unrecognised as being of *M. ginkgodens* identity. Moreover, some fresh specimens were washed ashore on Japanese beaches, and a single individual was recorded from Taiwan. Yet even though it may be widely distributed throughout the warmer regions of the tropical Indopacific, the small number of specimens and sightings on record imply that the Japanese beaked whale is either fairly rare or fairly elusive (or both!).[38,123]

Until recently, only 19 specimens had been recorded, but on 9 April 2003 a twentieth came to light, when an adult female, weighing almost 2 tons and measuring 15.6 ft long, was washed up dead on an Onaero beach in New Zealand's North Island. This was the first Japanese beaked whale ever recorded from New Zealand shores, and was only the third reported anywhere in the Southern Hemisphere. Its skeleton has since been stored at Wellington's Te Papa Museum, and its internal organs are held at Massey University in Palmerston North, to be used for research purposes.[123a]

COCHITO—A PETITE PORPOISE FROM THE GULF OF CALIFORNIA

The Japanese beaked whale was not the only new cetacean to be described in 1958. Back in 1950, an unusually small, sun-bleached porpoise skull, resting above the high tide line on a beach at Punta San Felipe in the Gulf of California, was sighted and brought back to the University of California by Dr. Kenneth Norris, one of the university's mammalogical researchers. At first it was merely assumed to be from an undersized specimen of the common porpoise *Phocoena phocoena*, but some fresh, similarly diminutive skulls were obtained from the Gulf of California in the next few years. After comparing these with each other and with known *P. phocoena* specimens, certain other morphological differences between the latter and the small ones were noted, so Norris and his Cornell University co-worker Dr. William McFarland decided that a new species was present. In 1958 they christened this petite porpoise *Phocoena sinus*.[124]

In general parlance, it is referred to as the Gulf of California porpoise, the vaquita ('little cow'), or the cochito—the name by which it is known among local fishermen, in whose nets it frequently becomes entangled. Dark grey on top, creamish-white underneath, the cochito attains a maximum length of only 5 ft, thus constituting the world's smallest species of porpoise. Unhappily, however, on account of what seems to be a comparably minute population size (estimated at only 100-300 individuals), a very limited distribution (apparently confined to the Gulf), and alarmingly high mortality incurred by accidental death in fishing nets, it would also appear to be among the rarest of all cetaceans. Following considerable publicity roused upon its behalf by the Defenders of Wildlife organisation, in 1985 the U.S. government officially classed the cochito as an endangered species, and it is categorised as critically endangered by the IUCN, but with few confirmed cochito sightings in recent years, its survival chances do not seem very promising.[38,103]

LIBERIICTIS—A MYSTERIOUS MONGOOSE FROM WEST AFRICA

By the early part of 1958, Heidelberg University ethnologist Dr. Hans Himmelheber had amassed an impressive collection of mammal skulls from northeastern Liberia. He had collected some of these himself, while undertaking research at native villages in that area; they had been retained from animals killed by the villages' hunters for food. Others had been sent to him by missionaries working in the area.[125]

His collection was examined by fellow Heidelberg University scientist Dr. Hans-Jurg Kuhn, who became particularly interested in eight large mongoose skulls that he was unable to assign to any known species. Consequently, while attending the fifteenth International Congress of Zoology, held at London during July 1958, he took the opportunity of visiting mammalogist Dr. R.W. Hayman of the British Museum (Natural History), where he showed him one of these eight anomalous skulls. Hayman readily discerned its distinct nature, and after later examining the other seven he announced that they were from a hitherto undescribed species, and one so far removed from all other mongooses that it deserved an entire genus all to itself.[125]

Designating the first skull that he had examined as the new mongoose's type (which had been collected some time between October 1957 and April 1958 at the village of Kpeaplay—the most northerly of two different villages bearing this same name), he christened the species *Liberiictis kuhni*.[125]

Structurally, the skulls of the Liberian mongoose compared most closely with those of the cusimanse mongooses (genus *Crossarchus*), but differed from them by virtue of their noticeably larger size, possession of an extra pair of upper and lower premolars, longer muzzle, and relatively smaller, weaker teeth. These latter characteristics suggested to Hayman that the Liberian mongoose may well be insectivorous, rather than a predator of more substantial animals such as rodents or reptiles.[125]

And so yet another taxonomically significant new mammal had become known to science—or at least its existence and its cranial features had become known. Science still did not know what it looked like in life—a situation that lasted for quite a time. Then in summer 1971, during an expedition to Liberia's northern Grand Gedeh County, two specimens were obtained near the town of Tar. One, an adult male, was captured in cutover High Forest by a local farmer on 7 July. The second, a juvenile female, was caught 22 days later in a burrow near to a termite's nest, again in cutover High Forest. Except for its pale-shaded neck (which bears a stripe along each side, running from ear to shoulder), the elusive *Liberiictis* proved to be principally dark brown in colour, and it had long claws—which would be of use in ripping apart termite mounds in search of the insects inside, assuming the validity of Hayman's postulated insectivorous diet for this species.[103,126]

In 1989 a male individual was captured alive in Liberia's Gbi National Forest, and was sent to Canada's Metro Toronto Zoo, in the hope that a captive breeding programme could ultimately be established, but unfortunately this was not accomplished.[113,126]

UFITI—THE FRIENDLY 'GHOST' OF NKATA BAY
One of the most controversial of the 20th century's mammalian discoveries involved a female chimpanzee named Ufiti. In August 1959, inhabitants of Nkata Bay, on the western shore of Malawi's Lake Nyasa, began to report sightings of a strange ape-like entity in the fringes of the adjacent forest. Such reports were readily confirmed, because the animal in question became very interested in the construction work that was taking place on a new bridge and road at the nearby Limpasa River, and stayed in the vicinity to observe the proceedings, so it was often seen. And as its amiable curiosity largely eclipsed its fear of humans, it could be closely approached.[18,127-8]

When questioned, the local westerners asserted that it was new to them, not previously known in the area, and the natives referred to it as *ufiti*—meaning 'ghost'. It was not a ghost, however, but a mature female chimpanzee—which came as a great surprise to zoologists, because chimpanzees had never before been recorded in Malawi. Indeed, the nearest colony on record was at least 480 miles northwest of Nkata Bay—in Tanzania's Nkungwe Mountains, on Lake Tanganyika's eastern shore.[18,127-8]

Ufiti—Nkata Bay's mysterious chimpanzee. (Loren Coleman)

In March 1960, a field expedition from the Rhodes-Livingstone Museum, headed by Drs. B.L. Mitchell and C.S. Holliday, travelled to Nkata Bay to observe and photograph Ufiti (as she had been nicknamed by then), as well as to obtain tape recordings of her vocalisations, and to study the prevailing ecology of the area.[18,127-8] The information gathered during that expedition was then sent to anthropologist Dr. W.C. Osman Hill, for his remarks and opinions, which in 1963 he documented within an article published by London's Zoological Society in a symposium of primate research papers.[128]

The photos and observations obtained during the expedition revealed that Ufiti, although definitely a chimpanzee, exhibited certain unexpected features. In view of her provenance, she should have been most similar in appearance to East African chimps—but instead, her completely black face, ears, hands, and feet, and also her short, dense coat, allied her more closely with western forms. Equally strange was the presence of a saddle-like area of pale grey fur across her back—a feature characteristic of mature male gorillas![128]

Prior to Hill's article, the predominant opinion among zoologists concerning Ufiti was that she must surely be just an escapee from captivity. However, the morphological features documented by Hill argued strongly against such an identity—inciting speculation that Ufiti represented a hitherto unknown taxon (subspecies?) of chimpanzee, native to Malawi and normally concealed in this country's dense forests, with Ufiti herself presumably being a wanderer, or an individual cast out of the population by its other members. Worth noting, as commented upon by Hill, is that the Nkata Bay area is well known for harbouring a number of animal and plant species more closely related to West African forms than to East African ones. Moreover, Hill later received accounts of chimpanzee-like creatures from Malawi that considerably preceded Ufiti's debut.[128]

Not everyone, however, was convinced by Hill's theory. In their *Mammals of Malawi* (1988), W.F.H. Ansell and R.J. Dowsett claimed that the mystery of Ufiti had been solved, and that she was nothing more than an escaped pet originating in Zaire (now the Democratic Congo). They stated that some years prior to their book's publication: "... the late Fr Tréguier, a White Father at a mission in the Misuku Hills, showed Dowsett a photograph which he had obtained from a colleague, Fr Rainville, of the animal [Ufiti] which was ... a household pet brought from Zaire at a time when many expatriates were leaving the country due to the troubled political situation". However, this apparently satisfactory solution is not quite as water-tight as it may seem. For as Ansell and Dowsett went on to say: "It is not known who brought it into Malawi or by what route".[128a] So how can anyone be absolutely sure that it did originate from Zaire? Moreover, it is unlikely that an individual chimpanzee can be conclusively identified from just a single picture anyway. Hence it is impossible to say with certainty that the chimp in the photograph truly was Ufiti.

Sadly, the truth as to whether Ufiti was a major discovery or just an escaped pet may never be known. In March 1964 she was captured, and sent to Britain's Chester Zoo, arriving there on 19 March. Unhappily, however, her health was found to be deteriorating rapidly; and so, to prevent her from suffering any further, on 23 April the zoo had no option but to put her down.[18,128a] With Ufiti's passing, the issue of Malawi's putative chimpanzee population was soon forgotten, so that over 50 years after her first appearance the friendly 'ghost' from Nkata Bay may still hold some surprises in store.

Staying with chimp classification: the common chimpanzee *Pan troglodytes* has been traditionally split into three subspecies. Following DNA analysis results released in July 1997 by a team of New York City anthropologists (featuring Prof. John Oates, Dr. Todd Disotell, and Katy Gonder among others), the existence of a fourth is now recognised. Their results revealed that a chimpanzee population existing in western Cameroon and southern Nigeria is genetically distinct from all three currently recognised subspecies. It has been dubbed *P. t. vellerosus*, thereby reviving a scientific name first applied back in the 1860s by John Edward Gray from the British Museum (Natural History) to a skin collected on Mount Cameroon by British explorer Sir Richard Burton.[128b]

RED GORAL—A RUG REVEALED ITS EXISTENCE
The red goral must assuredly be the only species of animal whose type specimen was very nearly a carpet! Gorals are goat-like relatives of the antelopes, and live in various mountain ranges of eastern and southern Asia. For many years, science only recognised the existence of a single type, grizzled greyish-brown in colour, and named *Nemorhaedus goral*, the grey goral, whose distribution extends northwards from Burma (now Myanmar) through northeastern China to Korea and the lower reaches of Siberia (though the Chinese, Russian, and Burmese representatives are nowadays split off by many researchers into two additional species—see below).

Yet as far back as 1863, as reported by naturalist Edward Blyth, the natives of Assam, Burma, knew of a quite different, smaller form of goral, whose coat was bright foxy-red in colour, and much softer in texture than that of the grey goral. Similarly, in 1912, the existence of a rufous-pelaged goral in the area of Sanga Chu Dzong in southeastern Tibet was substantiated by Lieutenant-Colonel F.M. Bailey, who learnt that the red-furred coats worn there by various of the local inhabitants had been made from the pelage of this scientifically-undescribed animal. During the next decade, a number of red gorals were shot in Assam's Mishmi Hills, but none was presented for inspection to any scientific establishment. Fortunately, this situation was rectified in 1931, when a specimen shot in northern Burma's Adung Valley by the Earl of Cranbrook was presented by him to the British Museum (Natural History), where it was examined by mammal specialist Captain Guy Dollman.[129]

It seemed that the mystery of the red goral would soon be resolved, because Dollman proposed to prepare a formal description of it, complete with an official scientific name. Surprisingly, however, he never carried out his promise, probably preferring to await additional specimens for absolute confirmation of the red goral's merit as a valid species—but no fresh material arrived.

Mounted red goral.

Hence the red goral was soon forgotten, its lone available specimen lingering in the British Museum still unnamed and undescribed—until 1960.[130]

During that year, the museum received a very attractive rug, made up of skins from a trio of rufous-coated mammals, which was sent to the museum by a Mr. H.L. Cooper, who wanted the skins to be identified. He reported that they were from three of the handful of red goat-like creatures that he and some native tribesmen had shot in the Mishmi Hills during 1922. Museum mammalogist Dr. R.W. Hayman compared the rug with the red goral skin donated by the Earl of Cranbrook, and recognised that they were all from the same species. In 1961, Hayman fulfilled the pledge that Dollman had made 30 years earlier—by publishing a full description of the new species, which he named *N. cranbrooki*.[130]

Three years later, science received its first opportunity to observe a living example of this species, when on 9 January 1964 a female red goral, caught near Lashio in Burma's Northern Shan State, arrived at Rangoon Zoo. An extremely agile animal, it was capable of substantial bounds and leaps, and in memory of its former mountainous home it chose not to sleep *inside* its hut but preferred instead to sleep *on top* of it, the roof of which was more than 5 ft above the ground.[131]

Worth noting is that as long ago as 1914, renowned zoologist Reginald Pocock had described a new species of goral, which he named *N. baileyi*, and which is generally referred to as the brown goral, because of the rich brown, ungrizzled shading of its pelage. Its description was based upon a single skin and skull collected on 3 July 1913 by Lt.-Col. Bailey at Dre, Yigrong Tso, Po Me, in southeastern Tibet.[132] No other specimen has ever been recorded. It was eventually demoted to the status of a grey goral subspecies, but when the existence of the red goral became established, some researchers wondered whether the brown form actually represented the latter species' summer coat phase. Hayman, however, closely compared red goral skins with the unique brown goral skin, and discovered various differences that ostensibly dismissed this possibility, so he concluded that the brown goral was a genuine species.[130]

Nevertheless, mammalogist Prof. Colin Groves more recently reported that new surveys of the western Yunnan mountains, lying between the respective type localities of the red goral and the brown goral, have found red goral there, thus linking the two types geographically. In the opinion of Groves and others, this therefore makes it fairly clear that, despite Hayman's belief, the red goral and the brown goral are in reality one and the same species.[133] This status has since become widely accepted, so the red goral is formally known now as *N. baileyi* (because this name was coined several decades before *N. cranbrooki*). In addition, two former subspecies of the grey goral are now classed as full species in their own right—the long-tailed goral *N. caudatus* (native to Russia, China, Thailand, and Myanmar), and the Chinese goral *N. griseus* (Myanmar, India, India, China, Vietnam, and possibly Laos).

Also deserving a brief mention here is the golden takin *Budorcas taxicolor bedfordi*, a relative of the gorals, which remained undetected by western science until 1911, when it was discovered in eastern China's Shensi and Kansu Provinces.[2]

BANANA BAT—A WINGED 'RHINOCEROS'!

Described in 1960, and named *Musonycteris harrisoni*, the Mexican banana bat is a small but instantly recognisable species—due to its unforgettable profile. The total length of this brown-furred bat scarcely exceeds

Leadbeater's possum. (Zoo Operations Ltd)

3 in, and in proportion to this its head would be relatively small—were it not for its extraordinarily long muzzle, so greatly elongated that it accounts for more than half of the skull's entire length. And as if to accentuate its muzzle's already bizarre appearance, perched incongruously at the very end is a pointed flap of skin, standing upright like a miniature rhinoceros horn just behind the bat's nostrils.[48,134]

Seemingly a rare species, and the only member of its genus, it was first discovered in 1958, when three specimens were captured within a grove of blooming banana trees, situated just over a mile southeast of Pueblo Juarez, Colima. Since then, others have been also recorded from the southwestern Mexican states of Michoacan and Guerrero.[48,134]

LEADBEATER'S POSSUM—RETURN OF THE NON-GLIDING GLIDER

Several squirrel species have a membrane of skin between forelimb and hindlimb on each side of the body. If, when leaping from a tree, these membranes are extended, the squirrels can sail through the air for considerable distances. Such species are termed flying squirrels (although they glide rather than fly). In Australia, the possum family contains some very close marsupial counterparts to these, known as flying phalangers—squirrel-like in general appearance, and with comparable gliding membranes.

The most familiar flying phalanger is a small species called the sugar glider *Petaurus breviceps*. In 1867, however, Sir Frederick McCoy, director of the National Museum of Victoria, described a species that, although very similar superficially to the sugar glider in many respects, was nonetheless immediately distinguished by its lack of gliding membranes. With a head-and-body length of roughly 6.5 in, a thin club-shaped tail of comparable length, and distinctive spatulate (shovel-shaped) toes, this densely-furred form, brownish-grey above, paler grey below, with a dark stripe running from forehead to tail base, was christened *Gymnobelideus leadbeateri*—sole occupant of a newly-created genus.[57]

Leadbeater's possum, as it came to be known, was named after John Leadbeater, a taxidermist at Victoria's National Museum who had discovered the very first examples a little while earlier. Two specimens had been collected in the Bass River Valley, 60 miles southwest of Melbourne in Victoria's South Gippsland. A third example from this same area was received by the museum in 1900, followed by one obtained in 1909 from the edge of the Koo-Wee-Rup Swamp (before it was cleared), 30 miles north of the Bass River.[57]

Except for a lone specimen discovered during 1931 in a collection of sugar glider skins and originating from East Gippsland's Mount Wills, nothing else materialised concerning Leadbeater's possum for over half a century, despite a number of extensive searches for it in eastern Victoria. The flying phalanger that couldn't fly (or even glide) seemed to have drifted into extinction, almost before science had even become aware of its existence.[57]

In 1959, naturalist Graham Pizzey began conducting a mammal survey in eastern Victoria's Healesville, Warburton, and Marysville ranges. On Easter Monday 1961, he was accompanied by H.E. Wilkinson from the National Museum and another colleague, on a trip to Cumberland Valley, 11 miles east of Marysville at an altitude of 3000 ft, and not previously visited by him during his survey. And there, alerted by a slight rustling noise on the trunk of a blackwood wattle tree close by, Pizzey and company briefly saw the first recorded specimen of a living Leadbeater's possum for 52 years.[135]

At first, they were so stunned that they could scarcely convince themselves that it really was a Leadbeater's. During that evening, however, after having driven 5 miles in the direction of Marysville, reaching a spot with mountain ash in abundance and known as Tommy's Bend Creek, they succeeded with the aid of a torch in observing another specimen, this time for a continuous period of about 10 minutes, as the light 'froze' it into immobility. With binoculars, Pizzey was able to check its salient features—club-shaped tail, spatulate toes, absence of gliding membranes. They all pointed to a single, undeniable identity—Leadbeater's possum. It was not extinct after all.[135]

Pizzey returned with colleagues to Tommy's Bend Creek five nights later, saw three specimens, took photos, and also collected one specimen, which was subsequently confirmed to be *G. leadbeateri*. Two months further on, a live individual was captured, which Pizzey was permitted to look after for a time, courtesy of the Fisheries and Wildlife Department. His priceless pet, which he named Jimmy, proved to be wholly nocturnal (which may explain at least in part how the species eluded detection for so long), and his spatulate toes functioned as effective suction pads, thanks to the large, damp pads on their undersurfaces.[135]

Since then, small colonies have been found in this same general area, and some specimens have also been discovered further afield. Indeed, by the 1980s, the total Leadbeater's possum population was estimated to have reached 7500 individuals, but from that peak it has since fallen to approximately 2000, and in February 2009 this species' only known habitat was swept by raging forest fires, destroying vast areas of bushland and scrub, thus imperilling its continuing survival. It is currently categorised as endangered by the IUCN, and there are no captive specimens.[48,135]

HUBBS'S UNDISCLOSED BEAKED WHALE

Hubbs's beaked whale was discovered by beaked whale expert Dr. J.C. Moore during his *Mesoplodon* researches of the early 1960s. While examining some skeletons obtained over the years from America's North Pacific coasts and hitherto classed as Stejneger's beaked whale *M. stejnegeri*, he realised that three of these clearly belonged to a quite different species, not previously documented scientifically. He also discovered that a specimen stranded alive in 1945 at La Jolla in California, and identified by biologist Dr. Carl Hubbs as an Andrews' beaked whale *M. bowdoini*, belonged in reality to the undescribed species (thus making the La Jolla specimen the earliest of this 'new' whale to have been preserved). And a beaked whale stranded off Japan in 1958 was shown to be yet another of its representatives.[38,136]

So in 1963, Moore christened this newly-detected species *M. carlhubbsi*, honouring its type specimen's initial researcher.[136] Measuring up to 17 ft long, and relatively similar in external appearance to *M. stejnegeri* (a feature no doubt responsible for the failure of scientists to recognise its separate specific status much earlier), Hubbs's beaked whale has one notably distinctive characteristic—the colour of its beak, which for much of its terminal portion is pure white, readily differentiating it from all other *Mesoplodon* species. Adult males also tend to exhibit extensive body scarring. By the end of 2009, a total of 31 strandings (including some recent cases from Hawaii) and one possible sighting for this elusive species had been recorded since its discovery.[38,136]

THE DWARF BLUE SHEEP OF BATANG

The blue sheep *Pseudois nayaur* is hardly the most aptly named of animals. Also known as the bharal (its Hindi name), it is not a true sheep (hence *Pseudois*—'false sheep'); in fact, it seems to be midway between sheep and goats, in terms of its habits and also its structural anatomy. As for its blue colouration (in reality, brownish-grey infused with slaty blue), this is present only in its first winter coat.

A mountain-dwelling species native to the Himalayas and western China, *P. nayaur* was long thought to be very distinct from all other living forms, the sole member of its genus. Then in 1934, during the second of three expeditions to the Tibetan plateau region, German zoologist Dr. Ernst Schaefer discovered an isolated population of blue sheep, in the Yangtze gorge near Batang (other specimens have since been recorded elsewhere in Tibet too), whose members were smaller in size and had proportionately shorter limbs than any previously recorded blue sheep. Three years later, Schaefer documented them as representatives of a new species, but did not give it a formal scientific name. This situation was not rectified until 1964, when noted mammalogist Dr. Theodor Haltenorth officially named it *P. n. schaeferi*, thereby classing it as a subspecies of the blue sheep.[137]

Popularly referred to as the dwarf blue sheep, it differs from its larger relative not only in terms of size but also by virtue of its drabber coat colouration, brightened only by a silvery sheen. Moreover, it exists in much smaller social groups.[137-8]

After taking these distinctive features into account, in 1978—41 years after Schaefer's original pronouncement—mammal taxonomist Prof. Colin Groves elected to restore the dwarf blue sheep to its former status, namely that of a valid species in its own right. Today, therefore, it is known scientifically as *Pseudois schaeferi* (although with recent molecular analyses showing differences between the two species to be slight, some workers have suggested that it should be reclassified as a subspecies of *P. nayaur*). Only around 200 individuals of this endangered mammal were known to exist in the year 2000.[138]

IRIOMOTE CAT AND DWARF PIG— REVELATIONS FROM THE RYUKYUS

By the end of the 1800s, zoologists were confident that all of the world's cat species had been described and named—but this belief received an unexpected challenge in 1965. That was the year in which Japanese author-naturalist Yukio Togawa (sometimes spelled 'Tagawa') learnt of a strange, unfamiliar cat form said to inhabit Iriomote—a tiny insignificant dot of an island in the Japanese Ryukyu chain, almost completely covered with dense mountainous rainforest, and little-explored by scientific researchers. By a fortuitous quirk of fate, work commitments led Togawa to visit Iriomote later that year, and once there he investigated its feline mystery, a creature referred to by the natives as the *pingimaya*. Togawa hoped to obtain specimens for formal identification, but was initially foiled by the inconvenient tendency of the natives to eat any such specimens that they encountered.[60,139]

Happily, however, two pelts and a skull finally found their way to the University of Okinawa, where they were studied by Togawa and by Ryukyu zoologist Prof. Tetsuo Takara, after which the specimens were passed on to the eminent Japanese mammalogist Prof. Yoshinori Imaizumi at Tokyo's National Science Museum. During a later visit to Iriomote, Togawa succeeded in obtaining three more skulls and a skin, which were once again sent

Iriomote cat. (Tadaaki Imaizumi/ Nature Production, Tokyo)

to Prof. Imaizumi. After studying these, together with some additional examples that included at least one living animal, in 1967 Imaizumi published an extensive description of the Iriomote cat, which he deemed to be so distinct from all others that he placed it in a brand-new genus, naming it *Mayailurus iriomotensis*.[60,140]

Little larger than a domestic cat, with greyish-brown fur, the Iriomote cat could easily be dismissed on first sight as nothing more than a feral domestic, and superficially resembles the common Asian leopard cat *Prionailurus bengalensis* too. However, its rounded ears with black backs spotted with white, the 5-7 longitudinal dark lines running along the nape of its neck to end on its shoulders, the distinctive vertical bands of spots on its body, and its 28-tooth dentition constitute just a few of the features that collectively distinguish it from both of these felids. Its tooth count is particularly unusual—on the one hand separating it from the normal 30-tooth dentition of most other Old World cats, yet on the other hand providing an unexpected correspondence with the mountain cat *F. guigna*, from Chile.[60,139-140]

This odd assemblage of characteristics inspired Imaizumi to postulate that the Iriomote cat was an extremely primitive species uniting the Old World's small cats with those of the New World. Later studies, however, suggested that its unusual dental complement arose independently of the Chilean mountain cat's, thus undermining its claim to fame as a feline 'missing link'. Nowadays, it is placed within the genus *Prionailurus*, containing the leopard cat, and recent analyses of its proteins confirm that its closest relative is indeed the leopard cat (in fact, some authorities class it as nothing more than a well-marked island subspecies of this felid). Nevertheless, it is an important find, wholly restricted as it is to a single tiny island with a total surface area of less than 113 square miles, but with no more than 100 individuals in existence here, the Iriomote cat is categorised by the IUCN as critically endangered.[60,139-141]

Nor is the Iriomote cat this island's only zoological distinction. In 1924, Japanese naturalist N. Kuroda self-published an article in which he documented a number of new mammals from the Ryukyus, including a form of dwarf wild pig from Iriomote and three other islands in this chain. He named it *Sus scrofa riukiuanus*, the Ryukyu dwarf pig, but it attracted little attention outside Japan, where in 1973, Prof. Yoshinori Imaizumi (who formally described the Iriomte cat in 1967) published a paper elevating it to the level of full species (though this has not been widely accepted). During an expedition to Iriomote a year later in 1974 to survey the Iriomote cat, however, this diminutive porcine form was also encountered and received wider publicity, as it proved to be only a little larger than the world's smallest species, the pygmy hog *Sus salvanius* (see p. 71).[142] Since 1982, it has been deemed by the IUCN to be vulnerable.

There has also been a more recent feline find on a tiny Japanese island. In winter 1989 it was announced that a new species of cat had been discovered on one of the two rocky Japanese islands of Tsushima, sited in Korea Strait, but this felid, the Tsushima cat or wildcat, is nowadays considered to be merely a subspecies of the leopard cat, albeit an endangered one.[143]

RETURN OF A LOST LEMUR
In 1875, Dr. Albert Günther of London's Zoological Society described a new species of Madagascan lemur, a small brown form with noticeably hairy ears. Its type specimen was a skin at the British Museum (Natural History), and it was christened *Cheirogaleus* [now *Allocebus*] *trichotis*, becoming known as the hairy-eared dwarf lemur. Two other skins were sent to the Paris Museum a few years later, but remained incognito until rediscovered there in 1956 by one of the world's leading lemur experts, Dr. Jean-Jacques Petter. Otherwise, the species was a complete mystery—until 1966, when a living specimen was obtained by some natives close to Mananara on Madagascar's eastern coast, and given to Dr. Petter's assistant, André Peyriéras.[144] More recently, in April 1989, Ruhr University zoologist Dr. Bernhard Meier discovered this species alive and well in lowland rainforest (Madagascar's very last surviving, extensive expanse), once again near Mananara. However, it is currently categorised by the IUCN as a critically endangered species, with an estimated total population of between 100 and 1000 individuals.[145]

Even more mysterious is the world's second smallest species of mouse lemur—the pygmy or western rufous mouse lemur *Microcebus myoxinus*. With a head-and-body length of 2.4 in, a tail length of 5.4 in, and weighing no more than 1.5 oz, it is no bigger than a slightly overweight house mouse. First described in 1852, this minute species was largely forgotten afterwards, and even those few specialists who were aware of it ultimately assumed that it had become extinct—until 1993, when a German scientific team rediscovered it in western Madagascar's Kirindy Forest.[146]

A RECLUSIVE RICE RAT
The hairy-eared dwarf lemur was not the only long-lost mammal that reappeared in 1966. Also making a belated comeback—but this time in the guise of a skull from a recently-deceased specimen—was the James Island rice rat *Oryzomys swarthi*. This naked-eared, mouse-like rodent from the Galapagos Islands had previously been known only from its type specimen and three others, all collected on James Island in 1906.[144]

WHITE-THROATED WALLABY— A NEW HOME IN NEW ZEALAND
In just over a year, no less than three supposedly long-extinct species of marsupial were rediscovered during the 1960s, the first of which was the white-throated (parma) wallaby *Macropus* (*Thylogale*) *parma*. A very handsome animal, dark brown on top, white underneath, with a black stripe running from its neck to midway

White-throated (parma) wallaby.
(Dr. Karl P.N. Shuker)

down its back, it was once plentiful in the Illawarra and Cambewarra mountainous areas of southern New South Wales. But as a result of its woodland habitat's wholesale clearance by man, its numbers rapidly dwindled. By 1932, this attractive mammal was considered extinct.[57,147]

In 1966, however, it made a reappearance that was particularly unexpected—due to the specific locality in which its reappearance took place, which was not in Australia at all, but instead in New Zealand, on a 500-acre island called Kawau, sited 30 miles north of Auckland. New Zealand is a country famed for having just two species of native mammal, both of which are bats. So how could the existence on a New Zealand island of an allegedly-extinct Australian wallaby be explained? For once, the answer was quite straightforward.[147]

In 1870, settlers had released several white-throated wallabies from Australia onto Kawau, just as they had earlier introduced many other non-native animals onto New Zealand's two principal islands. The wallabies had thrived, and multiplied, so that a healthy population now existed there (alongside those of four other wallaby species also brought here from Australia at various times).[147]

To safeguard this species' future, some of the island's white-throated wallabies have been sent to zoos around the world to initiate captive populations, just in case disease or some other threat should decimate the Kawau colony. In 1972, moreover, it was rediscovered on mainland Australia too, by G.H. Maynes, who located some notable forest-dwelling populations in an expanse of land stretching from the Hunter River to the Clarence River in northeastern New South Wales, thereby increasing its survival chances in the wild.[147]

BURRAMYS—A POSSUM FROM THE PAST
One day in August 1966, while residing at the Melbourne University Ski Lodge, high on the slopes of Mount Hotham in the Australian Alps of eastern Victoria, Dr. Kenneth Shortman (of Melbourne's Walter and Eliza Hall Institute) entered the lodge's kitchen, presumably in search of food. Whether he found any is not recorded, but something that he *did* find there, just behind the garbage pail and also looking for food, is most definitely on record—because it constituted one of the most extraordinary mammalogical discoveries of the 20th century.

In 1896, comparative anatomist Dr. Robert Broom disclosed that a few fragments of 20,000-year-old fossilised bone, extracted from some owl pellets that he had found in the Wombeyan Caves near to the New South Wales town of Burra, were from a hitherto undescribed species of extinct Pleistocene possum, which he named *Burramys parvus* ('small Burra mouse'). Its fossilised remains were later found at the Buchan

Burramys pygmy possum. (Fisheries & Wildlife Department, Victoria, Australia)

Caves in Gippsland, Victoria, too. In the years that followed, *Burramys* attracted widespread interest from palaeontologists, for being yet another representative of the diverse array of marsupial forms that had failed to survive into Recent (Holocene) times.[42,148]

At least, that was the assumption until August 1966, when Dr. Shortman's fateful visit to the ski-lodge's kitchen provided a very different storyline—because what he found there was a small, dormouse-like creature with brownish-grey upperparts, paler underparts, and a dark stripe running down each limb. It was evidently a possum, but he could not identify its species. Bemused, he captured the mystery mammal alive and unharmed, and took it to the Victorian Fisheries and Wildlife Department in Melbourne, where to everyone's amazement it was revealed by palaeontologist Norman Wakefield to be a living *Burramys*.[42,148]

It was cared for from then on by Dr. John Seebeck, who learnt a great deal concerning the habits of this 'living fossil', noting its principally nocturnal lifestyle, placid temperament, and its dietary preference for insects, apples, sunflower seeds, and honey diluted with water.[148]

Searches for more living specimens of *Burramys* (nowadays popularly termed the mountain pygmy possum) by Seebeck and others were unsuccessful at first, so that zoologists began to wonder if the ski-lodge kitchen's trespasser could be a marsupialian Mohican, the very last of its line. Happily, others were found in due course in the same area of Victoria, as well as a separate population in Kosciuszko National Park, New South Wales,, thereby securing on a firmer footing the modern-day existence of this highly significant species, which is presently categorised as critically endangered by the IUCN.[48,148]

DAY OF THE DIBBLER

Less than a year after the white-throated wallaby's reappearance and the resurrection of *Burramys*, yet another 'extinct' Australian mammal was unexpectedly discovered. The species in question this time was the freckled marsupial mouse *Parantechinus apicalis*, more commonly called the dibbler—its Aboriginal name. Up to 10 in long, with speckled grey fur and a distinctive white ring round each eye, this Western Australian mammal had last been sighted in 1884. It was thought to have preyed upon other mammals and small birds, but little was really known about its lifestyle—which was underlined by the surprising circumstances surrounding its rediscovery, 83 years later.[57]

In January 1967, wildlife photographer Michael K. Morcombe was near the Waychinicup River, east of Albany on Western Australia's southern tip, seeking a curious little creature called the honey possum *Tarsipes spencerae*, which lives almost solely upon nectar and pollen (not honey!) and seems to be the marsupial equivalent of a hummingbird. To trap some specimens temporarily for photographic purposes, Morcombe prepared some tiny cages that would catch the possums unharmed, and he fitted these over the flowers of the nectar-rich bottlebrush and banksia, which are among the honey possum's favourite flora.[149]

Normally, that is—because three weeks passed without any of the cages trapping a single possum. Then the weather grew hotter and drier, and one day Morcombe discovered that his cages had finally snared a couple of specimens—but they were not honey possums. To his amazement and delight, he saw that the two speckled marsupials sitting unharmed inside his cages were dibblers![149] Since then, other specimens have been recorded in southwestern Australia, including a sizeable population on the Jurien Bay islands of Boulanger and Whitlock by Dr. Chris Dickman during 1986-88. Nonetheless, Morcombe's couple remain particularly significant—not only because they were the ones that marked the species' revival, but also because they revealed a hitherto unobserved facet of dibbler lifestyle. In contrast to expectations, the dibbler is not predominantly carnivorous. Although it does devour any insects encountered on the blooms of large florescent bushes, its main interest is the flowers' nectar, especially in dry weather.[149]

PYGMY HOG—REDISCOVERING THE WORLD'S TINIEST WILD PIG

The 1970s began very promisingly for mammalian finds, with the rediscovery of a delightful little creature called the pygmy hog *Porcula salvania* (until 2007, it was known as *Sus salvanius*, but following a new genetic analysis of a large section of mitochondrial DNA from this species, it was reassigned to its own genus, *Porcula*, originally created for it back in 1847 but later abandoned). In keeping with its name, it is the smallest species of wild pig, with a total length of only 20-26 in, and a shoulder height of no more than 12 in. Once widely distributed in grassy swampland along the southern foothills of the Himalayas of Bhutan, Nepal, Sikkim, Bangladesh, and northeastern India, by the late 1950s it seemed to have totally died out—its disappearance blamed upon habitat destruction.[103,150]

Nevertheless, while science wrote it off as extinct, unconfirmed reports of its continuing survival persisted in various localities within its former range. These included a statement made by a forest ranger to noted Indian wildlife expert Edward Pritchard Gee that during the 1958-9 winter, pygmy hogs had been seen in Assam's Rowta Reserve Forest, on the northern bank of India's famous Brahmaputra River. This and other accounts

Dibbler—an elusive marsupial that vanished for 83 years, until its accidental rediscovery in 1967. (Dr. Christopher Dickman)

greatly intrigued amateur naturalist John Tessier-Yandell, who was living at that time in Upper Assam.[150]

By sheer good fortune, in 1967 his official work took Tessier-Yandell to the Brahmaputra's north bank—right in the heart of alleged pygmy hog territory. He lost no time in pursuing this species' trail, and two years later he learnt that a pygmy hog had definitely been killed not long before on an estate in nearby Mangaldai. Sadly, however, the specimen had been eaten afterwards and the remains discarded. Nevertheless, for Tessier-Yandell there was no longer any doubt concerning the pygmy hog's continuing existence. All that he needed now was conclusive evidence for scientists to examine and confirm.[150]

His investigations uncovered fresh pieces of encouraging information during 1970, so that it seemed only a matter of time before the pygmy hog would be formally rediscovered. And on 10 April 1971 that time did indeed arrive, when Tessier-Yandell received a telegram from colleague and fellow pygmy hog pursuer Dick Graves, requesting him to come to Mangaldai at once to identify some wild pigs that had been captured alive and which Graves felt sure were bona fide pygmy hogs. Nine animals were present—four young and two adult sows, plus one adult and two immature boars. Tessier-Yandell made his way to Mangaldai, and was thrilled to be able to verify Graves's belief—they were indeed pygmy hogs. Another memorable mammal had made a welcome return to the world list of living creatures.[150]

Since then, specimens have been captured for captive breeding programmes, and in recent years this distinctive if diminutive creature has been bred at Germany's Stuttgart Zoo and also in Japan. Moreover, the Durrell Wildlife Conservation Trust has not only launched a comprehensive conservation strategy and education programme for this species in its homeland but also established a highly successful captive breeding programme for it at the Pygmy Hog Research and Breeding Centre in Assam, thereby offering hope that

Pygmy hog at London Zoo in or around 1904.

the future of this critically endangered species may be secured.[103,150-1]

There are two pygmy-sized but nonetheless significant footnotes to this saga. Firstly, an unexpected bonus during the search for the pygmy hog was the capture of a hispid hare *Caprolagus hispidus*, another little-known species hitherto deemed extinct by some authorities.[150-1] Secondly, in 1977 four specimens of a new species of parasitic insect were taken off a pygmy hog in Darrang, northwestern Assam. It was later named *Haematopinus oliveri*, the pygmy hog sucking louse.[152]

KITTI'S HOG-NOSED BAT—
NO BIGGER THAN A BUMBLEBEE
What do a bumblebee and a bat have in common? Not a lot—until 1973, that is. During the early 1970s, Dr. Kitti Thonglongya—at that time Curator of Terrestrial Vertebrates at Bangkok's Centre for Thai National Reference Collections—became increasingly interested in Thailand's chiropteran (bat) fauna, and made a number of important discoveries in that field. The most notable of these comprised a series of exceptionally tiny bats collected by him in October and December 1973, from one or other of two limestone caves in the vicinity of Ban Sai Yoke's Forestry Station and the infamous River Kwai bridge, in southern Thailand's Kanchanaburi Province. Dr. Thonglongya's initial examination of these specimens convinced him that they represented something totally new, radically different from any species recorded before by science and meriting separate generic status at least.[152]

Thonglongya's opinion was shared by experts at the British Museum (Natural History); and so, working with British Museum colleague John Edwards Hill, he began to prepare a full-scale scientific description of this new species. Tragically, however, in February 1974 (only four months after he had collected the very first specimens) Thonglongya died, but Hill continued with their paper and published it later that year.[153]

The new bat was certainly very different from all others, exhibiting a bewildering mixture of morphological

Pygmy hogs—a colour engraving from 1853.

and anatomical features drawn from several quite separate chiropteran families. Accordingly, Hill deemed it necessary to create for it a completely new family (Craseonycteridae—'family of bats with mixed features'), a classification still accepted today. As for any vernacular name, two of the bat's most striking external attributes provided a couple of suitable versions nowadays in widespread use. One alludes to its remarkably small size: holder of the *Guinness World Records* title as the world's smallest species of mammal (and quite possibly the world's smallest warm-blooded vertebrate), with a body size no bigger than that of a large bumblebee and only weighing 0.07 oz, this miniscule mammal soon became known as the bumblebee bat. Equally, its pig-like snout earned it the alternative name of Kitti's hog-nosed bat. Other noticeable characteristics include the relatively large size of the flight membrane connecting its thighs, and its lack of a tail.[153]

Kitti's hog-nosed bat. (Jeffrey A. McNeely)

Following this species' discovery and description (in which Hill named it *Craseonycteris thonglongyai*[153]), attempts were made to find out how abundant, or rare, it was. Regrettably, only a handful of examples were located by the early 1980s, and its continuing survival seemed so uncertain that at the November 1984 General Assembly of the IUCN, Kitti's hog-nosed bat was named as one of the world's 12 most endangered animal species.[154] By 1986, however, events had taken a more promising turn; over 2000 specimens had been recorded in more than 21 caves, following an extensive search made by Surapon Duangkhae from Bangkok's Mahidol University, armed with geological maps to locate the precise type of cave most likely to house this species, and a high-tech electronic bat-detector.[155]

Even so, faced with the ever-present threat of habitat destruction, as well as with collection of specimens by locals to sell as souvenirs or to display as curios, the smallest but most significant bat species discovered during the 20th century is by no means out of danger, and requires full protection to ensure its long-term survival.[155]

In 2001, a single specimen of Kitti's hog-nosed bat was discovered in Mon State, Myanmar (formerly Burma), since when at least nine separate Burmese caves containing this species have been located. Interestingly, although individuals from the Thai and Burmese populations are morphologically identical, they exhibit different echolocation calls, but whether they are reproductively isolated (and hence deserving of classification as two separate species) has yet to be determined.[155a]

On 2 May 1948, a strange tailless fruit bat was caught by naturalist Angus Hutton on a remote coffee estate within the High Wavy Mountains in southern India's Western Ghats, but was not recognised to be a new species until 1972—when Dr. Thonglongya dubbed it *Latidens salimalii*, Salim Ali's fruit bat, thus creating a new genus for it, and honouring the renowned Indian zoologist Dr. Salim Ali. No other specimen was recorded until April 1993, when a flock was encountered on the same coffee estate by zoologists Nikky Thomas and Manoj Muni.[156] And another rare Asian bat, Ridley's leaf-nosed bat *Hipposideros ridleyi*, last reported in 1910, made a welcome flight back into being in 1975, when a colony of about 50 individuals was discovered in a forest near Malaysia's capital, Kuala Lumpur.[103,156]

CHACOAN PECCARY—AN ICE AGE RECLUSE
The modern-day existence of only two species of peccary, those pig-like mammals of tropical America, was an immutable and immortal zoological fact that had been faithfully reiterated within each and every zoological work for generations—until a third species was unexpectedly discovered in 1974. Nor was it 'just another species'.

During an extensive series of field studies carried out from 1972 to 1975 in the semi-arid Gran Chaco area overlapping northern Argentina, western Paraguay, and southeastern Bolivia, Connecticut University biologist

Chacoan peccary

The late Dr. Ralph Wetzel (left) during the early 1970s with a specimen of the newly-resurrected Chacoan peccary. (Dr. Ralph Wetzel)

Dr. Ralph Wetzel and his co-workers were at first puzzled to discover that the local inhabitants recognised the existence of *three* different types of peccary. The white-lipped peccary *Tayassu pecari* was referred to as *tâchycâti* or *tagnicate*, and the collared peccary *Pecari tajacu* as *cure-í* or *tayté-tou*, but the locals also spoke of a larger, mysterious form variously termed *tagua*, *pagua*, or, probably on account of its long ears, *curé-buro* ('donkey-pig'). After further enquiries, Wetzel succeeded in obtaining a series of skulls for all three peccary types from hunters' kills, and morphological comparisons of these duly proved that the 'donkey-eared' peccary was truly a totally separate type, distinct from the two previously-known modern-day species.[157]

Upon his return to the university, Wetzel made an even more surprising discovery. Although hitherto unknown in *living* form, the new peccary was not completely unknown to science. In 1930, the first specimen of a new fossil subspecies of Argentinian peccary had been described, and named *Platygonus carlesi wagneri*. Eight years later, Rusconi classified it as a species in its own right—*P. wagneri*. Since then, many other fossil specimens of this species have been found, but none of more recent date than the Pleistocene epoch (2 million to 10 thousand years ago), so it has always been catalogued as one of the many large mammals that died out during the Ice Ages. When he compared fossilised skulls of *P. wagneri* and present-day skulls of the 'donkey-eared' peccary, however, Wetzel perceived that they all unquestionably belonged to one and the same species. *P. wagneri* had survived the Ice Ages after all! His studies also led him to conclude that *P. wagneri* was really more akin to members of the genus *Catagonus* than *Platygonus*, so in 1975 he renamed it *Catagonus wagneri*, and it is nowadays commonly called the Chacoan peccary or tagua.[157]

Weighing 100 lb and standing more than 3 ft high, Rusconi's resurrected species is the second largest of the four present-day peccary species (a fourth species of living peccary was discovered during the early 21st century, see p. 276), and is less closely related to any of the other three than they are to each other. Except for its size, its most noticeable external features are its brown, bristly pelage, its faint collar of lighter hairs across its shoulders, its large head, and its relatively long limbs, tail, ears, and snout.[157] At first, its existence was known only from the Gran Chaco region of Paraguay and Argentina, but later reports from hunters in Bolivia's corresponding area exposed its presence there too. Yet, disturbingly, due to habitat destruction and to hunting by the local people for its hide and meat, the Chacoan peccary's continuing survival is becoming increasingly imperilled.[157-8]

This grave situation prompted Wildlife Conservation International (a division of the New York Zoological Society) to begin funding in 1983 an important conservation project in an attempt to prevent this species from becoming any rarer. Headed by zoologist Andrew Taber, the goal was to initiate and encourage positive conservation policies in the Chaco Basin region, in order to preserve not only the Chacoan peccary (currently categorised as endangered by the IUCN) but also the entire ecology of the area, which is home to many endemic species.[158]

The much-welcomed discovery or rediscovery of a major animal often has its ironic aspects too—and the formal exposure of the Chacoan peccary's present-day existence was certainly no exception. Following its 'official' return to the land of the living, news emerged that for a number of years *prior* to this, and wholly unbeknownst to science, its hide had routinely been used by New York furriers to trim hats and coats!

YELLOW-TAILED WOOLLY MONKEY—HOW A SOLDIER'S PET RESTORED A 'LOST' SPECIES

Whereas the common woolly monkey *Lagothrix lagothricha* is one of the best-known species of South American primate, its yellow-tailed relative *Oreonax* (formerly *L.*) *flavicauda* was renowned for a very great period of time as the rarest and most mysterious. Its type specimen had been obtained as long ago as 1802, by the famous explorer Alexander von Humboldt, who documented it a decade later, recording its provenance as the Peruvian province of Jaen. Nothing more was heard of this new monkey for over a century afterwards—until 1926, when three more were obtained by R.W. Hendee (an animal collector working on behalf of the Godman-Thomas expedition to Peru) at Pucatambo, in the Peruvian province of Rodriguez de Mendoza, Amazonas.[159]

It made a fleeting return to the scientific spotlight in 1963, when zoologist Jack Fooden revealed that two monkey specimens held at the American Museum of Natural History and obtained in 1925 from La Leija in Amazonas were yellow-tailed woollies, thereby boosting the number of recorded specimens from four to six. Even so, after that it again sank into scientific obscurity, and there it may well have remained indefinitely—had it not been for the interest that its history had aroused in Peruvian zoologist Dr. Hernando de Macedo-Ruiz (Head of Mammalogy and Ornithology at Lima's Natural History Museum), Harvard University zoologist Dr. Russell Mittermeier, and Lima resident B. Anthony Luscombe.[159]

In 1974, this trio of investigators organised and took part in a 12-day expedition to the area in which the six known specimens had been collected, and almost immediately achieved success. On only the second day, they encountered a local hunter carrying the skull and stuffed skin of an adult *O. flavicauda* that he had shot less than a week earlier. Evidently, the species was not extinct. In addition, he supplied them with three other skins and two skulls of specimens that he had killed for food—but their greatest success was still to come. On the very last day of their expedition, Dr. de Macedo-Ruiz and company were led by some children at Pedro Ruiz Gallo to the home of a soldier, whose current pet was none other than a juvenile *O. flavicauda*. The team rapidly purchased this unique animal—the first *living* yellow-tailed woolly monkey ever seen by scientists.[159]

The team's researches suggested that this species' natural habitat consisted of montane forests at 1665-8330 ft, primarily of Amazonian origin, but that its distribution range was extremely limited. Coupling this with the extensive destruction of its habitat through road development and provision of land for agricultural purposes, plus the additional threat to its survival posed by hunting, this rediscovered species was indisputably greatly endangered. Consequently, the team concluded its investigations by recommending the establishment of an official reserve within the monkey's distribution range—a recommendation that led to a formal campaign for conservation, and to the announcement in 1982 that a national park would indeed be created. And as an extra attempt to secure its future, a captive breeding programme for the yellow-tailed woolly monkey has been proposed at Lima Zoo. In November 2010, the discovery of a hitherto-unknown population of this species in Huanuco, Peru, was announced. It is presently categorised as critically endangered by the IUCN.[103,159-60]

THE NOVEL NINGAUIS AND OTHER MINISCULE MARSUPIALS

'Marsupial mouse' is something of a misnomer for the various species of tiny pouched mammal inhabiting Australia that are grouped together under this collective name, because their savage behaviour, insectivorous mode of existence, and sharp-muzzled features are far more reminiscent of shrews than mice. Pairing their diminutive size, moreover, with the extreme elusiveness characteristic of all predators, plus a predominantly nocturnal lifestyle, it can be appreciated that detailed observation and identification of these animals in the wild is not an easy task. Consequently, reports from time to time concerning the rediscovery of marsupial mice not seen for some years are neither uncommon nor unexpected—usually.

A notable exception occurred in 1975, however, when a series of such animals was scientifically described that not only constituted two *new* species, but also required the creation of a totally new genus for them. This was due to various cranial and dental features, not to mention their relatively short feet, which set them well apart from all previously known species of marsupial mouse.[161]

The two species were christened the Pilbara ningaui *Ningaui timealeyi* (from Pilbara in northwestern Western Australia) and the Wongai ningaui *N. ridei* (from Wongai in central Western Australia), by noted marsupial researcher Dr. Michael Archer of the Queensland Museum, who disclosed that the specimens representing them had been collected as far back as the 1950s in some cases, but had not been recognised until now as taxonomically distinct from all others. Just prior to the publication of Archer's paper, news services around the world reported the capture of a living ningaui (their generic name is nowadays used as their common name too), an adult female measuring just 4 in long, which had been caught unharmed in the Billiatt Conservation Park. Another female, this time with newborn babies (each measuring less than 0.5 in), was also captured, and all were safely transported to Adelaide's Institute of Medical and Veterinary Science for study.[161]

Incidentally, Archer's choice of generic name for these new species was particularly appropriate, because 'ningaui' is an Aboriginal term alluding to tiny mythological entities that are hairy, have short feet, hunt for food only at night, and eat all food raw—a perfect description of the new marsupial mice and their lifestyle.[161]

During the 1980s, several new species of marsupial mouse were discovered, including a third ningaui. Named the southern ningaui *N. yvonneae* in 1983, it hails once again from Western Australia.[162]

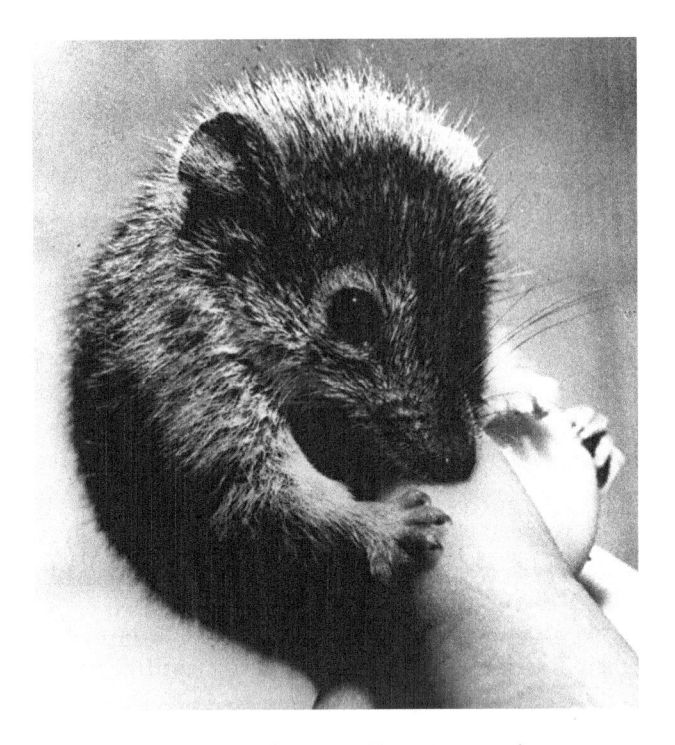

Ningaui. (Times Newspapers)

Back in 1975, marsupial mouse researchers were also celebrating the rediscovery of a very unusual 'lost' species. First made known to science in 1908 with the procurement of a single specimen at Pilbara in Western Australia,[163] the long-tailed dunnart *Sminthopsis longicaudata* has a tail more than twice as long as its head and body combined, ending in a conspicuous black tuft. After 1908, this distinctive species was not reported again until 1940, when a second lone individual was obtained at Marble Bar. In addition, a preserved specimen not previously identified as *S. longicaudata* was spotted in the National Museum of Victoria's collections.[57] Nothing more was then heard of this tiny marsupial with the tremendous tail for another 35 years, until 1975, when an adult female was obtained in Western Australia's Gibson Desert.[164] Six years later, two breeding pairs were captured alive here by members of the Western Australian Research Centre, and the species is now believed to be relatively widespread amongst this desert's rocky terrain. As for its extraordinary tail, researchers have surmised that it uses this to lure within range the insects upon which it feeds.[165]

The sandhill dunnart *S. psammophila* was discovered in 1894, when a single specimen was procured by the Horn Expedition in the desert near Lake Amadeus, in the proximity of the Northern Territory's famous Ayers Rock. Nothing else was known of this species until 1969, when farmer M. Andrews collected one on South Australia's Eyre Peninsula, about 650 miles southeast of Ayers Rock. Others have since been found even further southeast.[57,164]

INTRODUCING THE PROSERPINE ROCK WALLABY

Albeit to varying extents, by 1977 all seven species of Australian rock wallaby recognised at that time had sharply decreased in number, and some were greatly endangered. Hence the discovery of an eighth, totally new species was the very last thing that zoologists expected—yet this is precisely what happened that year. Indeed, if science had taken more notice of local testimony, it may well have occurred much earlier—because farmers at Proserpine, eastern Queensland, had often spoken about a strange form of rock wallaby in the area, but no-one had ever sought a specimen. Midway through 1977, however, a single individual of this mysterious marsupial was finally captured, and provided scientists with a great surprise.[166]

To begin with, it was rather larger in overall size than any of the known species of rock wallaby. Also, its tail was longer relative to body size, and, uniquely, was tipped with white. Examination of the living animal's external morphology was succeeded by comprehensive chromosomal and blood-protein comparisons, which demonstrated irrefutably that it was genuinely a species new to science. To ensure that this, the largest of all rock wallabies, did not suffer the marked decrease in numbers experienced by its smaller relatives, recommendations for its habitat's preservation were soon submitted to the Queensland State Government. Nevertheless, at the present time (2009), it is the only rock wallaby species to be categorised as endangered by the IUCN.[166]

As for its name, this novel marsupial simply became known as the Proserpine rock wallaby, after its type locality—in turn named after the Roman goddess Proserpina. And maintaining its nomenclatural link with classical legend, its scientific name was derived from Proserpina's Greek counterpart, because it was christened *Petrogale persephone*.[166]

BULMER'S FRUIT BAT—
THE FOSSIL THAT CAME TO LIFE!

In 1977, the *Australian Journal of Zoology* published a paper by New Guinea biologist James I. Menzies in which he described a new genus and species of large fossil fruit bat, of which 200 incomplete specimens had been excavated by archaeologist Dr. Susan Bulmer at a site in Papua New Guinea's Chimbu Province called the Kiowa rock shelter. The fossils were at least 9,000 years old, and the new (albeit long-extinct) bat was dubbed *Aproteles bulmerae*—'Bulmer's incomplete-at-the-front bat'. Its scientific name drew attention to its unique lack (among bats) of front teeth (i.e. incisors), and its closest living relative was another New Guinea species, *Dobsonia moluccensis*.[167] The discovery of Bulmer's fruit bat was of great interest to palaeontologists, representing a wholly new fossil genus, but within three years its name would become of equal if not greater interest to students of present-day wildlife too.

In 1980, another paper by Menzies appeared, and contained some stunning news. Not long after the publication in 1977 of his description of Bulmer's fruit bat, Menzies had examined a collection of current New Guinea animal specimens obtained two years earlier, and he had discovered within it two *modern-day* skulls and two isolated *modern-day* mandibles (lower jaws) of this species! In short, *A. bulmerae* was no longer an exclusively fossil form, supposedly extinct since the Pleistocene's close. Instead, it had plainly survived, unknown to science, right up to the present day. Somewhere in New Guinea, therefore, there should be living specimens of this resuscitated species.[168]

Naturally, Menzies had swiftly investigated the origin and history of this highly significant collection, and had learnt that it had been made by David Hyndman in 1975, whilst carrying out anthropological studies in western Papua New Guinea's Hindenburg Wall range. The *A. bulmerae* remains were from specimens that had been shot by bow and arrow in a huge cave nearby called Luplupwintem—which had contained numerous bats. Obviously, therefore, this cave would be the ideal locality to search for living specimens.[168]

And so, following Menzies's identification of the *A. bulmerae* material within his earlier collection, Hyndman had returned there in November 1977 for this specific purpose. To his horror, however, he had learnt upon arrival that, not long before, native hunters visiting the cave had either driven away or killed all but two members of the entire bat colony![168]

Happily, the species nonetheless survives. After being written off as extinct during the 1980s, Bulmer's fruit bat was rediscovered in May 1992. This was when 137 living specimens were recorded by Australian zoologist Dr. Tim Flannery and Papuan ranger-biologist Lester Seri in Luplupwintem. The only other recently-reported populations (both in Papua New Guinea) are from the vicinity of Herowana in Eastern Highlands Province and from the vicinity of Crater Mountain in Chimbu Province. *A. bulmerae* is now known to be the largest species of cave-dwelling bat in the world, and a traditional native taboo on killing this species, reinstated by the local Wopkaimin tribe who owns Luplupwintem, may help to safeguard it in the future, but at present it is categorised as critically endangered by the IUCN.[168]

FEA'S MUNTJAC—REDISCOVERING THE WORLD'S RAREST DEER

In December 1977, a female specimen of Fea's muntjac *Muntiacus feae* (named after zoologist Leonardo Fea), at that time the world's rarest deer, arrived at Bangkok's Dusit Zoo. It is a small dark-coated relative of India's barking deer. The only previous examples on record had been the species' type specimen (formally described in 1889) and one other, both originating from the borders of southern Burma (now Myanmar) and western Thailand. During the early 1980s, some more were captured and sent to Dusit Zoo—two females in 1981, followed by three males and three females from Xizang, Tibet, between February 1982 and April 1983, thereby establishing a small herd of this threatened species for captive breeding.[169-70]

Fea's muntjac. (Dave Powers)

One unexpected but exciting sequel to the Fea's muntjac story is that in March 1988, an old male muntjac was captured on the east slope of the northern sector of Gaoligong Mountain, in Gongshan County, within China's Yunnan Province, which was thought at first to belong to this species. Chromosomal analyses, conversely, divulged that it was in reality a separate species—one that had never been documented by science! A younger adult female, a juvenile female, and three skins all belonging to this same newly-disclosed species were later obtained too, all from Gongshan's Gaoligong and Biluo Mountains, and in 1990 it was formally described by Profs. Ma Shilai, Wang Yingxiang, and Shi Liming from the Kunming Institute of Zoology. They named it *Muntiacus gongshanensis*, the Gongshan muntjac (later suggestions that it was conspecific with the hairy-fronted muntjac *M. crinifrons* have not received widespread acceptance).[171]

Moreover, in 1982 mammalogists Prof. Colin Groves and Dr. Peter Grubb described a new muntjac, the Bornean yellow muntjac *M. atherodes*, from Borneo, which is nowadays considered to be a relict species. Prior to its discovery, only the common muntjac *M. muntjak* had been known from Borneo, which has larger antlers, and a different pelage colour.[172]

SULAWESI'S 'TREE-CLIMBING DOG' CLAMBERS BACK

One of the least-known of all viverrids (civets, genets) is the giant palm civet *Macrogalidia musschenbroeki*. A forest-inhabiting species measuring 5 ft in total length (a third of which constitutes its brown-and-white banded tail), it has a strikingly dog-like head, large feet, and sports a handsome coat of chestnut-brown fur dorsally, paler brown ventrally, with a reddish chest. Confined to the southeast Asian island of Sulawesi (Celebes), it was always presumed to be rare, and by the onset of World War II sightings had totally ceased.[173]

Consequently, in 1978 zoological circles greatly welcomed the news that a specimen had recently been spied and positively identified by Dr. John MacKinnon and Ir. Tarmuji while conducting a WWF survey in Gunung Ambang Reserve, located in the island's northern section. Equally exciting was the announcement that MacKinnon had actually succeeded in photographing it, as it sat in full view in a tree—the first time that the giant palm civet had ever been photographed alive in its natural habitat. The Sulawesi natives tell stories of a strange tree-climbing dog termed the *unguno bato* or *anjing hutan*, with hind limbs longer than its forelimbs. As MacKinnon has commented, this is evidently based upon sightings of the elusive *M. musschenbroeki* (which is currently categorised as vulnerable by the IUCN).[173]

LONG-FOOTED POTOROO— THE BIGGEST RAT KANGAROO

Smaller than kangaroos and wallabies, being closer in size to rabbits and hares, the Australian potoroos or rat kangaroos (genus *Potorous*) are characterised by their large canine teeth, short ears and hind legs, and rat-like form. Until 1967, only two species were known to exist (a third, the flat-faced potoroo *P. platyceps*, had died out a century earlier), but in June of that year a conspicuously large potoroo (an adult male) was caught in a dog trap in the forest southwest of Bonang, Victoria. It was initially assumed to be a long-nosed potoroo *P. tridactylus*, but when some other unusually large

Giant palm civet. (Dr. John MacKinnon/ World Wide Fund for Nature)

specimens were obtained at a nearby site named Bellbird (28 miles or so southeast of Bonang) in May 1968 and April 1978, suspicions that they may constitute something more significant were tested.[174]

Detailed morphological and chromosomal analyses were undertaken by biologists Drs. John Seebeck and P.G. Johnston, and by 1980 they had discovered differences in cranial and foot structure, and in blood proteins, as well as a totally distinct chromosomal complement—confirming that these rangy specimens really did constitute a new species. Its most obvious distinguishing feature gave rise to its scientific and common names—*Potorous longipes*, the long-footed potoroo.[174]

HELMET DOLPHIN— RECOGNITION AFTER 135 YEARS

In 1846, British Museum zoologist John Edward Gray described a new species of dolphin, based only upon a skull of unknown provenance from a dolphin specimen of equally unknown external appearance. Named *Delphinus metis*—then changed to *Stenella clymene*—this species was later reclassified as nothing more notable than a local variety of the spinner dolphin *Stenella longirostris*, and was omitted from lists of cetacean species for many decades afterwards. Then in 1975, dolphin researcher Dr. W.F. Perrin recognised that the skulls of two dolphins that had been obtained from the Texas Gulf coast so greatly resembled that of Gray's *S. clymene* (still housed at the British Museum) that these three skulls evidently belonged to the same species—but *which* species? Cranially, they corresponded most closely with *S. coeruleoalba*, the striped dolphin. Yet when photographs taken of the two Texan dolphins while still alive were examined, these showed that their body colouration and intricate pattern of markings were most similar to those of the spinner dolphin. Essentially, therefore, here were dolphins whose skulls resembled one species, yet whose external appearance resembled a second, totally separate species.[38,175]

As a result, Dr. Perrin suspected that *S. clymene*, with the British Museum skull as its type specimen and the Texan skulls as paratypes, was a valid species after all. In June 1976, another of these puzzling dolphins was obtained—it had beached itself at New Jersey's Ocean City. This specimen was carefully examined in 1977 at the Second International Conference on the Biology of Marine Mammals, held in Seattle, and its distinct nature was officially confirmed. So in 1981, Perrin and four colleagues fully redescribed *S. clymene*, combining a detailed anatomical report with an extensive account of its external morphology.[38,175]

Now known from Texas, Florida, New Jersey, the Caribbean, the mid-Atlantic, and West Africa, this re-elevated species exhibits a more complex pattern of swirling black, grey, and white markings along its flanks than the spinner dolphin, and is also distinguished from it by its shorter beak. Most characteristic of all, however, is the helmet-like series of markings on its beak and forehead. Consequently, it is widely referred to nowadays as the helmet dolphin.[38,175]

SUN-TAILED GUENON— FROM DINNER TO DEBUTANTE

In 1984, primatologist Dr. Mike Harrison returned home to Edinburgh from Gabon, West Africa, with a very beautiful monkey skin that had been left over from the dinner of a local hunter in central Gabon's secluded Forêt des Abeilles (Forest of Bees). The brilliant orange distal portion of its tail distinguished it from l'Hoest's guenon *Cercopithecus lhoesti*, the known species to which it bore the closest resemblance (but which in any event is limited to the eastern Democratic Congo). Enquiries by Harrison led to the discovery of some more skins, and a colleague, Rennes University zoologist Dr. Jean-Pierre Gautier, located a live specimen, kept as a pet in the same area of Gabon. Blood samples taken from this individual revealed that its species' chromosomal make-up differed from that of l'Hoest's guenon.[176-7]

The golden-tailed monkeys obviously comprised a new species, which Harrison duly described in 1988, naming it *C. solatus* ('sunstruck guenon'), emphasising the vivid colouration of its tail's tip, so that its common name became the sun-tailed guenon.[177-8]

This was the latest in a line of guenon surprises revealed during the 20th century. In 1977, what was originally deemed to represent a new species was found in central Zaire (now the Democratic Congo), when a Japanese traveller purchased a very strikingly patterned monkey skin from local hunters. Later named the salongo guenon *C. salongo*, it closely resembles the diana monkey *C. diana* of western Africa.[179] Central Democratic Congo is also the home of the dryas guenon *C. dryas*, formally described in 1932 by Dr. Ernst Schwarz and for a long time known only from its type specimen. In 1991, however, research was published that revealed the salongo guenon and the dryas guenon to be one and the same species with an age-related coat pattern. So the combined species is now known variously as the dryas guenon or the salongo guenon, with the scientific name *C. dryas*.[180]

And 1907 witnessed the description of an intriguing blackish-green species known as Allen's swamp monkey. Hailing once again from the forests of the Democratic Congo, as well as from those of the neighbouring Congo Republic, it was originally placed within the guenon genus *Cercopithecus* and named *C. nigroviridis*, but in 1923 Lang rehoused it within a genus of its own (renaming it *Allenopithecus nigroviridis*), on account of certain anatomical and behavioural similarities to the baboons. As it is the only species of *Allenopithecus*, Allen's swamp monkey is particularly important.[48,181]

THE QUEEN OF SHEBA'S GAZELLE

In the 1950s, while undertaking a mammalogical study in Yemen, Dr. Harold Hoogstraal collected many specimens, which were afterwards lodged in Chicago's Field Museum of Natural History. They included five skulls and skins from gazelles collected in mountains near Ta'izz, and which were assumed at that time to belong to the mountain gazelle's North Yemeni subspecies, *Gazella gazella cora*.[182]

In consequence, they attracted little further attention—until, more than three decades later, mammalogist Prof. Colin Groves and fellow researcher Dr. Douglas Lay re-examined them closely, comparing those characteristics of taxonomic importance with their equivalents in *known* specimens of the North Yemeni mountain gazelle. To their surprise, they found that the Hoogstraal material was decidedly different (especially with regard to the male skulls' unusually upright, almost straight horns). Indeed, it was apparent that these remains required formal description as the basis of a new species.

In 1985, this was carried out by Groves and Lay, who named the newly-exposed mammal *Gazella bilkis*, the bilkis or Queen of Sheba's gazelle—appropriate names, because the biblical kingdom reigned over by Bilqis, the Queen of Sheba, extended over much of what is now North Yemen. In or around that same year, a photograph was taken by Chris Furley of some gazelles living in a private collection, Al Wabra Wildlife Farm, in Qatar, which Groves claimed may be bilkis gazelles. However, this has never been confirmed, and no other specimens have been reported (despite a survey in the area in 1992), so at present this enigmatic antelope is, sadly, deemed extinct by the IUCN.[182]

MAKING A MONKEY OUT OF AN APE-MAN

Since the earliest times, people from many parts of China have been reporting sightings of strange hairy ape-men, some as tall as man himself (even taller in certain instances) and bipedal, others much smaller and given to running on all fours. The zoological identities of the larger forms, collectively termed the Chinese wildman or *yeren*, are still unknown, but in 1985 the mystery of the smaller version finally came to an end, when one of these elusive creatures was captured alive and unharmed.[183]

Caught near Anhui Province's Huangshan Mountain, and transferred to Hefei Zoo, it proved to be a macaque (i.e. related to the famous rhesus monkey *Macaca mulatta* and Gibraltar's barbary ape *M. sylvana*). However, it was sufficiently larger than any previously recorded to incite speculation that it may represent a new species, or at least a new subspecies of one of China's two species of stump-tailed macaque.[183] Chinese anthropologist Dr. Zhou Guoxing, a *yeren* expert, feels that the Huangshan Mountain specimen is of the same type as the unidentified giant macaque killed on 23 May 1957 near Zhejiang Province's Jiulong Mountain, and whose preserved hands and feet he examined in 1982. But further studies and specimens are required to resolve this issue satisfactorily.[184]

DENILIQUIN WOMBAT—BURROWING BACK FROM EXTINCTION?

In the 1890s, a population of wombats was observed near to the town of Deniliquin, in the Riverina district of New South Wales, Australia. They seemed to represent a species unknown to science, distinct from the northern hairy-nosed wombat *Lasiorhinus krefftii*, but before they could be investigated they disappeared, never to be seen alive again. In 1985, however, George G. Scott, a researcher with a special interest in wombat taxonomy, working at Canberra's Australian National Museum, learnt that just two years earlier a picnicker had reputedly seen a wombat in the Riverina area, a locality not frequented by common wombats. Investigating this claim, local naturalists discovered some wombat burrows there, which appeared to be freshly-made, but they did not spot any wombats. Other burrows have since been found, some containing hairs that have been shown to match very closely with those from a recently-uncovered taxidermic specimen—prepared from a wombat now known to have been collected at Deniliquin prior to the 1890s! Searches have continued since then, in the hope that these marsupial denizens of Deniliquin do survive, but none has so far been found.[185]

VANZOLINI'S SQUIRREL MONKEY

The year 1985 saw the description of the black or Vanzolini's squirrel monkey *Saimiri vanzolinii*—the first notably new South American monkey to be described and accepted as a valid species since the emperor tamarin in 1907 (p. 34). Named after Brazilian herpetologist Dr. Paul Vanzolini, it was found by Brazilian primatologist Dr. Marcio Ayres (from São Paulo's National Institute for Amazonian Research) in a very small area of rainforest between Brazil's Japura and Amazon Rivers. Its facial markings and darker fur colouration distinguish it from its neighbouring species, *S. sciureus*, but ally it with the Bolivian squirrel monkey *S. boliviensis*—even though that species is only found several hundred miles further southwest. The most likely explanation for such a zoogeographically anomalous affinity is that Vanzolini's squirrel monkey evolved from a population of *S. boliviensis* that became isolated from all others of its species when the latter's distribution range shrank during the last Ice Age.[186]

ONZA—HAS MEXICO'S LEGENDARY CAT FINALLY BECOME A REALITY?

For zoologists, 1986 opened in a decidedly dramatic manner—with the apparent procurement on 1 January of a legendary creature whose existence had been denied by science for centuries. That evening, Mexican ranger Andres Rodriguez Murillo surprised a very large cat close to his home in the valley behind Parrot Mountain, in Mexico's Sinaloa State. Fearing that it was a jaguar about to attack him, he shot it, but when he examined its body he found that it was neither a jaguar nor a puma—the only other large felid 'officially' existing in Mexico.[60,187]

It did resemble a puma superficially, but its limbs were longer and its body was more slender, giving it a cheetah-like outline; in addition, its ears were unusually big, and the inner surfaces of its forelimbs bore dark

Rodriguez onza—this adult female specimen, shot in 1986, has long been believed to be a bona fide example of Mexico's cheetah-like mystery cat . . . but is it? (International Society of Cryptozoology)

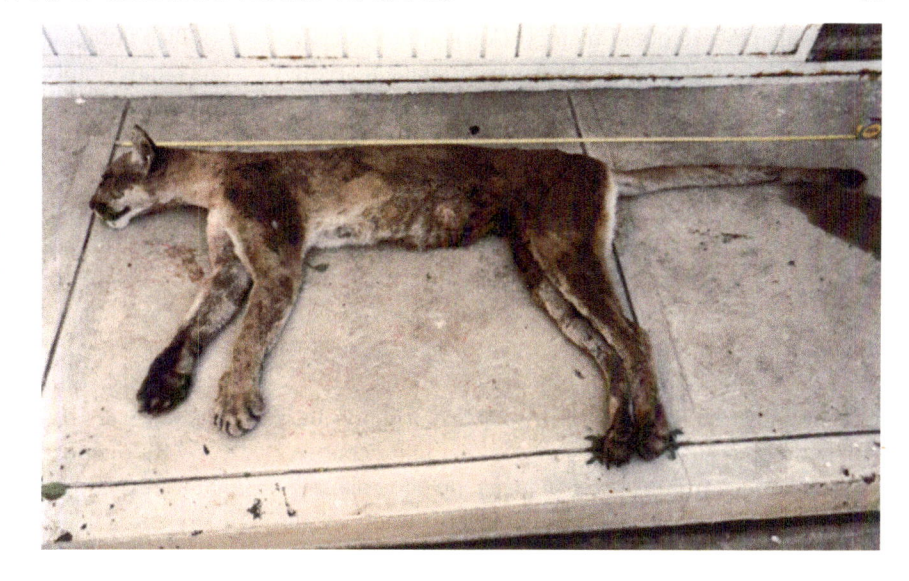

markings not possessed by pumas. As Rodriguez had little knowledge of wildlife, he contacted an expert hunter by the name of Manuel Vega to come over and identify it for him. When Vega arrived, he felt sure that he recognised the cat—identifying it as an onza, the fabled *third* large cat of Mexico.[60]

The onza's existence is supported by more than 300 years' worth of local eyewitness accounts (even including the testimony of visiting missionaries and Jesuit priests). Yet these had always been dismissed by zoologists as reports of poorly-seen or misidentified pumas.[60,187] Further details concerning the onza's early history are given in one of my previous books, *Mystery Cats of the World* (1989).[60]

Some 20th-century accounts of alleged onzas had actually been supported for a short time by complete specimens, though in every case these were somehow lost or destroyed. In spring 1926, for example, hunter-cowboy C.B. Ruggles trapped and killed a supposed onza southeast of Yaqui River in Mexico's Sonora State. After taking some photos of its carcase, and noting that it had very skinny hindquarters, with dark spots on the innerside of its limbs, Ruggles discarded it. A few years later, American naturalist J. Frank Dobie reported shooting an onza caught in traps set on Mexico's Barrancas de la Viboras; regrettably, its skin was subsequently devoured by bugs.[188]

In 1938, while in the company of renowned hunters Dale and Clell Lee from Arizona, Indiana banker Joseph Shirk shot an onza on Sinaloa's La Silla Mountain; photos show that, as with all of the others, it resembled an extremely gracile, long-limbed puma with big ears and unusual limb markings. Although much of its carcase was discarded, its skull was retained—only to vanish without trace when sent to a museum whose name, frustratingly, does not appear to have been placed on record. The dead body of a large unidentified felid that may well have been an onza was taken to Texas University in the late 1950s, but this cannot be traced either.[188]

J. Richard Greenwell, at that time the secretary of the International Society of Cryptozoology (ISC), who had a particular interest in the onza, succeeded in locating two onza skulls—one from a specimen shot in 1938 on La Silla Mountain by R.R.M. Carpenter while accompanied by Dale and Clell Lee (thus providing a remarkable parallel to the history of the Shirk specimen), the other from an onza shot by Jesus Vega (father of Manuel Vega) in much the same area sometime during the mid-1970s (this skull is now owned by rancher Ricardo Urquijo). Additionally, another onza investigator, Arizona hunter Robert Marshall (author of *The Onza*, the first book devoted to this subject) was successful in obtaining the incomplete skull (its lower jaw was missing) of an onza shot during the 1950s at Los Frailes, Sinaloa.[60,189]

More recently, around the time of the shooting of the Rodriguez specimen in January 1986, an onza was allegedly captured *alive*, and held for several days in captivity at a ranch in northern Sonora, where it was supposedly photographed too. Tragically, however, when no-one showed any interest in it, its owner shot it and threw its body away. As for the photographs, these have yet to make a published appearance. And in early 1987, yet another onza was reportedly shot in Sinaloa, this time by a wealthy Mazatlan businessman; but, true to form, its remains were not preserved.[190]

In short, with the exception of three skulls, physical remains of onzas have displayed a disconcerting tendency to disappear beyond the reach of scientists. The Rodriguez specimen, however, changed all of that—for instead of destroying its remains, the ranchers contacted Richard Greenwell. Following a complex series of interchanges, the precious specimen was transported to the Regional Diagnostic Laboratory of Animal Pathology, belonging to Mexico's Ministry of Agriculture and sited in Mazatlan, Sinaloa. There it was painstakingly studied and dissected by a biological team headed by American puma researcher Dr. Troy Best, who had been working alongside Greenwell during his earlier onza investigations. After the dissection, extensive samples of skeletal material, tissues, and blood were taken for examination and analysis in various U.S. research institutions, in a bid to uncover the onza's taxonomic status. Several different taxonomic identities have been offered over the years, but the principal four are as follows.[60,187]

The most exciting possibility is that it could be a currently undescribed species—one that quite conceivably descended from the typical puma, but later diverged from it to fill the ecological niche left vacant by the extinction 10,000 years ago of *Acinonyx trumani*—an extraordinary felid now known to have been a true American cheetah (it was first classed as a cheetah-like puma). Indeed, German felid specialist Dr. Helmut Hemmer initially mooted that the onza may actually be a surviving representative of Truman's cheetah, but after studying casts of two onza skulls and comparing them with fossil material from *A. trumani*, he changed his mind. However, regardless of whether it is even related to (let alone synonymous with) Truman's cheetah, if the onza is indeed a valid species in its own right it should be distinguishable from the puma via the biochemical and genetic analyses conducted by those researchers studying the Rodriguez specimen (see below).[60,187]

Another intriguing, frequently-raised possibility is that the onza is a naturally-occurring hybrid of puma and jaguar. No evidence for this identity was obtained, however, from examination of the Rodriguez specimen. In any case, hybrids of this type that have been bred in captivity bear no resemblance to onzas, appearing visibly intermediate between puma and jaguar instead.[60,187]

Two rather more conservative identities remain. The onza may simply be a mutant form of the puma, the result of a genetic aberration that has yielded an uncommonly slim, leggy variety. Its gracility is, indisputably, a *natural* facet of its appearance (rather than starvation-induced emaciation), because the Rodriguez specimen was found to possess adequate amounts of body fat. If not a genetic freak, the onza could be a separate subspecies of puma, kept apart from the Mexican puma *Felis (Puma) concolor azteca* by behavioural differences and dissimilar habitat preferences.[60,187] If the last-mentioned identity were the correct one, however, then once again (as with the possibility that it is a separate species) we would expect the onza to be distinguishable biochemically and genetically from other pumas. But is it?

Early biochemical tests had failed to uncover any characteristics differentiating the Rodriguez specimen from pumas, which would seem to suggest that onza and puma are very closely related. However, this conclusion stems from a fundamental assumption that, although widely accepted in the zoological (and especially the cryptozoological) community, has never been confirmed. Namely, that the onza specimen whose tissues provided these biochemical results, i.e. the Rodriguez specimen, really was an onza![191]

Is it conceivable, however, that Manuel Vega had been mistaken, and that this animal had merely been a malformed or infirm puma that only outwardly resembled a genuine onza? After reflecting upon this disturbing possibility for some time, in January 1998 I heretically aired it within an onza article of mine[191]—and now it seems that my suspicions may indeed have been justified. A few months after my article appeared, the much-delayed volume of the ISC's journal *Cryptozoology* covering the years 1993-1996 was finally published, and contained a report by a research team featuring Prof.

Onza investigator Robert Marshall (left) examining the Vega onza skull, held by its present owner, Mexican rancher Ricardo Urquijo (right). (International Society of Cryptozoology)

Stephen O'Brien, an expert in feline molecular genetics. The team's report revealed that after conducting comparative protein and mitochondrial DNA analyses using tissue samples taken from the Rodriguez onza and from specimens of known North American cat species, the results obtained for the Rodriguez onza were found to be indistinguishable from those of North American pumas.[191] Of course, this does not mean that all onzas are pumas, but how savagely ironic it would be for the most celebrated onza specimen not to be an onza after all.

Nevertheless, a solution to the riddle of the onza's identity may yet be in sight. On 15 April 1995, an alleged male onza was shot behind Parrot Mountain, this time by rancher Raul Jiminez Dominguez. Later that same day, after having been frozen, the onza's corpse was examined by two biologists from the Universidad Nacional Autonoma de Mexico, who took tissue samples away with them for electrophoretic analysis. The remainder of its carcase was preserved and dissected at Mazatlan for future study.[191]

No further information has been released concerning this specimen, but let us hope that it will eventually provide a precise, unequivocal answer to the long-standing question of the onza's taxonomic identity. Until then, Mexico's feline enigma will remain a cryptozoological controversy—a far cry indeed from its popular yet sadly premature image as an erstwhile mystery cat whose reality is no longer in doubt.

A BEWILDERMENT OF BAMBOO LEMURS

Until its rediscovery in 1972 by French zoologist Dr. André Peyriéras in a small forest near Kianjavato, situated 63 miles east of the town of Fianarantsoa in southeastern Madagascar, the greater bamboo lemur *Hapalemur* [now *Prolemur*] *simus* was believed to have been extinct since the 20th century's opening years. In 1985, French researcher Corine Dague discovered what seemed at first to be a second colony of this species, inhabiting a rainforest near Ranomafana, 28.5 miles east of Fianarantsoa. By comparing these individuals with those back at Kianjavato, however, Ruhr University zoologist Dr. Bernhard Meier showed in 1986 that they were not greater bamboo lemurs at all. On the contrary, they represented something even more exciting—a completely new species, never before recorded.[192]

Readily distinguished from the ruddy-grey greater bamboo lemur and from the smaller, grey bamboo lemur *H. griseus* by virtue of its fur's rich golden hue on its throat, cheeks, and eyebrows, as well as by its quite different alarm call, scent-marking glands, and high chromosome number, the new species was named *H. aureus*, the golden bamboo lemur. Even its feeding habits are different, as it lives exclusively upon the young shoots and branches of bamboo, whereas the other

Golden bamboo lemur. (Dr. Bernhard Meier)

two species also eat fruit and foliage from additional plant species. Two golden bamboo lemurs, a male and female, were captured alive for study, and were taken to Tsimbazaza Park, in the island's capital, Antananarivo. This species is currently categorised as endangered by the IUCN.[192]

GIANT STRIPED MALAGASY MONGOOSE—AN OVERLOOKED EUPLERID

Another mammalian surprise from Madagascar that came to light in 1986 was discovered not in any forest, but in a museum cabinet instead. Madagascar is home to several very strange-looking mongooses that are overtly unlike any found elsewhere in the world. One of these is *Galidictis fasciata*, the broad-striped Malagasy mongoose—named after the three wide horizontal bands of dark brown that run close to one another along each creamy-white flank, from just behind the creature's ear to the base of its tail. (Malagasy mongooses, incidentally, belong to a separate taxonomic family from true mongooses, and are nowadays known formally as

*Giant striped Malagasy mongoose.
(Grigory Morozov/Wikipedia)*

euplerids.) It was once thought to be the only member of its genus, but a second species was described in 1986 by zoologist Dr. W. Chris Wozencraft, from Kansas University's Museum of Natural History.[193]

During an in-depth study of Malagasy mongoose specimens in museums worldwide, Wozencraft had come upon two labelled as *G. fasciata* that differed conspicuously from other recorded examples of this species by way of their longer, sturdier, and broader skulls, their larger canine teeth and carnassials, and their widely-spaced, horizontal brown bands. As a result, Wozencraft designated these as the first recorded representatives of a new species, the giant striped Malagasy mongoose *G. grandidiensis*—named after the Grandidier collection of specimens from which its type had been acquired (a year later, in 1987, Wozencraft changed its name to *Galidictis grandidieri*). Its second representative specimen, the skin and damaged skull of an adult male, had been collected near to southwestern Madagascar's Lac Tsimanampetsotsa on 15 February 1929, and local villagers in that area subsequently reported the existence of an animal fitting the giant striped Malagasy mongoose's description, which was subsequently confirmed.[113,193]

CHAETOMYS—REAPPEARANCE OF A THIN-SPINED MISSING LINK

Variously deemed to be a rat with unusually spiky hair or a porcupine with unusually fine quills, the thin-spined porcupine rat *Chaetomys subspinosus* constitutes a perplexing taxonomic bridge between the spiny rats and the American porcupines, combining features of both families. Even its tail is tantalisingly transitional—whereas those of spiny rats are not prehensile, and those of American porcupines are fully prehensile, the tail of *Chaetomys* is semi-prehensile. This ambiguous but beguiling little animal, not exceeding 2.5 ft in total length, bristling with soft, flexible brown spines, and hailing from the drier forests of southeastern Brazil, was officially described as long ago as 1818. Yet it has been conclusively sighted only very rarely since; until quite recently, the last definite report came in 1952—accompanied by the first photograph ever taken of the species in the wild. Apart from various unconfirmed reports in later years, nothing more was heard of this intriguing rodent.[48]

In December 1986, however, during a specific search for the species in the vicinity of Valenca, in Bahia State, mammalogist Ilmar Bastos Santos was taken by a local hunter to a tree that contained two thin-spined porcupine rats, one of which was a pregnant female. Additional specimens were later sought, and it is hoped that a protected breeding area for this taxonomically significant species can be established.[194]

Koopman's tree porcupine *Coendou koopmani*, a prehensile-tailed New World porcupine, was formally described as a new species in 1992, from Brazil's Amazon lowlands. It is a very small, dark-spined species, no larger than a guinea-pig,[195] but was later found to be identical to *C. nycthemera*, a species described as long ago as 1818, which means that '*C. koopmani*' is now a junior synonym of the latter.

RETURN OF THE RHINOS

The possibility that animals as large as rhinoceroses have remained unknown to science for decades must seem very remote. Yet between 1986 and 1988, zoologists learnt of two separate cases in which this is precisely what had happened. The first announcement came in 1986, disclosing that up to five specimens of the hairy Sumatran rhinoceros *Dicerorhinus sumatrensis* had been found in Sarawak, a Malaysian state on the island of Borneo. Science had hitherto assumed that this species had died out in Sarawak by the end of World War II. Inhabiting a remote valley in the Ulu Baram area, they had first come to the notice of local inhabitants in 1983, and were reported at once to the Sarawak Forest Department's National Parks and Wildlife Office. It decided to keep the discovery secret for a while, however,

The Sumatran rhinoceros is Asia's only two-horned species. (Alan Pringle)

Javan rhinoceros—recorded in 1988 on the Asian mainland, the first such record since the 1940s. (Zoological Society of London)

to allow time for protection measures to be drawn up and implemented—assisted by Dr. Julian Caldecott of the Earthlife Foundation, who spotted Sumatran rhino tracks himself in August 1986.[196]

An even more spectacular revelation occurred just two years later—because this was when the Javan rhinoceros *Rhinoceros sondaicus*, the most critically threatened of all rhino species and supposedly confined totally to Java since the 1940s, was rediscovered on the Asian mainland. In November 1988, a female specimen was shot by a local tribesman in the jungles of southern Vietnam, about 80 miles northeast of Ho Chi Minh City (formerly Saigon). Its remains, together with those of a second shot specimen, were later examined and conclusively identified by Dr. George Schaller, Vice-President of the New York Zoological Society, who had been working at the time on a faunal survey with Vietnamese scientists in the jungles of southern Vietnam. In February 1989, Schaller also sighted some Javan rhino tracks, on the banks of the Dong Nai River (75 miles northeast of Ho Chi Minh City).[197]

Prompted by this dramatic find, the Vietnamese government established a Rhinoceros Conservation Group, and considered plans to extend the neighbouring Cat Tien National Reserve to encompass the entire region housing the rhinos (only a portion was contained within it at that time), which would be of great benefit to the long-term survival of this small but priceless population of a species ranked in 1984 as one of the world's 12 most endangered species of animal. Indeed, no living specimen of Javan rhinoceros had even been photographed in Vietnam—until May 1999. That was when one such individual made photo-history by walking through a laser beam that triggered the automatic cameras set up by a WWF survey in the Cat Tien National Park. Seven pictures of it were snapped in total. Current estimates indicate that less than 12 individuals still survive in Vietnam, all within the Cat Tien National Park, with another 40-50 in western Java's Ujung Kulon National Park. Hence the species remains critically endangered.[197]

Incidentally, it is often alleged that the Javan rhinoceros has never been maintained in captivity, but this is far from being true, with at least 22 captive specimens on record prior to the 20th century. Reproduced here is a photo of an Asian rhinoceros on show at London Zoo from 1874 to 1885 that was unquestionably a Javan rhino—as evinced by the characteristically scaly, mosaic-like patterning upon its hide, and especially by the fold of skin in front of its shoulder, which runs right across its back (in its close relative, the great Indian rhino *R. unicornis*, this fold does *not* run right across its back). In addition, as I learnt from John Edwards, of the Zoological Society of London, the large Asian rhino that had lived at Adelaide Zoo until 1907 had always been looked upon as a great Indian—until 1948, that is, when researchers revealed that it had actually been a Javan! Indeed, it is quite likely that a number of other supposed great Indians from the earlier days of zoological parks will ultimately be exposed as incognito Javans.

A BRONZE QUOLL AND A BLACK TREE KANGAROO

Australasia's answers to the mustelids (weasels, martens, and suchlike) and viverrids (civets and genets) found elsewhere in the world are the dasyures, traditionally constituting a quintet of brown-coated, white-spotted marsupial carnivores. Two are commonly referred to as quolls—New Guinea's *Dasyurus albopunctatus*, and eastern Australia's *D. viverrinus*—but in 1987 a third species of quoll was described, from the savannahs of southwestern Papua New Guinea, thereby increasing the total number of dasyure species to six.

Dubbed *D. spartacus* by Queensland Museum zoologist Dr. Stephen Van Dyck, its common name is the bronze quoll, on account of the deep bronze colouration of its pelage, dappled with notably small white spots (except for its tail, which is uniformly black). It can also be differentiated from the other dasyures by the extreme narrowness of its muzzle measured between the left and right lachrymal canals (channels in the skull for the tear ducts). It first came to light in the early 1970s, when five specimens were collected as part of a mammal survey in the Trans-Fly Plains of southwestern Papua New Guinea during 1972-3, but these were classed as western dasyures *D. geoffroii* until Van Dyck examined them and perceived their distinctness from known examples of the latter species.[198]

The description of a new dasyure from New Guinea in 1987 was followed in 1988 by the discovery on this same island of a very special tree kangaroo by Australian Museum zoologist Dr. Tim Flannery, assisted by ranger-biologist Lester Seri. Heavier than Matschie's tree kangaroo *Dendrolagus matschiei* (the biggest species then recognised), this new species is further delineated by its short face, thick jet-black fur, and unusually short tail (which is also characterised by a striking flash of orange at its base). The biggest mammal native to Melanesia, it inhabits the remote Torricelli mountain range in Papua New Guinea's West Sepik Province, and first came to Flannery's attention in 1985 when he purchased a strange black claw from a villager at Wigotei. He subsequently learnt from the locals there that it was from a large black tree kangaroo known to them as the *tenkile*.[199-200]

In 11 June 1988, the skin and skull of a juvenile specimen were obtained at a height of 4516 ft on Sweipini, a ridge just west of the Mount Somoro summit in the Torricelli Mountains. These became the type specimen of this newly-discovered species, which was formally described in 1990 by Flannery and Seri, who christened it *Dendrolagus scottae*. Like many other recently-disclosed mammals, Scott's tree kangaroo may be in grave danger of extinction. It is currently categorised as critically endangered by the IUCN, with an estimated total number of around 250 individuals, and a habitat of only about 10.5 square miles in area.[199-200]

LESSER BEAKED WHALE—A PRIZE FROM PERU

In 1988, the discovery of a new beaked whale was made public—the fifth *Mesoplodon* species to have been revealed during the 20th century. Its existence first became known on 2 February 1976, with the procurement of a partial cranium and lumbar vertebra by Smithsonian Institution mammalogist Dr. James G. Mead, close to the fish market in San Andres, Peru. Their dimensions suggested a total length of around 11.5 ft for the complete animal. Nine years later, on 1 May 1985, an immature female beaked whale was captured by fishermen about 30 nautical miles off Pucusana, Peru, and its entire skeleton, together with various of its major internal organs, were duly preserved for study.[201]

By late 1988, seven more specimens had been obtained, all from along the coasts of south central Peru, and consisting variously of complete carcases and selected portions of carcases. All were carefully examined and compared with other *Mesoplodon* material by Mead and two fellow cetacean researchers, Drs. Julio C. Reyes and Koen van Waerebeek (both from Lima's Centro Peruano de Estudios Cetológicos).[202]

Their studies confirmed that these (plus an additional Pucusana example, obtained in June 1989) belonged to a species that was as yet undescribed. This situation was rectified in January 1991 with the publication of their official description of this new *Mesoplodon* member, christened *M. peruvianus*, and commonly known as the lesser, pygmy, or Peruvian beaked whale. Uniformly grey above, paler underneath, this is the world's smallest species of *Mesoplodon* beaked whale, with a maximum recorded total length of only 3.72 m (just over 12 ft), and characteristically small teeth too, ovate in cross-section.[203]

GOLDEN-CROWNED SIFAKA—
BOTH CRYPTIC AND CONSPICUOUS!

Less than three years after the discovery of a new species of bamboo lemur in Madagascar came the description of a new species of sifaka from this island. Also known as sun worshippers on account of at least one species' habit of sitting upright with the palms of its hands stretched upwards, as if in supplication to the sun, sifakas are among the largest of lemurs, and derive their name from the sound of their call. Traditionally, only two species were recognised (in recent years, however, these have each been split into several more), but in 1974 mammalogist Dr. Ian Tattersall observed some odd-looking specimens in a dry forest near Daraina, in the Madagascan province of Antseranana. Tattersall was puzzled by their unique complement of short white fur, brightly-coloured, readily-visible golden crowns, and very large, well-furred ears—a combination of features that differentiated them from both of the known species of sifaka.[204]

Nevertheless, he did not consider at that time that they could represent a new species, and so the incident attracted no attention—until they were spied in the late 1980s by American zoologist Dr. Elwyn Simons, from the Primate Center at North Carolina's Duke University, who did recognise that they warranted separate specific status. In early 1989 he duly described the new species, naming it *Propithecus tattersalli*, in honour of

Kellas cat trapped in 1984 at Revack Lodge, northern Scotland. (Edward Orbell)

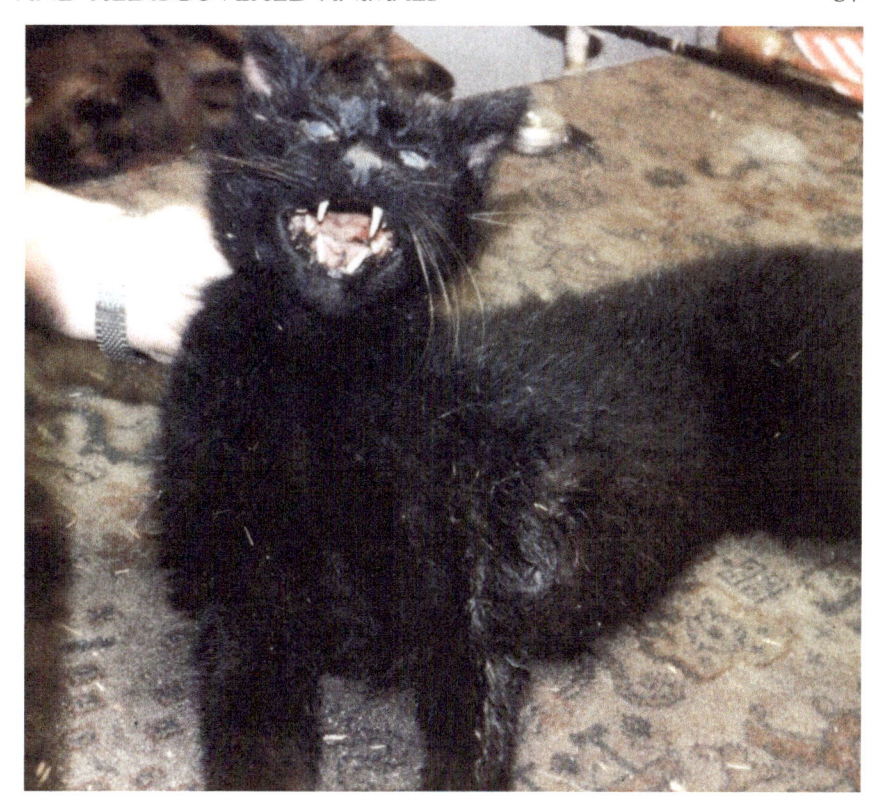

its original observer. Commonly called the golden-crowned sifaka, its greatest mystery for Simons was why it had not been discovered much earlier—because its golden crown makes it so conspicuous that it can be readily spotted in treetops up to half a mile away! Sadly, this distinctive species is critically endangered.[204]

Speaking of new sifakas: during October 1999, Dr. Ken Glander led an expedition from Duke University's Primate Center to Madagascar's Mahatsinjo forest, seeking a mate for the Center's Romeo—who happens to be a diademed sifaka *P. diadema*. On 10 October, the team captured a female (inevitably dubbed Juliet), plus a young male, which they took to Madagascar's Ivoloina Zoological Park for close examination. There, the team discovered that these two supposed diademed sifakas' fur was darker than Romeo's, and lacked the orange colouration which characterises Romeo's species of sifaka. The shape of the captured sifakas' faces also differed from Romeo's, and they sported distinctive white 'spectacles' of fur encircling their eyes. No diademed sifaka matching this description had previously been recorded, and Glander's team is now seriously considering the possibility that Juliet's version constitutes a wholly new, currently undescribed species of sifaka. Tragically, only a month after her capture, Juliet died.[205]

KELLAS CAT—A SHORT-LIVED 'SPECIES' FROM SCOTLAND

During the 1980s, several unusual fox-sized cats, very gracile with notably large fangs and bristly black fur sprinkled with white primary guard hairs, were found near the West Moray hamlet of Kellas and elsewhere in northern Scotland. The first to attract media attention was a specimen trapped in a fox snare at Revack Lodge in 1984. Unfortunately, its body was later lost, but its photograph is reproduced here. The media speculated that these distinctive felids, now known as Kellas cats, were a species new to science. In my book *Mystery Cats of the World* (1989) and my Kellas cat paper (1990), however, I predicted that they would prove merely to be introgressive (complex) hybrids of domestic cat and Scottish wildcat—an identity later confirmed by anatomical analyses at the Royal Museum of Scotland.[60,206]

LION TAMARINS AND ZEBRA MARMOSETS

The first major mammalian discovery to hit the headlines in the 1990s was made public on 21 June 1990—when, at a scientific conference devoted to those exceptionally handsome, golden-furred New World monkeys called lion tamarins, primatologists announced that a new, fourth species had recently been discovered in Brazil by biologists Drs. Lucia Lorini and Vanessa Persson. It is known as the black-faced lion tamarin, and was officially named *Leontopithecus caissara* after the caicaras (coastal fishermen who live and work on its island home of Superagui, south of São Paulo). It is currently categorised as critically endangered by the IUCN.[207]

In October 1992, Dr. Russell Mittermeier, of Conservation International in New York, and two colleagues formally described the Maués marmoset *Callithrix* [now *Mico*] *mauesi*, a new species marked with a hint of zebra-like stripes. Discovered on 15 April 1985 by Marco

Schwarz during a field trip to the region, it occurs on the west bank of the Maués-Açô River, in a remote portion of central Brazilian Amazonia.[208] This is in stark contrast to the black-faced lion tamarin, whose previously undetected existence astonished zoologists because its home is in one of Brazil's most heavily-populated regions.[207]

(This latter scenario was duplicated by the discovery of a new spiny bandicoot. Dubbed the Fly River spiny bandicoot *Echymipera echinista* and described in 1990, this slender-snouted marsupial was found in New Guinea's Western Province, an area that had already been extensively explored; it is currently known from two specimens.[209])

TATU BOLA—IN THE MARKET FOR REDISCOVERY

Equally unexpected was the mode of rediscovery featuring the Brazilian three-banded armadillo *Tolypeutes tricinctus*. Referred to locally as the *tatu bola* on account of its ability to roll itself up into an impregnable ball like a surrealistic armoured hedgehog, it is known only from the drought-blighted, semi-arid plains of northeastern Brazil. Here its talents offer it protection against natural predators—but not against man, partial to its chicken-tasting flesh. Indeed, as a direct result of this fatal quality of its flesh, the Brazilian three-banded armadillo had not been reliably reported since the early 1980s, and was feared to be extinct—until 1990, when a team of field researchers seeking Lear's macaw (see p. 142) in Bahia spotted five living adult specimens of the *tatu bola* being offered for sale at a local market.[210]

Needless to say, the team bought all five and transferred them to Minas Gerais Federal University at Belo Horizonte. Plans are underway to seek out more information concerning this endangered species' dietary requirements and reproduction, in order to instigate a captive breeding programme once further specimens have been obtained (in recent years, a specimen has been exhibited at Edmonton Zoo, Canada).[210]

WARTHOGS OF THE DESERT

During much of the 20th century, zoologists had only recognised one living species of warthog, the common warthog *Phacochoerus aethiopicus*. A century earlier than that, conversely, the warthogs of South Africa's Cape Province and the Orange Free State had been deemed by some to constitute a separate species, the Cape warthog, but this had died out by the end of the 1800s. Following his studies on the few specimens preserved in museums, however, ungulate researcher Dr. Peter Grubb disclosed in 1991 that a widely-separated population of desert-dwelling warthogs in Somalia and northern Kenya belongs to this ostensibly 'lost' second species (which may also exist in Eritrea, Djibouti, and Ethiopia), and that the latter is truly of discrete taxonomic status.[211]

Bearing in mind that the South African representatives had all gone, Grubb proposed 'desert warthog' as a suitable common name to replace 'Cape warthog'. As for its scientific name, however, researchers discovered that this restored species held a claim to the name *Phacochoerus aethiopicus* which predated that of the common warthog. According to the rules of nomenclatural precedence, therefore, the desert warthog was duly christened *P. aethiopicus*, and the common warthog was renamed *P. africanus*.[211]

THE BAY CAT REBORN—IN BORNEO

On 4 November 1992, the Sarawak Museum in Borneo received a very unexpected feline specimen. The animal in question was comparable to a domestic cat in size, but sported an extremely long tail and a dark chestnut-red pelt. It had been captured alive several months earlier by native hunters on the border of Sarawak and Kalimantan (Indonesian Borneo), and had been retained by them ever since. Tragically, however, its protruding bones and wasted muscles bore stark if silent testimony to the neglect that this once-handsome creature had suffered during its enforced confinement, and it died shortly after reaching the museum.[212]

An ignoble end to a noble animal, but its life had not been without significance. On the contrary, its very existence had resurrected an entire species from more than 60 years of 'official' extinction, because this starved, emaciated individual belonged to one of the world's poorest-known species of mammal—*Catopuma* (*Badiofelis*) *badia*, the Bornean bay cat.[212]

In 1856, pioneering evolutionist Alfred Russell Wallace sent to the British Museum (Natural History)

Brazilian three-banded armadillo.
(Chris Stubbs/Wikipedia)

Bay cat—one of the world's least-known species of wild cat, unexpectedly rediscovered in 1992.

the skull and the severely torn, fragmented skin of a subadult male cat collected near Kuching in Sarawak. It was provisionally classed as a flat-headed cat *Felis (Prionailurus) planiceps*, but its nearly uniform reddish-brown coat, exceedingly long tail, and relatively small body size influenced its subsequent reclassification as a juvenile Temminck's golden cat *Catopuma temminckii*. However, a detailed examination of its skull revealed that this perplexing cat was almost fully grown. Clearly, therefore, it could not be a Temminck's golden cat either, and it did not correspond at all with any other known species.[213]

Nevertheless, the specimen was in such a poor state that its principal investigator, John Edward Gray, was reluctant to base the description of a new species solely upon it, preferring instead to await the procurement of some additional, complete specimens—but he waited in vain. At last, when almost 20 years had passed by without a single example having been obtained, Gray relented, and on 19 May 1874 he designated Wallace's unique if unsatisfactory specimen as the holotype of *Felis badia*, the Bornean bay cat.[213]

Since then, only a handful of further bay cat specimens have been collected. Perhaps the most interesting of these was an adult male, obtained from the vicinity of the Entoyut River in 1894 by animal collector Charles Hose. Its preserved skin was deposited the following year at the British Museum (Natural History). Unlike all previously and subsequently documented skins of the bay cat, however, the pelage of Hose's specimen was grey, not red. This indicates that the bay cat is dimorphic, i.e. occurring in two wholly discrete colour phases, red and grey—thus providing yet another parallel with Temminck's golden cat.[212]

Of course, external appearances can sometimes be deceptive, and after studying the bay cat's cranial features felid expert Reginald Pocock deemed this bemusing felid distinct enough to warrant its own genus. Accordingly, in 1932 he rechristened it *Badiofelis badia*.[212]

Yet whereas it is true, for instance, that this species' skull is higher and rounder than that of Temminck's golden cat, and that its first upper pre-molar is abnormally small and has only a single root, in general the bay cat is sufficiently similar to *C. temminckii* for many taxonomists to have treated it merely as a miniature island version of the latter species, presumably descended from a population of *C. temminckii* that became permanently isolated when Borneo separated from other land masses towards the close of the Pleistocene epoch, approximately 10,000 years ago.[212]

Blood and tissue samples derived from the erstwhile captive specimen obtained in November 1992 were earmarked for DNA and other genetic analyses, in order to tease from them their taxonomic secrets, and ultimately unmask the real identity of the bay cat, And indeed they did, revealing that the bay cat was a valid species in its own right, closely related to but taxonomically distinct from *C. temminckii*, with the two diverging from a common ancestor roughly 4 million years ago (and therefore long before Borneo became separated as an island from other Sunda Shelf isles). Consequently, this species is now known as *Catopuma badia*.[214] (Having said that, in 2006 a taxonomic revision of the carnivorans suggested that both the bay cat and Temminck's golden cat should be rehoused within the genus *Pardofelis*, thereby allying them with the marbled cat *P. marmorata*.)

In 1998, the bay cat was filmed in the wild for the first time, and again on 27 July 2003. During the late 2000s, a camera trap located in the northwestern part of Sabah's Deramakot Forest Reserve snapped a single photo of a male bay cat, thereby expanding the northern portion of this species' known distribution range. In total, 15 separate sightings of bay cats were recorded in Kalimantan, Sabah, and Sarawak but not in Brunei, between 2003 and 2005, and in 2005 a captive individual was photographed by observer Jim Harrison in Sarawak. The bay cat is currently categorised as endangered by the IUCN, so it is hoped that continued attempts will be made to observe and scientifically monitor this feline enigma in its natural domain, because its lifestyle and biology are virtually unknown at present. Based largely upon unconfirmed sightings, native testimony, and early collection data, all that can be said is that it apparently frequents rocky limestone outcrops and dense tropical forests, and has a reputation among local tribes for being vicious.[212]

Hardly an exhaustive dossier for a species first brought to scientific attention more than 150 years ago, but at least the zoological world now knows that the bay cat still exists!

DEBUT OF THE VU QUANG OX

At the beginning of the 1990s, the very idea of discovering any major new species of mammal anywhere in the world had long since become distinctly unfashionable within the scientific community. But beginning in 1992 and continuing throughout the 1990s, a previously-unexplored stretch of Vietnamese forest terrain called Vu Quang and the remote Annamite Mountains in which it is ensconced, forming a natural border between northern Vietnam and Laos, proved to be a zoological treasure trove—unfurling spectacular new creatures on a scale unprecedented since the early 1900s. Little wonder, then, that Indochina, and Vietnam in particular, remains to this day the focus of intense scientific scrutiny . . . and astonishment.[215]

Yet because all of these remarkable new (and rediscovered) species are ones that were already familiar to the locals (i.e. ethnoknown), technically they are not just zoological but cryptozoological finds. In recognition of this, on 1 May 1997 the Vietnam National University of Hanoi established a pioneering new sub-department, dubbed the Vietnam Cryptozoic and Rare Animals Research Center, in the university's Teachers Training College. This centre's founding director was Prof. Tran Hong Viet, who was also chairman of the university's zoology department.[215]

Miraculously escaping defoliation and bombing during the Vietnam War, what is now the Vu Quang Nature Reserve consists of dense mountain rainforest encompassed by protective rocky cliffs. It was here, during a field trip in May 1992, that a scientific team led by WWF representative Dr. John MacKinnon came upon three strange pairs of horns (one pair with part of the skull still attached) in the homes of local hunters. The horns were long and dagger-like, similar to those of the African and Arabian oryxes—but there was no Asian mammal species known to science that had such horns. Conversely, the locals knew the creature well, referring to it as the *saola* (also used quite commonly nowadays as an English name for it, and translating as 'spindle horn') or *son duong* ('mountain goat').[216]

Consequently, the Vietnamese participants of this expedition made four return visits to the region in search of additional specimens of this seemingly undiscovered species, and succeeded in procuring over 20 specimens—including three complete skins, one of which was later preserved as a taxiderm exhibit. These important remains were examined by a number of mammal experts in Britain and Australia, and subjected to DNA analyses in the United States and Denmark, which confirmed that they were indeed from a hitherto undescribed species—one so distinct from all others, in fact, that it required a new genus.[216]

Although it appears to warrant classification within the ox subfamily of bovid ungulates (with DNA analyses variously suggesting an apparent affinity to the nilghai or to the anoa), and has therefore been dubbed the Vu Quang ox, some authorities consider that this major new mammal may comprise a taxonomic bridge between the oxen and the antelopes, as implied externally by its oryx-like horns. Certain others, conversely, favour a link with the goat subfamily. In June 1993, MacKinnon and his Vietnamese colleagues formally described their sensational discovery, christening it *Pseudoryx nghetinhensis*.[48,216]

A female Vu Quang ox, held in captivity at Lak Xao, Laos, in 1996—the first adult specimen of this dramatically new species to be examined by scientists. (William Robichaud)

Standing 32-36 in at the shoulder, with a total length of 5.5-7 ft, and weighing around 220 lb when adult, the Vu Quang ox is comparable in size to Sulawesi's lowland anoa *Bubalus depressicornis* (p. 37), but its horns (up to 20 in long) and slender neck are lengthier. Its fine-haired coat is deep brown in colour, and its face is decorated with bold black and white markings. There is also a blackish-brown narrow dorsal stripe running onto the short tail, which bears a black tassle, and a whitish stripe on the outer rump. One of this species' most distinctive features is the pair of slit-like maxillary glands, located in front of its eyes on its cheeks, which it uses in marking territory, and which are covered by skin flaps that can uniquely be flipped open and shut under the animal's conscious control.[48,216]

Despite its very belated scientific debut (bearing in mind its large size), the Vu Quang ox has a sizeable distribution range in Vietnam, encompassing over 20 different localities inside the provinces of Nghe an and Ha tinh along Vietnam's border with Laos, comprising a total area estimated at more than 1540 square miles. It inhabits montane forest, never entering neighbouring agricultural lands, but ventures into the lowlands during the winter, when it is hunted by the local people. They state that it travels in small groups, usually containing two or three animals, seeking the leaves of bushes and fig trees on which it browses.[216]

According to local testimony, less than a decade ago as many as several hundred specimens may have existed within its known distribution range in Vietnam, but sightings have decreased during subsequent years, and with the combined threat of hunting and habitat destruction jeopardising its future survival the Vu Quang ox is currently categorised as critically endangered by the IUCN. In order to ensure its survival the Vietnamese Ministry of Forestry has increased the Vu Quang Nature Reserve from 16,000 to 60,000 hectares (i.e. from roughly 60 to 230 square miles), and has prohibited logging here.[216]

In June 1994, the WWF announced the exciting news that the zoological world had been eagerly awaiting for the past two years—a living Vu Quang ox had been captured! The first ever seen by scientists, it proved to be a juvenile female specimen—a diminutive, delightfully Bambi-esque creature, aged 4-5 months old, standing scarcely 2 ft tall and weighing only 40 lb, with tiny horns, and nervous, timid demeanour.[217]

She had been captured by a local hunter in the Khe Tre watershed forest just outside Vu Quang, and was transferred to the roomy botanical gardens of Hanoi's Forest Inventory and Planning Institute (FIPI), where scientists planned to study her before releasing her back into the wild. In July, a second, slightly older Vu Quang ox was captured, this time a male—raising the possibility of initiating a future captive-breeding programme for their species. Tragically, however, it was not to be. By mid-October 1994, both animals had died—one from liver and bladder complications, the other from diarrhoea.[217]

However, during that same year, reports of Vu Quang oxen were also filtering out from Vietnam's western neighbour, Laos, indicating that this species inhabited both sides of the Annamite mountain range. In January 1996, this assumption was sensationally confirmed with the capture by villagers just outside Laos's Nakai Nam Theun Reserve of a living adult female specimen—the first adult Vu Quang ox to be examined by scientists. Tragically, it transpired that this was one of five specimens encountered, but the other four had all been killed by hunting dogs. The lone survivor was taken to a military-run menagerie at the village of Lak Xao (also spelt 'Lac Sao') in Laos's Khammouane province (which has famously introduced two other new species to the scientific world—see the muntjacs entry following), where she was found to be pregnant. Sadly, she died before giving birth, but lived long enough, and in apparent contentment, within captivity to permit scientists to study her.[218]

In late May 1998, another pregnant adult Vu Quang ox was captured, but on this occasion in Vietnam. After being studied in captivity for a week, she was released in Vietnam's Thua Thien-Hue province—the first time that a specimen had been returned to the wild following capture. And on 31 October 1998 yet another momentous episode in the *Pseudoryx* saga took place—when staff at the Pu Mat Nature Reserve in Vietnam's central Nghe An province announced that a set of automatic infra-red camera traps laid by a team led by Mike Baltzer from Fauna and Flora International had snapped two pictures of this elusive species.[219]

These were the first photos ever obtained in its native habitat of a wild Vu Quang ox—unique portraits of the biggest new species of land mammal to have been revealed since the kouprey more than 50 years earlier. A zoological sensation in every sense, it is true to say that the Vu Quang ox is the okapi of modern times, for it is unquestionably the greatest cryptozoological triumph since the latter beast's own extraordinary unveiling back in 1901.

MUNTJACS, MUNTJACS, EVERYWHERE!

The Vu Quang ox *Pseudoryx nghetinhensis* was even more than a major zoological find in its own right. It was also the first in a veritable stampede of significant new and rediscovered ungulates to be disclosed in Vietnam and neighbouring countries during the 1990s—a mass mammalogical unveiling on a scale wholly unprecedented in modern times. Several of these constituted overlooked or forgotten species of muntjac (barking deer).

While still pursuing information about *Pseudoryx* in Vu Quang during 1993, the MacKinnon team of researchers also heard tell of a strange deer that seemed equally unfamiliar at that time to the zoological world. Judging from 19 trophies, comprising skulls and frontal portions of preserved heads, found once again in the homes of local hunters during March 1994, it was evidently a type of muntjac—but no ordinary one.[220]

The antlers were four times as long as those of the largest muntjac species then known to science, and whereas those of other muntjacs had long bases (pedicels) and short tines, the mysterious giant version's had short bases and long tines. Questioning the hunters, MacKinnon learnt that 15-20 specimens of this outsized muntjac had been trapped locally between July 1993 and January 1994, indicating that it was fairly common.[220]

Genetic examination by Dr. Peter Arctander at Copenhagen University of skin samples from the preserved heads confirmed that it was indeed a new species, and due to its remarkable antlers the giant muntjac was deemed distinctive enough to warrant the creation of a brand-new genus for it. Accordingly, in 1994 a zoological team that included two Vietnamese workers, MacKinnon, and Arctander published a paper in which

The world's largest barking deer, the giant muntjac eluded scientific detection until as recently as 1994. (William M. Rebsamen)

this latest mammalogical sensation was formally dubbed *Megamuntiacus vuquangensis*—'the giant muntjac of Vu Quang', though not exclusively Vu Quang (or even Vietnam), as it turned out.[220]

As with the Vu Quang ox, the procurement of skulls and preserved specimens of the giant muntjac (also termed the large-antlered muntjac) was ultimately succeeded by the discovery of a living specimen. Indeed, at much the same time that MacKinnon's team was collecting *M. vuquangensis* hunting trophies in Vietnam, British conservationists Robert Timmins and Tom Evans were doing the same in its western neighbour, Laos, within the Nakai Nam Theun reserve. In addition, however, they also unexpectedly chanced upon a live giant muntjac—an adult male, sharing a pen with two female common muntjacs *Muntiacus muntjak* in the earlier-mentioned military-owned menagerie at Lak Xao, which was visited by them during February 1994.[221-2]

Reportedly captured near Ban Thabak in Khammouane province, this specimen confirmed that the giant muntjac is easily the largest currently-known species of muntjac in the world—a third longer and twice as heavy as any other species. Its coat was brown and grizzled in colour, with a black patch on top of its notably broad, almost triangular tail, and its skull was conspicuously larger than that of other muntjacs.[48,221-2]

Later in 1994, Timmins, Evans, and two other colleagues prepared a formal description of the giant muntjac, in which they proposed christening it 'Muntiacus kaysoni'—in honour of Kaysone Phomvihane, a former prime minister of Laos.[222] However, the paper by MacKinnon, Arctander, and company, in which they had named this species *Megamuntiacus vuquangensis*, was published first—hence their name for it took nomenclatural precedence.

Investigations by conservationist Dr. George Schaller from New York's Wildlife Conservation Society (WCS) and Yale University's Dr. Elisabeth Vrba, during which a dead female specimen and seven sets of antlers were also obtained in Laos, revealed that the giant muntjac occurred over a stretch of 280 miles along the southeastern Laos slope of the Annamite mountain range (the topographical demarcation line between Laos and Vietnam) from the Chat River south to the village of Dakchung. It may also occur in northern Cambodia, again along the border with Vietnam. Principally frequenting old-growth evergreen forests at elevations above 300 ft, the giant muntjac currently appears to be quite abundant, but it is extensively hunted by local tribes, and its habitat is also threatened.[223]

In their paper, Schaller and Vrba presented the results of DNA studies, which in their view indicated that the giant muntjac was not sufficiently distinct from other species to warrant being placed in a genus to itself. Moreover, while in captivity the Laos menagerie's male had mated with the female common muntjacs, yielding three hybrid offspring—which also indicates a close taxonomic affinity between the giant muntjac and *Muntiacus* muntjacs. Consequently, in 1997 it was renamed *Muntiacus vuquangensis* (with *Megamuntiacus* abandoned as a valid separate genus).[222-3]

Ironically, it would seem that the giant muntjac had actually been brought to western science's attention long before its official 1990s disclosure, but had not been recognised as something new. Austrian zoologist Dr. K. Bauer has recently revealed that within his book *Les Grands Animaux Sauvages de l'Annam* (1930), Fernand Millet, a longstanding forestry official in Indochina, documented various Annamite hunting trophies consisting of the antlers from an exceptionally large form of muntjac—and in the accompanying photograph, some of the diagnostic features of *M. vuquangensis* can be clearly seen.[223-4]

In January 1995, Timmins returned to the Laos military menagerie, accompanied by Schaller and his conservation team, to observe further the adult male giant muntjac. To their great surprise, however, they discovered that this specimen now shared its enclosure not only with the female common muntjacs, but also with a much smaller, darker male muntjac—one whose species appeared wholly unfamiliar. However, it did correspond with descriptions given by local hunters to Schaller of a highly elusive, rainforest-dwelling, blackish muntjac while he had been conducting a WCS wildlife survey in Laos's Annamite region during December 1994 and January 1995.[225]

The new Laos captive's coat was very dark in colour, and its limbs were black, as was its chin—rendering it unique among muntjac forms. It also sported a tuft of orange hair between its antlers. Hunting trophies from this seemingly new species were obtained, and in 1999 a study comparing diagnostic DNA characters of different muntjac species was published within the *Journal of Mammalogy* by a research team featuring Dr. George Amato (from the American Museum of Natural History) and Schaller. Their study revealed that three specimens of the unidentified Laos muntjac had identical DNA sequences with an obscure, largely-forgotten species known as Roosevelts' muntjac *Muntiacus rooseveltorum* (named after both Teddy and Kermit Roosevelt, hence the use of the plurals 'Roosevelts'' and '*rooseveltorum*' in its names).[226]

This long-overlooked deer had previously been known only from its type specimen. This was an immature male collected at Ban Muangyo in northern Laos on 16 May 1929, formally described as a new species three years later by mammalogist W.H. Osgood, and retained by the Chicago Field Museum of Natural History. Here it remained, unique and unexplained, an anomaly popularly assumed to represent a now-extinct species—until it was examined by Amato, Schaller, and colleagues during their investigation of the mysterious Laos muntjac, and duly resurrected by them as a valid species that was still very much alive.[226]

The next novel Indochinese muntjac to hit the cryptozoological headlines first came to scientific attention in April 1997, during research conducted in the Truong Son mountain range of the west Quang Nam province, central Vietnam, by an international team backed by the World Wide Fund for Nature (WWF) and the United Nations Environment Programme. Known locally as the *sam coi cacoong* (translating as 'deer living in dense forest'), it was a dwarf species of muntjac weighing no more than 35 lb and only half as big in size as the familiar common muntjac. Even its antlers were minuscule, no longer than a thumbnail. Subsequently dubbed the Truong Son or Annamite muntjac, its existence was made public in August 1997, by which time this newly-unveiled deer was represented by 17 skulls and two tails, obtained from village hunters. DNA was extracted, and studied at Copenhagen University by Arctander. His studies confirmed its status as a separate species in its own right, and in 1998 it was formally described by a research team that included Arctander, MacKinnon, and Amato, who christened it *Muntiacus truongsonensis*.[227]

In best cryptozoological tradition, however, this 'new' species is of course a familiar creature to the local hunters, who revealed that it was black in colour, and hunt it for its meat. Its most distinctive features are the canine teeth of the female, which are almost as long as those of the male (in most muntjac species, the female's are much shorter).[227]

So far, everything seemed straightforward concerning these lately-disclosed Indochinese muntjacs—until November 1999, that is, when I received an e-mail from Robert Timmins that cast an intriguing shadow of doubt upon the popularly-held school of thought (as documented by me above) concerning their classification. Within his e-mail,[228] he wrote:

> The taxonomy of the *M. rooseveltorum* / *M. truongsonensis* group of muntjacs is rather confused at the moment, and all recent publications contain various inaccuracies in this respect. The animal I photographed [i.e. the small dark muntjac sharing an enclosure with the giant muntjac when Timmins visited the Laos military menagerie in January 1995], if it is actually one of these two species (not a certainty), is most likely *M. truongsonensis*, and quite unlikely to be *M. rooseveltorum*.

It would appear, therefore, that the last word has yet to be spoken after all in relation to the identity of the Laotian captive giant muntjac's unassuming little companion. Indeed, nowadays the general consensus is that *M. rooseveltorum*, *M. truongsonensis*, and two other recently-revealed Indochinese muntjac species, *M. putaoensis* and *M. puhoatensis* (see below), form a clade dubbed the *M. rooseveltorum* species-complex, with uncertainty as to just how many valid species it truly contains.

Meanwhile, returning to little muntjacs: the smallest species of muntjac currently known anywhere in the world first came to scientific attention in 1997, during a visit (spanning 23 February-29 April) by American conservationist Dr. Alan Rabinowitz to Myanmar (formerly Burma). Ten skulls and one freshly-killed adult female specimen of a tiny form of chestnut-coated muntjac were collected by Rabinowitz between the villages of Alanga and Shinshanku in northern Myanmar. On account of its extremely diminutive size, standing just 20 in at the shoulder and weighing no more than 25 lb, the village hunters called this deer the leaf muntjac. For they alleged that it could actually be wrapped up inside a single leaf from *Phrynium cadellianum*, a local species of zingiber tree.[229]

Comparative taxonomic studies by Rabinowitz, Amato, and Dr. Mary G. Egan involving the leaf muntjac's DNA, cranial characteristics (including its distinctive, unbranched antlers), and external morphology convinced them that it constituted yet another new species. Thus in 1999, within their official description, they dubbed it *Muntiacus putaoensis*, after Putao, which is the most northerly town in Myanmar and the closest reference point for this species. Its nearest relative appears to be the Truong Son muntjac of Vietnam.[229]

During February 2003, a quartet of Indian scientists announced in the journal *Current Science* that they had recently found evidence in the form of many skulls that this diminutive deer also exists in the dense forests of Arunachal Pradesh, within an area close to Namdapha Tiger Reserve. Emphasising the significance of their unexpected discovery, in their published report the team states: "This, perhaps, represents the only addition, so far, to the ungulate [hoofed mammal] fauna of the Indian subcontinent in the last century".[229a]

Yet another recently-revealed mini-muntjac from Indochina is the Pu Hoat muntjac, first made known to science from a single skull with attached antlers, obtained during October 1997 in Que Phong, within Vietnam's Nghe An province, by Le Trong Trai, a senior zoologist at Vietnam's Forest Inventory Planning Institute. DNA analyses conducted at Copenhagen University on this material indicated that the muntjac that it came from represented a distinct species, which was named

Muntiacus puhoatensis by Trai that same year. Its discovery was made public in early 1998. According to reports at that time, the Pu Hoat muntjac was also the world's smallest muntjac (but these reports preceded those announcing the discovery of the leaf muntjac in Myanmar), and is dark brown dorsally, with short unbranched antlers, and thick brown and yellow hair covering its head. Later material, including two almost-complete specimens, has since been collected. As noted earlier, however, it is now categorised within the *M. rooseveltorum* species-complex and thus may not be a separate species after all.[230]

HEUDE'S PIG AND SCHOMBURGK'S DEER— A COUPLE OF BELATED SURVIVALS?

With so many mammalian discoveries occurring in Indochina, it seemed likely that some rediscoveries would also be recorded from this cryptozoological oasis. And sure enough, hot on the hooves of Roosevelt's muntjac came another notable revival—and possibly two.

The classification of wild pigs is still a highly controversial subject, with opinions concerning the number of existing valid species varying greatly from one specialist to another. Back in 1892, for instance, based upon a description of two skulls from Vietnam published by Père Pierre-Marie Heude, a Jesuit naturalist-priest in Shanghai, a very mysterious Vietnamese relative of the warty pig *Sus verrucosus* was christened *Sus bucculentus*. But as neither skull seemed to have been received by a scientific establishment, *S. bucculentus* subsequently vanished from the zoological species list.[231]

While in Laos during January 1995, however, Dr. George Schaller learnt that the native hunters sometimes killed pigs that corresponded with the priest's description of *S. bucculentus*—a long-snouted, warty-faced form with a distinctive yellow and rust-red coat. Moreover, when he returned home to New York, Schaller brought back not only news of the small dark deer now deemed to be Roosevelts' muntjac but also a skull from a juvenile male specimen of this 'lost' species of warty pig, which he had obtained at the village of Ban Ni Ghiang on the Nam Gnouang, plus a sample of its meat.[232]

DNA was extracted from muscle tissue attached to the skull for taxonomic analysis by Dr. George Amato, who duly confirmed that the meat was from neither of the region's two known species of pig. In addition, one of Père Heude's two original specimens was uncovered by Australian National University mammalogist Prof. Colin Groves in July 1996, ensconced yet unregistered, but bearing an identification in Heude's own handwriting, within the collection of the Beijing Institute of Zoology, part of China's Academia Sinica. After more than a century, Heude's long-forgotten species, nowadays commonly termed the Vietnamese warty pig, was AWOL no longer, and its rediscovery was documented by Groves, Schaller, Amato, and Laotian forestry worker Khamkhoun Khounboline in a *Nature* paper on 27 March 1997.[233]

However, although it is now known to survive, *S. bucculentus* has lost much of its zoological significance, due to a recent reassessment of its taxonomic status. On 17 March 2010, in response to my request for information concerning a number of enigmatic Asian ungulates, I received the following illuminating data from Prof. Colin Groves concerning the Vietnamese warty pig:

> [Re] *Sus bucculentus*. In October last year, I was able to go back to the Beijing collection for a few days, and study all of the Indochinese pig specimens from the old Heude collection. I regret to say that Heude appears to have selected what in effect is one extreme of the range of variation to describe as a distinct species! In other words, the type specimen of *Sus bucculentus* does not appear to be anything but what is the one-and-only wild pig from Indochina (and in fact I can't find any difference between the Indochinese wild pig as a whole and that from Burma and indeed western China).[233a]

Exit *S. bucculentus* as a valid species!

Meanwhile, the second potential mammalogical rediscovery in Indochina since Roosevelt's muntjac is still officially pending. Closely related to the Indian barasingha *Cervus duvauceli* (indeed, classed as a subspecies of it by some workers), Schomburgk's deer *C. schomburgki* was a little-known species that may have occurred as far north as Yunnan and Laos, but was definitely recorded only from the swampy plains of

Schomburgk's deer.

Schomburgk's deer—rare early 1900s photo.

Thailand, principally within the Chao Phraya Basin. The adult stag was a magnificent animal, standing 3.5 ft tall at the shoulder, with a rich chocolate-brown coat, and spectacular antlers that were extraordinarily elaborate and highly-branched, but for which it was extensively hunted by the locals.[20,48,234-5]

Schomburgk's deer also suffered greatly from the destruction of its habitat by cultivation, forcing it into Thailand's junglelands, which were unsuitable for a species adapted to open lands with sparse tree growth, and where it soon dwindled in numbers. The last confirmed specimen was shot in September 1932, since when this impressive species has been deemed extinct. Remarkably, it had never been seen alive in the wild by Europeans, and only a few specimens reached Western zoos, notably Berlin Zoo during the early 1900s.[20,48,234-5]

In February 1991, however, a pair of Schomburgk's deer antlers was spotted and photographed by United Nations agronomist Laurent Chazée while visiting a Chinese medicine shop in a remote area of Laos. Moreover, the shop's owner informed him that the antlers were from a deer killed just the previous year (1990), within a forested region in a nearby district of Laos.[236] Although far beyond the Thai confines of this species' known distribution (and also outside its favoured habitat range), as hunting is prohibited in this forested region of Laos it is possible that deer have sought refuge and lingered here, unmolested, while others elsewhere (including Thailand) were killed. To date, no further evidence for the survival of Schomburgk's deer has been forthcoming, but the remoteness of the area and the many zoological surprises emerging with remarkable regularity within this region of southern Asia offer hope that one day its modern-day existence will be unequivocally confirmed.

GIANT LORISES AND VIKING DEER

Even more controversial than the survival of Schomburgk's deer in Laos are the following couple of semi-cryptic Vietnamese mammals.

During a visit to Vietnam in late 1994, Doug Richardson, London Zoo's assistant mammal curator, was walking around the new animal market in Hanoi on 4 December when he spotted (and photographed) a very unusual primate inside a large bird cage. Although it superficially resembled a slow loris *Nycticebus coucang*, it was noticeably bigger (with a head and body length of at least 20 in), and its fur was very light cream instead of this species' pale brown shade. Could it simply have been an exceptionally large, aberrantly coloured specimen—or did it represent a species new to science? When its animal dealer owner was asked how much he would sell it for, he quoted a price of $1,100; such a high price may be an indication of the rarity of this still-mysterious form of loris.[237-8]

In 1994, Vietnamese biologist Nguyen Ngoc Chinh visited Pu Mat, just north of Vu Quang, in search of Vu Quang oxen. He didn't find any, but returned instead with local reports of a strange deer known to hunters as the *quang khem*—'slow-running deer'. One hunter had also given him the skull of a *quang khem*, which was very unusual, on account of its bizarre antlers—for these were nothing more than primitive unbranched spikes that bore a startling resemblance to the horns on a Viking's helmet![238-9]

Technically, this odd-looking deer had actually been discovered three decades earlier—but no-one had realised! Shortly after Chinh's findings, MacKinnon spotted some *quang khem* skulls in a box of bones at Hanoi's Institute of Ecology and Biological Resources— bones that had been collected as long ago as the late 1960s, but which had not previously been examined or sorted. DNA samples were sent to Arctander at Copenhagen University, who was unable to match them with the DNA of any known species. Nevertheless, the elusive Vietnamese slow-running deer has still to be scientifically described and named.[238-9]

In an e-mail of 15 December 1999, Prof. Colin Groves mentioned to me that he hadn't heard anything more concerning this deer from Arctander or anyone else. He remained unsure of its likely zoological identity, being "unable to decide whether it was just sambar with undeveloped antlers (i.e. very young or very old, 'going back'), or a sort of paedomorphic sambar. Certainly the evidence indicated *Cervus* (*Rusa*)". Clearly, therefore, whatever it does prove to be, the *quang khem* is one mystery deer that is not a muntjac.[239]

Heartened by his serendipitous find at the Institute, MacKinnon made another visit there not long afterwards—and uncovered what seemed to be yet another new but hitherto-overlooked species! In the very same box of skulls that had contained those of the slow-running deer, he spied an unfamiliar pair of antlers that, according to the testimony of Vu Quang hunters, were from a dark-coated species termed the *mangden* or black deer. Yet its description did not correspond with any deer known at that time. Could this enigmatic form be one and the same as the subsequently-unveiled Truong Son muntjac?[238-9]

HOLY GOAT!—IT'S PSEUDONOVIBOS!

Most enigmatic, and controversial, of all, however, is *Pseudonovibos spiralis*, assuredly the least-known, most elusive member of the Indochinese cornucopia of recently-revealed ungulates—always assuming, of course, that this hoofed mystery beast ever existed at all....

Its convoluted scientific history began in earnest on a market stall in southern Vietnam's Ho Chi Minh City—for that is where, in early 1994, German zoologist Dr. Wolfgang Peter, visiting from Münster's Zoological Gardens, spotted a strange pair of horns that were unlike any that he had seen before. They were approximately 18 in long, heavily spiralled, blackish in colour, and their upper portions were greatly splayed—so that they bore more than a passing resemblance to a somewhat strange pair of motorbike handlebars![48,238,240]

Although he didn't purchase them, Peter did take some photographs. And when he and his colleagues back home in Germany and elsewhere around the world were unable to assign these mystifying horns to any known species, he and fellow German zoologist Dr. Alfred Feiler, from Dresden's State Museum of Natural History, paid several further visits to southern Vietnam. Here they succeeded in uncovering eight pairs of these peculiar horns, one pair becoming the type specimen for the formal description (published by Peter and Feiler later in 1994) of their still-unseen owner as a new species, housed within its own, brand-new genus. Incredibly, this was the third new genus of large ungulate to be described in just over a year, following on from *Pseudoryx* in 1993 and *Megamuntiacus* in 1994.[48,238,240]

Conversations with locals in the Vietnamese districts of Kon Tum, Dac Lac, and Ban Me Thuot revealed that they were familiar with this creature, which they call the *linh duong*—sometimes translated as 'holy goat'. Scientifically, however, it is *Pseudonovibos spiralis* ('spiral-horned false kouprey')—emphasising its spiralled horns, and their deceptive similarity in shape to those of the Cambodian wild ox or kouprey *Bos (Novibos) sauveli*, which itself remained concealed from scientific detection until 1936.[48,238,240]

This, however, is only half of the *Pseudonovibos* saga. At much the same time that Peter and Feiler were discovering its horns in Vietnam, Norway-based zoologist Dr. Maurizio Dioli was visiting northeastern Cambodia's Mondulkiri and Rattanakiri provinces when he purchased two pairs of unusual spiral horns at a market. Each pair was attached to a portion of skull, and seemed to resemble the horns of a juvenile female kouprey.[241]

Upon closer observation, however, Dioli found that the skulls' sutures were completely fused—conclusive proof that the animals had been adults, not juveniles. Moreover, whereas those of female koupreys are smooth and markedly oval in cross section, Dioli's horns bore very pronounced rings, were almost perfectly circular in cross section, and were more widely splayed. Clearly, then, these were not from a kouprey. Nor did they match those from either of the other two species of wild cattle known in Cambodia—the gaur and the banteng. Indeed, they did not correspond with the horns of any animal documented by science.[241]

When Dioli made enquiries, he learnt from local Cambodian hunters that these mystifying horns belonged to a large bovine beast that they call the *kting* (or *kthing*)

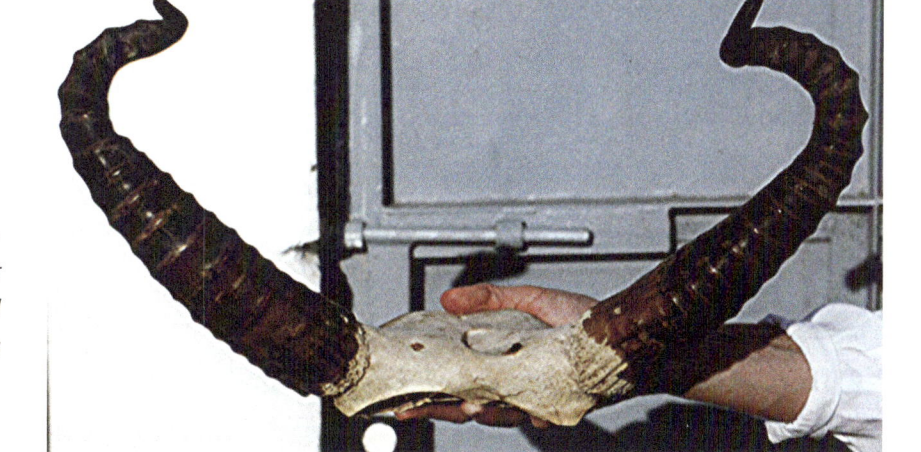

Discovered in 1994, the inordinately elusive holy goat Pseudonovibos spiralis is still known to science only from its horns. (Dr. Maurizio Dioli)

voar. This name translates as 'wild cow with vine-like horns', referring to their rings and curved shape.[241]

Judging from hunters' accounts collected by Dioli and also, more recently, by the Cambodia National Tiger Survey, the *kting voar* weighs 440-660 lb, stands 3.5-4 ft at the withers, and is said to be somewhat bovine in basic form. However, it is taller and more slender than a banteng or domestic cow, and has sambar-like legs, as well as a well-developed coat, which is variously claimed to be uniformly greyish-black or dark red in colour. Very shy, rare, fleet-footed, and agile, given to standing on its hind legs to browse off leaves on trees, it lives in small family groups amid the region's mountainous dipterocarp forests. Evidently, therefore, although it has successfully eluded scientific detection, this reclusive animal is no stranger to the region's people, thus making all the more interesting their second, alternative name for it—*kting sipuoh*, or 'snake-eating wild cow'![240-1a]

It is not unique for primitive native folklore to incorporate fanciful beliefs regarding herbivorous ungulates consuming serpents—Indian tribes tell similar stories concerning the ibex-like markhor *Capra falconeri*. Although intriguing, these curious claims have no scientific foundation.[241]

The hunters alleged that when the *kting voar* devours snakes, these reptiles bite its horns, creating their rings, and imbuing them with venom. This supposedly bestows the horns with medicinal properties against snake bite. Consequently, as soon as a *kting voar* is killed by a hunter, its horns are removed and used for making venom antidote. This is prepared by burning the horns in a fire—hence few survive to be sold at the markets as trophies.[241]

Ling—is this unidentified animal from a 17th-Century Chinese encyclopaedia one and the same as the mysterious Pseudonovibos?

Not long afterwards, Dioli learnt of Peter and Feiler's investigations, and he recognised that the horns of their Vietnamese holy goat matched those of his Cambodian *kting voar*. These two mysterious mammals were one and the same—both belonged to the newly-named species *Pseudonovibos spiralis*. Moreover, Dioli revealed that two horns supposedly from a young female kouprey that were collected in southern Vietnam as long ago as 1929 and donated to the Kansas Museum of Natural History are actually those of *Pseudonovibos*.[241]

Researches have suggested that *Pseudonovibos* had once been most common in Vietnam, but has been so heavily hunted there that today it survives predominantly across the border in Cambodia. Despite being known to the western world now for over 15 years, however, one major mystery remains unsolved. Scientists have yet to spy a living *Pseudonovibos*—apart from native testimony, therefore, we still do not know for sure what it looks like![241]

Having said that, however, a certain antiquated Chinese encyclopedia may offer a unique clue, as revealed by Drs. Alastair A. Macdonald and Lixin N. Yang. Entitled *San Cai Tu Hui*, compiled by Wang Chi and Wang Si Yi, and published in 1607, it contains a drawing and short piece of accompanying text concerning a sturdy horned creature known as the *ling*. According to this encyclopedia, the *ling*: ". . . looks like a goat but is larger. Its horns are round and have pointed tips". There is also a fanciful account of how it uses its horns to hang from trees at night to sleep. The animal depicted in the drawing does not call to mind any known species of ungulate—except, that is, for its horns, whose shape and ribbed pattern, as acknowledged by Macdonald and Yang, do recall those of *Pseudonovibos*.[242]

Moreover, in 1999 a team of German zoologists, which included Feiler and also Dr. Ralph Tiedemann from Kiel University, published a short communication documenting the results of some mitochondrial DNA sequence analyses featuring DNA extracted from *Pseudonovibos* horn fragments and compared with corresponding DNA sequences from a range of other bovid ungulates in an attempt to ascertain its taxonomic affinities. These analyses revealed that *Pseudonovibos* did seem to be more closely related to goats than to antelopes or to cattle (despite its generic name), but in a recent e-mail Dr. Tiedemann informed me that his team would be publishing more extensive genetic comparisons at a later date.[243]

And in his e-mail to me of 15 December 1999, Prof. Colin Groves revealed that Australian zoologist Dr. Jack Giles of Taronga Zoo in Sydney had recently visited Vietnam, where he had been shown an old black-and-white photo of a local hunter sitting upon a dead *Pseudonovibos*! Frustratingly, however, the photo was of such poor quality that little detail could be discerned, other than the fact that the animal was not particularly large. Making matters even worse, the hunter was perched upon the dead beast's head, thereby obscuring any distinguishing facial or cranial features that may have been visible.[239]

Even more tantalising are reports from early January 1995 documenting the recent capture of a still-unidentified mystery mammal in central Vietnam during December 1994. An immature female specimen, it was caught alive near the village of A Luoi in the central Vietnamese province of Thua Thien-Hue, more than 180 miles southeast of Vu Quang. Referred to by its captors as a *tuoa*, it died shortly afterwards, and was eaten before its body could be scientifically examined. Its mother had also been captured, but escaped. The calf weighed 36 lb and was said not to be a Vu Quang ox, resembling a goat instead, with a roundish head, long ears, horns, stout body, and a black and white coat patterned with buff and grey patches. According to quotes attributed to Hanoi University zoologist Prof. Ha Dinh Duc, it seemed to be different from any bovid species known scientifically in Vietnam.[244]

Nothing more has been heard about the *tuoa*, but in view of its morphological description, could this cryptic creature be one and the same as *Pseudonovibos*? If so, how ironic, and tragic, that the only complete—and living—specimen to come within reach of modern-day science found its way into a local cooking pot instead!

Moreover, in his above-noted e-mail to me, Robert Timmins opined that it may already be too late for this most elusive Indochinese hoofed debutante: "It's looking like *Pseudonovibos* has disappeared like the rhinos for a perceived medicinal value to its horns".[228]

However, we should not—indeed, cannot—forget that during the 1990s Indochina has hosted an unparalleled spectacle of mammalogical revelations, and with research continuing here the 21st century may well witness many more surprises in this cryptozoologically rich and still far from well-explored region. There must surely be hope, therefore, that one of these surprises will be the long-awaited discovery of living specimens of *Pseudonovibos*—or will it . . . ?

In January 2001, a team of French biologists including Drs. Arnoult Seveau, Herbert Thomas, and Alexandre Hassanin published a pair of startling, highly controversial papers, in which they claimed that *Pseudonovibos* is non-existent—a forgery. They based their claim upon the results of two separate studies of *Pseudonovibos* material. In one of these, they sequenced two DNA markers from the bony cores of four sets of *Pseudonovibos* horns, and compared them with the equivalent genetic markers in Vietnamese domestic cattle. In the second study, they conducted a histological examination of the keratin in six *Pseudonovibos* frontlets. The results of the DNA study revealed that the markers from the *Pseudonovibos* material were a perfect match with those from the Vietnamese domestic cattle. And the keratin study exposed the *Pseudonovibos* frontlets to be nothing more than domestic cattle horns whose keratin sheaths had been skilfully manipulated by heat treatment, followed by twisting and trimming, to create the distinctive spiral, heavily-ridged horns characterising *Pseudonovibos*.[244a]

Yet although there can be little doubt that these particular specimens are indeed fakes, there is currently no evidence that any of the several other sets of *Pseudonovibos* horns on record (including this species' type material) are also fraudulent. Hence the French team's bold statement that *Pseudonovibos* is not a new animal and its scientific name should be abandoned is premature, to say the least. Kansas University mammalogist Prof. Robert M. Timm has recently published an extensive paper on *Pseudonovibos*, in which he and fellow mammalogist Dr. John H. Brandt documented two sets of *Pseudonovibos* trophy horns procured by two western big game hunters in Vietnam during 1929. Following the appearance of the French team's claims, Timm averred that he had no doubt *Pseudonovibos* is a valid taxon, having uncovered various overlooked records from the 1880s and 1950s that documented a mysterious spiral-horned bovine beast ostensibly synonymous with *P. spiralis*.[244a]

Moreover, in a separate paper a team of Eastern European scientists announced that their phylogenetic analyses of nearly-complete 12S mitochondrial rDNA sequences for this enigmatic creature and a number of other bovids indicate that *P. spiralis* is a valid species belonging to the buffalo subtribe (Bovina), and should be placed between the Asiatic buffaloes *Bubalus* and the African buffalo *Syncerus*.[244b]

Personally, I consider it possible that the answer to the riddle of whether *Pseudonovibos* truly exists is that this enigmatic beast is a real but extremely rare species, so rare that procurement of its much-prized, supposedly snake-repelling horns even by locals is extremely difficult—which has in turn led to the deliberate preparation of copies for use in rituals. In other words, some of the preserved horns on record are indeed fakes, yet were created not to fool, but merely to act as substitutes for the real thing. This is also the opinion of Prof. Colin Groves, though as he noted to me, if the type material for *Pseudonovibos* is examined and is also shown to be fake, then regardless of whether this animal does exist, the name '*Pseudonovibos spiralis*' must be abandoned, and every remaining specimen must be examined to see whether any genuine material does exist.[244a]

In any event, Indochina's many other extraordinary new animals offer those who seek undiscovered creatures a very potent weapon for warding off criticism. If confronted in the future by cynics attempting to pour scorn on the cryptozoological concept of major new animals still awaiting discovery, there is now a very simple but effective means of responding—just whisper "Vu Quang", and watch them change the subject very hastily!

A MYRIAD OF MARMOSETS AND OTHER SOUTH AMERICAN PRIMATES

Not content with the spectacular herd of radically new ungulates revealed during the 1990s in Indochina, science has also been going ape about the equally impressive (and unexpected) troupe of totally new primates unfurled during that same decade in South America.

Following on from the black-faced lion tamarin *Leontopithecus caissara* and the Maués marmoset *Mico mauesi* (p. 87), the next couple of previously overlooked primates were triumphs of taxonomy rather than foundlings in the field (or rainforest). After conducting a detailed study of a trio of long-tailed, densely-furred monkeys known as titis, Chicago mammalogist Dr. Philip

Black-capped dwarf marmoset (young specimen). (Dr. Marc van Roosmalen)

Adult female Acarí marmoset—named after Brazil's Rio Acarí. (Dr. Marc van Roosmalen)

Hershkovitz proclaimed in 1990 that these constituted four separate species, not three, and he named his newly-created, fourth titi *Callicebus hoffmannsi*—Hoffmanns's titi. (Since then, continued taxonomic splitting has created many additional new titi species.) A year later, Brazilian researcher Dr. M. de Vivo published an extensive revision of the classification of marmosets, in which he distinguished a new species, Hershkovitz's marmoset (honouring Dr. Hershkovitz), formally dubbed *Callithrix* [now *Mico*] *intermedia*.[245]

The recognition of new species by the taxonomic splitting of existing ones is a relatively common occurrence. Far rarer, particularly with such ostensibly well-documented animals as monkeys, is the discovery of new species in the wild—which is why the following three examples received such attention from the international primatological community.

In 1992, Dr. Stephen F. Ferrari from the Federal University of Pará, Brazil, and fellow researcher Dr. M.A. Lopes announced that during a recent survey of monkey distribution around Humaitá on the Rio Madeira in western Brazilian Amazonia, they had discovered a hitherto unknown species of marmoset lacking the prominent ear tufts that certain other marmosets possess. Due to its head's inky-coloured fur, it was christened *Callithrix* [now *Mico*] *nigriceps*, the black-headed marmoset. Weighing under 1 lb, this diminutive debutante is of notable zoogeographical significance, because although it was often seen, it may well be confined to an area scarcely exceeding 3860 square miles. If true, this is one of the smallest (and hence most vulnerable) ranges for any Amazonian primate.[246-7]

Adult male Satéré marmoset.
(Dr. Marc van Roosmalen)

Prince Bernhard's titi (male).
(Dr. Marc van Roosmalen)

Rather more substantial in body size (weighing about 4.4 lb) is the Ka'apor capuchin monkey, whose existence was also revealed in 1992, this time by Dr. Helder L. Queiroz from the Emilio Goeldi Pará Museum. He had discovered it while conducting fieldwork in eastern Brazilian Amazonia's Maranhão state, assisted by the Urubu-Ka'apor Indians. One of them owned an adult female specimen as a pet—inspiring Queiroz to name this new species of capuchin *Cebus kaapori*, after his local helpers.[247-8]

The most familiar species of capuchin, *Cebus apella*, the black-capped or tufted capuchin, is characterised by a distinctive tuft of hair on top of its head. Conversely, the Ka'apor capuchin, like other species, has no such tuft. Yet whereas it appears to be separated geographically from its fellow untufted capuchins by distances exceeding 250 miles, it inhabits the same region as its tufted relative. However, the Ka'apor capuchin is much less common, and less robust too.[247-8]

The next simian revelation from South America was the Satéré marmoset, which was named *Callithrix* [now *Mico*] *saterei* by researchers Dr. José de Sousa e Silva and Dr. Mauricio de Almeida Norohona in 1998 within the journal *Goeldiana*.[249] Its discovery was made public in June 1996 by primatologist Dr. Russell Mittermeier, the president of Conservation International in Washington. Like the Ka'apor capuchin, it was named after an Indian tribe—who engagingly refer to it as the *zip*, and share its jungle domain in the central Amazonian area between the Rio Madeira and Rio Tapajós.[250]

This squirrel-sized marmoset is distinguished by its unpigmented facial skin, its pale mahogany-coloured body fur, and also by the presence of strange, fleshy appendages on the genitalia of both sexes, for which zoologists have yet to discover a function. Fortunately, however, they should have plenty of time to disclose these structures' secrets—unlike the previous new South American monkeys discovered in the wild during this decade, the Satéré marmoset seems to be fairly common and is not threatened by deforestation or hunting.[250]

When discussing this marmoset, Mittermeier stated that he would not be surprised if another five new species of monkey were found in South America by the year 2000, as there are still large unexplored areas in this vast, verdant continent.[250] In fact, Brazil-based Dutch

Juvenile female Kamayurá spider monkey. (Dr. Marc van Roosmalen)

primatologist Dr. Marc van Roosmalen appears to have already surpassed this estimate—in 1999, he claimed to have discovered no fewer than 14 new species of South American primate![251]

Working for the National Institute for Amazon Research in Manaus, Brazil, where he has also established a primate orphanage, his epic revelations first hit the headlined in early January 1998, when press reports announced that after more than a year's searching amid the Amazon rainforests, Dr. Roosmalen had succeeded in discovering four new species of monkey. He had also found a hitherto-undescribed species of tree porcupine, and had spied but not captured what may prove to be a remarkable number of other significant new mammals (p. 276, and p. 283).[252-3]

His quest had begun in 1996, incited by his discovery of a previously-unknown species of monkey now termed the black-capped dwarf marmoset, which he formally described in 1998, naming it *Callithrix humilis*. No bigger than a mouse, this is the world's second smallest species of marmoset (only the pygmy marmoset *Cebuella pygmaea* is smaller), and first came to Roosmalen's attention on 16 April 1996, when a native backwoodsman in the town of Novo Aripuanã showed him a milk can that contained a young marmoset, chirruping like a furry grasshopper.[252-4]

Quite apart from its minute dimensions (its body was only 4 in long, its tail was 10 in long, and it weighed a feather-light 5.6 oz), the little animal's black crown, broad white border of fur fringing its face (disappearing when adult), exposed ears, grey-green fur, and black tail collectively distinguished it from all other known species of marmoset, as Roosmalen realised when examining it in May 1996.[252-4]

However, this curious little marmoset's precise origin was unclear, so Roosmalen decided to seek out its species' undisclosed homeland, hidden somewhere in the forests bordering the 2000-mile-long Rio Madeira. Armed with a camera and snapshots of this recently-revealed marmoset, he showed pictures of it to the local Indian tribes that he encountered during his travels. Often they would claim to know where specimens could be found, and sometimes they even brought a few to him. Invariably, however, these specimens were found not to belong to the species that he was seeking—but Roosmalen was not disappointed. On the contrary, he was ecstatic, because four different specimens brought to him each proved to be a representative of a species hitherto unknown to science![252-4]

One is the manicore marmoset, living in an area by the Rio Manicore, 300 miles south of Manaus. With an adult head-and-body length of 9 in, plus a 15-in tail, this naked-eared mini-monkey has a silvery-white upper body, orange-yellow belly, grey head cap, black tail, and weighs about 12 oz. In a *Neotropical Primates* paper published in 2000, Roosmalen formally christened it *Callithrix* [now *Mico*] *manicorensis*.[252-3]

Another member of this once-cryptic quartet was a new titi monkey, locally termed the *zog-zog*, which may derive from the vocal duets aired when pairs of its species are establishing their territory. Greyish-brown, with a head-and-body length of 16 in, a slightly longer tail,

Adolescent female Manicore marmoset. (Dr. Marc van Roosmalen)

Black-tailed dwarf porcupine —a new species of South American tree porcupine discovered at Lake Matupiri. (Dr. Marc van Roosmalen)

and weighing 2.2 lb, the *zog-zog* is characterised by its distinctive reddish-orange belly and beard, as well as by a noticeable white spot at the tip of its tail. Roosmalen plans to dub this new species 'Callicebus [now *Mico*] aripuanensis', the after the lower Rio Aripuanã, around whose confluence with the Rio Madeira in the Amazon Basin is the previously little-explored locality where he has made so many of his remarkable discoveries.[252-3]

The remaining two species are *Callithrix* [now *Mico*] *acariensis*, a small marmoset from the interfluvium region of the Rio Acarí and Rio Sucundurí, with a snowy-white upper body and belly, a grey back with a stripe running down to its knees, and an orange-tipped black tail; and 'Ateles aripuanensis', a new spider monkey, from the lower Aripuanã region, still awaiting formal description and naming. The Acarí marmoset superficially resembles the Satéré marmoset, but it has black cheeks (the Satéré marmoset's are pink) and a distinctive white marking shaped like an inverted 'V' upon and above its nose.[253]

Roosmalen has also differentiated a second new spider monkey, which he has dubbed 'Ateles kamayurensis' (after the Kamayurá tribe) and is from the Upper Xingú area of the Mato Grosso. There were another two new titis as well, the size of small cats and both described in 2002. One is Prince Bernhard's titi *Callicebus bernhardi*, named after Prince Bernhard of the Netherlands, and hailing from the interfluvium of the Rio Aripuanã and Rio Madeira. This species is readily distinguished by its bright orange sideburns, chest, and inner limbs, its reddish-brown back, and its white-tipped black tail. The other is Stephen Nash's titi *C. stephennashi*, a silver-furred species with black brow and red sideburns and chest, which was named after Conservation International illustrator Stephen Nash, and is endemic to the eastern bank of Brazil's Purus River. In addition, there were two proposed new species of tamarin, plus a new subspecies of bearded saki, two new subspecies of uakari, and two new tamarin subspecies.[253]

As for Roosmalen's new porcupine, this was discovered on the left bank of Brazil's Rio Madeira by Lake Matupiri. Just like its freshly-revealed monkey compatriots, it too is decidedly diminutive—a dwarf tree porcupine, with a black tail, pink nose, and deceptively fluffy pale-hued fur concealing its armoury of sharp yellow spines. Mammalogists Drs. Robert S. Voss and Maria N.F. da Silva named it the black-tailed dwarf porcupine *Coendou* [now *Sphiggurus*] *roosmalenorum* (after Roosmalen and his son Tomas van Roosmalen) in 2001. Moreover, a second new porcupine species, *C.* [now *S.*] *ichillus*, the streaked dwarf porcupine, inhabiting the Amazonian lowlands of eastern Ecuador, was also described and named in the same paper.[252, 254a]

Meanwhile, the hideaway habitat of the black-capped dwarf marmoset, which was the object of Roosmalen's quest in the first place, had also been uncovered by him—in a small triangular stretch of land east of the middle Rio Madeira, west of the lower Rio Aripuanã, from its mouth to the Paraná Capimtuba, and east of the lower Rio Mataurá and the Rio Uruá. Here, and in particular within a pair of morototó trees in the backyard of local inhabitant Damiao Lisboa Pereira, he saw a number of these small monkeys feeding.[252]

Roosmalen considers this marmoset to be his most important primatological discovery to date, on account of its unique combination of morphological, physiological, and behavioural features. For instance, although it possesses claws (rather than nails) like other marmosets, and also has marmoset-like teeth, its ear tufts grow out of a different portion of the ears from that which gives rise to ear tufts in other marmoset species. Also, black-capped dwarfs only give birth to a single baby whereas typical marmosets usually give birth to twins. And in a troop of black-capped dwarfs, several different females will give birth at any one time, yet in other marmoset species the usual rule is for one dominant female to be the only member of the troop to breed. Moreover, although marmosets are normally viciously territorial, the black-capped dwarf is not.[252-4] Indeed, it was subsequently deemed so distinct from all other marmosets that it was rehoused in an entirely new genus, *Calibella*, created specially for it. Consequently, this remarkable species is now known formally as *Callibella humilis*.[253]

Nor was this the end of Roosmalen's astonishing success in discovering new mammalian species (which incited *Time Magazine* to elect him, very deservedly, as one of their six Heroes for the Planet in the Wildlife Category, within their issue for 28 February 2000), as will be seen in this present book's 21st century section.

Moreover, during his intensive studies of the area around the Aripuanã-Madeira confluence, Roosmalen has uncovered evidence for exceptionally high biodiversity—possibly the highest ever recorded anywhere in the world, thereby offering hope indeed for further zoological revelations during the coming years.[253]

DINGISO—THE FORBIDDEN TREE KANGAROO OF IRIAN JAYA

One of the world's most distinctive species of mammal remained undiscovered by science until as recently as 1994, but a few scattered clues betraying its existence had turned up four years earlier. In November 1990, while visiting the Sudirman (Maokop) Mountains in Irian Jaya (the western, Indonesian half of New Guinea), wildlife photographer Gerald Cubitt encountered a hunter from the local Dani tribe, who offered to sell him as a pet a young tree kangaroo, predominantly black in colour but with distinctive white markings on its chest, face, and tail. Although he declined the offer, Cubitt did photograph the appealing little animal, but when he returned home to Britain and attempted to identify it he was very surprised to discover that it did not correspond with any species currently known.[255]

Exhibit #2 in this curious case of the unexplained tree kangaroo took the form of a tree kangaroo jawbone, which had been given in 1990 to Australian Museum zoologist Dr. Tim Flannery (later director of the South Australian Museum) by Lani tribesmen at Billingeek, a rock shelter near the settlement of Kwiyawagi in the Sudirman mountain range. The jawbone eluded Flannery's attempts to identify its species—as did Exhibit #3. This was a piece of tree kangaroo fur, black with a flash of white on its chest, that had originally been fashioned into a Lani war bonnet, which Flannery had bought during his 1990 visit to Kwiyawagi.[199]

Perplexed by these enigmatic trinkets and Cubitt's photos, Flannery was keen to return to Irian Jaya to resolve the mystery of the black-and-white tree kangaroo, and in late spring 1994 he received the opportunity to do so, courtesy of an invitation to talk about wildlife to the local community at Tembagapura, situated a mere 75 miles west of Kwiyawagi. When he reached Tembagapura, accompanied by Indonesian biologist Boeadi from Bogor's Zoology Museum and Sydney University anthropologist Dr. Alexandra Szalay, Flannery lost no time in showing the local Moni and Dani tribespeople the photos of the young mystery tree kangaroo, and asking them if they knew of it. Both tribes confirmed that they did.[199,256]

The Dani referred to it as the *nemenaki*, which they hunted for its meat, and did not look upon as being anything special. A very different scenario emerged, however, when the Moni hunters were questioned, because

Young specimen of the dingiso—a major new tree kangaroo from Irian Jaya. (Dr. Tim Flannery)

they revealed that it was an important animal in their cultural beliefs—so much so that they had three different names for it.[199,256]

In common Moni parlance, it was the *dingiso* (translating as 'large black game animal', which has now become its accepted English name too), and also the *mayamumaya* ('the one who has received the face of a man)'. However, it also had a third, sacred name—*bondegezou*, or 'man of the alpine forest', which was a term of deep respect used by Bibida Moni, for whom the harming, hunting, and eating of this creature was strictly forbidden. The reason for this, as Flannery later discovered, is that the Moni believe the dingiso to be the ancestral being from which they are descended.[256]

Armed with ample information that it did indeed exist here, Flannery was now anxious to obtain a specimen of what seemed increasingly likely to be a bona fide new species. Days of failure followed, but finally, in late May, while seeking at altitudes of around 14,850 ft, one of the local Dani hunters working with Flannery's team brought back some partial skins and bones of two specimens that had recently been eaten. A few days later, another Dani hunter returned, carrying on his shoulders a complete, albeit dead, specimen, sitting upright—photos of which appeared in press reports worldwide during July.[199,256-7]

And in late October 1994, Flannery returned to Tembagapura to see his first living dingiso—after being alerted by a phonecall that a young male had been discovered hiding in a disused machinery shed at Tembagapura's mining site. Once he had examined, filmed, and tagged it, Flannery released the half-grown dingiso, nicknamed 'Ding', back into the wild.

Up to 4 ft long and possibly weighing up to 33 lb, the dingiso is a sizeable, singularly remarkable creature. Superficially resembling a panda-patterned koala, it is instantly distinguished from all other species of tree kangaroo by its eyecatching pelage. In the adult, this consists of a black back, a white chest and belly, a white half-collar across its throat, and a varyingly-marked black-and-white tail, usually with a white tip. Its black face is highlighted by a distinctive white star-shaped mark on the centre of its brow, and a pair of white lines encircling its muzzle's base (although not definitely present in juveniles).[199,256]

Equally noticeable are its short bear-like muzzle, its very short tail, and in particular its limbs, which are more slender than those of other tree kangaroos, and betray its extremely surprising ecological secret. With blatant disregard for zoological nomenclature, the dingiso is a tree kangaroo that spends much of its time on the ground! Typical, arboreal tree kangaroos have very thick, robust limbs, needed to support them when they leap down from trees to the ground. As the dingiso

Dingiso. (Dr. Tim Flannery)

is primarily terrestrial, however, it does not perform these dramatic downward leaps, and therefore does not require such burly limbs.[199,256]

In 1995, Flannery, Boeadi, and Szalay formally described the dingiso, which they christened *Dendrolagus mbaiso*, 'the forbidden tree kangaroo', emphasising the Moni's refusal to hunt this species—'mbaiso' translates as 'forbidden' in the Moni's language.[238,256]

In fact, the dingiso probably owes its survival to this traditional religious taboo, because it is very common throughout the high elevation mossy forest along the western slopes of the Sudirman Range where the Moni live. Indeed, as a direct consequence of the protection afforded to it by their cultural system, this marsupial is so tame that it can be readily approached and even picked up, as it has never learned to fear humans, its only protestation being a loud, piercing whistle. Conversely, in the southern Sudirman slopes, where it is hunted by the Dani and other tribes, the dingiso is much rarer and shyer.[199,256]

Missionary Jon Cutts, working with the Moni, has been instrumental in preserving their taboo against hunting the dingiso, and it is to be hoped that his sterling efforts will be perpetuated by others, and the Moni themselves, in the future, to ensure the continuing well-being of this delightful newcomer to the zoological world.[199]

It may seem strange that a species as striking, abundant, and tame as the dingiso could elude scientific detection for so long, but its habitat is exceedingly rugged, cold, mist-enshrouded, and virtually impenetrable, which dissuaded earlier expeditions to New Guinea from penetrating its domain. Moreover, only in recent years has Indonesia permitted detailed explorations of Irian Jaya by outside teams, which is why, even today, this vast jungleland remains one of the least-known territories, zoologically-speaking, in the world.

Moreover, there may be a similarly noteworthy sequel still to take place on New Guinea's smaller, neighbouring eastern island of New Britain. In 2007, John Lane, principal scientist at Chico Environmental Science and Planning in Chico, California, led a six-week expedition to New Britain, and while there his team not only reported a number of new species of butterfly, frog, snake, and fish, but also sighted some tree kangaroos. They duly filed a report of their findings to the Nature Conservancy, and Lane was later informed that the presence of tree kangaroos on New Britain was unexpected and may even involve a new species.[257a]

Aware of the conservation publicity value of discovering a major new mammal here (this area is at risk from loggers and palm-oil encroachment), in July 2009 Lane returned to New Britain with a team of explorers, to track and possibly even capture a specimen of this intriguing creature in order to obtain DNA samples, so the eventual release of Lane's latest findings report will no doubt be of great interest.[257a]

A WILD AND WOOLLY REDISCOVERY

Ever since its discovery in 1888, the woolly flying squirrel *Eupetaurus cinereus* has remained one of the world's least-known mammals. Native to the Himalayas, it was believed extinct since 1924—until its unexpected rediscovery during summer 1994 in Pakistan's Sai Valley made belated headlines around the world in spring 1995. Its resurrection was the culmination of two years of diligent searches by New Yorkers Peter Zahler (a freelance writer) and Chantal Dietemann (a college maths teacher). And at the end of their search, it was a disembodied forepaw, not to mention two local sellers of crystallised *E. cinereus* urine(!), that led them to their long-awaited quarry—in the form of a female woolly flying squirrel squatting in a sack, which was given to Zahler and Dietemann by the urine purveyors in return for their promised US $150 finder's fee.[258]

For a species hitherto lurking amid the shadows of scientific obscurity, the woolly flying squirrel is an uncommonly impressive animal—the world's largest squirrel, standing 2 ft tall, sporting an eyecatching 2-ft-long tail, and able to glide up to 100 ft through the air. After studying their furry captive, Zahler and Dietemann released it back into the lofty cave, high upon a steep and relatively inaccessible slope, inside which it had earlier been captured.[258]

RETURN OF A RAT KANGAROO

Not surprisingly, Gilbert's potoroo *Potorous gilberti* has traditionally been viewed as one of Australia's least-known mammals, represented by just two specimens that had been sent to the British Museum (Natural History) by 19th century Australian naturalist John Gilbert. He had discovered this small, superficially rodent-like species of rat-kangaroo in 1840, and reported that it lived amid the dense thickets and rank vegetation bordering swamps and running streams within the King George's Sound area of Western Australia. Predation by foxes and domestic cats introduced by man into Australia from the Old World, hunting by the locals, and encroachment upon its habitat by farmers and their livestock, however, were severely threatening its survival, and the last confirmed report of this inoffensive creature was in 1869. Since then, Gilbert's potoroo has been listed as one of zoology's 'classic' extinct mammals, whose future rediscovery seemed a wholly unrealistic possibility.[20,259]

Woolly flying squirrel.

In 1994, zoology student Elizabeth Sinclair from the University of Western Australia was seeking specimens of a small wallaby known as the quokka, at the Two Peoples Bay nature reserve just east of Albany, on Western Australia's southern coast, which famously hosted the rediscovery of two supposedly long-extinct species of bird during the early 1960s (see p. 134). She was using special, non-harmful traps to capture some living specimens during her research, but after eight days of failing to catch anything she was startled to discover that her traps had captured two smaller creatures that were definitely not quokkas. With short black tail, long pointed muzzle, grey fur, long nails, and a black stripe running from nose to brow, they bore an uncanny resemblance to the long-demised Gilbert's potoroo—which is not surprising, because that is precisely what they were.[259]

One of the two specimens was an adult female with young in her pouch; the other was a juvenile male. After formal confirmation of their identity had been made, the male was tagged with a radio-transmitter and released back into the wild, but the female was retained for further study. Follow-up trapping programmes in the Mount Gardiner area of Two Peoples Bay succeeded in capturing three additional specimens—an adult male, another juvenile male, and another adult female. Against all the odds, Gilbert's potoroo had somehow survived, but was clearly exceedingly elusive—so much so that it had totally evaded detection via a similar scientific trapping programme instigated here back in 1975.[259]

With specimens to hand, scientists have since been investigating the precise taxonomic status of Gilbert's potoroo using genetic analyses, as some researchers believe that it may not be a separate species in its own right, but merely a subspecies of the three-toed potoroo *P. tridactylus*. Whatever its identity may prove to be, however, there is no longer any question of its existence—after 125 years, Gilbert's potoroo is extinct no more, although it remains one of Australia's most endangered mammals.[259]

ANOTHER TRIO THAT TURNED UP DOWN UNDER

Cuscuses are Australasian marsupials resembling sluggishly animate teddy bears, but with the bonus of a long tail, and a new recruit to their membership was described in 1995 by Australian Museum zoologist Dr. Tim Flannery and Indonesian biologist Boeadi from Bogor's Zoology Museum. Christened *Phalanger alexandrae* in honour of Sydney University anthropologist Dr. Alexandra Szalay, and taxonomically differentiated from the closely-related *P. ornatus*, it inhabits Gebe Island, in the North Moluccas. Its large size and bright red shoulders collectively distinguish it not only from these

Gilbert's potoroo.

islands' other cuscuses but also from those elsewhere in Australasia. Back in 1987, Flannery had described another distinctive new species of cuscus. Formally dubbed *P. matanim* ('matanim' is its local name), the Telefomin cuscus had been discovered two years earlier, but remains today one of the rarest of all possums. Grey, noticeably plump, and known only from the Telefomin area of western Papua New Guinea, it appears to be a primitive, relict species, with no close relatives.[260]

September 1996 marked the rediscovery of the central rock-rat *Zyzomys pedunculatus*, one of Australia's relatively few species of native non-marsupialian mammal. Two females and a male were captured alive by National Park volunteers constructing a walking track in the West MacDonnell Ranges, about 70 miles west of Alice Springs. Approximately 1 ft long in total length, this rodent possesses the decidedly reptilian ability to shed part of its notably fat, hairy tail if attacked, just like various lizards. It is indigenous to the MacDonnell, James, and Davenport Ranges, plus the Granites and the Napperby Hills, all in Australia's Northern Territory, and was last collected in 1960.[164,261]

Several additional living specimens were captured since September 1996, including four during autumn 1997, by researchers from the Northern Territory Parks and Wildlife Commission, who believed that they may have discovered a small isolated colony. They also hoped to establish a captive breeding colony of this vulnerable species, within the safety of the Alice Springs Desert Park, and in 2001 specimens were recorded in 14 different locations. Tragically, however, following drought and a bush-fire in the West MacDonnell Ranges in 2002, no specimen has been reported, leading the IUCN to categorise this species as critically endangered, possibly extinct.[261]

Another important mammalian event Down Under was the rediscovery on mainland Australia in 1995 (but not made public until mid-1996) of the eastern quoll *Dasyurus viverrinus*. This weasel-like carnivorous marsupial is the size of a small domestic cat, and its dark fur is dappled with large white spots. Until two separate confirmed sightings were made by scientists working in New South Wales, however, it had last been recorded on the mainland in 1964, and had thus been believed to be confined to Tasmania.[262]

BAHAMONDE'S BEAKED WHALE

It's not every day that a new species of mammal estimated to be as big as an elephant is discovered, but this was the conclusion drawn by cetologists after finding a strange partial skull of a subadult beaked whale on Chile's Robinson Crusoe Island in June 1986. Possessing at least three unique characteristics, including an exceedingly broad rostrum (beak), it was deemed sufficiently distinct from all other beaked whales by Drs. Julio Reyes, Koen van Waerebeek, Juan Carlos Cárdenas, and José Yáñez to be classified by them as a wholly new species. In 1995, they named it *Mesoplodon bahamondi*—Bahamonde's beaked whale, in honour of renowned Chilean marine biologist Prof. Nibaldo Bahamonde, who founded Robinson Crusoe Island's marine research station. It constitutes the sixth new species of *Mesoplodon* beaked whale to have been discovered and described in the 20th century[263] . . . or at least it did.

In a *Marine Mammal Science* paper of 2002, the authors announced that mitochondrial DNA and morphological examinations of the Robinson Crusoe Island skull had revealed that '*M. bahamondi*' is merely a junior synonym of *M. traversii*—a name coined way back in 1874 by British Museum zoologist John Edward Gray for a whale now known as the spade-toothed beaked whale.[263a] In any event, Bahamonde's beaked whale has now been consigned to taxonomic history, but a related mystery is still unresolved—the external appearance of *M. traversii*, which remains unknown, as no complete specimen has ever been documented.

FALSE POTTOS AND REDEFINED BUSHBABIES

The potto *Perodicticus potto* is a small nocturnal primate from western Africa, allied to the little-known angwantibo *Arctocebus calabarensis*, but closely resembling (and also closely related to) the lorises of Asia, and possessing only the merest stump of a tail. It was first brought to scientific attention as long ago as 1704, since when it has generally been classified as a single species. In 1996, however, primatologists were startled to learn that the potto had a taxonomic traitor in its midst. Or, to put it another way—all pottos are not the same!

During the 1990s, Pittsburgh University mammalogist Dr. Jeffrey H. Schwartz had been studying the potto's anatomy, and was examining some skeletons at Zurich University when he came upon two specimens that seemed rather peculiar. Originating from Cameroon's tropical forests, one was an almost complete adult female skeleton, the other a subadult skull and mandible. On first sight, they had appeared similar to other specimens of this species. However, when Schwartz conducted some detailed comparative analyses with them and other potto material, he found himself confronted by ample anatomical discrepancies—not least of which were the specimens' small size, longer premolars, and in the case of the near-complete specimen, the lack of the true potto's distinctive spiny vertebrae, plus the presence of a lengthy, conspicuous tail—to convince him that these two were impostors, representing a hitherto undescribed species.[28,264]

Moreover, they were sufficiently different from the potto to necessitate the creation of an entirely new genus. Throughout the 20th century, only a handful of newly-disclosed species of mammal have required fresh genera to accommodate them. Yet here was another one.[264]

Schwartz's remarkable new primate, which he considers to be more primitive than the true potto, was revealed to the scientific world via a paper published in early 1996. He formally christened it *Pseudopotto martini*, Martin's false potto—thus emphasising its deceptive resemblance to the true potto, and also honouring Dr. R.D. Martin, the director of Zurich University's Anthropological Institute and Museum.[264]

Needless to say, the finding of so significantly novel a species as this in such unforeseen circumstances is noteworthy in itself, but what makes it even more so is the extraordinary realisation that somewhere amid the seclusion of West Africa's jungles was a radically new mammal that no zoologist had ever knowingly set eyes upon (and apparently had not even been reported by the native people) prior to Martin's unfurling of *Pseudopotto* via the two museum skeletons. Strictly speaking, therefore, the false potto is not even a creature of cryptozoology, as its discovery was not prefaced by any anecdotal testimony.

During the late 1990s, however, this exceptionally reclusive mammal was at last identified in the field, by primatologist C. Wild. Even so, the biggest irony of all concerning the false potto was the discovery that the near-complete skeleton came from of a specimen that had actually lived for a time, many years ago, at Zurich Zoo—yet no record apparently exists of what this unique captive looked like![48]

The classification of prosimians (potto, bushbabies, lorises) has undergone dramatic change during the 1990s due to extensive, combined taxonomic, ethological, and ecological studies conducted in the laboratory and the field. One prominent zoologist at the forefront of this wholesale taxonomic revision is Dr. Simon K. Bearder from Oxford Brookes University, heading the Nocturnal Primate Research Group, whose intensive field and university studies of bushbabies (galagos)—the most comprehensive ever undertaken in relation to these small, shy primates—have so far resulted in the identification of several new species, thereby raising the total number recognised from 11 in 1986 to 17 by 1999, and currently (in 2011) a total of 19.[265]

These lately-defined species include *Galago gabonensis*, *Galagoides granti*, *G. orinus*, *G. udzungwensis*, *G. rondoensis*, and *Euoticus pallidus*, all of which have been variously delineated by DNA profiles, outward morphological traits, vocalisations, and other distinguishing characteristics. Moreover, Dr. Bearder believes that the number of species will increase still further as his team's work continues—and not just bushbaby species either. During the 1990s, some zoologists looked favourably upon the taxonomic splitting of the potto into three distinct species within the genus *Potto*—a number that in Bearder's opinion will also rise once these intriguing primates attract the same degree of research attention as bushbabies. Currently, however, only a single potto species is recognised, split into four subspecies.[265-6]

In addition, there are some very odd records on file of unidentified African prosimians resembling unexpectedly large bushbabies that are greatly deserving of investigation too. For instance, during June 1985, in the centre of Senegal's Casamance Forest, naturalist Owen Burnham spied a mysterious creature resembling a giant bushbaby. It was the size of a half-grown cat, with pale grey fur, and was accompanied by two or three babies. More recently, similar animals have been reported from the Ivory Coast and Cameroon.[266-7]

LET'S HEAR IT FOR A NEW BRITISH PIPISTRELLE!

Perhaps the most unlikely location for the uncovering of a new species of mammal is the exceedingly well-explored island of Great Britain. Nevertheless, in March 1996, Drs. John D. Altringham and Kirsty J. Park from Leeds University together with Dr. Gareth Jones from Bristol University announced that Britain's common pipistrelle bats, traditionally grouped together within the single species *Pipistrellus pipistrellus*, most probably belong to two separate species, as betrayed by the frequencies of their ultrasonic squeaks, used for echolocation and mate-attraction purposes.[268]

The team discovered that whereas some British colonies of pipistrelle were emitting squeaks whose frequency was close to 45 kHz, others were emitting squeaks close to 55 kHz, with no colonies emitting intermediate-frequency squeaks, and no colonies containing both 45 kHz bats and 55 kHz bats. This situation was later found to be mirrored in mainland Europe too. As these squeaks specifically attract mates belonging to their own respective colonies, this effectively isolates the two phonic types of pipistrelle from one another reproductively—thereby fulfilling the fundamental requirement for creating two taxonomically distinct species.[268]

Specimens of these two phonic types can be distinguished from one another genetically too, via comparative analyses of mitochondrial DNA sequences. Other delineating factors have also been disclosed by studies of their respective habitat preferences (55 kHz pipistrelles are more riparian, showing greater tolerance of river pollution than do their 45 kHz counterparts), maternity roosts (those of 45 kHz pipistrelles contain significantly fewer individuals than those of 55 kHz pipistrelles), diet, cranial features, and some subtle outward characteristics (for example, the 45 kHz pipistrelles have a dark face band and a more pointed snout, whereas the 55 kHz pipistrelles are somewhat paler, leaner, and smaller).[269]

As these collectively yield unequivocal evidence for the reality of two hitherto-cryptic but nonetheless valid pipistrelle species, which Dr. Jones and colleagues believe may have diverged from one another 5-10 million years ago, zoologists have since been contemplating which one of these two species should retain the original name *Pipistrellus pipistrellus*, and which one should be given a separate name. Addressing this nomenclatural dilemma, in 1999 Dr. Jones and co-worker Dr. Elizabeth M. Barratt from the Institute of Zoology in London proposed that the then-obsolete name *Pipistrellus* [originally *Vespertilio*] *pygmaeus*, coined in 1825 by W.E. Leach for a small pipistrelle specimen from southwestern England that was initially deemed to represent a new species, should be revived and applied to the 55 kHz pipistrelle (with 'soprano pipistrelle' as its suggested common name), and *P. pipistrellus* be retained for the 45 kHz pipistrelle. This suggestion has been duly adopted.[270]

Another cryptic chiropteran, whose disclosure dated back to March 1995 but was not publicly revealed until autumn 1996, is among the world's smallest species of fruit bat or flying fox. It was brought to scientific attention during a faunal survey of Mount Iglit, on the Philippine island of Mindoro, by the University of the Philippines' Wildlife Biology Laboratory.[271]

UNVEILING A NEW BROCKET FROM BRAZIL

Some of the world's least-known deer are the small South American brockets (*Mazama* spp.). In 1992, an odd-looking specimen at São Paulo's Sorocaba Zoo attracted attention from Brazilian zoologists, as it seemed different from the four species of brocket previously recorded (recently taxonomic splitting has created four more), including Peru and Bolivia's dwarf brocket, *M. chunyi*, described in 1959. Chromosomal analyses were conducted, and in autumn 1996 this strange brocket was confirmed as a hitherto unknown species. It had been caught in Capão Bonito City, containing fragmented portions of rainforest. Similar brocket skins derived from there have since been found in the São Paulo Museum, where they had been mislabelled as *Mazama bororo*, the small red brocket.[272]

RETURN OF THE BARBARY LION—
OR A CASE OF MISTAKEN IDENTITY?

One of the largest and most impressive lion subspecies was the Barbary lion *Panthera leo leo*—the great shaggy-maned lion that confronted gladiators and Christians in the terrifying spectacle of the Roman arena. Native to North Africa, the Barbary lion measured up to 10 ft in total length, weighed as much as 500 lb, and was characterised by its huge mane, which covered almost half of its entire body. Described by early 1900s zoologist Dr. Richard Lydekker as being dusky ochre in colour, this hirsute mass was very dense, long, and extended to the middle of the lion's back as well as underneath its body.[20,234,273]

Yet even this mightiest of the mighty leonine lineage was doomed to fall when pitted against human aggressors, and a male specimen shot in the Atlas Mountains of Morocco during 1922 is traditionally claimed to be the very last Barbary lion—but was it?[20,234,273]

In July 1996, the worldwide media reported the unexpected discovery by South African vet Dr. Hym Ebedes of eleven lions within a zoo in Addis Ababa, Ethiopia, that allegedly bore much more than a passing resemblance to the supposedly demised Barbary lion. And indeed, it is known that Barbary-lookalikes that may have been the genuine article had formerly been maintained by Ethiopian ruler Emperor Haile Selassie within his royal palace's zoo.[273-4]

Moreover, by the end of 1996, news reports had also appeared concerning another three putative Barbary lions, this time discovered within an abandoned circus in Maputo, Mozambique, by workers from a London-based charity called Animal Defenders. These were rescued and transported by Animal Defenders to a new home at Hoedspruit Research and Breeding Centre for Endangered Species, in South Africa.[273-4]

However, a closer reading of these reports, and scrutiny of the accompanying photos of the supposed Barbaries readily reveal some serious discrepancies. As mentioned above, the Barbary lion allegedly sported a dusky ochre mane, yet the media reports persisted in claiming that this extinct subspecies possessed a black mane, one that extended not only onto its back but also down its chest and belly. This description much more readily calls to mind a second great, late subspecies of lion—*Panthera leo melanochaita*, the Cape lion, formerly indigenous to South Africa's Cape Colony, but which became extinct through over-hunting in c.1865. Similarly, photos of Akef (later renamed Giepe), the only male member of the Mozambique trio, clearly reveal that his mane is dark (especially ventrally), though admittedly not very extensive.[20,234,273,275]

So are these recently-revealed specimens truly Barbary lions—or could they perhaps be resurrected Cape lions instead, as suggested by a few zoologists since the two stories broke in 1996? A hair sample taken from the male Mozambique lion was compared visually with hair from a stuffed Barbary at London's Natural History Museum and was said to resemble the latter closely, but only genetic comparisons can offer a satisfactory answer.[276]

Until then, as noted by Doug Richardson, London Zoo's mammal curator, even if the recently-revealed lions do possess some Barbary characteristics, this does not make them Barbary—or, as the case may conceivably be, Cape—lions. Indeed, quite possibly they are neither—for as Richardson also pointed out: "If you put any Kenyan lion in a temperate climate it will grow a black mane and a belly mane".[276]

Clearly, life with the lions, at least as far as issues regarding taxonomic identification are concerned, is not as cosy as media reports would have us believe. Nevertheless, in 2005, mitochondrial DNA research revealed that an African lion housed at Neuwied Zoo in Germany is not of sub-Saharan origin according to its mitochondrial lineage. Hence this specimen may indeed be a genuine Barbary lion descendant.[276a]

A MOUSE THAT GOES FISHING

On 28 November 1997, Dr. Paulina D. Jenkins from the British Museum (Natural History) and Dr. Adrian A. Barnett from London's Roehampton Institute published the official description of an extraordinary new species of mouse—one that catches fishes. This bizarre rodent, less than 10 in long and formally dubbed *Chibchanomys orcesi*, is wholly confined to Ecuador's Las Cajas Plateau, and is made all the more remarkable due to the fact that it is virtually blind, and thus catches its prey not by sight but by touch. It uses its long sensitive whiskers to detect subtle movements made by resting fishes at night, and then snatches them from their aquatic abode in icy streams and rivers with its front paws.[277]

Yet how it can actually discern and differentiate the resting fishes' relatively gentle movements among the violent perturbations of the water current within these streams and rivers remains a mystery. After all, to quote the apt words of Barnett: "With all that swirling water, locating such food by touch must be a bit like trying to home in on a piccolo playing in a heavy metal concert".[278]

Named after Prof. Gustavo Orcés, a pioneering Ecuadorean mammalogist, *C. orcesi* is not the only known species of fishing mouse; several other species had previously been described, and are collectively termed ichthyomyines. However, it is only the second member

Barbary lion—does this supposedly vanished big cat still exist in captivity?

of the genus *Chibchanomys*, it and lives at higher altitudes (up to 13,000 ft) than the others. As might be expected from its lifestyle, Orcés's fishing mouse has thick water-repellent fur, broad powerful hindfeet for propulsion through its watery habitat, and is so streamlined that its external ears are completely hidden within its head's dense pelage.[277]

This intriguing little animal also has a memorable history, which began on 22 August 1981 when Barnett and some colleagues collected two adult male specimens (one of which ultimately became this species' type specimen) at a height of approximately 11,600 ft during the Oxford Expedition to Las Cajas. Initially misidentified as *Anotomys leander*, a species of fishing rat, it was subsequently differentiated from the latter rodent at the British Museum (Natural History). However, Barnett's specimens were then loaned for several years to a colleague elsewhere, and although several others were later collected during the 1980s this mystifying piscivorous mouse remained resolutely undescribed.[279]

And then came spring 1994, when a new documentary concerning Andean wildlife was screened at the Mammal Society's conference. Its viewers were startled to see footage of a strange little aquatic rodent inhabiting an area where no such species was supposed to exist. It was, of course, the still-unnamed Orcés's fishing mouse, whose long-awaited description finally occurred three-and-a-half years later.[279]

UNMASKED MICE AND SPINY SPLITTING

Although Australia's most famous mammals are marsupials, it also possesses a notable number of native rodent species. Tragically, however, many of these are currently in danger of extinction, and some appear to have already died out. One species traditionally placed in this latter category was the Alice Springs mouse *Pseudomys fieldi*, known only from a single damaged specimen collected in the Alice Springs region of the Northern Territory during 1895. At the beginning of 1998, however, Australia's Federal Environment Minister, Senator Robert Hill, made an unexpected but most welcome announcement—this seemingly lost species was still alive, because it had been found to be one and the same as another, supposedly distinct species, the Shark Bay mouse or djoongari *Pseudomys praeconis*.[164,280]

Formerly occurring on mainland Australia too but extinct there since European colonisation, this scrub-inhabiting grizzle-furred rodent is nowadays confined entirely to Bernier Island in Western Australia's Shark Bay, and is itself classed as vulnerable. By being conspecific with the Alice Springs mouse, however, its existence thereby removes one name from Australia's list of vanished fauna. It is now the focus of action by the Australian government for ensuring its protection, involving the implementation of a national recovery plan for this "two-in-one" species.[164,280]

Almost as famous as its marsupials are Australia's monotremes—those extraordinary egg-laying mammals comprising the platypus and the echidnas (spiny anteaters). One species of echidna, moreover, is native to New Guinea—Bruijn's long-nosed echidna *Zaglossus bruijni*. Long deemed to constitute a single, distinctive species, in 1998 it was formally split into three separate species via a statistical and non-metric analysis conducted by zoologists Dr. Tim Flannery (now director of the South Australian Museum) and Prof. Colin Groves of the Australian National University. Two of these are *Z. bruijni*, occurring west of the Paniai Lakes, and *Z. bartoni*, occurring in the central cordillera between the Paniai Lakes and the Nanneau Range, plus the Huon peninsula.[281]

The third is a newly-established species dubbed *Zaglossus attenboroughi* (after British television naturalist David Attenborough). The smallest *Zaglossus* species, it is presently known only from its type specimen, collected in Western new Guinea's Cyclops Mountains during 1961, and is possibly extinct. Having said that, in 2007 a month-long expedition to the Cyclops Mountains by a scientific team from the Zoological Society of London's Evolutionarily Distinct and Globally Endangered (EDGE) programme observed burrows and tracks possibly produced by this species. They also learnt from native tribesmen here that sightings of the animal itself had been made as recently as 2005.[281a]

KEEPING UP WITH THE COATIS

Zoologists have traditionally recognised four species of coati—those long-nosed, lengthy-tailed relatives of the raccoons. Three are 'true' coatis, classed within the genus *Nasua*, but the fourth occupies a genus to itself, *Nasuella*. Known as the mountain coati *Nasuella olivacea*, it has a stockier body and a thicker, shorter tail than the true coatis, and occurs only in Columbia.[282]

On 19 October 1999, however, in a personal interview conducted by *Exotic Zoology* editor Matthew Bille, Peruvian zoologist Dr. Peter Hocking revealed that he had recently discovered a new, second species of mountain coati. In 1998, a farmer gave him a dead specimen of a strange-looking mountain coati that derived from cloud-forests in the Peruvian state of Apurimac. And in 1999, Hocking found two more specimens, but these were both alive—living in a local zoo. At the end of 1999, he was working with mammalogist Dr. Victor Pacheco from Lima's San Marcos University, but to date the supposed new species that these three specimens represent has still not been formally named and described.[282]

WEASELING OUT AN INCA SURPRISE

Staying in South America: at the end of 1999, mammalogist Dr. Louise H. Emmons from the Smithsonian Institution formally described a significant new species of rodent, based upon a specimen that in April 1997 had been quite literally placed at her feet—by an Andean weasel. While conducting a research trip to a hitherto-unexplored region of Peru's Vilcabamba Mountains, at an altitude of over 2000 ft, she unexpectedly encountered an Andean weasel that had just killed a large rat-like rodent. Startled by her presence, the weasel dropped its prey and fled, and when Emmons examined the dead rodent she realised that she had never seen anything like it before. Subsequent research back at the Smithsonian Institution confirmed that it was wholly new to science—a member of the abrocomid (chinchilla rat or chinchillone) family of rodents, but sufficiently distinct from all previously-recorded members to warrant the creation of a brand-new genus for it.[283]

Arboreal, powerful in build, weighing approximately 2 lb, and equipped with large claws, the new species is pale grey in colour, with a white streak running from its head to its snout. It has been formally dubbed *Cuscomys ashaninka*, the Asháninka arboreal chinchilla rat, by Emmons, honouring the ancient Inca town of Cusco, which is close to where its type specimen was hastily deposited by the weasel, and also the region's native Asháninka people.[283]

Emmons believes that the closest relative of *Cuscomys* may be a reputedly long-extinct species of large arboreal abrocomid known as the Inca tomb rat *Cuscomys oblativus* (formerly *Abrocoma oblativa*)—because it was apparently kept as a pet by the Incas in nearby Machu Picchu (northwest of Cusco). Some Inca tomb rats have been found inside human tombs within this city, suggesting that they may have been deliberately buried with their deceased owners. Moreover, the wholly unexpected discovery in this area of a new mammalian species as distinctive as *Cuscomys* has led Emmons to speculate that perhaps the Inca tomb rat may not be extinct after all—that it too might still survive here, awaiting formal rediscovery by future expeditions.[283]

COIMBRA'S TITI—ANOTHER BRAZILIAN REVELATION

Still in South America: just as they began, the 1990s ended with the news that this continent had revealed to scientists yet another new species of monkey. A new species of titi this time, it is distinguished from others by its zebra-striped lower back as well as its black brow and ears, and was discovered, remarkably, in a region of northeastern Brazil's Sergipe State that has been almost denuded of trees by deforestation. It has been named *Callicebus coimbrai*, Coimbra's titi, in honour of Brazilian primatologist Dr. Adelmar Coimbra-Filho.[284]

SPOTTING SOME CIVETS

In 1997 (*not* 1986, as stated in some media reports), a new, strikingly handsome species of civet was described by a team of Russian and Vietnamese scientists. This was based upon a series of skins and skeletons, some obtained back in the late 1950s and early 1960s in Tonkin and Annam, Vietnam. Named *Viverra tainguensis*, the Tainguen civet, after the Tay Nguyen

(Tainguen) Highlands where some of its material had been collected, its pelage superficially resembled that of the somewhat larger Indian civet *V. zibetha*. However, the Tainguen civet's bold stripes and spots were more striking, and its skeleton possessed a number of distinguishing anatomical characteristics.[285]

No living specimens, however, had been sighted, and nothing more was heard about this elusive species, until April 1999. That was when the Cambridge-based conservation charity Fauna and Flora International released to the world's press an excellent photo of a live Tainguen civet, vividly depicting its eyecatching pelage markings and boldly banded tail, which had been recently snapped by their infra-red photographic equipment in the Na Hang Nature Reserve in Tuyen Quang province, northern Vietnam. The civet's identity had been confirmed by two members of the original team that had described this species—Dr. Pham Trong Anh from Hanoi's Institute of Ecology and Biological Resources, and Dr. Viatcheslav Rozhnov from Moscow's Severtsov Institute of Ecology and Evolution. Ironically, the photographic equipment responsible for snapping this significant picture had been set up not to seek evidence for the Tainguen civet's existence in the reserve, but rather to confirm the survival here of the extremely rare Tonkin snub-nosed monkey *Rhinopithecus avunculus*. More recently, DNA research has shown that this mysterious civet should be reclassified within the painted civet subspecies of the Indian civet, i.e. as *V. z. picta*.[285]

Worth noting here is that two specimens of an unusual, strangely-coloured civet initially believed to be another hitherto-undescribed species and originating from Lao Cai province in the far northwest of Vietnam were documented in 1995 by Dr. Pham Nhat from the Forestry University in Xuan Mai. In an e-mail of 7 November 1999, however, Dr. Vern Weitzel of the United Nations Development Programme based in Hanoi informed me that these skins were actually from a known civet species, but had been unusually prepared, causing their fur to change colour.[286-7]

And still on the subject of strange southeast Asian civets: in June 1986, tropical agriculturalist Tyson Hughes visited the little-explored Indonesian island of Seram (Ceram) in the Moluccas group. During his work there for the VSO, he received reports from the local inhabitants concerning several unidentifiable beasts. The most tangible evidence obtained by him for any of these, however, was the furry tail of a mysterious mammal, about 18 in long and encircled by a series of dark rings. According to the native people, it was from a beast that was half-dog and half-cat, and when he presented it to the WWF's Indonesian branch for examination, officials there suggested that it might be from an unknown cat. Judging from the natives' claim regarding the creature from which this tail had originated, however, a viverrid, specifically a civet, may be a more likely identity, as their description is reminiscent of the giant palm civet (see p. 78), indigenous to the nearby island of Sulawesi.[287]

Moreover, in 2009, a reclassification of Sri Lanka's golden palm civet *Paradoxurus zeylonensis* created three additional species, hitherto dismissed as mere morphs. These are: the golden wet-zone palm civet *P. aureus*, the golden dry-zone palm civet *P. stenocephalus*, and the Sri Lankan brown palm civet *P. montanus*.[287a]

... AND A RABBIT WITH STRIPES

It is both fitting and perhaps inevitable that the last new mammal to be documented in this section of the present book occurs in Vietnam—the country that has yielded more remarkable mammalian revelations than any other during the closing years of the 20th century.

Between December 1995 and February 1996, the carcases of three freshly-captured specimens of a mystifying striped rabbit were observed on sale at a meat market in Ban Lak, a rural town in Laos, by Laos-based British conservationist Robert Timmins (see also p. 279). The rabbits were very distinctive in appearance, with short ears and tail, a red rump, and dark brown or black stripes on their face and back. The only previously-known species of striped rabbit was *Nesolagus netscheri*, an extremely rare species found only on the island of Sumatra (more than 1000 miles south of Ban Lak), and whose stripes were longer. Consequently, Timmins lost no time in bringing this hitherto-unknown Laotian version to the attention of rabbit expert Dr. Diana J. Bell and geneticist Dr. Alison K. Surridge, both based at the University of East Anglia in Norwich.[288]

Examination of dead specimens and analysis of their DNA by these researchers confirmed that the Laos striped rabbit was a species new to science, which in 2000 was duly christened *Nesolagus timminsi*, honouring its discoverer. As announced in a *Nature* paper of 19 August 1999, their studies also revealed that although related to the Sumatran striped rabbit, the Laotian version has been separated from it for at least 8 million years. Moreover, workers in Vietnam's Pu Mat Nature Reserve, in the Annamite Mountains on the Vietnam-Laos border, subsequently photographed living specimens of the new striped rabbit—now known as the Annamite striped rabbit—using automatic cameras, thereby demonstrating that it existed both in Laos and Vietnam. And in early 1998, automatic cameras placed by Fauna and Flora International in the Kerinci Seblat National Park confirmed the continuing existence of the Sumatran striped rabbit too—a species so rare that it had only been spied in the wild once since 1916 and had last been collected in 1929.[288]

Male Congo peacock. (A. J. Haverkamp)

Male Congo peacock. (A. J. Haverkamp)

THE BIRDS

Resurrected Takahes and Peacocks From the Congo

In 1948, a discovery was made in New Zealand that shook the ornithological world out of its usual comatose condition in an incredible manner—no less than the discovery (or rediscovery) of a bird that had vanished, a bird that had, for the last fifty years, been believed to be extinct. It was, to give it its full title, the Notornis or Takahe (*Notornis mantelli*), and the whole history of this bird is one of the most fascinating in the annals of ornithology.

GERALD DURRELL—*TWO IN THE BUSH*

The story of the discovery of this bird begins with a single feather which I found on the hat of a native at Avakubi in the Ituri Forest in 1913. It was a secondary wing-quill, rufous with regular blackish barring, and from its form and texture seemed to be that of some gallinaceous bird. But after comparison with feathers of many birds of that group, I found myself unable to identify it, nor could any of the many friends to whom I showed it. So I laid it away in safety, but never forgot it. . . .

DR. JAMES P. CHAPIN—'THE CONGO PEACOCK',
In: COMPTE-RENDU DU IX CONGRÈS ORNITHOLOGIQUE INTERNATIONAL

Some of the 20th century's most celebrated zoological debutantes and restorées have been of the feathered variety. These have featured in such ornithological revelations as the finding of a veritable phalanx of flamboyant new pheasants, and the debut of a bewildering passerine from Peru that continues to defy all attempts to classify it; the reappearance of a trio of long-lost Australians all within the space of just a few months, as well as the most welcome revival of several 'classic' extinct species; the long-hoped-for (but little-expected) return of a multicoloured wanderer from New Zealand, and the astounding resurrection of a Bermuda seabird previously believed extinct for more than three centuries; plus the unfurling of a zoogeographically-anomalous partridge-lookalike in Tanzania, the successful pursuit of a very special Brazilian macaw missing for over 120 years, and the extraordinary history of a pair of dusty taxiderm specimens, neglected and forgotten for years, which became the greatest ornithological discovery of the century.

FEARFUL OWL—SMALL BUT SUFFICIENT

The most spectacular avian discovery during the 19th century's closing years was *Pithecophaga jefferyi*, the Philippine or monkey-eating eagle, native to four of the Philippine islands. A magnificent but nowadays greatly endangered, harpy-reminiscent species with a mighty 7 ft wingspan, it was discovered on the island of Samar by English naturalist John Whitehead, and was formally described in 1896 by William Robert Ogilvie-Grant[1]—but was not seen in living form beyond the Philippines until 1909, when a live specimen was exhibited at London Zoo.

Compared to such an impressive species as this, the Solomon Islands' fearful owl *Nesasio solomonensis*, one of the most interesting new birds to be discovered at the beginning of the 20th century but with a total length of

The Philippine eagle.

only 15 in, may not seem particularly noteworthy on first sight. Certainly, its barred upperparts, wings, and tail, and its paler, creamy underparts, are rather bland; only its thick, startlingly white eyebrows are likely to attract a second glance. However, first appearances can often deceive.[2-3]

Viewed more closely, its beak is seen to be disproportionately powerful and sturdy for such a modest-sized owl, and its talons are also exceptionally formidable. The explanation for these unanticipated features seems to be that the fearful owl, despite its small size, is the Solomon Islands' ecological counterpart to the generally huge, exceedingly powerful eagle owls found elsewhere in the world. In keeping with this status, it preys upon such sizeable creatures as possums and large birds, thereby fully meriting its common name.[2-3]

The only member of its genus, the fearful owl was first discovered on the island of Santa Isabel (Ysabel), by Albert S. Meek, but it also occurs on Choiseul and Bougainville. Formally described in 1901 by Ernst Hartert, it inhabits low and hill forests.[2-3]

ROTHSCHILD'S PEACOCK PHEASANT— AN UNEXPECTED 'MISSING LINK'

Despite their name, peacock pheasants are not hybrids of peacocks and pheasants. In fact, they constitute a genus, *Polyplectron*, of very distinctive grey or brown pheasants, whose males possess a resplendent peacock-like tail, liberally adorned with mirror-resembling eye-spots (ocelli), which they raise and spread open during their mating displays.

By the end of the 19th century, five species were known to science, and ornithologists were confident that there were no more of these large, showy species still awaiting detection. The error of this assumption was decisively laid bare in January 1902, when a team of Bornean hunters employed by animal collector John Waterstradt obtained six peacock pheasants (four males, two females) that did not belong to any species already known to science. Procured in the State of Ulu Pahang within the central Malay Peninsula, they proved most interesting, as their species appeared to be transitional between the most primitive species of peacock pheasant and the four more advanced ones.[4]

The most primitive, the bronze-tailed peacock pheasant *P. chalcurus*, has a long pointed tail, and lacks the reflective ocelli decorating the tail and wings of the other species—whose tail feathers, moreover, are broad with rounded tips. Accordingly, the bronze-tailed was initially placed within its own genus, as *Chalcurus chalcurus*. As for the new species, this became known as Rothschild's peacock pheasant in honour of Lord Walter

Rothschild (a prominent figure in early 20th century zoology). However, whereas its males possessed the long pointed tail characteristic of the bronze-tailed, their tail and wings were embellished with the ocelli exhibited by the four advanced species.[5] So how was it to be classified?

When Lord Rothschild described this new species in 1903, he deemed the shape of its tail to be more significant than the presence of fully-formed ocelli, so he christened it *Chalcurus inopinatus*—its specific name recalling both its discovery and its equally unexpected status as a 'missing link' in the peacock pheasant lineage's evolutionary progression. Eventually, however, this latter attribute became obscured, on account of a name-change. Taxonomists ultimately decided that all six species (by 2010, this had become eight via the taxonomic splitting of certain existing species) were sufficiently closely related to be included within the single genus *Polyplectron*, so Rothschild's species is known today as *P. inopinatum*.[4-5]

Notwithstanding their striking plumage and chicken-sized build, peacock pheasants are not easy to observe in the wild, because their favoured habitat consists of dense, inaccessible jungle, where they seek cover in the most impenetrable vegetation at the slightest provocation—thus explaining the success of *P. inopinatum* at eluding scientific detection for so long. In fact, it is possible that there is at least one more species *still* awaiting scientific discovery. The *alovot* is a seemingly undescribed pheasant-like bird that allegedly dwells only within the densest forests of Sumatra. Those who have been fortunate enough to glimpse it assert that it is chicken-shaped and also chicken-sized, with a comb-like crest in some cases (present only in the males?) and dark brown plumage dappled with lighter spots. This description compares with certain of the more advanced species of peacock pheasant.[6]

WAKE ISLAND RAIL—DESCRIBED AND DESTROYED IN UNDER 45 YEARS

One of the lesser-known casualties of World War II, this species has the tragic distinction of having been described and destroyed all within the first half of the 20th century.

It was a diminutive, flightless, bar-breasted member of the rail family—a multifarious assemblage of birds that includes such familiar species as the moorhen, coot, water rail, and corncrake. *Gallirallus* [originally *Hypotaenidia*] *wakensis* was described in 1903 by Lord Rothschild from a series of ten specimens obtained by a Japanese vessel in 1892 on Wake Island, a tiny mid-Pacific dot of land that constituted this species' only known home.[7] Nevertheless, it seemed to be in no danger of extinction, and attracted little attention—until 1941, when its island was occupied by a Japanese battalion during World War II.

Wake Island rail. (Errol Fuller)

Food became ever scarcer on Wake Island as the conflict progressed, year after year, until the soldiers faced outright starvation. Inevitably, they ate whatever they could find—including the defenceless rails. By the close of 1945, every single bird had been killed and eaten. World War II had ended—but so too had the life of an entire species.[8]

MIKADO PHEASANT— BEAUTY IN BLUE FROM FORMOSA

One of the world's most beautiful pheasants was unknown to science prior to 1906. In that year, British ornithologist Walter Goodfellow noticed two very long and attractive tail feathers, deep steely-blue in colour and banded across with thin white stripes, in the head-dress of a hill-native from Formosa (now Taiwan). They clearly belonged to some type of pheasant, but none with which Goodfellow was familiar. Upon questioning the native, he learnt that they had been obtained from a bird that the native had killed on the island's Mount Arizan. Goodfellow was able to purchase the feathers, which, as he had suspected, did not match those of any pheasant recorded by science. Hence they became the type of a new species, described later in 1906 by William Robert Ogilvie-Grant, who named it *Calophasis mikado*, the mikado pheasant.[9]

Mikado pheasant. (World Pheasant Association)

Not long afterwards, Goodfellow succeeded in obtaining a whole skin, of a female mikado pheasant, on the Racu-Racu Mountains. Other complete specimens, females and males, were subsequently procured too, and in 1907 Lord Rothschild produced the first scientific description of the entire bird. Later studies revealed that this ornate inhabitant of Taiwan's steep conifer-bearing slopes was closely related to the long-tailed *Syrmaticus* pheasants, so in 1922 William Beebe renamed it *Syrmaticus mikado*, thereby allying it with such comparably elegant species as Elliot's, Humes's bar-tailed, and the copper pheasant.[5,10]

As with most pheasants, the female mikado is fairly drab, its plumage composed of soft, nondescript shades of brown and fawn. In contrast, the male is extremely eyecatching. Much of its head and body shares the dark blue colour of its tail, which at 20 in or so accounts for just over half of its total length of 3 ft. Intruding upon this predominantly blue colour scheme, however, are two white bands and numerous shorter stripes on each wing, white banding across its tail, an expanse of deep metallic green on the lower part of each shoulder, black spotting on its breast and back, and a bright red patch of skin around each eye.[5,10]

Confined solely to Taiwan and hunted widely in the past by this island's natives, the mountain-dwelling mikado is one of the rarer pheasant species. Fortunately, it breeds well in captivity, thereby offering a means of perpetuating it, and of supplementing its wild populations via reintroduction programmes using captive-bred birds, which has proven successful in recent times.[5,10]

African green broadbill.

AFRICAN GREEN BROADBILL— AN OUT-OF-PLACE ODDITY

In 1908, at an altitude of roughly 6670 ft in the Itombwe Mountains of what is now the Democratic Congo (formerly the Belgian Congo, and later Zaire), bird collector Rudolf Grauer collected a very pretty, principally green-plumaged bird with bright blue cheeks, throat, and chest. Grauer's discovery came as an appreciable surprise to ornithology, for although his bird was first thought to be an aberrant flycatcher, it proved to be a hitherto unknown species belonging to the taxonomic family Eurylaimidae. This family's members are primitive perching birds popularly known as broadbills, which were previously believed to be exclusively Asian in distribution. In honour of its discoverer, and in recognition of its superficial similarity to the green *Calyptomena* broadbills from the southeast Asian Sunda Isles, in 1909 Lord Rothschild named it *Pseudocalyptomena graueri*, the African green broadbill.[11-12] Bird enthusiasts were naturally most interested in this African anomaly, and eagerly awaited further news of it—but none arrived. It would be another 20 years before a second specimen appeared.

One morning in 1929, after vainly seeking this 'lost' species day after day during an American expedition to the Itombwe Mountains, Alan Moses was sitting down, taking a rest from his search and despairing of ever sighting his elusive quarry. Luckily, he happened to glance up from his gloomy reverie just at the right moment—for one of these mysterious, almost magical African green broadbills was perched on the branches just above his head! He was unable to catch it, but he now knew that its species did still exist, and by the end of the expedition a few specimens had been acquired, verifying its rediscovery.[12]

In 1967 it was discovered in Uganda, near the Bwindi Swamp in the Impenetrable Forest, and a population has also been revealed in the mountains west of the Democratic Congo's Lake Kivu, but it remains an endangered species and is categorised as vulnerable by the IUCN.[12]

More recently, *P. graueri* has lost its unique status as Africa's only species of broadbill—and in a rather unexpected way. The genus *Smithornis* contains three African species of dull, brown-coloured bird, which were all known to science before *P. graueri*, but were assumed to be flycatchers—until they were closely re-examined during the 1960s, whereupon they were found not to be flycatchers at all, but broadbills.

Even more remarkable is the very recent DNA-driven revelation that the broad-billed sapayoa *Sapayoa aenigma*, a small olive-plumaged species native to lowland rainforests in Panama and northwest South America, originally housed in an exclusively New World taxonomic family of perching birds known as manakins, and formally described by Ernst Hartert in 1903, is not a manakin at all but seems instead to be most closely related to the exclusively Old World family of broadbills![12a]

ROTHSCHILD'S MYNAH— BALI'S ONLY ENDEMIC BIRD

On 24 March 1911, while participating in the second Freiburger expedition to the Moluccas, avian expert Dr. Erwin Stresemann collected an adult female of a quite exquisite species of crested starling at Bubunan, on the northern coast of Bali. Except for the black edge to its tail and its black wing tips, its plumage was an immaculate snowy white. In contrast, its unfeathered legs and feet were pale grey, its bill was brownish-yellow, and a conspicuous patch of bright blue skin encircled each eye. A new species, related most closely to the mynahs, it was unique to Bali—moreover, it is this island's *only* endemic bird.[13]

In 1912 it was officially described by Stresemann, who created a new genus for it, and named it *Leucopsar rothschildi*, as a token of his gratitude to Lord Rothschild for permitting him to spend such a considerable time during his ornithological researches at the magnificent natural history museum at Tring, founded and owned at that time by Rothschild.[13]

Rothschild's mynah. (Dr. Karl P.N. Shuker)

The highly attractive appearance of Rothschild's mynah ensured its rapid rise to fame as a popular cage bird, but by being restricted to such a tiny island as Bali (no more than about 2000 square miles in area, of which only the Bubunan portion is inhabited by the mynahs) it is, unavoidably, a species with a small population size. As a result, the depletion of its numbers in the wild by local trappers supplying birds to zoos, aviculturalists, etc, ultimately transformed it into an endangered species— so much so that in 2001 the wild population had reached an all-time low of just six birds and is categorised as critically endangered by the IUCN.[14]

Happily, however, it breeds well in captivity, enabling its numbers to be built up. At Jersey Zoo, for instance, 185 mynahs had been bred by mid-1990 from an original group of just four. Conservationists hope that captive-bred specimens released onto Bali will boost the wild population to its former level, before this mynah became a popular aviary species.[15]

CRESTED SHELDUCK—DOES IT STILL SURVIVE?

The crested shelduck *Tadorna* [originally *Pseudotadorna*] *cristata* is a renowned mystery bird that might well have become known to western zoologists much earlier than the 20th century if they had paid more attention to oriental works of art—because this extremely attractive and unmistakable species was frequently portrayed in Chinese paintings and tapestries, as well as in antiquarian Japanese illustrations and tomes dating back to the early 18th century. Obviously it must have been common in those days, but by the time that it had first engaged scientific interest it was virtually extinct.[10]

The first specimen recorded by science was a female obtained in April 1877 near Vladivostok, but this was dismissed by zoologist Philip Sclater as a hybrid of the falcated teal *Anas falcata* and the ruddy shelduck *Tadorna ferruginea*. In the closing weeks of 1913 (or 1914, according to some accounts), a pair of specimens was collected at the mouth of western Korea's Kun-Kiang River (the female was later lost), followed in December 1916 by a female obtained on Korea's Naktong River. By now, this unusual waterfowl had attracted formal scientific interest at last, and the Naktong River bird became the species' type, officially described and named in 1917 by Nagamichi Kuroda. In summer 1924, a second male was collected, once again at the mouth of Korea's Kun-Kiang River.[10,16-17]

The male of this species is particularly striking, by virtue of its predominantly tricoloured plumage. Its chin and crown, lower neck and upper back, breast, tail, and wing flight feathers are dark greenish-black, as is its long, diagnostic drooping crest. These all contrast sharply with the grey, finely-barred plumes of its face, upper neck, lower back, and underparts; and also with its rufous flanks and shoulders—slightly darker than its deep pink feet and bill. The only intrusion upon this three-shaded colour scheme is the whiteness of its wings' covert feathers. The female is basically a washed-out, faded counterpart of the male in terms of colouration, with the addition of a distinctive white ring around each eye.[16]

Except for the lost Kun-Kiang female, all of the above specimens were preserved (unlike, tragically, a trio of individuals obtained in northwestern Korea in 1917), which is just as well—because no additional example has ever been procured since. Much more alarming,

A pair of crested shelduck (male is on the left).

however, is the extreme scarcity even of *sightings* since 1924. In late March 1943, an alleged sighting of two individuals was recorded from South Korea's Chungchong Pukto Province. More than 20 years later, on 16 May 1964, V.I. Labzyuk and Yu. N. Nazarov saw three birds, a male and two females, associating with a flock of harlequin ducks on a rocky island southwest of Vladivostok in the Rimski-Korsakov archipelago. The drake's characteristic tricoloured plumage was readily visible, and easily distinguished from the harlequins' equally distinctive appearance. This (or another) specimen was also seen with a female a few days later, on a nearby island, enhancing the report's veracity and significance.[10,17]

Of particular note was a sighting that featured no less than six of these birds (including two males), spied at the mouth of North Korea's Pouchon River in March 1971. Two birds were reported from eastern Russia in 1985. Moreover, in recent times a number of unconfirmed sightings in northeastern China were reported during a survey of Chinese hunters, and a Chinese forestry worker claimed that he had unknowingly eaten two such birds in 1984. All of this offers tentative hope to ornithologists that the exotic crested shelduck does indeed still survive in viable numbers[17]—a species that might easily have been bred in captivity and thereby granted a secure future, if only science and the western world had recognised its existence and its validity as a genuine species a little earlier.[10,17]

INACCESSIBLE ISLAND RAIL— A FLIGHTLESS BIRD FROM ATLANTIS

Inaccessible Island is a tiny islet of the Tristan da Cunha group, sited in the south Atlantic roughly midway between southern Argentina and South Africa, and would have little claim to fame, were it not for a very peculiar member of its avifauna. The species in question is a miniscule rail, only 5 in long (little larger than a newly-hatched chicken), and with such tiny, poorly-formed wings that it is totally flightless, making it the world's smallest extant species of flightless bird. Its habit of scampering swiftly through the island's wide expanses of dense tussock grass thus makes it seem to the casual observer more akin to a mouse than to a bird. This illusion is enhanced by its strange feathers, which are decomposed (i.e. atrophied) and hair-like. Its upperparts

Inaccessible Island rail. (Dr. Ian Best/Denstone Expeditions Trust, courtesy of Michael Swales)

are reddish-brown, its underparts are dark grey, and its belly, flanks, and wing-covert feathers bear paler bands.[18]

A remote spot, not readily reached, Inaccessible Island was well-named. Due to its inaccessibility, its diminutive rail (found nowhere else in the world) escaped scientific attention until 1923, when the Reverend H.M.C. Rogers, resident chaplain on Tristan da Cunha, collected some skins of it in response to a request made by the Shackleton-Rowett Expedition's naturalist, a Mr. Wilkins. The expedition had visited the island group a little earlier, and had heard the locals speak about the tiny 'island hen' of Inaccessible, but had been unable to travel there to seek it out.[18]

On 5 July 1923, two of the skins collected by Rogers arrived at the British Museum (Natural History), and were described that same year by Percy Lowe, who named the new species *Atlantisia rogersi*. 'Atlantisia' alludes to the belief by some workers that the Tristan da Cunha islands are remnants of the fabled sunken continent of Atlantis. Previously little-studied, in the 1980s *A. rogersi* was the subject of a detailed field survey by South African researchers Drs. M.J. Fraser and W.R.J. Dean, and Dr. I.C. Best from Bahrain.[18]

TWO PHEASANT FINDS IN INDO-CHINA

During an expedition to Indo-China in 1923, aviculture expert Dr. Jean Delacour rediscovered a previously obscure species of blue pheasant, when, aided by ornithologist Pierre Jabouille, the Resident of the Quangtri Province (in what is now Vietnam), he induced the natives to capture no fewer than 22 living specimens of Edwards's pheasant *Lophura edwardsi* in the province's back hills. Until then, this species had been known to science from just four skins, sent in 1895 to the Paris Museum from Quangtri by French missionary Père Renauld.[5,19]

Delacour's specimens established Edwards's pheasant in captivity in the west, where it has been bred successfully, ensuring its survival—which is just as well. Following the collection of a single live bird in 1928, this reclusive species, recorded only from Vietnam, eluded all attempts to locate further specimens, prompting many ornithologists to fear that it had become extinct in the wild. Happily, however, this was not the case, as confirmed in August 1996 with the capture of a mating pair by villagers in central Vietnam's Bach Ma National Park, who took them to the Forest Guard Station in Phong Dien district.[5,19]

Yet even more remarkable than his rediscovery in 1923 of Edwards's pheasant was Delacour's discovery, during the very same expedition, of a completely new pheasant—when a living pair of birds ostensibly belonging to a totally unknown species was sent to him from

Imperial pheasant. (Jean Howman)

the limestone mountains of Donghoi and northern Quangtri, in northern Annam, on the Vietnam-Laos border. These two specimens were transported safely back to France, where they bred, eventually giving rise to a large number since distributed to aviaries and parks worldwide—all of which is very fortunate, because no other living specimens of this species obtained in the wild have ever survived the journey back to the west.[5,19] In 1990, an immature male specimen was trapped in the wild by a rattan collector, and a second was captured in February 2000.

Described by Delacour and Jabouille in 1924, it was originally named *Hierophasis imperialis*, the imperial pheasant, but was later renamed *Lophura imperialis*, stressing its affinities with the *Lophura* species—known collectively as the gallopheasants ('chicken pheasants'), and including such attractive species as the silver pheasants, firebacks, and kalijs.[5,19]

Up to 30 in long, with a wide tail and short crest, the male imperial pheasant is a dark, midnight coalescence of black and very deep blue, encroached upon only by a bright scarlet, twin-lobed wattle on each side of its face, crimson-coloured legs, pale bill, and brown-tipped wings. The latter features are shared by the female, but she lacks the male's blue-black colouration, replacing it with chestnut-brown shades above, becoming lighter below.[5,19]

The extreme rarity of this mysterious pheasant had long puzzled aviculturalists, but it was not until 2003 that the remarkable solution to this riddle was finally revealed. In that year, a detailed paper was published in which its authors announced that morphological comparisons, hybrid experiments, and DNA analyses had collectively confirmed the imperial pheasant not to be a valid species at all. Instead, it was a rare but naturally-occurring hybrid of two other Vietnamese gallopheasants—Vo Quy's pheasant *L. hatinhensis* (see p. 136) and the silver pheasant *L. nycthimera*.[19a]

BAKER'S 'LOST' BOWERBIRD

In 1928, Rollo H. Beck, a bird collector from the American Museum of Natural History, but with little experience of New Guinea's avifauna, struck a decisive blow for gifted amateurs everywhere. Upon his return home from a collecting trip to that island's Mandated Territory, he presented the museum with three specimens of a thoroughly resplendent—and wholly new—species of bowerbird. Of comparable shape and size to the common starling, its principally black plumage was set aflame by a fiery-feathered cape flowing over its shoulders, a bright golden-yellow band across each wing, and a vivid scarlet crown above its ebony-plumed face. According to Beck, the three specimens had been collected in the vicinity of Madang, a coastal town in the Adelbert Mountains' Astrolabe Bay.[20]

The following year, their species was officially described, and named *Xanthomelus* [later renamed *Sericulus*] *bakeri*, Baker's regent bowerbird[20]—after which it vanished entirely. For 30 years, searches within its type locality failed to find a single specimen of this magnificent new bird, so that it became one of the greatest enigmas in modern-day ornithology.

In 1956, however, Prof. Thomas Gilliard, a major authority on bowerbirds and birds of paradise, set out to discover its secret homeland. After establishing that it did *not* exist near Madang, he learnt from Beck's widow that in actual fact the three specimens collected by him had been obtained at a spot some distance away from Madang, to the west. And sure enough, when Gilliard arrived there, he sighted several birds, the first examples of Baker's regent bowerbird seen in the wild by a scientist since its discovery back in 1928.[21]

RETURN OF THE RELICT GULL

On 24 April 1929, a single individual of what seemed to be a new subspecies of the Mediterranean gull *Larus* [now *Ichthyaetus*] *melanocephalus* was collected by K.G. Söderbom at Tsondol on the Etsin River of northern Inner Mongolia. Two years later, it became the type specimen of *L. m. relictus*, the relict gull. For many years afterwards, the Etsin River bird was also its *only* specimen—despite several searches by ornithologists, no other example of this new gull was found. (Ironically, a second specimen *had* been collected, on 9 April 1935 in China, but remained unrecognised for what it was, a relict gull, in the Zoological Institute of Leningrad's Academy of Science until belatedly identified in 1971.)[14,22]

As a result, some authorities eventually suggested that the relict gull was not a separate taxonomic form at all, but merely a hybrid between two already-known species—the great black-headed gull *L.* [now *Ichthyaetus*] *ichthyaetus* and the brown-headed gull *L. brunnicephalus*. During the early 1960s, however, this identity was disproved, and the mystery of the relict gull finally solved, when it was rediscovered in small breeding colonies. In 1967, around 100 pairs were nesting on Baroon Torey Nor, one of Transbaikalia's series of Torey lakes.[14,22]

Nowadays classed as a full species in its own right as *Ichthyaetus relictus*, the relict gull has since been found to breed on Kazakhstan's Lake Alakul too as well as in several Mongolian localities and one in China, with a total population estimated at no more than 10,000 birds, resulting in its categorisation as vulnerable by the IUCN. Specimens have been collected from as far afield as this species' Etsin River type locality, the Yellow Sea's coastlands, and northern Vietnam.[14,22]

GIANT PIED-BILLED GREBE—
HOW TOURISM TERMINATED THE POC

The grebes constitute a taxonomic order of primitive aquatic birds whose members include such well-known European species as the dabchick and the great-crested grebe. In America, one of the most familiar species is the common pied-billed grebe *Podilymbus podiceps*, and in 1929 ornithologist Ludlow Griscom discovered that it had a giant-sized flightless equivalent, peculiar to Lake Atitlán in the southwestern highlands of Guatemala. Accordingly, he named it *P. gigas*. Its holotype had been collected as long ago as 1862, but had not been considered to be anything special, so it had remained undescribed until Griscom had chanced upon it in the Dwight collection of bird specimens at the United States National Museum.[23]

Once numbering up to 100 pairs, since the 1960s the giant pied-billed grebe's story has been one of calamitous, unrelenting decline, its numbers dwindling remorselessly year by year. This is principally due to a voracious species of fish called the large-mouthed bass—introduced into Lake Atitlán to attract angling-inclined tourists, but wreaking havoc upon the lake's unique

Giant pied-billed grebe. (Dr. Anne LaBastille)

ecosystem by devouring its supplies of smaller fishes and crabs, the staple diet of the grebes. Depletion of its reedbeds by native Indians who used the reeds to weave mats for another lucrative tourist demand also threatened the grebes (known locally as *pocs*), which nested among the reeds. The outlook for the species' continuing survival seemed grave. Tragically, the reality was even worse.[24]

One of the *poc*'s principal researchers and campaigners on its behalf was Dr. Anne LaBastille, who recognised that the only way to safeguard this species' remaining individuals would be to capture them all and transfer them to a safer locality elsewhere. When a colleague, Dr. Laurie A. Hunter of North Dakota State University, attempted to do this, however, at least a third of the total number *flew* away—an event that filled her with despair, for an extremely significant reason.[24]

The giant pied-bill was *flightless*—therefore, the specimens that had flown away could not have been genuine *pocs*. Instead, they were nothing more than misidentified common pied-bills! Worst of all, when Drs. LaBastille and Hunter investigated the lake's entire grebe population they found that only two or three individuals were giant pied-bills; *all* of the rest were imposters, belonging to the common, flying version, or were hybrids of the latter with the *poc*. In short, while science had been blissfully consoling itself that at least 30 or so of the *poc*, *P. gigas*, still existed, this species had actually declined to the point of no return, undercover of its smaller, more common relative, *P. podiceps*. In 1991, within her definitive book *Mama Poc*, Dr. LaBastille announced that there were no longer any giant pied-bills living. The *poc* was extinct—another irreplaceable species irredeemably lost.[24]

A PRECIOUS PARTRIDGE

The year 1932 saw the description of *Arborophila rufipectus*, the Sichuan hill partridge—a small, sturdy bird with a huge, gaudy, fan-like crest. Limited solely to four counties within southwestern China's Sichuan Province, for the next three decades it was known only from its type, a male. During the 1960s, however, Prof. Li Guiyan obtained a series of specimens, enabling him to prepare the first scientific account of the female's morphology. Yet even today it remains one of China's rarest species of bird, categorised as endangered by the IUCN.[25]

CONGO PEACOCK—
AN ORNITHOLOGICAL DETECTIVE STORY

A head-dress belonging to a native tribesman from the former Belgian Congo (later Zaire, now the Democratic Congo), a pair of old, forgotten museum exhibits set aside from the main collection, and the keen, retentive memory of a dedicated scientist. These were the principal components of a complex detective story spanning more than 20 years, but climaxing in the mid-1930s with what is widely acclaimed to be the most important ornithological discovery of the 20th century.

Congo peacock (male, with female in background). (Alan Pringle)

In 1913, Dr. James Chapin was in the Belgian Congo's Ituri Forest, taking part in an okapi expedition. An ornithologist from the American Museum of Natural History, he had been collecting bird specimens for the museum, and as the region's native hunters wore elaborate head-dresses of feathers he decided to obtain a selection of these too. Most of their feathers were from brilliantly-coloured cuckoo-related birds called touracos, whose species Chapin was readily able to identify. A certain Avakubi head-dress, however, contained one particular feather (*not* two, as many books erroneously claim) that puzzled him a great deal. A secondary wing-quill of rufous background colour, overlain with regularly-spaced black bars, its texture and form suggested to Chapin that this was from some type of gallinaceous bird (i.e. pheasant, grouse, quail), but he was unable to assign it to any known species from the Ituri region. He was so intrigued by it that he bought the complete head-dress and rigorously investigated its mysterious plume's possible origin.[26-7]

In the Congo, Chapin learnt that such feathers came from a bird known locally as the *mbulu*—but he was unable to discover whether or not this name merely referred to a breed of domestic chicken. Back in America at the museum, his assiduous examination of the feather confirmed that it was certainly from a gallinaceous bird, but all attempts at ascertaining the latter's identity met with failure. And so, albeit reluctantly, Chapin carried on with his official work and locked the enigmatic quill away in a drawer. Fortunately, however, he never forgot about it, nor about the mysterious *mbulu* from which it had originated.[26-7]

By 1936, Chapin's continuing interest in the avifauna of the Belgian Congo had ultimately led him to compile a definitive work on the subject, and his researches for it had taken him to the foremost scientific establishment concerned with that country—the Congo Museum at Tervueren, Belgium. It was here, while browsing through its ornithological collection, that he caught sight of two old, dusty, and totally neglected stuffed birds, mounted on a board and standing on top of a tall cabinet, placed aside from the main collection—due at least in part to their poor state of preservation. To Chapin, however, they were the answer to a prayer—because even from a distance he could very readily see that one of these birds, more rufous in colour than its darker partner, sported wing quills identical to the strange plume that he had obtained 23 years earlier in the Avakubi head-dress![26-7]

Needless to say, Chapin immediately made full enquiries concerning the origin and identity of those stuffed birds, and learnt from the museum's director, Dr. Henri Schouteden, that they had been part of a collection of taxiderm specimens kept for some years at the Brussels office of the Kasai Trading Company. According to a faded label attached to their board, these two strange birds were nothing more than imported, immature specimens of *Pavo cristatus*, Asia's familiar blue peacock (thereby providing another reason why they had been kept apart from Tervueren Museum's collection of *native* Congolese avifauna). However, the merest glance at them was more than sufficient for Chapin to recognise that this identification was totally incorrect, and that in reality they did not resemble any species of bird known to science. In a later study of them, Chapin described the pair as follows:[27]

> The blackish individual had large spurs, and was plainly an adult male. Its whole back and rump were blackish, glossed with dull dark green. The base of the neck, chest, lesser wing-coverts, and tips of retrices [tail plumes] had brighter violet reflections. There was an upright crest of narrow black feathers, and just in front of that, in the middle of the crown, a curious patch of short whitish bristles. The rufous bird appeared to be an adult female of the same species, and its back was glossed with brilliant metallic green.

But what could they be? An affinity with pheasants and peacocks seemed most likely from their appearance; but, zoogeographically at least, such a notion must surely be nonsense, as it was well known that pheasants and peacocks did not occur in Africa. Schouteden favoured the possibility that they represented a peacock-chicken hybrid, derived from outside Africa. During lunch with some old Congolese acquaintances a little while later, however, Chapin casually mentioned these puzzling birds and the Avakubi feather, and learnt to his delight that one of his friends knew the *mbulu* well. Furthermore, this friend stressed that it was neither a domestic form nor a hybrid, but was an elusive species that dwelt deep within the most secluded part of the Ituri Forest.[26-7]

This momentous piece of news galvanised Chapin and Schouteden into speedy action—contacting ornithologists worldwide, collating all available information regarding the mysterious *mbulu*, and sending out requests to Congolese people to forward all news of future *mbulu* sightings to the Tervueren Museum. A great deal of fresh data eventually emerged, so that there could no longer be any doubt. The *mbulu* was unquestionably a species new to science, and, in flagrant defiance of all previous concepts, it was a bona fide African pheasant, most closely related to the Asiatic peacocks.[26-7] Consequently, on 20 November 1936 Chapin formally named it *Afropavo congensis*—'African peacock from the Congo' or, more simply, the Congo peacock—with the Tervueren Museum's male taxiderm specimen as its type. All that now remained for him to accomplish was the acquisition of some specimens directly from its native habitat, in order to confirm the anecdotal evidence that he had been amassing.[28]

On 19 June 1937, Chapin set off to Africa to achieve this, but on arriving at Stanleyville he learnt from local vet Dr. T. Els that a pair of these birds had already been obtained at Ayena, not far from Stanleyville, and were now in his possession, preserved via formalin injection. Chapin also received a letter from a correspondent, Dr. Pierre Dyleff, who informed him that four specimens were awaiting his arrival at Angumu. Chapin's persistence had paid off—one small, bewildering feather had led to the discovery of a very beautiful, radically new species, represented by several specimens awaiting formal scientific attention.[26-7]

Interestingly, close examination of these new ones revealed that the male Congo peacock actually bears a vertical tuft of long white bristles, positioned just in front of the small crest of black feathers. In the original, poorly-preserved Tervueren specimen, this tuft of bristles had either been worn down or had broken off at its base, leaving behind only the "curious patch of short whitish bristles" that had puzzled Chapin when he had first seen it.[26-7]

In-depth anatomical studies have confirmed that despite lacking their magnificent, ocellated train, the Congo peacock is truly most closely allied to the Asiatic peacocks, although it seems to reflect a much more primitive stage in peacock evolution. Unlike many of the reclusive denizens of little-explored jungles, this highly significant species has become a familiar sight in the western world over the years, because several have been successfully maintained in various zoos, including those of New York, London, Antwerp, and Rotterdam[26]—a far cry from the days when its very existence was completely unsuspected, with its ambassadors left to gather dust and disinterest on top of a lofty museum cabinet.

CABANIS'S TANAGER—THE CASE OF THE MISSING CAGE-BIRD

The finch-like tanagers of the Americas are among the world's most brilliantly-coloured birds, and hence are very popular cage-bird species. With its grey-blue neck and crown, bright azure back and rump, and blue-bordered wings and tail, a particularly attractive representative of this group is Cabanis's tanager *Tangara cabanisi*, a 6-in-long species named after German ornithologist Jean Cabanis, and once lost for over 70 years. It was first described in 1868, by Dr. Philip Sclater of London's Zoological Society, from a single skin obtained at Costa Cuca, near Quetzaltenango in western Guatemala. Nothing more was heard of this species until 1937, when it was unexpectedly rediscovered on Mount Ovando, Chiapas, in Mexico, by Florida University bird specialist Dr. Pierce Brodkorb. Apart from one other Chiapas sighting in 1943, it vanished again, but during the 1970s small flocks were seen on several occasions at El Triunfo, Chiapas. There are currently believed to be as many as 2500-10,000 individuals in existence, though their numbers appear to be in decline; this species is classified as threatened by the IUCN.[12,29]

Cabanis's tanager—depicted in a painting from 1868.

ZAVATTARIORNIS—A STRANGE STARLING, OR A CURIOUS CROW?

In 1938, the Italian ornithologist Dr. Edgardo Moltoni described a species of bird destined to become one of ornithology's greatest anomalies.[30] Indeed, in more recent years this singular species and the equally extraordinary Congo peacock have been said "to represent the two most remarkable ornithological discoveries made in Africa this [i.e. the 20th] century".[12]

Named *Zavattariornis stresemanni* by Moltoni (honouring Italian zoologist Prof. Edoardo Zavattari and German bird specialist Dr. Erwin Stresemann), it had been discovered only a few months earlier, in the thorn-bush and acacia grasslands at Yavello (Javello) in southern Ethiopia's Sidamo Province, where it builds large dome-shaped nests at the tops of acacia.[30] Subsequent investigations have disclosed that it is entirely confined to an area of just over 2300 square miles, and although once common there, its rapidly-decreasing population means that it is currently categorised as endangered by the IUCN.[12,30-2]

Except for its rather slender starling-like bill, *Zavattariornis* is ostensibly crow-like in basic outline, and jackdaw-sized. Its relatively drab plumage—consisting of white underparts and forehead; grey upperparts, chest, and flanks; and blue-black wings and tail—is quite comparable to an American species of crow known as Clarke's nutcracker *Nucifraga colombiana*, but *Zavattariornis* is distinguished by the bright blue patch of bare skin ringing each eye[12,30-2] (as in Rothschild's mynah, p. 119).

Whereas its general appearance, therefore, is quite unremarkable, studies of its finer morphological and anatomical details have provided some fundamental surprises regarding its taxonomic affinities—or lack of them. Although *Zavattariornis* had been classed by Moltoni as a corvid (member of the crow family) and afterwards became known as the Ethiopian bush-crow, it differs markedly from all other corvids in a variety of different ways. Moreover, certain aspects of its behaviour are more reminiscent of starlings (such as the frequently observed emergence of *three* adult birds from a single nest, indicating assisted rearing of fledglings), as are its parasitic biting lice or mallophagans (and closely-related host species do often possess closely-related parasites). Similarly, it has been spied consorting with various species of starling when feeding (on terrestrial insects).[12,30-2]

Inevitably, therefore, *Zavattariornis* has incited much debate regarding its classification, with avian researchers variously supporting a corvid allegiance, favouring its inclusion within the starling family, or even suggesting that it should be housed within a family all to itself.[31-3] Today it is usually classed as a highly aberrant corvid—a categorisation openly acknowledged, however, to stem more from considerations of convenience than certainty.

RIBBON-TAILED BIRD OF PARADISE—THE TALE OF A TAIL

Not all birds of paradise are large, brilliantly-coloured species. Some, particularly the manucodes, are relatively small with dark plumage, much less conspicuous and

Ribbon-tailed bird of paradise (male). (Errol Fuller)

thus more readily overlooked than their bigger, flamboyant relatives. Yet the most recently discovered bird of paradise was *not* of this smaller, less noticeable, drab type, but was actually one of the family's most spectacular members. Its discovery's history is no less extraordinary.

In his book *Papuan Wonderland* (1936), Jack G. Hides recorded seeing some pairs of a very strange bird of paradise to the west of New Guinea's Mount Hagen. The male form was particularly eyecatching, because its tail constituted just a single pair of incredibly long and slender, ribbon-like feathers, 3 ft in length and creamy-white in colour. These two feathers contrasted sharply with the bird's very much smaller, predominantly dark-plumaged body, and they frequently flicked as they trailed behind it in flight. Not recognising the species, Hides instructed one of the police officers accompanying him on his expedition to shoot a male specimen and obtain its exceptional tail feathers for study.[34]

Learning about that incident, New Guinea explorer-naturalist Frederick Shaw Mayer decided to find out more about this remarkable bird, which did not seem to correspond with any species known at that time. By 1938 he had discovered that it could be found approximately 80-100 miles west of Mount Hagen, and that its amazing tail plumes were often worn in the hair of native tribesmen from this region. In August 1938, a missionary gave him two such feathers obtained from one of these natives, which were subsequently passed to Dr. C.R. Stonor, who designated them as the type specimen of a brand new species. Officially describing it in 1939, he named it *Astrapia mayeri*, in honour of the man responsible for bringing it to scientific notice. As for its common name, its immensely long tail feathers—3.5 times the combined length of its head and body—ensured that it would be referred to ever afterwards as the ribbon-tailed bird of paradise.[21,35]

Shortly after Stonor had received the tail feathers from Mayer, three complete specimens, collected by two explorers in the forests west and northwest of Mount Hagen, were sent to Australian Museum zoologist Dr. Roy Kinghorn. The species has since been recorded from Mount Giluwe too, and living examples have been exhibited at Sydney's Taronga Zoo.[21]

Three other new birds of paradise have also been unfurled during the 20th century. One, closely related to the ribbon-tail, is the Huon astrapia, discovered in the Rawlinson Mountains of New Guinea's Huon Peninsula. Described in 1906 by F. Foerster, who named it *Astrapia rothschildi*, the male of this species, just over 2 ft long, is very beautiful. It has a glossy blue crown; a nape of reddish-copper, transforming into shimmering cerise across the forepart of its back and over the short cape of feathers above its shoulders, but darkening into velvet black overlain with a viridescent sheen upon its back's lower reaches; a transverse band of bronze across its chest, separating its shiny blue-black chin and upper breast from its gleaming green lower breast and abdomen; lustrous black wings; and a very long, broad tail principally black but surfaced with glistening purple.[21,36]

Also native to the Rawlinson Mountains is the Huon six-wired bird of paradise, dubbed *Parotia wahnesi* in 1906 by Lord Rothschild. It is the most recently revealed *Parotia* species—all of which bear a sextet of slender, wire-like feathers, three on each side of their head just behind the eye, and each one terminating in a paddle-shaped racquet.[21,36] And in 1911, a crow-like bird of paradise, with a pair of yellow facial wattles and a smaller pair of blue wattles sprouting from the base of its bill, was described from western New Guinea's Mount Goliath. Due to its very short tail and kinship with an already-known species called the paradigalla, it was named *Paradigalla brevicauda*, the short-tailed paradigalla.[21,37]

ARCHBOLD'S BOWERBIRD—SEPARATED FROM ITS CRESTS FOR ELEVEN YEARS

Named after Richard Archbold (an expert on New Guinea birds), Archbold's bowerbird *Archboldia papuensis* is the most recently described bowerbird species. The only member of its genus, it was discovered in 1939 within the alpine forests of western New Guinea's Oranje Mountains by Austin L. Rand, who described it in 1940. The male specimens collected by him were quite large, principally black-bodied birds, with similarly-shaded heads—which made the male specimen obtained from a Mount Hagen native on 12 June 1950 by Prof. E. Thomas Gilliard very worthy of note. Although apparently belonging to the same species as Rand's birds, it was instantly differentiated by its two striking golden crests—one on its forehead, the other on its nape. Obviously, therefore, Archbold's bowerbird existed in two manifestly distinct subspecies. Consequently, the crestless version represented by Rand's specimens was dubbed *A. p. papuensis*, and Gilliard's double-crested equivalent was christened *A. p. sanfordi* (after Dr. Leonard C. Sanford, a trustee of the American Museum of Natural History). Moreover, some authorities nowadays class these subspecies as separate species in their own right.[21,38]

PRINCE RUSPOLI'S TOURACO AND OTHER ETHIOPIAN ENIGMAS

Touracos constitute a taxonomic family of gaudy-plumaged birds endemic to the African tropics and traditionally categorised with cuckoos. Principally green (occasionally blue) in colour, these crow-sized species are most famous for possessing two plumage pigments (turacin and turacoverdin) peculiar to themselves; no other species in the entire animal kingdom produces either of them.

In 1896, a specimen of a very beautiful but wholly unknown species of touraco was uncovered at the Genoa Museum. It had been obtained close to Lake Abaya in southern Ethiopia, and was one of a collection of birds amassed by Prince Ruspoli of Italy. Befitting its royal connection, its plumage was the quintessence of avian grandeur—its pastel-green neck and underparts combining with the deeper green of its back and wing coverts to yield a viridescent backdrop for its startlingly scarlet crest and wing primaries, its white face and throat, jet-black rump, and vivid purple tail. In honour of its eminent discoverer, the new touraco was named *Tauraco ruspolii*[39]—after which it promptly sank back into ornithological obscurity, because for over 40 years nothing more was heard of this beautiful new species.

Happily, however, Prince Ruspoli's touraco proved merely to be elusive rather than extinct (as had been feared by some authorities), because in 1942 it was rediscovered, inhabiting juniper woodland at Arero, about 60 miles east of Yavello, again in southern Ethiopia. Five specimens were collected by C.W. Benson, and since then further sightings have been reported periodically—but as the species appears to be totally restricted to a plot of juniper woods and evergreen foliage covering no more than 10 square miles, its population size is inevitably small, and it is now categorised as vulnerable by the IUCN.[10]

Other finds made by Benson in southern Ethiopia during that same period included a new species of swallow—the white-tailed swallow *Hirundo megaensis*, whose type specimen was an adult male that he had collected on 10 September 1941, 10 miles north of Mega; and three specimens of the Teita falcon *Falco fascinucha*, collected at Yavello—and hence several hundreds of miles north of this species' type locality of Teita, Kenya (where the only two previously-recorded specimens of this short-tailed raptor had been obtained in 1895).[31]

CAHOW—THE SEABIRD RESURRECTED FROM THREE CENTURIES OF EXTINCTION

In the mid-1940s, a small, unassuming species of seabird called the Bermuda petrel or cahow *Pterodroma cahow* made what must assuredly be the most sensational comeback on record for any bird during the 20th century. Its historic saga began more than 300 years earlier, in 1609, when Bermuda witnessed the arrival of the first permanent population of settlers from Great Britain. As with human settlement throughout the world, an unwelcome stowaway soon accompanied them ashore—the black rat. Its presence in this particular instance was even more disastrous than usual, because it multiplied so rapidly on Bermuda that it literally ate the settlers out of house and home, causing a tremendous famine that may well have resulted in these people's wholesale starvation. This, however, did not happen, because they found a highly valuable native source of food—the affable, ground-dwelling cahow. By 1621, hardly any were left, and even the passing of a specific law to safeguard the few survivors appeared to have come too late. Within a few years, the cahow seemed to have vanished. Almost three centuries were to pass before the first indication of its possible perpetuation emerged.[8,10]

Close to the principal island of Bermuda is a group of tiny islets called the Castle Harbour Islands. While visiting one of these in 1906, L.L. Mowbray collected a petrel that was shown ten years later to be a cahow! This became the species' type specimen—because in spite of its former abundance, the cahow had never been scientifically described or named.[8,10]

Having tantalised ornithologists with this undeniable proof that somehow, against all the odds, it had survived the slaughter waged during the 17th century, the cahow abruptly disappeared again. No additional conclusive evidence of its existence surfaced for almost three decades—until on the evening of 8 June 1935 a young bird was taken to biologist Dr. William Beebe on Bermuda; it had been sent by the lighthouse keeper. A third cahow, which had killed itself by colliding with a telegraph wire on Bermuda, was obtained in June 1941, but there was still no sign of a breeding colony. The young bird from 1935 was proof enough that breeding *was* still taking place—but where?[8,10]

The answer came towards the end of World War II, when in March 1945 the cahow's secret breeding locality was discovered in the Castle Harbour region by ornithologist Frederick Hall, who found it by accident, during the construction there of a United States Air Force base. This fortuitous find was followed up in early 1951 by cahow seekers Robert C. Murphy and Louis Mowbray (L.L. Mowbray's son), who noted 18 pairs.[8,10,40]

After 330 years, the cahow had risen like a maritime phoenix, and has received full protection ever since. Its numbers remain vulnerably low—no more than approximately 250 birds in 2005—but its breeding chances have been greatly improved by introducing it to the ecologically-restored Nonsuch Island in the Bermuda chain, and by the attaching to its nesting burrows of artificial, cahow-sized entrance holes. These permit ready access to the cahows but prevent entrance by the slightly larger red-tailed tropic birds, thereby saving it from the threat of competition with the latter species for nesting sites.[8,10,14]

Just four years after Hall's rediscovery of the cahow as a breeding species, a previously unknown relative was also revealed, when in 1949 Murphy's petrel *P. ultima* was formally described; its type specimen had been collected on the Polynesian island of Oeno.[41] One of the least-known of all seabirds, it is only rarely sighted, and its eggs were spied for the very first time as recently as March 1990. That find was made during a highly successful search for this species, led by ornithologist Peter Harrison, on the uninhabited rock tower of Marotiri, in a southeastern Pacific chain of islands called the Australs. Several hundred pairs of Murphy's petrel were discovered on the tower,[42] but the search had an even more exciting sequel. On the following day, the expedition sailed to Rapa, Marotiri's closest island neighbour, and there Harrison discovered a hitherto undiscovered form of storm petrel. One of the largest known to science and initially thought to be a totally new species in its own right, it proved to be a new subspecies of white-bellied storm petrel *Fregetta grallaria*, and was subsequently christened *F. g. titan*.[42]

TAKAHE—THE NEW ZEALAND WANDERER'S RETURN

The year 1948 closed with the zoological world celebrating another of the 20th century's most spectacular rediscoveries—a large flightless bird with multicoloured plumage, native to New Zealand's South Island, and known as the takahe. Its name is of Maori origin, and translates as 'the wanderer', which is particularly apposite, because the takahe had been wandering in and out of scientific obscurity for more than a century.

In 1847, New Zealand naturalist Walter Mantell excavated some bones at Waingongoro and Wanganui in North Island. They proved to be from a sturdy, flightless species of rail, distantly allied to the familiar moorhen and coots, but most closely allied to a collection of larger, brilliantly-coloured species known as purple gallinules (genus *Porphyrio*) with representatives distributed all over the world. Although previously unrecorded by science, Mantell's rail was well known to the Maoris of North Island, who had hunted it in earlier times and referred to it as the *moho*, but stated that it was no longer seen. Its bones were described the following year by British palaeontologist Prof. Richard Owen, who named this new but seemingly extinct species *Notornis mantelli* (in more recent times, however, it has been renamed *Porphyrio mantelli*, as separate generic status does not seem to be warranted).[43]

In 1849, Mantell's rail reappeared, in an unexpected locality and in a most unexpected way. A *living* specimen was captured—at Duck Cove on Resolution Island, one of the southwestern islets of South Island, by a dog belonging to a seal hunter camped there. Its skin was purchased by Mantell, and was sent to the British Museum; the remainder of it supplied the sealer with an unexpected addition to his menu![8,10,26,44]

It transpired that this species was known to the South Island Maoris as the *takahe*, and unlike its North Island counterpart it evidently still survived. Two years later, another South Island specimen was caught, this time in Deas Cove, opposite Secretary Island in Thompson

The takahe's rediscovery in 1948 was a sensational ornithological triumph. (J.L. Kendrick/Department of Conservation, Wellington, New Zealand)

Sound (and is now housed in New Zealand's Dominion Museum). After this, however, the takahe vanished for 28 years, and was believed extinct, but in December 1879 a rabbit catcher captured one not far from the southernmost end of South Island's Lake Te Anau; it was later purchased by the Dresden Museum (but was 'lost' during World War II).[8,10,26,44]

After making comparisons between specimens of the South Island takahe and the North Island moho, German zoologist Dr. Adolf B. Meyer decided that the two were sufficiently distinct for the takahe to merit separate specific status. And so in 1883, with the Dresden example as its type specimen, he named it *N. hochstetteri*, in honour of explorer Ferdinand von Hochstetter, who had searched for it in vain during its previous 28-year period of 'extinction'.[45] Later studies, conversely, overturned Meyer's ruling, demoting the takahe to subspecific level (i.e. as *N. mantelli hochstetteri*), because the only major difference between takahe and moho appeared to be the slightly longer length of the moho's legs. Indeed, some researchers even deny the takahe subspecific distinction.[8] (Having said that, however, in 1996 Dr. S.A. Trewick published an examination of osteometric data obtained from fossil and modern-day takahe bones in which he concluded that moho and takahe were indeed distinct species; he even claimed that they had evolved from flying ancestors wholly independently of one another, rather than from a common flying ancestor.[45a])

Although the moho was known to science via skeletal remains, its appearance in life was (and still is) unrecorded, but if it was anything like the takahe it must have been quite magnificent—because the takahe certainly is. Standing 1.5 ft high, and of stocky build like a small turkey, its extremely soft, silky plumage is a rich pageant of colour—commencing with dark indigo-violet upon its head, neck, and underparts, transforming into shimmering ultramarine over its shoulders and wings, and deep jungle-green across its back, but mellowing into paler, tawny-olive shades upon its rump and its tail (which also has a tuft of pure-white undertail coverts). As a final flourish, its legs and much of its extraordinarily massive beak are coral-pink, ripening into scarlet at its beak's shield-like base.[8,10,26,44]

One might expect that such a visually arresting bird as this would be difficult to overlook—yet the takahe's history reveals only too readily that it is singularly adept at concealing itself. No fresh conclusive sightings emerged until autumn 1894, when a truly unique event took place—the capture of a *living* specimen on North Island!

This was the one and only time that a moho appears to have been caught since its bones were first uncovered by science in 1847. It was procured by surveyor Norman Carkreek, and its skin and feathers were retained for many years afterwards by Roderick A. McDonald at his homestead in Horowhenua. This historic specimen could have finally provided science with the

Takahes—a trio of New Zealand's famously no-longer-lost wanderers.

long-awaited opportunity to document the external appearance of North Island's moho, hitherto believed extinct, but tragically it was eventually lost without any record of its morphology ever having been made.[46] The only moho seen alive by a Westerner was gone—not even death, it seemed, could prevent a specimen of this elusive bird from vanishing.

As no other modern-day moho has ever been collected or even sighted, all subsequent reports given here apply solely to its South Island equivalent, the takahe—whose name is generally used nowadays for the entire species.

On 7 August 1898, the moho's southern counterpart re-emerged yet again from obscurity, when a takahe was caught alive by the dog of Donald Ross, on the shore of Middle Sound, Lake Te Anau. For the first time, the entire specimen—skin, skeleton, and internal organs—were preserved, and reached Dunedin's Otago Museum, where it ultimately became known as 'the last of the takahes', because, true to form, the takahe disappeared once more, but for much longer this time than ever before.[8,10,26,44]

By 1948, almost 50 years had passed without a single takahe having been collected. Admittedly, there were many rumours relating to supposed sightings of takahes, but these were never confirmed, and so avian researchers finally declared that it must surely be extinct this time. Not everyone, however, was fully convinced. Dr. Geoffrey Orbell, a physician from Invercargill, was one such person. He collected a great deal of information concerning the takahe's possible existence, including much testimony from Maoris inhabiting the area around Lake Te Anau, where specimens had been captured in the past. He learnt from these sources that in the mountains around this lake's eastern shores another large lake existed, whose valley was the traditional homeland of this reclusive bird but was currently undocumented by westerners. Such was its connection with the takahe that it was known to the Maoris as *Kohaka-takahea*—'the takahe's nesting place'.[8,10,26,44,47]

Inspired by this optimistic and stimulating news, Orbell led a small expedition to the newly-revealed lake's valley in April 1948, but although he discovered some signs that indicated the presence of takahes, no birds were seen. Even so, it was promising enough for Orbell to feel justified in leading a second expedition there seven months later.[8,10,26,44,47]

On 20 November 1948, he and his companions were trekking through the valley, across a clearing carpeted in snow-grass, when—in what must have seemed to them to be an incongruously casual manner, given the occasion's scientific significance—the first takahe conclusively reported for half a century stepped out into view just ahead of them, and straight into ornithological history. Once again, this amazing bird had fooled everyone—thriving in blissful ignorance of the fact that it was supposed to be extinct, and frequenting a locality not previously known by science to exist. By the end of 1948, the team had succeeded in netting two specimens (alive and unharmed), which they meticulously filmed and observed before releasing them again; they also saw a third, which eluded capture.[10,26,44,47]

In 1949, Orbell led yet another expedition to this valley, and as a result of sightings made during that visit he estimated that at least 20 breeding pairs lived here, with additional takahes in another valley close by. The whole region was swiftly declared a protected zone, with all visitors strictly prohibited unless they had received full governmental permission to enter it. More recently, a captive breeding programme was initiated, to ensure that this species survives even if some dire event should overtake the wild stock, and it has also been introduced onto the island bird sanctuary of Tiritiri Matangi, just northeast of Auckland, where I was very privileged and delighted to view takahes at close range in the wild when I visited this island in November 2006. For the moment, therefore, the takahe and its known native range, appropriately christened Takahe Valley, seem safe.[10,14,26,44,47]

Let us hope that continued preservation and good fortune will guarantee that the future never witnesses a fresh encore of the takahe's famed disappearing act, and that New Zealand's wanderer has, instead, returned for good.

AFRICAN BAY OWL—A ZOOGEOGRAPHICAL ENIGMA

The bay owl *Phodilus badius* is a widely-distributed Asian species whose range extends from northern India to Indonesia. Taxonomically, however, it stands aloof, sufficiently different from even its closest relatives, the barn owls, to require a subfamily all to itself—or at least that was the situation until the early 1950s. One of the most unexpected ornithological discoveries of the 20th century occurred in 1951, when a previously unknown species of bay owl was found, but not in Asia—instead, in Africa!

The discovery in Africa of a species belonging to a group traditionally looked upon as exclusively Asian is not new (as already shown with the African green broadbill and the Congo peacock), but is nonetheless very unusual—and made even more so in this particular instance by virtue of the remarkable fact that only two specimens of this decidedly out-of-place owl have ever been obtained. Its type was procured at Muusi, in a grass clearing at an altitude of 8100 ft within the Itombwe Mountains of eastern Zaire (now the Democratic Congo), and differed from the Asian bay owl by way of its somewhat darker plumage, smaller feet, and flattened bill. Officially described in 1952, its cryptic species was dubbed *P. prigoginei*.[48]

A sighting of an owl that was almost certainly an African bay owl was made in 1974, at the Rwegura Tea Estate in Burundi, and the calls of a possible specimen were tape-recorded in Rwanda's Nyungwe Forest during 1990, thus suggesting that the species is not totally confined to the Democratic Congo. In 1996, a female African bay owl was captured alive within montane gallery forest in a remote region of eastern Zaire called the Massif, situated in the extreme southeast corner of Itombwe Forest, by a team from the Wildlife Conservation Society and the Zaire Institute for Nature Conservation. Otherwise, however, this zoogeographical enigma is no better known scientifically than it was on the day after its discovery, 49 years ago.[12,48]

The Itombwe Mountains also house a number of other rare and scantily-known species of bird first described in the 20th century. These include: the African green broadbill (1907—p. 118), forest ground-thrush *Zoothera oberlaenderi* (1914), Chapin's flycatcher *Muscicapa lendu* (1932), Rockefeller's sunbird *Nectarinia rockefelleri* (1932), Schouteden's swift *Schoutedenapus schoutedeni* (1960), and Albertine owlet *Glaucidium albertinum* (1983).[49]

JAMES'S FLAMINGO—OUT OF SIGHT (AND REACH) FOR 70 YEARS

During an expedition in 1886 to the salt lakes beyond the Atacama Desert within the high, southern Andean mountain range that constitutes the Chile-Bolivia border, its participants not only observed the already-known Andean flamingo *Phoenicoparrus andinus* and Chilean flamingo *Phoenicopterus chilensis* but also discovered a previously unknown member of this family. In honour of the expedition's sponsor, British businessman Berkeley James, it was christened *Phoenicoparrus jamesi*, James's flamingo. Slightly smaller than the other two species, the new flamingo was further differentiated by its predominantly orange-yellow bill (with only the tip black), the black 'mask' encircling its eyes and extending to the base of its bill, and the rich rosy colour of its legs.

Specimens were collected and sent to many major museums worldwide—after which James's flamingo became a notable enigma. Due to the almost inaccessible nature of its type locality, other expeditionary teams declined to look for it there, and searches made elsewhere failed to uncover any evidence of its existence. Furthermore, as the museum specimens began to fade with the passing decades, even references in books

James's flamingo—out of sight (and reach) for 70 years. (Alan Pringle)

to its diagnostic colours became confused and inaccurate. Eventually, many ornithologists feared that it must be extinct; some even dismissed it as an unimportant variety of one or other of the other two species.[50]

In 1956, however, both of these hypotheses were disproven, when American explorer A.W. Johnson and a bold team of Chileans scaled the southern Andes' hostile peaks to reach the very highest lakes in the range. There they found James's flamingo, still living, and corresponding precisely with the original descriptions of the specimens obtained in 1886, thus vindicating its classification as a valid species. In January 1957, and a little further to the north, Johnson and company reached the salt lake of Laguna Colorada ('Red Lake') in southwestern Bolivia, and encountered this long-lost species' nesting grounds. Later visits revealed the presence of several thousand breeding birds there. More recently, it became a successful breeding species at Berlin Zoo; and specimens have been exhibited at the late Sir Peter Scott's world-renowned Wildfowl Trust at Slimbridge, in Gloucestershire, England.[50]

Some day there may be a momentous sequel to this exotic bird's discovery and rediscovery. The local Andean Indians inhabiting its provenance unhesitatingly distinguish the three species of flamingo living there—referring to the Chilean flamingo as *guaichete*, the Andean as *tococo*, and James's as *chururo*. In addition, they claim that a fourth type of flamingo also exists there, which they refer to as the *jetete*.[51] Could there be an unknown species of flamingo still awaiting discovery in this remote, little-explored region? After all, the nesting grounds of James's flamingo—extensively populated by a species neither small nor inconspicuous—remained undiscovered until little more than 55 years ago.

SEYCHELLES SCOPS OWL—MISSED IN THE MISTS FOR HALF A CENTURY

One of the world's least-known owls, the Seychelles scops owl *Otus insularis*, a handsome russet-plumaged species characterised by its lengthy, unfeathered legs, is found only on the tiny island of Mahé. Like so many other indigenous forms of Seychelles wildlife, its numbers fell rapidly following man's increasing interactions with this island group's ecology. Some (though not all) avian researchers believe that man's introduction of the South African barn owl onto Mahé provided *O. insularis* with severe competition for food. In any event, by the beginning of the 20th century it had become gravely endangered, with what was ultimately thought to be the last record of a living specimen occurring in 1906.[10,12,52]

No further news of the Seychelles scops owl emerged, and it was eventually deemed to be extinct—until 1959, when it was unexpectedly rediscovered by French ornithologist Philippe Loustau-Lalanne amid the

Seychelles scops owl.

remote, mist-shrouded mountains of Mahé's south-central region. At first, only a few pairs were believed to exist, but later searches and estimates led to a revised figure of around 80 pairs, in secondary forest across one third of the entire island. Moreover, shortly after its initial rediscovery, scientists learnt that a single specimen had in fact been collected in 1940, but had failed to attract any attention. Happily, the Seychelles scops owl no longer seems to be in decline; indeed, its nocturnal and highly reclusive lifestyle may have been responsible for at least part of its supposedly 'critically endangered' status, but it is still classed as endangered by the IUCN.[10,12,52]

Less than a year earlier, on 7 September 1958, a new species of scops owl had been discovered, and in another Indian Ocean island chain. This time the locality was La Convalescence, on Grand Comoro of the Comoro Islands, and the species was *Otus pauliani*, the Comoro scops owl.[53] For a long time, this small, mysterious bird was known only from its type specimen—a male, collected by the British Ornithologists' Union Centenary Comoro Expedition—but others were heard calling in 1983. Moreover, in September 2005 this elusive species was found to be abundant on the southeastern flanks of Mount Karthala, an active volcano on Grand Comoro.[12,53]

In June 1992, another elusive Comoro owl, the whistling Anjouan scops owl *Otus capnodes*, was unexpectedly rediscovered by biologist Dr. Roger Safford. It had not been reported since its discovery in 1886.[54]

PUERTO RICAN WHIP-POOR-WILL—WINGING ITS WAY FROM PREHISTORY TO PRESENT DAY

A relative of the well-known American whip-poor-will and Britain's elusive nightjar, a short-winged form of whip-poor-will was known to inhabit the forests along Puerto Rico's north and south coasts during the late 19th century and early 20th century. But as it closely resembled the American species, *C. vociferus*, it did not receive any scientific attention. Moreover, after a reliable sighting of some such birds by naturalist Alexander Wetmore, made on 23 December 1911 in the vicinity of the Insular Experimental Station at Rio Pedros, it disappeared, and was considered extinct.[8,10,14]

A few years later, Wetmore excavated some fossil whip-poor-will bones, several thousand years old, within the Clara and Catedral Caves near Morovis, Puerto Rico, and recognised that they represented an undescribed species, which in 1919 he dubbed *Setochalcis* [later becoming *Caprimulgus*] *noctitherus*. Shortly afterwards, he studied a dead specimen of the lost modern-day form of Puerto Rican whip-poor-will, and realised that the present-day and the prehistoric types were one and the same.[8,10,14]

As the species was assuredly extinct now, however, the matter seemed to be of academic interest only—until March 1961, when a strange call was tape-recorded by Cornell University researcher George B. Reynard during fieldwork on Puerto Rico. Although clearly a whip-poor-will cry, it did not match that of any known type—thus agreeing with native Puerto Rican testimony that their whip-poor-will had sounded quite different from other species. Using an amplified version of the recording as a lure, later that year Reynard enticed within collecting range a male whip-poor-will, belonging to the supposedly demised Puerto Rican species! Populations have since been located in at least three different forests within the island's southern section, but it remains critically endangered.[8,10,14] *C. noctitherus* is sometimes classed as a subspecies of *C. vociferus*, but its distinctive voice is sufficient for many to treat it as a full species.

EYREAN GRASSWREN— VIRTUOSO OF THE VANISHING ACT

Within the space of just a few months in 1961, three of Australia's most famous species of 'extinct' bird were sensationally rediscovered.

The Eyrean grasswren *Amytornis goyderi*, an attractive, warbler-related species with a long tail held high in the air, has acquired the reputation of being one of the world's most elusive birds—and for very good reason. Roughly 6 in long, with brown upperparts overlain by short white streaks, darker wings and tail, buff underparts, white breast, and a thick sparrow-like bill, this evanescent inhabitant of spinifex grasslands was discovered in 1875 by F.W. Andrews, during the Lewis Expedition to South Australia's famous Lake Eyre and its environs. Two specimens were caught, ultimately reaching the British Museum (Natural History), after which it made the first of its celebrated disappearances. No further report of Eyrean grasswrens occurred for over half a century—until 1931, when a single pair was sighted in the area where the original couple had been collected. After the 1931 record, however, the species vanished again, this time for 30 years.[55]

On 3 September 1961 it was rediscovered once more, when two adult birds and a nest containing two young were spied at Christmas Waterhole on the Macumba River, near Lake Eyre North, by an ornithological team from Victoria.[56] On 2 August 1966, during the West-East Crossing Expedition to the Simpson Desert, north of Lake Eyre, naturalist Keith Davey saw a pair of grasswrens 37 miles northwest of this desert's Poeppel's Corner. He was sure that they were Eyrean grasswrens— if so, this is its earliest Northern Territory record.[55]

Another surprise was recorded in August 1976, when Ian May discovered that the Eyrean grasswren was actually quite abundant on the sandhills east of Poeppel's Corner and west of Eyre Creek, as well as at the aptly-named Birdsville (just inside eastern Queensland). True to form, however, when those same areas were visited the very next year the birds had largely vanished, with just a few remaining near Poeppel's Corner. In the hope of establishing a captive breeding programme to safeguard this unpredictable species, a pair was captured in 1977 and maintained thereafter at Sydney's Taronga Zoo.[55]

At the time of the Eyrean grasswren's reappearance in 1961 only seven grasswren species were recognised, but in 1967 Normal Favaloro collected some specimens of a pale-coloured form that proved to be an eighth (by 2010, ten grasswren species were recognised)—described by him a year later and christened *A. barbatus*. Popularly termed the grey grasswren, and inhabiting southwestern Queensland and northwestern New South Wales, it was first reported as long ago as 1921, but no specimen was obtained. In 1942, Favaloro saw one by the Bullvo River, but the form as a whole was dismissed by other authorities as just a pale mutant variety of the western grasswren *A. textilis*. Twenty-five years later, however, Favaloro succeeded in proving them wrong.[57]

And just a year after that, a third species of 'lost' grasswren reappeared, when in 1968 the black grasswren *A. housei* was positively sighted for the first time since its discovery in 1901.[58]

NOISY SCRUB-BIRD— SOUNDING OUT ITS RETURN

The lyrebirds, with their ornate tails and talent for vocal mimicry, are among the most famous of Australia's diverse avifauna. Less spectacular, and hence less well-known, are their smaller, superficially thrush-like relatives, the scrub-birds. Two species exist—the rufous *Atrichornis rufescens*, and the noisy *A. clamosus*, the latter of which staged another widely-extolled ornithological coup of 1961, by returning from almost a century of 'extinction'.

The noisy scrub-bird was officially described in 1844 from Drake's Brook, Waroona (south of Western Australia's capital, Perth), and was later discovered at various localities along Australia's southwestern coast, including Torbay and Albany. A long-tailed, short-winged form, 8.5 in long, with brownish-red upperparts, white chin and underparts, plus a distinctive brown throat-band, it earned its common and scientific names from its extremely loud (albeit melodious) voice. This generally provided would-be observers with their only clue to its presence, because its plumage blended very effectively with the thick swampside foliage in which it usually remained hidden from view.[8]

Although it seemed most common around Albany, the noisy scrub-bird was never an abundant species, as its preference for swampland and running streams limited its population size. By the 1880s it had become exceedingly rare, and after a lone specimen had been collected in Torbay in 1889 by A.J. Campbell, it was believed to have become extinct.[8]

Australian ornithologist Vincent Serventy was quite stunned, therefore, when on Christmas Eve 1961 he received a phone-call from a newspaper reporter asking him to comment upon the noisy scrub-bird's alleged rediscovery just a few days earlier! Anxiously, Serventy requested full details, and learnt that on 17 December naturalist Harley Webster had clearly heard its unmistakable call emerging from some thick scrub-land on a small headland close to Mount Gardner at Two Peoples Bay, about 25 miles east of Albany. Moreover, it transpired that just a month earlier, on 5 November, P.J. Fuller and Charles Allen had actually *seen* a noisy scrub-bird in this same locality, giving them the first conclusive sighting of its species for 72 years. Serventy later journeyed with Webster to this spot, and discovered a small colony of the birds.[59]

Yet no sooner had this species returned to the land of the living than it straight away seemed set to plummet back into extinction—due to some plans to build a small town on the edge of its very limited and only known area of distribution! Thankfully, however, the plans were abandoned—due in no small way to royal intervention in the form of a personal plea by H.R.H. Prince Philip of the United Kingdom for the area to be left undisturbed. And by 1967, a zone of 13,600 acres, which totally contained this species' known distribution range, was declared a faunal reserve, providing added security for the noisy scrub-bird's continued survival.[10,59]

Noisy scrub-birds.

WESTERN BRISTLEBIRD—ANOTHER TURN-UP AT TWO PEOPLES BAY

Also making a dramatic comeback in 1961, and once again at Two Peoples Bay, was the western bristlebird *Dasyornis longirostris*. '*Dasyornis*' ('hairy bird') and 'bristlebird' both refer to the cluster of stiff, hair-like feathers around the bill of all three species belonging to this Antipodean genus of warbler-like passerines, whose ground-dwelling members are predominantly brown-coloured, and roughly 6 in long, with lengthy, broad-ended tails and pale underparts.

The western bristlebird was discovered in 1839 by John Gilbert on the Swan River, near Perth (though none has been recorded from there since), and was described and named in 1840 by John Gould. In the later 1800s it was collected at King George's Sound, and was reported in the early 1900s from Wilson's Inlet—until a fire destroyed this latter colony in 1914. Except for a single specimen collected in 1945 at Two Peoples Bay, nothing more was heard of this species for almost half a century. Then in April 1961, small numbers were rediscovered at Two Peoples Bay, and later at King George's Sound too. In 2005, the known breeding population was estimated at 300-450 pairs.[8,14]

NEW AND REDISCOVERED OWLS

The Nduk eagle owl *Bubo vosseleri* (sometimes classed as a well-delineated subspecies of Fraser's eagle owl *B. poensis*) is a large, striking species with bright brown upperparts, silky white underparts overlain by slender brown cross-barring, and an ochre-yellow face with prominent ear-tufts. It was formally described in 1908 by Reichenow, basing his account upon an adult specimen obtained at least two years earlier from Amani, in northeastern Tanzania's Usambara Mountains. The species was named after zoologist Dr. Julius Vosseler of Tanzania's Biological Institute in Amani, who had obtained a young individual and had then sent it to the Berlin Museum on 15 October 1906. For more than 50 years, these were the only specimens of the Nduk eagle owl available to science. A bird that may have been of this species was sighted near Amani on 20 December 1930, and on 6 September 1931 (flying in the daytime); otherwise, there were no records from the wild either.[12,14,60]

Then on 28 April 1962, Dr. G. Pringle, Director of Amani's East African Institute of Malaria, was visited by Gabriel Joseph, who had brought with him a young owl that he had found in the high forests nearby. Pringle was fairly sure that it was a Nduk eagle owl, and after caring for it over the next few weeks he passed it on to London Zoo, where his provisional identification was duly confirmed.[60] Since then, other specimens have been obtained—from the late 1970s there have been three more individuals displayed at London Zoo, and a number of confirmed sightings in the field. As far as is known, however, this species is confined to Tanzania's Eastern Arc mountain ranges (including the Usambaras and Udzungwas), and appears to have declined in numbers as a result of forest destruction within its already restricted habitat.[12]

In 1965, the type specimen of a new species of scops owl was collected at Kenya's Sokoke Forest, and in 1992 this species was recorded from Tanzania's East Usambara Mountains too. Named *Otus ireneae* when described in 1966 by eminent ornithologist Dr. S. Dillon Ripley,[61] the Sokoke scops owl appears to be more common than *B. vosseleri*, but is similarly threatened by forest clearance.[3,12,61]

RUFOUS-HEADED ROBIN—AN EASTERN ENIGMA

Few species of striking appearance are so elusive that they can remain hidden even from their local human neighbours, let alone the scientific world—which makes the rufous-headed robin rather special. In July 1905, English explorer Alan Owston was travelling through central China when some of his animal collectors working in the Tsin-Ling Mountains of Shensi Province captured and brought to him three specimens of a small but extremely beautiful bird. Although it obviously belonged to the thrush family, it was instantly set apart from all known species by its fiery-coloured head, contrasting markedly with its sombre, slate-grey back, greyish-white underparts, brown wings, and black-bordered pure-white throat.[10,62]

Owston's three examples of this exquisite little thrush were sent to Lord Walter Rothschild's renowned bird collection at Tring Museum, and after studying them Ernst Hartert named their species *Larvivora ruficeps*. Its highly attractive appearance made it a much sought-after bird by other ornithologists visiting the Tsin-Ling Mountains, but no-one succeeded in obtaining—or even seeing—any other specimens. Most surprising of all, not even the local people could offer any help, because they had never seen it before either![10]

Decades rolled by, with additional studies of the three original examples leading to the species' reclassification as a member of the robin genus, so that it was referred to thereafter as *Erithacus ruficeps*, the rufous-headed robin. Otherwise, it was largely forgotten.[10]

The date was now 15 March 1963. That evening, during a session of bird study and ringing at the peak of Mount Brinchang in west-central Malaya's Cameron Highlands, ornithologist Dr. Elliot McClure sent his assistant to inspect their nets, to see if they had captured any birds for ringing purposes. When the assistant returned, he was carrying, very carefully, a living specimen of a very small but gorgeous robin-like bird that McClure was unable to identify. Offsetting its drab brown and grey body was the brilliant orange colouration of its head. Recognising its worth, if not its species, but nonetheless unwilling to kill such a beautiful little bird for collection purposes, McClure elected to take some close-up colour photos of it, while hand-held. He also weighed, measured, and ringed it, and afterwards released it back into the wild.[63]

Eager to identify his unexpected find, McClure sent descriptions of it to fellow ornithologists far and wide, and its photo appeared in several publications. As a result of these efforts, its identity was eventually ascertained—it was the elusive *E. ruficeps*, a discovery that was doubly startling. Not only had a near-mythical bird reappeared, but in addition no-one had ever expected it to occur as far away from central China as west-central

Malaya (areas separated by more than a thousand miles).[63]

Another facet of this exceptionally secretive species' existence had thus been disclosed, but the world has yet to uncover anything else concerning it—its lifestyle, behaviour, population size, nest, and much more all variously remain totally or virtually unknown. Only its continuing existence is certain—three specimens were recorded in 1985 at Jiuzhaigou, Sichuan, and six singing males were sighted in 1987. In modern times, this mystifying little bird has been renamed yet again—it is now *Luscinia ruficeps*, thus allying it with the nightingale, bluethroat, firethroat, and rubythroats among others, all of which have been recategorised taxonomically as flycatchers rather than thrushes.[64]

VO QUY'S PHEASANT—MAKING A MEAL OF A MYSTERY BIRD

Vo Quy's pheasant *Lophura hatinhensis* (also known as the Vietnamese pheasant) is the most recently discovered and also the least-known of all pheasant species—only two specimens had been recorded prior to the 1990s. The first of these, the species' type, was collected in 1964 by the late Do Ngoc Quang in the vicinity of Son Tung and Ky Thuong, in Vietnam's Nghe Tinh Province. A male, with predominantly royal blue plumage offset by a white crest and long white tail feathers and also by a bright red area of bare skin around each eye, it was preserved afterwards as a mounted taxiderm exhibit, and is currently held at Hanoi's Institute of Ecology and Biological Resources. The second specimen, another male, was obtained close by in 1974, by Troung Van La; sadly, however, it was not preserved. Others were spied in nearby valleys, but none was collected. In 1975, this species was formally described (in Vietnamese) by biologist Prof. Vo Quy of Hanoi University, who had encountered it in the wild during his youth, and now perceived it to be a species awaiting scientific recognition.[65]

For quite a time, however, it seemed as if such recognition would be of little practical worth, because after an absence of confirmed sightings for several years zoologists began to suspect that Vo Quy's recently-described pheasant had died out. Then in February 1990, a team of investigating Vietnamese ornithologists made a startling discovery. Far from being extinct, *L. hatinhensis* was being regularly trapped and eaten by local farmers! Needless to say, the scientists immediately requested the farmers to bring to them alive and uninjured any future specimens that came their way, and soon afterwards two males and one female (the first time that a female, brown in plumage, had been seen by zoologists) were duly handed over. These were sent to Hanoi Zoo, where they were duly placed on public display;

Vo Quy's pheasant. (Dr. Karl P.N. Shuker)

since then, a thriving captive breeding population has been established here.[65] Interestingly, a single male imperial pheasant *L. imperialis* was also caught alive—the first seen for many years—but tragically it died the following day[65] (see also p. 121). In 1999, the first specimens seen in Britain arrived at the Cotswold Wildlife Park. Others have since been exhibited at Chester Zoo and, most recently, at Dudley Zoo.[65]

Despite the relatively recent date of Vo Quy's pheasant's 'official' discovery, at least one notable western scientist may have encountered it many years earlier. In 1920, Prof. Jacques Berlioz (then the director of the Laboratory of Ornithology at Paris's National Museum of Natural History) saw a pheasant in Vietnam that he could not identify with any species previously recorded by science.[66] Could it have been a Vo Quy's pheasant? If not, it must have been an imperial pheasant (as this form was not formally discovered until 1923)—unless, of course, there is a further species of Vietnamese pheasant *still* awaiting detection?

GREATER YELLOW-HEADED VULTURE— A BELATED BIRD OF PREY

Not all new birds come to light in the field. The year 1964 saw the scientific debut of a new species of American bird of prey, when the greater yellow-headed vulture *Cathartes melambrotus* received its long-overdue official description, based upon a specimen originating from Kartabo in Guyana. Ironically, this 'new' vulture had actually been known to ornithologists for many years, and was already well-represented in museums all over the world, but unfortunately it had been confused in the past with the *urubitinga* subspecies of *C. burrovianus* (nowadays termed the lesser yellow-headed vulture), so that its true identity as a valid species in its own right had not previously been realised.[67]

A TALE OF TWO RAILS—ONE SAVED, ONE LOST

Sometimes treated as a distinctive subspecies of the slate-breasted rail *Lewinia* [formerly *Rallus*] *pectoralis* rather than as a separate species, for a long time the Auckland Islands rail *L. muelleri* was known only from its type specimen, described in 1893. Then in 1966, a single individual was captured alive near to a rubbish dump on Adams Island, a member of the Auckland group, by the southern party of the joint Dominion Museum-D.S.I.R. expedition visiting these islands from 14 January to 14 February. It was taken to the New Zealand Wildlife Division's Native Bird Reserve and aviary at Mount Bruce, near Masterton, and was later observed and photographed there by naturalist Anthony Whitten during work for his Churchill Fellowship. A small chestnut-coloured rail whose underparts are boldly striped in transverse black and white bands, for a time it was the only confirmed representative of its species on record since 1893; similar birds had been sighted on Adams Island during the 1940s, as well as on nearby Ewing Island during the 1940s and 1960s, but none was captured for identification. Fortunately, others were subsequently found there, and in 1993 this long-lost rail was also discovered alive and well on nearby (and in-aptly-named!) Disappointment Island. At present (2011), their species' total population is estimated at just over 2000 individuals.[8,68]

In 1973 a single specimen of the Fijian barred-wing rail *Nesoclopeus* (=*Rallina*) *poecilopterus* was unexpectedly encountered on Viti Levu's Nadrau Plateau. Prior to then, ornithologists had believed that this large brown-and-grey species had been exterminated on its island homes of Viti Levu and Ovalau some time around 1890, by cats, rats, and mongooses—all introduced there in earlier days by man.[8,14]

Kakapo.

Fijian barred-wing rail.

Normally, the rediscovery of a long-lost species offers it a new lease on life. With *N. poecilopterus*, however, the ultimate tragedy is that unless further specimens also exist its solitary representative can do nothing more for it than provide it, upon death, with a new extinction date. In short, although it had a living representative as recently as the 1970s, the Fijian barred-wing rail was already *effectively extinct* (aka *functionally extinct*).

When employed precisely, this term describes any species whose last surviving examples are, for various reasons, unable to perpetuate it by reproduction. Such reasons can include: the existence of just one single living specimen (as with *N. poecilopterus*), or the existence of several specimens but all of the same sex; or the existence of specimens of both sexes but too old or diseased to mate and/or to produce viable offspring.

A once-common species that has suffered a devastating decline via persecution at the hands of man and at the jaws of his canine, feline, and rodentine entourage is New Zealand's owl parrot or kakapo *Strigops habroptilus*. In the late 1970s, only 12 living kakapos were known—all males. Happily, however, this species was rescued from the limbo of effective extinction in 1981, when a female was found on Stewart Island, followed later by others; as of February 2010, 122 living individuals were known.[69]

ALDABRA BRUSH-WARBLER—
TINY BIRD, TINY KNOWN EXISTENCE

Discovered in 1968, and described a year later, the Aldabra brush-warbler *Nesillas aldabrana* had one of the tiniest distribution ranges of any bird species—about 24 acres of coastal vegetation on Malabar Isle in the Seychelles atoll of Aldabra.[70] As an inevitable consequence, it was recognised to be one of the world's rarest bird species. Following its discovery, no further specimens were recorded until 1975, when six individuals, all males, were ringed and photographed by Robert Prys-Jones of the British Museum (Natural History). By 1983, only a single individual, a male, was known to exist, but as this warbler's terrain—profusely pitted with razor-edged cracks and crevices in the underlying coral—was not conducive to easy exploration, ornithologists hoped that others may exist undetected. Sadly, however, this proved not to be the case, and in 1986 the Aldabra brush-warbler was confirmed extinct, which meant that its very existence had been known to science for a mere 18 years. Its extinction may have been due to the introduction here many years previously of cats, rats, and goats.[14]

FOUND AND LOST IN A DECADE—
THE WHITE-EYED RIVER MARTIN

Another equally mysterious and seemingly lost bird is, or was, the white-eyed river martin. It was unexpectedly discovered in early 1968, when nine specimens were trapped during a series of nettings in reed beds at Bung Boraphet, a very large man-made lake in central Thailand. Some were brown-plumaged juveniles, others were dark glossy-green adults—which were additionally distinctive in appearance thanks to their two extremely long and racquet-tipped central tail feathers, plus their unusually stout legs and beak (for a species of swallow or martin), white rump, and very large, white-ringed eyes.[70a]

As readily recognised by eminent Thai zoologist Dr. Kitti Thonglongya, this was unquestionably a species new to science—indeed, the only known species that in any way recalled it was, remarkably, an exclusively African species known as the African or Congo river martin *Pseudochelidon eurystomina*. Consequently, when he formally described this anomalous Asian version later in 1968, he duly christened it *P. sirintarae* (honouring a Thai princess), though in more recent years some researchers have rehoused it in a genus of its own, as *Eurochelidon sirintarae*.[70a]

During the next few years, some additional specimens were obtained and sighted, but the last reliable record was a field observation at Bung Boraphet in 1978. Apart from two possible further sightings in Thailand during the 1980s, and a possible observation in Cambodia during 2004, no more reports, not even unconfirmed ones, have emerged. Consequently, some investigators have speculated that perhaps the specimens obtained in Thailand were vagrants, and that this seemingly lost species' true provenance has yet to be discovered. That may be true, but with every year that passes with no additional reports of any kind being revealed, it looks ever more likely that Asia's enigmatic river martin has simply become extinct.[70a]

PARDUSCO—A CONUNDRUM
FROM THE CLOUDS

In June 1973, two male specimens of a small, warbler-like bird were collected in Peru's Carpish Mountains during an ornithological field exploration organised by the American Museum of Natural History. Some additional specimens of this species, one that proved to be new to science, were obtained there during 1974-5. It is known to the locals as the pardusco; but as far as ornithologists are concerned, a more appropriate appellation for it would be the word 'paradox'. With mostly plain brown upperparts and tawny-olive to tawny-ochraceous underparts, the pardusco superficially resembles the wood warblers or parulids, an exclusively New World family of songbirds. However, it exhibits two primitive, highly perplexing features that complicate this classification.[71]

One of these is the presence of the hypoglossus anterior, a muscle lost in all ostensibly similar birds (i.e. those that possess nine primary wing feathers). The other is its unusual ceratohyoideus muscle, which, very unexpectedly, takes its origin not only from the lateral, but also from the medial, surface of this bird's ceratobrachiale bone. When its anatomical characteristics were assessed *in toto*, the resulting combination was so bewildering that its researchers, Drs. George Lowery and Dan Tallman, were quite unable to assign the pardusco with comfort to *any* existing taxonomic family of birds! In July 1976, they christened this singular species *Nephelornis oneilli* ('O'Neill's bird of the clouds')—in honour of Dr. John P. O'Neill, a leading expert on Peruvian birds, and in recognition of the pardusco's habitat, consisting of cloud-covered forests. Presently categorised hesitatingly as a highly aberrant tanager, its taxonomic affinities to other birds remain controversial.[71]

PO'O-ULI—WHAT WAS THAT MASKED BIRD?

Deep within a remote forest on the northeastern slope of Haleakala, a volcanic peak on the Hawaiian island of Maui, a species of bird hitherto unknown to science was discovered by ecology students Jim Jacobi and Tonnie L.C. Casey in July 1973. If external appearances were anything to go by, measuring a mere 5.25 in, with nondescript brown upperparts, cream underparts, a black

Pardusco. (Dr. John P. O'Neill)

face mask, and a short pointed bill for pecking tree bark in search of insects, it hardly seemed a likely candidate for stirring up great excitement within ornithological circles. Yet no assumption could have been further from the truth, because this new species, soon named *Melamprosops phaeosoma* ('brown-bodied black-face') and known locally as the po'o-uli, proved to be a member of one of the world's most extraordinary bird families—Drepanididae, the Hawaiian honeycreepers.[72]

Apparently descended from a single, generalised finch-like species, the Hawaiian honeycreepers consist of more than 30 modern-day species, each with its own characteristic bill shape for use within its own, exclusive ecological niche. Thus, there are certain species with sharp bills or sickle-shaped ones for various insect-capturing techniques, some with parrot-shaped bills for seed-eating, and others with long slender ones for nectar-sipping. Collectively, these remarkable birds, endemic to the Hawaiian archipelago, probably constitute the world's most spectacular example of adaptive radiation—the evolutionary development and divergence of many species, from a single ancestral one, to occupy many ecological niches. Their especial scientific significance thereby makes it all the more tragic that, within the last 130 years, at least a third of all honeycreeper species have become extinct—due to such factors as over-hunting by natives for their plumes, predation by introduced vermin, and their vulnerability to diseases carried by introduced bird species, not to mention over-collection of specimens for museums by excessively zealous bird collectors.[72]

In the wake of such appalling extinctions, the unexpected detection of a totally new species (so distinctive that it required a new genus) was therefore extremely exciting, a most welcome addition to a family more associated with decimation than with discovery.

Unhappily, all is not good news, because the po'o-uli, seemingly confined to the upper Koolau Forest Reserve on Haleakala, is critically endangered—by the end of 1997, its total population had apparently fallen to as few as three individuals. One of these, a male, was captured on 9 September 2004 and taken to the Maui Bird Conservation Center in an attempt to initiate a captive breeding programme, but it died shortly afterwards, before a mate could be found for it, and the other two individuals have not been seen since 2004 either. The po'o-uli, is fully protected by United States federal law from disturbance, capture, and killing, but whether (always assuming that it does still exist) it can survive the menacing presence of tree-climbing rats and exotic avian diseases is very much a matter for conjecture.[8,14,72]

The po'o-uli is the most recently-discovered Hawaiian honeycreeper.

A LIKING FOR REMOTE LAKES—
THE HOODED GREBE

In 1974, a hitherto-unknown species of grebe was unexpectedly discovered on the Laguna de los Escarchados, a remote mountain lake in southwestern Argentina, when Argentinian biologist and film-maker Maurice Rumboll shot a specimen of what he initially assumed was either a silvery grebe *Podiceps occipitalis* or a white-tufted grebe *Rollandia rolland*, but which, when subsequently examined, proved to belong to a wholly unfamiliar, scientifically-undescribed species. Later that same year, Rumboll formally named it *Podiceps gallardoi*, the hooded grebe (after its striking black, white, and orange head plumage), and this long-overlooked bird has since been discovered at several other equally remote lakes elsewhere in Argentina too, as well as in neighbouring Chile.[72a, 73a]

KABYLIAN NUTHATCH—
AN ALGERIAN SURPRISE

The Kabylian nuthatch *Sitta ledanti* is a small dainty species of tree-dwelling passerine bird that was first made known to science in 1975—when one very small population was discovered in a relict group of conifers on the summit ridge of Djebel Babor, in Algeria's Little Kabylie range. With only 20 or so pairs in total, this noteworthy species—Algeria's only endemic bird—was in a very vulnerable position, especially as its only known habitat, the surrounding forest, was threatened by summertime overbrowsing by cattle and goats. Due to appreciable ornithological concern for this newly-disclosed species' survival, however, the entire Djebel Babor area was duly declared a national park, and the number of pairs here is now estimated at around 80 in total. In June 1989 a second population was found—consisting of 350 birds in Algeria's fully-protected Taza National Park, at Jijel—and two further populations, adjacent to the Taza birds, were later discovered in the Tamentout and Djimla Forests. Even so, the total number of Kabylian nuthatches may not exceed 1000 birds, and it continues to be categorised as endangered by the IUCN.[14,73, 73a]

XENOGLAUX AND XENONETTA—
A STRANGE OWL AND A STRANGER DUCK

On 23 August 1976, trekking through cloud forest on the Andes' eastern slopes in northern Peru, Louisiana State University Museum ornithologists Drs. John P. O'Neill and Gary P. Graves saw a tiny brown owl that seemed totally unlike any previously documented species. What made it so distinct were its long and fragile facial filaments, which extended beyond the edge of its head in a delicate feathery fringe, and the very long and striking bristles at its bill's base that grew upwards to yield a fan-like crest between its eyes. The owl's species became known as the long-whiskered owlet, and comparative studies confirmed that it was indeed new to science. Weighing no more than 2 oz, this engaging little bird seemed most closely related to those comparably minute species the pygmy owls (genus *Glaucidium*) and (like them) was without ear-tufts, but it also lacked feathers on its tarsi and feet. It was thus allocated a genus to itself, *Xenoglaux* ('strange owl'), and its full scientific name, *X. loweryi*, honours Dr. George H. Lowery, director of the Louisiana State University Museum.[74]

Following its discovery, however, precious little was seen or heard of *Xenoglaux* for the next 30 years; two birds were mist-netted in 1978, a third in 2002, and that was all—no further sightings or records of any kind were documented. Consequently, there were even fears that this odd little owl had become extinct—until February 2007, when it was encountered on three separate occasions during the day and its calls recorded at night by American Bird Conservancy researchers and rangers working in the Area de Conservación Privada de Abra Patricia-Alto Nieva.[73a]

Moving from strange owls to stranger ducks, the Campbell Island flightless teal *Xenonetta* [=*Anas*]

Peru's long-whiskered owlet.
(Dr. John P. O'Neill)

nesiotis—formerly endemic to Campbell Island itself (about 380 miles south of New Zealand)—was not formally described until 1935, and by 1944 it was presumed extinct, wiped out on its island homeland by introduced brown rats. There was an unconfirmed report in 1958, but nothing more until 1976—when a tiny population of this brown-plumaged, yellow-billed duck with reduced wings was positively identified on Dent Island, a small islet near Campbell Island, by the New Zealand Wildlife Service. Between 1984 and 1990, Dent Island's highly-vulnerable population was removed and taken into captivity for safety and breeding purposes by New Zealand's Department of Conservation. This has proven very successful, and thriving populations have since been re-introduced into the wild, including back onto Campbell Island (after its entire 200,000-strong population of rats had been killed). Some authorities nowadays deem the Campbell Island flightless teal to be nothing more than a well-marked subspecies of the New Zealand flightless teal (a slightly larger, darker-billed bird), and thus refer to it as *Anas aucklandica nesiotis*; others prefer to retain its separate specific status.[75]

WHITE-WINGED GUAN— RIGHT BIRD, WRONG HABITAT

The white-winged guan *Penelope albipennis* is a large pheasant-like bird, whose type specimen was obtained on 18 December 1878 on northern Peru's Countess Island—ringed coastally with mangroves but bearing tall dry forests at its centre. Two more were procured a short time later, again from northern coastal regions of Peru—after which nothing else was heard of this species. As mangrove forests predominated in the provenances of all three specimens, such areas were scoured meticulously in future decades, but to no avail.

Then in the mid-1970s, a local inhabitant of northwestern Peru informed his neighbour Gustavo del Solar (an agriculturalist and longtime guan seeker) that he knew of a concealed locality in which this 'vanished species' still survived. Yet this locality consisted of a deep pass on the western Andes' foothills—very different from the white-winged guan's alleged coastal habitat. Nonetheless, accompanied by Peruvian avifauna expert Dr. John P. O'Neill, del Solar elected to investigate this secluded area, just in case.[76]

And on 13 September 1977 (just a few months short of a century after the species' original discovery) their perseverance was rewarded by a conclusive sighting there of a white-winged guan. Shortly afterwards, its hideaway was visited by Peruvian zoologist Dr. Hernando de Macedo-Ruiz, in the company of del Solar, and also bird photographer Heinz Plenge who succeeded in taking the first pictures of living white-winged guans in their *natural* habitat. This, incidentally, proved not to be coastal mangrove vegetation at all, but the thickets of the dry, interior forests instead.[76]

Indeed, it was this confusion that seems to have been the prime reason for the guan's previous successes at evading detection—its would-be detectors had been looking in entirely the wrong habitat! (As pheasant expert Dr. Jesus Estudillo Lopez later commented: "It was as ridiculous as looking for camels in the Arctic".) Also, based upon conversations with the local people Dr. O'Neill later estimated that this species may number in the several hundreds, so extreme rarity could not be offered as an explanation for its evanescence either. Having said that, however, the white-winged guan is currently categorised as critically endangered by the IUCN on account of its fragmented distribution and a total population estimated in 2011 at approximately 350 birds.[76]

MAGENTA PETREL AND CHATHAM ISLAND TAIKO—TWO BIRDS IN ONE

The rediscovery of the magenta petrel *Pterodroma magentae* was cause for double celebration by zoologists, because it solved not just one but two ornithological mysteries.

Until 1978 it was known only from a specimen taken at sea just south of Pitcairn Island in the south Pacific, in 1867. Several years later, some geologically-recent skeletal remains discovered on Chatham Island (500 miles west of Pitcairn) were shown to be from a petrel species not recorded before by science, and which became known as the Chatham Island taiko. No living specimen was taken, and according to the islanders' testimony the taiko had died out around 1914 (although unconfirmed sightings had continued into the 1940s).[8,14]

The two petrel species unexpectedly became one in 1964, when, after comparative studies of the taiko remains and the single specimen of magenta petrel, Dr. W.R.F. Bourne declared that these two lost forms belonged to one and the same species—though it would

Campbell Island flightless teal.

naturally be desirable to have some more magenta petrels to hand for further comparisons to check this conclusion. And on 1 January 1978, 111 years after the latter's sole representative had been acquired, two more magenta petrels were indeed obtained (with a third merely observed), by David Crockett, though not on Pitcairn Island, but instead from the taiko's homeland, Chatham Island![8,14,77]

Photographs and measurements taken of the two captured specimens fully confirmed Bourne's belief—'magenta petrel' and 'Chatham Island taiko' were nothing more than two different names for the same single species. Afterwards, the birds were released back onto Chatham, within the dense bush forest in which they had been captured. Today, the taiko is confined entirely to this island's forested Tuku valley system, with a total population estimated at 100-150 birds, and is classed as critically endangered by the IUCN.[8,14,77]

LEAR'S MACAW—EVERYONE KNEW OF IT, BUT NO-ONE KNEW WHERE IT CAME FROM!

Another mysterious species rediscovered in 1978 was Lear's macaw *Anodorhynchus leari*, a smaller, turquoise-headed relative of the spectacular hyacinth macaw *A. hyacinthinus*. Its existence first became known to science in 1831, when Victorian bird painter and nonsense-rhymes writer Edward Lear painted a macaw of unrecorded origin that he believed to be a hyacinth macaw but which was later recognised to be a separate species, and was named in honour of him (though some authorities also refer to it as the indigo macaw). Despite having been represented in aviaries worldwide since 1831, it remained a major conundrum to ornithologists for over a century—because no-one knew where these captive specimens had actually been caught. Not even their *country* of origin, much less their precise provenance, was known. Indeed, this species might even be extinct in the wild—always assuming that it *was* a valid species, and not a hybrid of the hyacinth macaw and the closely-related glaucous macaw *A. glaucus*, as some researchers were beginning to suggest.[78]

In 1964, the late Dr. Helmut Sick, a German-born Brazilian ornithologist, began an intensive programme of searches for this mysterious missing macaw in a bid to solve its riddles once and for all. It was a programme that would take 14 years before success arrived, but arrive it did. On 31 December 1978, he spied three Lear's macaws in a little-explored area of Brazil's northeastern Bahia region, called the Raso de Catarina. And in January 1979 he sighted a flock of about 20, proving that it was not a hybrid form. These turned out to be part of a population numbering just over 100 birds. Moreover, in June 1995, a team of Brazilian biologists discovered a second population of Lear's macaw, several hundred miles from the first one, consisting of 22 birds on a nesting cliff. In 2009, the total wild population was estimated at around 1000 birds.[78]

In 1990, a single specimen of Spix's macaw *Cyanopsitta spixii* was spotted at a site in northern Bahia by an ornithological expedition sponsored by the International Council for Bird Preservation (ICBP). This small, grey-headed, blue-bodied species was first discovered as long ago as 1819 by Austrian naturalist Johan Baptist von Spix, but it remains one of the world's least-known parrots. Spix's macaw had hitherto been written off as extinct in the wild, and only about 85 specimens presently exist in captivity—most in private collections. In March 1995, following months of painstaking preparation by the Spix's Macaw Recovery Committee, a female captive specimen was released into the wild near to where the male had been observed, in the hope that they would breed. Sadly, however, she disappeared just seven weeks later, and the lone male died at its original Bahia site in October 2000, since when Spix's macaw has been presumed extinct in the wild. As noted, there are believed to be around 85 individuals in captivity, and captive breeding successes have been achieved,

A painting of Spix's macaw from 1931.

Edward Lear's painting from 1831 of the mysterious macaw species that was subsequently named after him following its initial misidentification as a hyacinth macaw.

despite some contention arising between various parties, but the species remains critically endangered.[14,79]

A FLUTTER OF FLIGHTLESS BIRDS

In 1965, the Japanese Ryukyu Islands' southernmost member, Iriomote, offered up to science a previously unknown species of cat (p. 68), plus a rediscovered dwarf pig in 1974 (p. 69). In 1981, it was the turn of the Ryukyus' principal member, Okinawa, to provide a zoological surprise—this time yielding the scientific description of a new, virtually flightless species of rail.

Christened *Rallus* [now *Galirallus*] *okinawae*, inhabiting the island's northernmost portion, Yambaru, and roughly equal in size to an adult chicken, the Okinawa rail is a particularly attractive, colourful species, thus making its belated zoological detection, in 1978, all the more surprising. Needless to say, it is well known to the local people of this area, who call it *yanbaru kuina*. The Okinawa rail's olive-brown wings provide a subtle backdrop for its dark blue underparts, handsomely barred with white, and for its light orange legs and long slender bill. Its throat and face are black, decorated with a broad white stripe running backwards towards the neck from the rear edge of each of its red eyes, and brightened by its head's bluish-grey crown.[80]

Speaking of island rails: an unconfirmed sighting of the New Caledonian woodrail *Gallirallus* [=*Tricholimnas*] *lafresnayanus* occurred in 1984. This species was believed until then to have become extinct in the 1930s, but judging from local testimony it seems possible that small numbers do exist here, probably in the island's mostly inaccessible mountain forests.[81] (This may also be true of the New Caledonian lorikeet *Charmosyna diadema*, currently known only from two specimens—one now lost—obtained prior to the 1860s[81].) Having said that, a faunal survey on New Caledonia in 1998 failed to uncover any conclusive evidence of the woodrail's survival, either from hunters or from fieldwork, but many locals still believe that it persists, and it is presently categorised by the IUCN as critically endangered (as is the New Caledonian lorikeet) rather than extinct.

A second interesting flightless bird described in 1981 was the white-headed steamer duck *Tachyeres leucocephalus*. Native to a small stretch of coast around the Golfo San Jorge in Argentina's southern Chubut and northern Santa Cruz Provinces, like other steamer ducks it derives its name from the steamship-like sound that it makes when swimming.[82]

YELLOW-FRONTED GARDENER BOWERBIRD— NEVER BEFORE SIGHTED IN THE WILD BY A WESTERNER

The yellow-fronted gardener bowerbird *Amblyornis flavifrons* is a species that has attained near-legendary

New Caledonian woodrail.

status within ornithological circles as the ultimate 'lost bird'. Its plumage is principally rufous-brown, and provides a suitably sedate background against which to admire its immense crest of gleaming gold, billowing over its shoulders to stunning effect—or at least this would be the effect if anyone were ever fortunate enough to observe a living specimen, a feat that ornithologists have until quite recently been inordinately unsuccessful at achieving.[21]

To begin with, its official scientific discovery did not take place in its native New Guinea homeland. Instead, it became known by way of three (possibly four) dead specimens bought, for scientific study, at various of the major plume markets of Europe. Its striking appearance not only warranted its classification as a distinct new species (described in 1895 by Lord Walter Rothschild), it also inspired at least a dozen scientific expeditions in search of living examples—but none was obtained.[21] In fact, the species was not even *sighted*—a sorry saga of failure that seemed destined never to end in success, until 1981.

Following a previous visit during October 1979, in January 1981 California University zoologist Dr. Jared Diamond was exploring the uninhabited Foja (Gauttier) Mountains in Irian Jaya (the western, Indonesian half of New Guinea) when he became the first westerner to see a living yellow-fronted gardener. Indeed, he saw not just one but about 22 of them, and estimated that up to a few *thousand* birds may exist there![83]

He even observed its courtship, involving the building of an elaborate turret-like bower by the male, who decorated it with fruit and flowers before attempting (unsuccessfully) to entice a female into it for mating, offering her a large blue fruit in return for her favours.[83]

All in all, the yellow-fronted gardener's history of mystery seemed finally to be at an end, but then came a fickle quirk of fate that enabled this bewitching bird to

The yellow-fronted gardener's rediscovery in 2005 was commemorated in the philatelic world by this magnificent Indonesian miniature sheet, from the author's collection. (Dr. Karl P.N. Shuker)

retain a measure of mystique after all. Needless to say, after encountering the gardeners Dr. Diamond lost no time in taking plenty of photographs of them, and of their mating ritual and bowers—only to lose them all, together with his camera, when his boat later capsized in a river! Except for the memory of his observations, all that he had left were some tape-recordings that he had made of their calls, which ranged from noises similar to the sounds of chopping wood and paper rustling to a motley cacophony of screeches, clicks, whistles, and croaks.[83] Bearing in mind Diamond's misfortune, could they have been the bowerbird equivalent of laughter?

Happily, however, the yellow-fronted gardener's continuing existence was finally confirmed beyond any shadow of doubt two-and-a-half decades later, when it was observed and filmed in the Foja Mountains by an international team of 11 scientists co-led by Dr. Bruce Beehler of Conservation International, which spent a month exploring this remote region during November and December 2005.[83a]

OWLS AND O-OS

In 1981, the cloudforest screech owl *Otus marshalli*—an attractive new species with long white eyebrows—was described from the Peruvian mountains.[84]

Also in 1981 was a probable sighting of the laughing owl or whekau *Sceloglaux albifacies*, one of New Zealand's only two species of true owl, and last recorded in 1914, when a confirmed sighting was made in the Mount Richmond State Forest Park. In 1985, moreover, two very frightened American tourists with no prior knowledge of the whekau heard what they described as "the sound of a madman laughing" while camping in a forest near the small South Island village of Cave, and which a New Zealand ornithologist subsequently claimed was a perfect account of the characteristic 'manic laughing' cry of this species. Yet in spite of extensive searches initiated by these reports, no fresh evidence has emerged so far in support of the whekau's survival.[85]

A much more tangible event of ornithological importance in 1981 was a certain oddly-named bird's equally odd reappearance. Based upon their call (and pronounced 'oh-oh'), 'o-o' is the onomatopoeic common name for a quartet of extremely attractive species of bird traditionally classified as belonging to the honeyeater family, Meliphagidae (but reassigned in 2008 to a family of their own, Mohoidae, following DNA analyses), and confined solely to the Hawaiian Islands. With velvety-black plumage decorated by elegant epaulettes of primrose-yellow, and by similar flourishes upon their long forked tails (and also along the edges of

New Zealand's elusive whekau or laughing owl.

A pair of o-o-aa or Kauai dwarf o-o, all-too-briefly restored from extinction in 1960.

their wings in some species), the o-os were among the most ornate members of their archipelago's avifauna. Tragically, however, their plumes were prized by the native Polynesians, their habitats were swiftly invaded and destroyed by western species introduced by man, and their skins were looked upon as valuable additions to the collections of western museums.

Originally, each major island had its own unique species of o-o, but one by one they disappeared—the Oahu o-o *Moho apicalis* by 1837, Bishop's o-o *M. bishopi* of Molokai by 1904, the exceptionally beautiful Hawaiian o-o *M. nobilis* by 1934, and the Kauai o-o *M. braccatus* (also called the o-o-aa or dwarf o-o) shortly afterwards. The last-mentioned of these was rediscovered in 1960, at Kauai's Alakai Swamp, but since two hurricanes hit the island in 1977 and 1987 respectively, there have been no further sightings of it, and it is presently categorised once again as extinct by the IUCN.[86]

In addition to these species, there was one recorded sighting of an o-o on the island of Maui too, made by bird collector Harry Henshaw in June 1901.[87] But as no further reports emerged from here, it was eventually dismissed as a misidentification of some other bird.

In 1981, however, Henshaw's opinion was finally vindicated, when an American ornithologist called Mountainspring had a clear sighting of an o-o on the northeastern slope of Maui's extinct volcano Haleakala, during a survey of bird migration in the Hawaiian Islands region, between the dates 1 August and 30 November. Most interesting and unexpected of all was the disclosure that this newly-revealed o-o was *not* a new species, unique to Maui, but seemed instead to be a representative of the Molokai species, *M. bishopi*—the first time that a single species of o-o has been recorded from two different islands (subfossil remains of this same species have also been found on Maui). Sadly, however, no further sightings have been recorded either on Maui or on Molokai, so Bishop's o-o is currently categorised once more as extinct by the IUCN.[88]

AN ALBATROSS FROM AMSTERDAM ISLAND

Amsterdam Island is an extremely small, subtropical island, one of the French Southern and Antarctic Territories, sited midway between South Australia and South Africa in the Indian Ocean's southern reaches. It is also the only known breeding ground of a very large, white-faced, dark-bodied species of albatross officially described by science in 1983. Also distinguished from other species by their striped, black-tipped bills, the first specimens were discovered there in 1978. When ornithologists ascertained that the total population consisted of only 30-50 birds, however, they decided very commendably to refrain from taking any for formal description, delaying the latter task until a specimen could

Bishop's o-o. (Errol Fuller)

become available via natural causes. This occurred in 1982, when a dead albatross was found on the island and became the type specimen of the new species—*Diomedea amsterdamensis*, the Amsterdam albatross. Even today, the total population is thought not to exceed 130 individuals, including 80 mature birds, and the species is duly categorised as critically endangered by the IUCN.[89]

MACGILLIVRAY'S PETREL—HEADING FOR REDISCOVERY THE LITERAL WAY

Prior to the 1980s, MacGillivray's (Fiji) petrel *Pseudobulweria* [=*Pterodroma*] *macgillivrayi* was one of the world's most mysterious seabirds, a small dark-grey species known only from a single fledgling collected on the Fijian island of Gau (=Ngau) in October 1855 by naturalist John MacGillivray while serving aboard 'HMS Herald'. Its preserved remains have been housed ever since at the British Museum (Natural History). In 1983, 128 years later, naturalist and Fijian resident Dick Watling began what was to become a year-long pursuit of this 'lost' species. One memorable evening in May 1984, his quest came to a successful end, when he rediscovered MacGillivray's petrel—in a singularly unexpected manner.[90]

He had been seeking petrels not only during the day but also through the night—using powerful flashlights to lure onto the ground any that may be flying nearby. That evening, one petrel had apparently become so dazzled by the lights that it crash-landed—right on top of Watling's head! And so it was that a dazzled petrel and a dazed Dick Watling restored another missing species to life. After carefully examining and identifying the bird—which seemed none the worse for its head-to-head collision—Watling released it. He has since been engaged in estimating on Gau the likely population size of MacGillivray's petrel, which is currently believed to number less than 50 individuals in total. Between 1984 and 2007, 17 reports were recorded of grounded birds on Gau (some of which died), and the only sightings of this species at sea were made off Gau in May and October 2009. It is categorised as critically endangered by the IUCN.[90]

RED SEA CLIFF SWALLOW—LOSING THE LOST

Also of ornithological note during 1984 was the discovery of a new species of swallow—the Red Sea cliff swallow. It was aptly christened *Hirundo perdita* ('lost swallow'), as the locality of its breeding grounds is unknown. Moreover, its type specimen (found dead at Sanganeb lighthouse on an islet off Port Sudan on 9 May) seemed to have been an off-course individual that had somehow been caught up in a widescale emigration of Palaearctic bird species. No other specimen of *H. perdita* has been recorded, although it most likely exists in the Red Sea hills of Sudan or Eritrea, so for the time being the lost swallow is once again lost. However, it is possible that various unidentified swallows sighted in Lake Langano and in Awash National Park in the Rift Valley in Ethiopia belong to this highly elusive species.[91]

A COUPLE OF CRYPTIC PARROTS FROM SOUTH AMERICA

Until 1985, the last time that a new species of parrot had been identified in the western hemisphere was in 1914. However, in 1985 two were revealed within weeks of one another.

During a field trip to the mountain forests of Ecuador's El Oro Province in 1980, ornithologist Dr. Robert Ridgely caught sight of a green parakeet that was clearly a member of the genus *Pyrrhura* ('fire-tailed'), an event that he was quick to recognise as being distinctly odd for two important reasons. Firstly, *Pyrrhura* parakeets were not supposed to exist in this area; secondly, its combination of red and blue wing-patches, red crown, and maroon tail distinguished it from *all* species of *Pyrrhura*, regardless of locality. He was unable to follow up his unexpected sighting until 1985, when a return expedition succeeded in obtaining a series of specimens for formal study. Moreover, he learnt that a single specimen of identical appearance was actually contained within the British Museum's collection of birds at Tring; it had been procured in 1939, but had never been classified or described. In 1988, this species was christened *P. orcesi*, and is popularly known as the El Oro parakeet.[92]

The second new South American parrot from 1985 was the Amazonian (Manu) parrotlet—spied for the first time by Charles Munn, in eastern Peru's Manu National Park. A particularly small species with a powder-blue crown, but otherwise predominantly grass-green in colour, it proved to be a member of the genus *Nannopsittaca* (occupied until then by only a single species), and when it was formally described in April 1991 it was christened *N. dachilleae*, in memory of a famous environmental journalist, the late Barbara D'Achille.[93]

JERDON'S COURSER—ABSENT-WITHOUT-LEAVE FOR 85 YEARS

Not since 1961 had there been such a momentous year for the reappearance of long-lost birds—within just a few weeks in 1986, two of the world's 'classic' extinct birds were rediscovered, followed by the restoration of a third greatly-lamented 'missing' species too.

The first species to re-emerge was Jerdon's courser *Cursorius* [now *Rhinoptilus*] *bitorquatus*, a dainty, long-legged plover-like bird with buff plumage and two characteristic bands of white—one lying like a necklace around its throat, the other located a little lower, extending down from its shoulders to lie across its breast. First recorded in 1848 by Dr. T.C. Jerdon, it appeared to be restricted to the Pennar and Godavari river valleys in India's Andhra Pradesh state, and was very rarely seen at the best of times—but after a reliable sighting made in 1900 by Howard Campbell near Anantpur, it seemed to have vanished completely. Several searches made in later decades failed to find any sign of it, so ornithologists regretfully added Jerdon's courser to the all-too-long list of extinct modern-day birds.[8,86,94]

In the 1960s, the renowned Indian ornithological expert Dr. Salim Ali of the Bombay Natural History Society headed what would prove to be another unsuccessful quest for this species, and a further expedition by the society in the mid-1970s fared no better. Yet even in the face of continuing failure Dr. Ali was reluctant to accept that Jerdon's courser was extinct—if only because there seemed no good reason at all why it should be. It had not been persecuted by man (a refreshing change!), its habitat had not been disrupted, and it seemed to have few if any serious predators.[94]

So in summer 1985, concentrating their efforts upon the Pennar Valley, Dr. Ali and the Bombay Natural History Society launched yet another search, during which they cross-examined the local people in depth, and distributed coloured paintings of the species in poster form. Eventually, the party received its first ray of hope—three different shikaris (native hunter guides) each informed

Jerdon's courser. (Errol Fuller)

Ali and colleagues that he had recently seen just such a bird; the team even learnt that it is referred to locally as the *kalivi-kodi*.[94]

On 12 January 1986, more than eight decades of pursuit ended—in success! One of the three shikaris who had reported seeing the species obtained unequivocal support for his claim—a living specimen of Jerdon's courser, captured that same evening in scrub jungle near his home in Cuddash District. On 15 January, it was examined by Bharat Bhushan, an ornithologist from the Bombay Natural History Society, who confirmed its identity. The society hoped that the courser could be successfully maintained alive in captivity, but sadly it died just four days later. However, its skin was preserved, and other living specimens were soon discovered, but none was captured this time. Instead, the society recommended that the area containing them should become a protected zone, and the Andhra Pradesh forest department agreed, thereby safeguarding the future of this significant species. News of its remarkable rediscovery was afterwards hailed by Dr. Nigel Collar, research director of the International Council for Bird Preservation (ICBP), as being 'the bird conservation highspot of the 1980s.'[94]

IVORY-BILLED WOODPECKER— REPRIEVES FOR THE VAN DYCK OF BIRDS?

With a total length of 20 in, and extremely striking black-and-white plumage highlighted in the male by a brilliant scarlet crest (black in the female), the ivory-billed woodpecker *Campephilus principalis* is a magnificent sight, one that inspired the celebrated bird painter John James Audubon to hail it as the 'Van Dyck of birds'. Two subspecies are recognised—North America's *C. p. principalis* and Cuba's *C. p. bairdii*.[8]

Although widely distributed throughout its range in the forests of the southeastern United States during the 19th century, the American ivory-bill was never a common bird in any given area, due to the specificity of its habitat requirements. Only extremely mature forests were suitable, and each breeding pair of birds required a territory of at least 2000 acres. As a consequence, the clearance of huge expanses of woodland by the developing timber industries towards the end of the 1800s sounded the death-knell for this extremely vulnerable bird—the relentless destruction of its habitat driving it inexorably towards extinction. By the late 1930s, the total population of the entire American subspecies was estimated to number less than two dozen individuals.[8]

Since then, quite a number of eyewitness reports have been documented, but due to the superficial similarity of the ivory-bill to the slightly smaller but much more common pileated woodpecker *Dryocopus pileatus* the chances are that few were valid. Of these latter few, however, two are particularly noteworthy. John V. Dennis, a leading woodpecker expert in the States, sighted an ivory-bill in the forests of Texas's Neches River valley on 10 December 1966 (the first positive Texas record since 1904), and again on 19 February 1967 (this time in the company of fellow ornithologist Armand Yramategui). Subsequent forays convinced Dennis that a few pairs (perhaps as many as ten) existed in the area.[95]

Two years later, on 4 April and again on 15 April 1969, animal sculptor-artist Frank Shields sighted an ivory-bill in a tree on his land (containing a large tract of forest) at Interlachen, Florida. The 15 April sighting was made at a distance of no more than 80 ft, and Shields, familiar with the species' distinguishing markings from his professional studies of animal colouration for his work, unhesitatingly identified the bird as an ivory-bill—but the best evidence was yet to come. On 11 June, once again on his land, Shields discovered a single, very striking black-and-white feather. From its precise size, shape, and markings, he was able to identify it as an ivory-bill's wing feather—specifically, one of the smaller, inner primaries adjoining the secondaries.[95] Since these encounters, however, no additional supportive evidence for ivory-bill survival in the U.S.A. was forthcoming . . . until the close of the 20th Century, that is.

In April 1999, while hunting turkeys in woodlands within Louisiana's Pearl River Wildlife Management Area, student David Kulivan from Louisiana State University (LSU) had a close-range sighting, lasting several minutes,

Ivory-billed woodpecker.

of two extremely distinctive woodpeckers that he feels sure constituted a pair of ivory-bills. They were bigger and with more white on their wings than the pileated woodpecker, a species familiar to him. And whereas one had a red crest, the other's crest was black (both sexes of pileated woodpecker have a red crest). Also, their bills were creamy-white (the pileated woodpecker's is greyer). And when they finally flew away, they gave voice to a call likened by Kulivan to the sound of a toy trumpet—yet another ivory-bill characteristic. He later informed his zoology tutor, Prof. Vernon Wright, and signed a statement for him describing his sighting.[95a]

Wright was impressed enough to send a copy of it to Prof. Van Remsen, curator of birds at LSU's Museum of Natural Science, who, after rigorously interviewing Kulivan, was also impressed, deeming it to be the most credible sighting reported to LSU during the past 20 years. However, a subsequent 30-day search of the area beginning on 17 January 2002, featuring Kulivan and trained ornithologists, failed to espy the birds.

Nevertheless, it was not wholly without potentially significant incident. At 3.30 pm on 27 January, four of the expedition's six members heard a series of loud double raps on an undisclosed tree trunk amid the dense oak woodland surrounding them—identical to the powerful, distinctive, rhythmic tattoo produced exclusively by this equally distinctive species as a means of communication. No other American woodpecker makes this sound, echoing like gunshot, nor does any other produce the eyecatching feeding-associated bark peeling that the team also encountered here.[95a]

So could the ivory-bill still exist in the United States after all? Buoyed by these findings, the team's spokesman, Prof. David Luneau of Arkansas University, stated that he now believed it does indeed still exist. Sadly, however, the team's findings would seem to have raised false hopes. Analysis of the tantalising double-rap sounds revealed that they were merely distant gun shots, whose reverberations sounded to human ears like drumming. Nevertheless, Cornell University ornithologist Dr. John Fitzpatrick, a leader of the search, still believed that ivory-bills may linger undiscovered in parts of the Pearl River forest beyond the area previously searched.[95a]

During an intensive year-long search in the Cache River and White River national wildlife refuges of Arkansas in 2003-04, conducted by the Big Woods Conservation Partnership, led by Cornell University's Cornell Laboratory of Ornithology and the Nature Conservancy, the team made several sightings of a bird that clearly resembled an ivory-bill—culminating on 25 April 2004 with a four-second video film of a male bird taking off from a tree trunk.[95b]

During frame-by-frame playback, the extensive, diagnostic white on the trailing edges of its wings coupled with white on its back that readily distinguish the ivory-bill from the pileated woodpecker could be readily perceived. Loud double-raps, also characterising the ivory-bill, were heard on three separate occasions by members of the team, and other, independent recordings detected bird calls that sounded like a tin horn. Although only a single bird has been seen at any one time as yet, the vast wilderness of this locality is such that it seems reasonable to assume there are others, especially as the Cornell team has so far focused its efforts on only a relatively small proportion of the total 850 square miles of forest here.[95b]

Nevertheless, because no additional sightings have been reported, some ornithologists have become sceptical that the bird seen and filmed was truly an ivory-bill, and further searches by the Cornell team has achieved no success in locating ivory-bills. As a result of this continued failure, in October 2009 the team suspended future searches, but in February 2010 team leader Ron Rohrbaugh announced in the scientific journal *Nature*:

John James Audubon's celebrated painting of ivory-billed woodpeckers.

"We don't believe that recoverable populations of Ivory-billed Woodpeckers persist in those places throughout the species' range that have received significant, systematic search effort. It is possible that a small population of birds exists in heretofore unsearched or under-searched habitats".[95b]

In September 2006, meanwhile, a research team comprising ornithologists from the University of Windsor and Auburn University released a report containing several sightings, numerous drumming recordings, and suggestive foraging signs considered by its authors to be evidence that ivory-bills still exist in the cypress swamps of the Florida panhandle. During 2009, a thorough search of this area was undertaken in the hope of obtaining further sightings and, preferably, photographic and DNA evidence that could be analysed to determine conclusively whether this most enigmatic of species does truly survive, but nothing was found. At present, however, the IUCN categorises the ivory-bill as critically endangered rather than extinct. And in March 2011, Naval Research Laboratory scientist Dr. Michael Collins published a paper that provided positive analysis of audio recordings consistent with those of the ivory-bill alongside discussions of the videos in which they appeared, and which showed a putative ivory-bill in flight.[95c]

Meanwhile, its Cuban counterpart had been suffering a similar fate, due again to habitat destruction, so that by the 1970s the continued existence of this subspecies was equally open to question—which in turn meant that the entire species could well be extinct. To the delight of ornithologists everywhere, however, and as subsequently publicised extensively worldwide, this dismal prospect was triumphantly repudiated on 16 April 1986. This was when, after weeks of unconfirmed reports and all-too-fleeting glimpses of what *might* have been an ivory-bill, an unequivocal sighting of a male specimen in flight was recorded, and at a distance of only 18 ft.[96]

The eyewitness responsible for this historic observation was, by very good fortune, another woodpecker expert—Dr. Lester L. Short, from the American Museum of Natural History. Dr. Short had been participating in a Cuban-led search for the ivory-bill in the island's northeastern Guantanamo province, and made his sighting in a hilly pine forest there called Ojito de Agua. (It should be noted, however, that although receiving far less media publicity internationally, a very brief but clear sighting of an ivory-bill had been made here a month earlier, on 13 March, by Cuban ornithologist Alberto Estrada.)[96]

Conversely, two extensive expeditions in 1991 and 1993 led by ornithologist Martjan Lammertink failed to find any ivory-bills or evidence of their existence in this species' lately-revealed Cuban hideaway either. The last positive sighting here, of a female specimen in flight about 645 ft away, had been in 1987, and although there have been some unconfirmed reports since then, many ornithologists nowadays believe that the Cuban subspecies probably became extinct in 1990.[96]

GURNEY'S PITTA—UNSEEN FOR 34 YEARS

Gurney's pitta *Pitta gurneyi* is an exceedingly beautiful ground-dwelling species whose multicoloured plumage elegantly combines brown wings, barred yellow underparts, and a bright turquoise tail, with a black-and-gold face, and a shimmering blue crown. Once common in the lowland forests of southern Thailand and Myanmar (formerly Burma), its habitat has suffered such long-standing and wide-ranging devastation at the hands of farming and human settlement that this exquisite species had not been positively sighted since 1952, and at the end of 1985 it was formally classed as extinct. Happily, in June 1986, four years of persistent searches by Uthai Treesucon and Philip Round (from Bangkok's Mathidol University Wildlife Research Centre) paid off handsomely, when they succeeded not only in observing, but also in photographing, Gurney's missing pitta. They even discovered a nest with fledglings. Plainly, therefore, the species was still breeding. So its two rediscoverers subsequently sought to persuade the Thai government to declare as a protected area the section of forest in which their pitta sightings were made, in the hope of securing this bird's safety from the combined threat of farming, logging, and trapping for the animal trade market. Sadly, however, Gurney's pitta remains endangered.[97]

In 2003, however, a team of conservationists drawn from a combination of different societies announced the very welcome discovery of specimens in four additional sites, but not in Thailand—in Myanmar instead. This is the first record of Gurney's pitta in Myanmar for 89 years, but these recently-located specimens are currently threatened with the risk of habitat loss due to forest clearance for logging. Happily, additional territories housing this species have since been discovered in Myanmar, which will be of great benefit in assisting this once-lost species' survival.[97a]

In 1988, Schneider's pitta *P. schneideri*, a species exclusive to Sumatra, was rediscovered in rainforest on Gunung Kerinci, the island's highest mountain. Its last previous sighting had been in 1977.[98]

Returning to 1986, this year continued its compelling record for avian comebacks with the revival of the golden-naped weaver *Ploceus aureinucha*. Last reported in 1926, several specimens were sighted in the Ituri Forest of Zaire (now the Democratic Congo) by three zoologists studying one of the 20th century's most famous newcomers, the okapi.[99]

HELMETED WOODPECKERS AND HOODED ANTWRENS

In February 1987, one of Brazil's most striking birds reappeared, via a confirmed sighting of a female helmeted woodpecker *Dryocopus galeatus* in western São Paulo. Not too dissimilar from its famous relative the ivory-bill, but easily differentiated by its smaller size and heavily barred underparts, this species has also been reported spasmodically from eastern Paraguay, but with less conviction, thereby making the São Paulo sighting—by an ornithologist called Edwin—the first reliable record of the helmeted woodpecker for several decades.[100]

Another long-lost bird from Brazil made headlines in 1987, when a pair of black-hooded antwrens *Myrmotherula erythronotos* was observed in September by members of the Rio de Janeiro Birdwatchers Club, at a secret locality within the remains of what had once been an extensive forest in Rio de Janeiro State. A small, ebony-headed, chestnut-backed member of an exclusively New World family of birds, it had previously been known only from museum specimens collected a century earlier, from a region of tropical forest between Rio de Janeiro and Campos that was largely destroyed in later years—thus making this species' rediscovery all the more unexpected and important.[100]

MADAGASCAN SERPENT-EAGLE— REDISCOVERED BY AN EAGLE-EYED EXPEDITION

Summer 1988 saw the rediscovery after 58 years of the Madagascan serpent-eagle *Eutriorchis astur*, when Drs. B.C. Sheldon and J.W. Duckworth, two members of an expedition from Cambridge University, England, spotted a very distinctive bird hunting along a river valley beneath the canopy of Madagascar's dense, humid, northeastern forests. During their 45-minute period of observation, they could clearly discern the bird's curious hood-like crest of feathers—a crest that succinctly distinguishes *Eutriorchis* from all other birds of prey on the island. Moreover, on 23 February 1990 herpetologist Dr. Chris Raxworthy found the decomposed body of a *Eutriorchis* by a trail running between the villages of Iampirano and Ranomena II in the Ambatovaky Special Reserve's rainforests. He collected its skull and three primaries, which were formally identified by ornithologist Peter Colston of Tring Museum. Most exciting of all: on 14 January 1994, a team of field biologists working in northeastern Madagascar for the Idaho-based Peregrine Fund captured a living serpent eagle in the Masoala peninsula. After photographing it, ringing it, and attaching a transmitter to it in order to monitor its movements, they released the eagle back into the wild.[101-2]

Fairly small, measuring no more than 23-26 in (much of which constitutes its very long tail), and with heavily barred underparts, the Madagascan serpent-eagle was last recorded conclusively in 1930, and the 1990 skull is the first intact example in any museum.[101]

Summer 1988 also witnessed the refinding of São Tomé's dwarf ibis *Bostrychia* [=*Lampribis*] *bocagei* (last seen in 1928), following a visit to this very small West African island shortly before by an ICBP team.[102]

NIGHT PARROT—BROUGHT BACK TO LIFE BY A DEAD SPECIMEN

Not only nocturnal but also virtually flightless, the night parrot *Pezoporus* [formerly *Geopsittacus*] *occidentalis* is a very unusual, ground-dwelling species that was common in arid areas throughout Australia in the 1800s. During the 20th century, however, sightings were very scarce, most during its early years, and substantiated by the collection of just one specimen—in 1912 at Western Australia's Nichol Spring. Consequently—yet with surprising disregard for the fact that four specimens were reliably spied at Cooper's Creek in South Australia's

Night parrot—one of Australia's most elusive birds.

far northeastern section in 1979—the night parrot has been considered by most authorities to be one of this continent's *former* inhabitants. In late October 1990, however, during a drive along a road near Queensland's Mount Isa, Australian Museum ornithologist Walter Boles stepped out of his vehicle to observe a rare bird, and discovered an even rarer one lying dead at his feet. It was a recently-killed night parrot, one that had met its death in a collision with a car or some similar vehicle. A veterinary pathologist from Sydney, who later examined this stunningly serendipitous find, verified that it had been dead for no longer than a year at most, and quite probably for as little as three months.[103]

An incredibly fortuitous discovery had thus restored the night parrot back onto the list of living Antipodean fauna. In 2005, three live individuals were reliably sighted by biologists Robert Davis and Brendan Metcalf near Minga Well, in Western Australia's Pilbara region; and a year later rangers in southwestern Queensland's Diamantina National Park found a dead juvenile specimen that had flown into a barbed wire fence and decapitated itself.[103]

SÃO TOMÉ GROSBEAK—A MAJOR REDISCOVERY ON A MINOR ISLAND

Hot on the heels of their success in uncovering its dwarf ibis in 1988, ICBP ornithologists revisiting the tiny West African island of São Tomé in 1991 scored an even greater triumph when they observed a pair of small finch-like birds in a forest here near the Rio Xufexufe in the island's southwestern portion—for these belonged to São Tomé's endemic species of grosbeak, last seen alive in 1890!

Rusty-brown in colour, with a massive bill, the São Tomé grosbeak *Neospiza concolor* is one of the world's least-known birds. It was discovered in 1888, when Francisco Newton collected two specimens in forests near Angolares on the Rio Quija, on the island's eastern coast. It was next reported two years afterwards, but despite several specific searches for it since then, it had never been spotted again—until its unexpected rediscovery 91 years later by David Sargeant and Tom Gullick. Following São Tomé's settlement by the Portuguese in 1483, great expanses of lowland forest were destroyed to make way for coffee plantations, implying that habitat loss was the principal reason for this species' apparent extinction. Clearly, however, in spite of such desecration the grosbeak has managed to survive, though the total population has remained a very small one, probably numbering less than 50 birds, and must be very vulnerable to any further dangers. Consequently, it is presently categorised as critically endangered by the IUCN.[104]

The early 1990s saw a number of other noteworthy resurrections within the bird world too. Hitherto known only from its type specimen, collected in 1937, the Cocha antshrike *Thamnophilus praecox* was refound in 1990 when a team from Philadelphia's Academy of Natural Sciences observed three pairs on the shores of Imuya Cocha, in Ecuador's Rio Lagarto drainage. In August 1991, the Santo Martin starling *Aplonis santovestris*, endemic to Vanuatu's Espiritu Santo and not seen by any of the expeditions that have visited this island since 1961, was spotted in cloudforest on Mount Santo.[105]

A drake netted alive on 29 August 1991 by a fisherman at Lake Alaotra, in northeastern Madagascar, proved to be a Madagascan pochard *Aythya innotata*, a species of duck restricted almost entirely to this lake and not spied since 1960. In November 2006, moreover, a flock of nine adult pochards and four recently-hatched ducklings were discovered at a volcanic lake in a remote area of northern Madagascar, and 25 mature pochards were observed here in 2008. It is currently listed as critically endangered by the IUCN.[105]

In 1993, the four-coloured flowerpecker *Dicaeum quadricolor* was rediscovered, in a small patch of limestone forest on the Philippine island of Cebu, where it was last seen 80 years previously. This small passerine

Madagascan pochard.

species was later discovered in three other sites too, and its present population is estimated to be 85-105 birds in total.[105]

UNMASKING THE TYNE PETREL
Since 1988, several strange storm petrels with dark rumps and forked tails have been spied each summer in northeastern England, and have sometimes even been ringed. Dubbed Tyne petrels, they were believed for a time to constitute a species new to science. However, DNA fingerprinting techniques revealed in 1992 that they were Swinhoe's petrels *Oceanodroma monorhis*, an eastern Asiatic species not previously recorded from the North Atlantic.[106]

THE INVISIBLE KIWIS
One specimen of the brown kiwi outwardly looks very much like another—but their genetic profiles tell a very different story. Native to the upper two-thirds of North Island and to several regions of South Island, in 1993 the results of genetic testing using two independent techniques revealed that at least three separate, valid species of brown kiwi were present. The reason why they had not been differentiated from one another before is that their genetic differences are ones that do not noticeably affect their external morphology, i.e. their taxonomic characteristics are 'invisible'. The three species constitute the North Island brown kiwi *Apteryx mantelli*, and two different species in South Island.[107]

The most widespread of the lately-defined South Island species has been dubbed the tokoeka ('tokoeka' is the southern Maori name for the kiwi) *A. australis*. It is in turn split into four separate subspecies, which collectively occur in Fjordland, the Southern Alps, and also on Stewart Island, and are said to be bigger sometimes than the North Island species, but not all kiwi experts agree with this. One of these subspecies, the Haast tokoeka, inhabits the rugged mountains behind Haast in New Zealand's Southern Alps. In summer, the Haast tokoeka stays in the high subalpine tussock grasslands, but in winter it may venture down to the lowland forests.[107]

The third, and most intriguing, member of this recently-unmasked trio is the Okarito brown kiwi or rowi *A. rowi*, confined to a small expanse of lowland forest just north of Franz Josef, on South Island's western coast. What makes this species (formally described and named in 2003) so interesting is that genetically it seems to be more closely related to the North Island brown kiwi than to any of its fellow South Island brown taxa. Close observation reveals that its plumage is slightly greyer than that of the other South Island brown kiwis, and it sometimes has white facial feathers, but these morphological characteristics are not readily discernible.[107]

XENOPERDIX—AN OUT-OF-PLACE PARTRIDGE GENUS FROM AFRICA
Africa's contribution to Phasianidae, the partridge and pheasant family, consists of the Congo peacock *Afropavo congensis*, the stone partridge *Ptilopachus petrosus*, and numerous species of superficially partridge-like birds called francolins, but no real partridges—or at least that was the belief of ornithologists until 1991. In that year, Dr. Lars Dinesen led a team of Copenhagen University ornithologists to the little-explored evergreen forests of

Four-coloured flowerpecker.

Udzungwa forest partridge—African in origin, but Asian in taxonomic affinity. (Philippa Foster)

Tanzania's remote central Udzungwa Mountains to study their avifauna. While there, the team spied several specimens of a baffling partridge-like bird that was readily distinguishable from all known species of francolin. Remarkably, it seemed most similar in overall appearance to the hill-partridges—a group of *Arborophila* species native to the Orient![108]

Studies of collected specimens confirmed this zoogeographical anomaly's taxonomic affinity with the latter birds. However, it was very distinct even from these. Consequently, in 1994 it was formally dubbed *Xenoperdix udzungwensis*, the Udzungwa forest partridge—sole occupant of a brand-new genus. Just like the Congo peacock (p. 123), *X. udzungwensis* is a relict species, surviving today in a little-disturbed, environmentally stable pocket of East Africa, but possibly descended from some ancestral species that may have formerly inhabited an area stretching from Africa to the Far East, at a time in the distant past when these regions were connected by a temporary land-bridge.[108]

A second population, discovered in Tanzania's Rubeho Mountains in 2000, was subsequently shown, following a review of its birds' molecular and morphological characteristics, to constitute a second species in its own right. In 2005, this new species was christened *X. obscurata*, the Rubeho forest partridge.[108a]

ON THE WING OF A MYSTERY—THE NECHISAR NIGHTJAR

On the evening of 3 September 1990, Dr. Roger Safford, a member of a visiting conservation team from England's Cambridge University, was driving along a rough track in an area of treeless grasslands within southern Ethiopia's Nechisar Plain National Park, when his torch's beam revealed a very squashed bird carcase in the road just up ahead. Stopping, he discovered that it had once been a nightjar, but the only portion that could be salvaged for inspection was its left wing. Five years later, his find proved to be a notable ornithological highlight of the 1990s when in August 1995 a team of experts at Tring Museum, the ornithological branch of the British Museum (Natural History), announced that its deceased owner belonged to a female or immature specimen of a hitherto unknown nightjar species. For in their opinion, the distinctive white 'flash' marking on the carpal region of the wing, and the wing's large size (1 ft long) readily distinguished it from all previously described nightjars.[109]

Commonly referred to as the Nechisar nightjar, it was formally dubbed *Caprimulgus solala*—'single-wing nightjar'. This is a very apt name, because for a number of years, the solitary wing found by Safford remained the only physical evidence for this highly elusive bird's existence. In 2009, however, an expedition to the Nechisar Plain National Park led by ornithological field-guides author Ian Sinclair briefly spotted an adult male Nechisar nightjar on the very first evening there (like other nightjars, it is nocturnal), and it was seen several more times during the next few nights.[109]

Previously, some researchers had expressed doubt that it was even a valid species, noting that some nightjar species exhibit great variation in plumage characteristics.

However, the adult male's much larger size than any other nightjars in the region, coupled with its wings' huge white carpal flash markings, provide strong evidence for supporting the validity of separate specific status for *C. solala*.[109-110]

Taxonomic tribulations also surround another lately-unveiled nightjar. In 1994, Frederico Lencioni-Neto described *Chordeiles vielliardi*, a new species of nighthawk (closely related to the *Caprimulgus* nightjars). Named in honour of the famous ornithologist Jacques Vielliard, and known commonly as the Bahia nighthawk after the Brazilian state in which it had been discovered, it had first come to scientific attention when a small group of specimens was observed in dry caatinga vegetation at Manga, near the village of Queimadas on the left bank of the Rio São Francisco. Since its description, however, some ornithologists have opined that this nighthawk has been misclassified, suggesting that it is actually a member of the genus *Nyctiprogne* rather than *Chordeiles*.[110-11]

Yet another notably elusive nightjar is Franklin's nightjar *Caprimulgus affinis*. Unreported for several years, it was rediscovered in 1999 when a specimen of this expertly-camouflaged bird happened to ruffle its feathers somewhat noisily while a group of birdwatchers was close by, at Keerampara Grama Panchayat in Kerala, India. They were also able to film it, and even discovered a couple of nests here.[112]

BULO BURTI BOUBOU SHRIKE—
DELINEATED BY ITS DNA

In 1991, a unique specimen made a unique claim to taxonomic fame—for this was the first time in ornithological history that a new species of bird was described and named using only DNA and a few feathers from a captured, subsequently released specimen as type material.

This singular episode began in late August 1988, when ornithologist Edmund F.G. Smith, from Merseyside, England, was visiting central Somalia, and happened to espy an unfamiliar-looking boubou shrike in hospital grounds on the edge of Bulo Burti, a town on Somalia's Shabeelle River. He came to this area again in October 1988, and most weeks thereafter when visiting the hospital he spotted the strange shrike there. Although it resembled wing-striped forms of the familiar, widespread tropical boubou *Laniarius aethiopicus*, the Bulo Burti specimen differed from these by having a long broad eyebrow-stripe of yellow, white underparts becoming yellow on the sides of the throat and breast, a mottled area with yellow feather-tips across the hindcrown, and a noticeably deeper bill.[112a]

Feeling sure that it represented a species new to science, but concerned that as he never saw more than one specimen its species may be near extinction, Smith contacted Dr. Nigel J. Collar at the International Council for Bird Preservation (ICBP) for advice on how to proceed. He was urged to document it fully via photographs, video film, and tape recordings, and to capture it alive for blood samples, but not to kill it. Following these suggestions, Smith duly caught this singular shrike and documented it thoroughly, as well as collecting blood samples and a few feathers moulted by the bird while in captivity. Unfortunately, due to civil unrest in Somalia at that time, the shrike could not be released afterwards, and was instead taken to Germany by J.-U. Heckel, a colleague of Smith, where it remained until Heckel returned to Somalia in March 1990. By then, however, Bulo Burti had little remaining habitat suitable for the release of the bird, so Heckel took it to a well-protected patch of riverine woodland lower down the Shabeelle, in the Balcad Nature Reserve. Here, on 23 March 1990, following a period of acclimatisation, the shrike was released.[112a]

Morphological studies based upon Smith's visual and audio records of the bird were conducted by Smith and two colleagues. These were supplemented by DNA research undertaken at Copenhagen University by Dr. Peter Arctander, involving comparative analyses of base sequences from the cyt-b gene of mitochondrial DNA extracted out of feather quills from this specimen, and from museum specimens and live specimens of various known shrike species. The results all supported Smith's original opinion that this bird represented a new species, and discounted possibilities that it was merely a hybrid or a colour morph of some already-known species. Accordingly, in 1991, Smith, his colleagues, and Arctander officially named its species *Laniarius liberatus* ('freed shrike'), the Bulo Burti boubou shrike.[112a]

Although this classification was subsequently criticised by some researchers, who were concerned that there was no type specimen available, it has recently been vigorously defended by Dr. Collar.[112a] And certainly, with such advanced DNA-based discrimination techniques as those already available, let alone those likely to be developed in the near future, this method of taxonomic determination offers a very laudable (and accurate) method of investigating new species that may also be exceedingly rare and therefore potentially at risk of extinction from the mode of specimen collection traditionally deemed normal when procuring zoological material on which to base an identification.[112a]

Having said all of that, it is now scarcely relevant to the Bulo Burti boubou shrike, however, because, with dramatic irony, in 2008 another research team published their own molecular analyses in relation to this and other *Laniarius* shrikes—which revealed that *L. liberatus* was synonymous with the Somali boubou *L. erlangeri*, and therefore not a separate species after all. Bye bye, then, to the Bulo Burti boubou.[112a]

OLD BIRDS IN NEW BUSHES

Quite a few supposedly-extinct birds turned up alive and well during the mid-1990s. The exceedingly elusive, aptly-named invisible rail *Habroptila wallacii* had not been reported on its Indonesian island home of Halmahera for more than 40 years. Then in 1995, this large slaty grey-blue species with browner wings and bright red beak and legs was finally sighted here in a swamp by an ornithological team working jointly with BirdLife International and the Indonesian PHPA. The team also visited the island of Sulawesi (Celebes), where they repeated their success by rediscovering two other 'vanished' species. These were the Lompobattang flycatcher *Ficedula bonthaina* (not seen since 1931), and, in October 1998 on Sangihe Island (just north of Sulawesi itself), the cerulean paradise flycatcher *Eutrichomyias rowleyi* (a bright-blue species known from a skin obtained in 1874, and one sighting in 1978).[113]

Equally notable were the rediscovery in 1994 amid evergreen forest in Chu Lang Sin, southern Vietnam, of the grey-crowned crocias *Crocias langbianis*, a babbler last seen in 1937[114]; and the Pohnpei mountain starling *Aplonis pelzelni* on its eponymous Micronesian island home, when on 4 July 1995 ornithologist Donald W. Buden obtained a dead specimen that had been shot by a local guide in 1994, the first recorded since 7 March 1956. To date, however, no further specimen of this starling has been recorded, so its species may now be extinct, and is currently categorised as critically endangered by the IUCN.[115]

SIZING A SHEARWATER UP (OR DOWN?)

The year 1995 also the official description of the Mascarene shearwater *Puffinus atrodorsalis*, a black-backed seabird related to the familiar Manx shearwater *P. puffinus*. Unlike some of the earlier-noted avian arrivals documented here, this species seems to be quite common, having been recorded from several different locations, as far apart as Israel and the coasts around east Africa and Madagascar. Whether or not it is a valid species, however, is another matter, as some ornithologists have subsequently suggested that its description was in reality based upon juvenile specimens of Audubon's shearwater *P. lherminieri*.[110,116]

ACROBAT BIRDS AND ANTWRENS—NEW GENERA FROM THE NEW WORLD

Recently-revealed species so different from their closest relatives that they require the creation of entirely new genera are hardly commonplace—which is why a certain warbler-sized species with pink legs that spends most of its time upside-down has attracted so much interest both in ornithological circles and among the world's media.

Frequenting the forest canopy providing shade for the cocoa fields of southeastern Bahia in coastal Brazil, the graveteiro (Portuguese for 'twig-gatherer') came to scientific attention in November 1994, when it was spied by Dr. Paulo Sergio Fonseca. An experienced birdwatcher, but unable to identify this small black-and-grey bird with any species already known from that region (or elsewhere), and intrigued by its habitual yet unusual preference for running to and fro along the underside of twigs in search of termites and other insects, Fonseca made two more excursions here with colleagues to confirm his tiny feathered acrobat's profound taxonomic significance. Thus, when the graveteiro was formally described by Dr. José F. Pacheco, Bret M. Whitney, and Luiz A.P. Gonzaga from Rio de Janeiro's Federal University in November 1996, they dubbed it *Acrobatornis fonsecai*—'Fonseca's acrobat bird'.[117]

Graveteiro. (Philippa Foster)

Ironically, although the graveteiro itself had hitherto avoided scientific detection, its nests have been readily visible for many years. A member of the ovenbird family (Furnariidae), *Acrobatornis* constructs huge nests of tightly interlaced twigs and sticks, containing one or more internal rooms that the bird enters through long tunnels. And it just so happens that a number of trees containing these eyecatching edifices stand along Highway BR-101, which runs through the graveteiro's distribution range and is plentifully supplied with traffic every day. Yet until the sharp-eyed Dr. Fonseca brought this species to attention, its nests had been consistently misidentified as those of a related ovenbird already known to science.[117]

Even more ironic, however, is the tragic fact that the recently-revealed graveteiro may soon be extinct. This is because the cocoa fields' shade trees on which it lives are being cut down and sold by cocoa farmers desperate to earn money, due to the wide-scale depletion of the cocoa fields themselves by an accidentally-introduced but lethal fungal infection. Hence Dr. Pacheco and colleagues hope that the Brazilian government will turn the graveteiro's known distribution range into a reserve, in order to protect what remains of its habitat.[117]

Many media accounts alleged that the graveteiro was the first new genus of Brazilian bird to have been described during the 1990s. In reality, however, any claim that *Acrobatornis* may have laid to this status had already been brushed aside a few months earlier. In June 1996, Brazilian zoologists Drs. Marcos R. Bornschein, Bianca L. Reinhert, and D.M. Teixera announced that a new genus of antbird had been discovered amid swampland in southern Brazil's Paraná State. Christened *Stymphalornis acutirostris*, the Paraná antwren is just under 6 in long, and is characterised by its narrow beak, long tail, and the distinctive structure of its vocal apparatus. Once again, however, this ornithological newcomer's continuing survival is at risk—due to human encroachment upon its habitat.[118]

AN OVERLOOKED OWLET AND OTHER SHY OWLS

One of the most remarkable avian comebacks must surely be that of the forest spotted owlet *Athene blewitti*. Previously known only from nine specimens all collected during the 1800s, this mysterious Asian species had been traditionally looked upon for much of the 20th century as one of ornithology's 'classic' extinct birds, just like three other equally elusive Indian species—the Himalayan mountain quail *Ophrysia superciliosa*, pink-headed duck *Rhodonessa caryophyllacea*, and Jerdon's courser *Rhinoptilus bitorquatus*.[8]

Unseen since 1868, the mountain quail currently remains in the land of the non-living, at least as far as official records are concerned. So too does the pink-headed duck, technically extinct since 1942 (though with a sizeable dossier of unconfirmed post-1942 sightings on file). Jerdon's courser, however, was sensationally rediscovered in 1986, after having last been sighted in 1900 (see p. 148). And eleven years later, 113 years after its 'final' confirmed record was made in 1884, the forest spotted owlet also reappeared—following an extensive trek by a trio of American researchers during November 1997 in search of evidence that it did still exist.[119]

Led by Dr. Pamela C. Rasmussen from Washington D.C.'s National Museum of Natural History, their quest was fuelled by a desire to end a long-running controversy regarding this distinctive 8-in-high owl, readily identified by its banded wings and underparts. In 1914, Colonel Richard Meinertzhagen (who had already acquired cryptozoological immortality as the discoverer in 1904 of the African giant forest hog *Hylochoerus meinertzhageni*, see p. 31) claimed to have rediscovered the forest spotted owlet—and he presented, as supposed proof of his claim, a stuffed specimen of this species.[119]

However, when Meinertzhagen's specimen was closely examined, it was found to have been stuffed via a unique, highly distinctive method devised and utilised exclusively by one particular scientist who had collected and preserved a forest spotted owlet specimen during the 1880s—a specimen that had subsequently gone missing from London's Natural History Museum and had not been seen again . . . until 1914![119]

Accordingly, Meinertzhagen's claim was discredited, as were many of his other ornithological records in 1993, but the owlet incident was sufficiently intriguing in itself to inspire Rasmussen and her team to set out for India's Maharashtra Province, hoping to resolve whether this largely-forgotten bird really did still survive. And survive it does—if only just. After 12 days of unsuccessful searching, the team encountered two specimens on two different days, but in the same wooded zone near Shahada, northeast of Bombay, which were duly photographed and videoed. They may represent a breeding pair, as they were certainly of opposite sexes—according to Rasmussen, two specimens of the same sex would not tolerate each other within the same hunting area. Since then, a number of individuals have been documented in several different sites across central India.[119]

Forest spotted owlet. (Philippa Foster)

Worth recalling from this present book's first edition, *The Lost Ark* (1993), is that some tantalising support for the forest spotted owlet's continuing survival was obtained back in 1968. This was when a photograph was taken of a small, short-winged, medium-brown owl with a conspicuous dark brown bar across its throat, sighted near Nagpur in Madhya Pradesh, India.[120]

Staying with rediscovered owls: in 1994, a captive specimen of the Madagascan red owl *Tyto soumagnei*, related to the familiar barn owl but only recorded once since 1934, was found in the town of Andapa and had been caught more than 180 miles north of all previous sightings. In August 1996, another specimen was recorded, this time in Madagascar's Zahamena Reserve, by a team of Conservation

International biologists conducting a species inventory there, and other sightings have been documented elsewhere in the island's southeastern region in more recent years.[121]

Several new species of owl have also been revealed lately. These include two new pygmy owls. *Glaucidium parkeri*, the subtropical pygmy owl, was discovered on the eastern Andean slopes of Ecuador and Peru. It was formally described in 1995 by Dr. Mark B. Robbins, collection manager for birds at Kansas University's Natural History Museum, and fellow ornithologist Dr. Steve N.G. Howell.[122] The second, also described by Robbins, together with Dr. Gary Stiles from Colombia's National University, and as recently as 1999, is the cloud-forest pygmy owl *G. nubicola*. This notoriously reclusive, tiny species, most readily distinguished by its song, inhabits the Pacific slopes of the western Andes' dense cloud forests in southern Colombia and northern Ecuador.[123]

Three new species of scops owl were described and named in 1998. The minuscule Indonesian island of Sangihe, near Sulawesi, is home to the widely-distributed yet long-overlooked Sangihe scops owl *Otus collari*, named by Drs. Pamela C. Rasmussen and Frank R. Lambert after renowned avian conservationist Dr. Nigel J. Collar of BirdLife International. Its type specimen is a female specimen retained in the State Natural History Museum of Braunschweig, Germany, which was collected as long ago as January 1887. Other belatedly recognised museum specimens include one deposited at Leiden in 1866, a second deposited there in May 1985, and one deposited at Dresden in 1871.[110,124]

Dr. Rasmussen's unrivalled success at unveiling owls during the 1990s also includes the Nicobar scops owl *Otus alius*, which she formally described from the only two known specimens of this new species. These were collected in 1966 and 1977 respectively on India's Great Nicobar Island.[125] Completing the *Otus* trio is the Mohéli scops owl *O. moheliensis* from Mohéli Island in the Comoros group. Existing in two separate colour morphs (one rufous, one brown), it currently has an estimated population of 400 individuals, which inhabit the last small section of forest still surviving on this island.[126]

SNIFFING OUT A SNIPE
As documented earlier, New Zealand's Campbell Islands are home to a rare, eponymous flightless teal, *Anas* [=*Xenonetta*] *nesiotis*, which was searched for in November 1997 by a Department of Conservation team from New Zealand. While seeking teal on a tiny rocky outcrop known as Jacquemart Island, however, one of their tracking dogs sniffed out a strange-looking snipe that none of the team could identify. Other specimens were duly sought, and about ten were discovered in total. When these were examined, the team realised to their excitement that they represented a form wholly new to science.[127]

Although a significant ornithological find, this unexpected bird is merely a new subspecies (albeit a well-marked one) of the New Zealand or subantarctic snipe *Coenocorypha aucklandica*—and was formally dubbed the Campbell Island snipe *Coenocorypha aucklandica perserverance* in February 2010. Consequently (as this section of the present book amply demonstrates!), it is certainly *not* "the first new bird species discovered since 1930", which was erroneously claimed in several media accounts at the time. It is named after Captain Frederick Hasselburgh's sealing brig *Perseverance*, which reached Campbell Island in 1810 but unfortunately also brought Norway rats with it when it was wrecked there in 1828; these nearly wiped out the island's avifauna, including its unique snipe.[127]

CHEERY NEWS ABOUT THE CHERRY-THROATED TANAGER
In 1870, a handsome black-and-white finch-like bird with startling cherry-coloured throat plumage was collected by naturalist Jean de Roure on the north bank of the Rio Paraíba do Sul near Muriahié in southeastern Brazil's Minas Gerais state. It became the type specimen of the cherry-throated tanager *Nemosia rourei*, which was not heard of again until 1941, when a flock of eight was observed near Itarana in the Limoeiro-Jatiboca area of Brazil's Espirito Santo state by Brazilian ornithologist Dr. Helmut Sick on 8 August. Since then, the only hint that this attractive little bird may still exist came in 1992, courtesy of a noteworthy yet unconfirmed sighting at Nova Lombardia Biological Reserve in Espirito Santo.[14,128]

Six years later, however, its survival was verified when up to four cherry-throated tanagers were observed, photographed, and even tape-recorded on 22 and 24 February 1998 by a team of six Brazilian researchers that included Dr. José Fernando Pacheco. The encounters took place on private land in the mountains of Conceição do Castelo municipality, the first lasting 20 minutes, the second 75 minutes. And on 12 September 2003, this elusive species' presence was also confirmed in Caetés Forest, 12.5 miles north of Vargem Alta.[128]

TAPPING INTO A NEW TAPACULO
Tapaculos constitute an exclusively Neotropical taxonomic family of small, ground-living perching birds—Rhinocryptidae, deriving its name from the moving flap or operculum that hides these birds' nares (nostrils), probably as a protection against dust. Several new species were discovered during the 1990s, but the most publicised of these was *Scytalopus iraiensis*, variously known as the lowland tapaculo or tall-grass wetland tapaculo. Its discoverers were none other than Drs. Bornschein and Reinert, who were responsible for bringing the Paraná antwren to scientific attention in 1996.[110,129]

Also from Paraná, southeastern Brazil (as well as from the highlands of Minas Gerais, further north in Brazil), this grey-black *Scytalopus* species was formally described and named in 1998, and is superficially similar to another tapaculo living in forests just a few miles away. However, *S. iraiensis* can be differentiated by its distinctive song, which is what first attracted its discoverers' attention, and is confined to marshland.[129]

A RAIL AND A ROBIN

In July 1998, ornithologists publicly announced the discovery of a new species of rail belonging to the genus *Gymnocrex* in the Talaud Islands of Indonesia. Formally dubbed the Talaud rail *Gymnocrex talaudensis* in 1998, the closest relative of this distinctive red-necked, green-winged species is the rare (or, at the very least, exceedingly elusive) bald-faced rail *G. rosenbergii*, also from Indonesia but this time inhabiting the large island of Sulawesi.[14,130]

A month later, graduate student Pamela Beresford from the American Museum of Natural History attended the 22nd International Ornithological Congress, held at Durban, in South Africa. While there, she disclosed that while participating in a WWF-sponsored study in the Central African Republic during November 1996, she and fellow team members had collected what appeared to be a new species of forest robin. Its distinguishing features are its yellow stomach, and its bright reddish-yellow throat and upper breast; those of other forest robins are white. In 1999, it was officially named *Stiphrornis sanghensis*, the Sangha forest robin.[131]

A ROTHSCHILDIAN REVELATION, AND THE ANTPITTA OF RIDGELY

One of the world's most mysterious birds, Rothschild's parakeet *Psittacula intermedia* was known for many years from only seven specimens, which had been sent to England between 1895 and 1907 from India, but were of unknown specific provenance. Consequently, many ornithologists eventually began to speculate that it must be extinct. Since the early 1980s, however, a few additional specimens have appeared each year in the bird markets of India, thereby raising hopes that it thrived after all, and that its secret homeland in the wild would eventually be found.[14]

Sure enough, Karthikeyan Vasudevan from the Wildlife Institute of India in Dehra Dun announced in 1998 that Rothschild's parakeet was one of ten ostensibly lost or vanished vertebrate species that had been rediscovered during a major two-year zoological survey of India. Completed in 1998, the survey had scoured India's southern rainforests, its Himalayan foothills, and its western desert scrub. However, the saga of Rothschild's parakeet's long-awaited rediscovery, far from having drawn at last to a close, would soon feature another revelation, and of a very different kind this time.[132]

In 1999, Dr. Pamela C. Rasmussen (rediscoverer of the forest spotted owlet) and BirdLife International conservationist Dr. Nigel J. Collar published an extensive paper analysing the seven original specimens of *Psittacula intermedia*. Its scientific name derived from the fact that this species is intermediate in appearance between males of the plum-headed parakeet *P. cyanocephala* and slaty-headed parakeet *P. himalayana*. This had led some ornithologists in the past to suggest that Rothschild's parakeet might simply be a hybrid of these two species, but no serious investigation of such a possibility had been conducted.[133]

Following their detailed examination of the seven specimens, however, as well as various alleged specimens alive in captivity, and historical records, Rasmussen and Collar concluded that *P. intermedia* was undoubtedly nothing more than the result of interbreeding between plum-headed and slaty-headed parakeets—thus explaining this bird's legendary rarity.[133]

And 1999 saw the long-awaited description and naming of a very distinctive new species of antpitta, which Neotropical ornithological expert Dr. Robert Ridgely and Ecuadorean colleague Lelis Navarrete had first heard, then later spied, back in November 1997 amid the mountains of southern Ecuador. At 10 in long, and characterised by its noticeable white facial markings and bark-like hooting cry, the Jocotoco antpitta *Grallaria ridgelyi* is the world's second largest antpitta species. Named after their superficial outward resemblance to the unrelated Old World pittas, antpittas are ground-dwelling species closely related to the New World antbirds, ovenbirds, and tyrant flycatchers.[134]

ONCE INCOGNITO IN INDONESIA AND INDIA

In 1985, a bright rufous-coloured owl was collected in Dumoga-Bone National Park, in northern Sulawesi, Indonesia, by F.G. Rozendaal. When examined, it was assumed merely to be an unusually red specimen of the ochre-bellied hawk-owl *Ninox ochracea*. A more recent re-examination, however, found that it constituted a valid species in its own right. Formally described by owl specialist Dr. Pamela Rasmussen in 1999, it was named *N. ios*, the cinnabar hawk-owl.[135]

Also in 1999, the discovery of a new monal pheasant was announced. Frequenting India's Arunachal Pradesh state in the Himalayas on the borders of Tibet and Myanmar, it resembles Sclater's monal *Lophophorus sclateri*, but its tail is white rather than chestnut. In 2004, this latest monal was formally classified and named as a new subspecies of Sclater's monal *Lophophorus sclateri*, and is known as the Arunachal Pradesh monal *L. s. arunachalensis*.[136]

THE REPTILES AND AMPHIBIANS

Hairy Frogs and Dragons of Komodo

I sat and gazed at them as though they had been beings from another world. Their casual identification of the picture, coming so unexpectedly, had quite startled me, for the drawing depicted a creature that I had long wanted to get hold of, perhaps the most remarkable amphibian in the world, known to scientists as *Trichobatrachus robustus*, and to everyone else as the Hairy Frog.

GERALD DURRELL—*THE BAFUT BEAGLES*

There was a rustle of dry leaves a little way up the shady ravine and then, without further warning, a prehistoric monster stepped out into the sun. No photograph or film can capture the threat of a dragon in the flesh, in search of food. He stood with his broad head held high, tasting the air with a forked yellow tongue, looking it seemed directly at me with a dark cold basilisk eye.

DR. LYALL WATSON (*describing his first sighting of an adult Komodo dragon*)—*EARTHWORKS*

Frogs with filamentous 'fur' and dragon-like lizards sometimes exceeding 10 ft in length are just two of the species listed on the roll-call of extraordinary reptiles and amphibians that first entered the zoological catalogue, or achieved notable re-entries within it, during the course of the 20th century. This select company of herpetological history-makers also includes: a frog as large as certain African antelopes, and another that hatches its eggs within its own stomach; a legendary New Zealand lizard on public display for more than a century before it was recognised to be a dramatically new species undescribed by science; two giants from the Seychelles that were sensationally resurrected from over 150 years of 'official' extinction; the world's longest toad and its most venomous frog; and a zoologically heretical tortoise—unnervingly fleet of foot, with a shell as flat and flexible as a pancake.

HAIRY FROGS AND GOLDEN FROGS

At the end of the 19th century, missionary-naturalist George Latimer Bates was stationed in Rio Muni (now Equatorial Guinea) on the Benito River, north of western Africa's Gaboon River, amassing a collection of animal specimens for the British Museum (Natural History). Among those that he obtained from the Benito River itself were some very peculiar frogs. Approximately 4 in long, and greenish-brown in colour with black markings, they superficially resembled ordinary pond frogs—except for one quite extraordinary feature. Their flanks and thighs seemed to be covered with hair![1,2]

Closer observations revealed this 'hair' to be a mass of fine skin filaments (dermal papillae), richly supplied with blood vessels, which drifted loosely when the frogs were submerged in water but clung against their bodies like matted fur when they were on land. Inevitably, this new species became known as the hairy frog, a soubriquet mirrored in its scientific name—describing it in 1900, George Boulenger dubbed it *Trichobatrachus robustus* ('sturdy hairy frog').[1,2]

This species has since been found elsewhere in western Africa, and scientists have learnt that only males have 'hair'—and only during the breeding season. During this

Hairy frog (male). (Dr. Jordi Sabater Pi)

season, the male needs to take in more oxygen than usual, in order to fuel its concomitant increase in physiological activity, but this need cannot be met by its lungs alone, because they are only poorly developed. However, frogs are also able to absorb oxygen directly through their skin, and in the hairy frog the success of this process is greatly increased by its 'hair', because the total surface area of these filaments equals that of the frog's entire body, i.e. they provide it with twice as much surface area of skin through which to absorb oxygen. Hence its lungs' inadequacy is compensated for by the effective performance of its 'hair' as accessory gills.[2]

Also worthy of note is the long claw present on the second, third, and fourth toes of each foot in this species. These are not ordinary claws; instead, each actually constitutes the tip of its toe's skeleton, protruding through the toe's skin (though it can be retracted back within this skin if need be). They enable the frog to cling to rocks, but can be used very productively in self-defence, as many animal collectors have discovered to their cost (see, for example, Gerald Durrell's delightful book *The Bafut Beagles*).[2]

A second distinctive frog described in 1900 was *Mantella aurantiaca*, Madagascar's fittingly-named golden frog. One of the most beautiful amphibians in the world, this 2-3-in-long species has an iridescent golden-yellow skin, bestowing upon it the guise of a living gem when seen squatting upon the rich green foliage of its forest habitat. It is also one of the very few species of frog to practise internal fertilisation; with most other frogs, the eggs are fertilised externally, as soon as they are ejected by the female. Like many amphibians, however, it is threatened with extinction, and is currently categorised as critically endangered by the IUCN.[3]

TEMPLE TURTLES AND PANCAKE TORTOISES
The year 1903 witnessed the scientific description of the yellow-headed temple turtle *Hieremys annandalii*, requiring the creation of a new genus. Since time immemorial, this large freshwater species has been greatly revered in Thailand, so much so that specially-erected, pool-encompassing Buddhist temples exist here to house it—making western science's ignorance of it until the 20th century all the more surprising.[4-5]

Also described in 1903 was a most unusual species of tortoise, unexpectedly fleet-footed and the sole member of another new genus. Inhabiting rocky regions of Tanzania and Kenya at altitudes exceeding 3300 ft, *Malacochersus tornieri* is aptly called the pancake tortoise, because its shell is almost flat, and is also surprisingly pliable, unlike the hard, rigid carapace of other tortoises. The shell's shape and flexibility enable its owner to make good use of low rocks and very narrow crevices in which to hide if threatened by predators, thereby exploiting a means of escape unavailable to related species.[4,6]

Pancake tortoises. (Frankfurt Zoological Garden)

GOLIATH FROG—WORLD'S LONGEST FROG
The hairy frog was not the only distinctive new frog discovered by George Latimer Bates during his West African forays at the turn of the 19th century. At Efulen, in Cameroon, he captured a specimen of what proved to be the world's longest frog. Measuring 10 in long, olive-brown in colour on top (with small dark spots on the body and irregular cross-bars on the limbs), yellowish-white underneath, and with a notably flattened head, it became the type specimen of what is nowadays known as the goliath frog. In 1906, this spectacular species was officially named *Rana goliath* by herpetologist George

Boulenger, but later studies showed that it was sufficiently different from other *Rana* frogs not just in size but also in anatomy to warrant its own genus, so it was renamed *Conraua goliath*.[7]

Since the discovery of its type specimen, even larger individuals have been recorded. According to *The Guinness Book of Records*, an enormous specimen was caught in Cameroon's River Sanaga during April 1989; it measured 14.5 in and weighed 8 lb—thus comparable in size to dik-diks and certain other small antelopes![8] One might expect that frogs of such extraordinary dimensions as these would have an appetite for sizeable prey. However, as confirmed by goliath frog specialist Dr. Jordi Sabater Pi, although this species will indeed eat small rodents and amphibians its principal prey consists of crabs and insects.[9]

THE DRAGONS OF KOMODO

Among the most notable new lizards from the early 1900s was the web-footed gecko *Palmatogecko rangei*—a semi-transparent consumer of termites, indigenous to southwestern Africa's Namib Desert. One of only two known members of its genus, it was described in 1908.[4,10] Another discovery was Weber's sailing lizard *Hydrosaurus weberi*—a large, basilisk-like species of aquatic agama native to the Moluccan islands of Halmahera and Ternate, and described in 1911 by Thomas Barbour.[4,11] None, however, could compare with the reptilian revelation that was about to take place on a hitherto-obscure Indonesian island called Komodo.

Once upon a time, giant reptiles were thought of only as creatures of the distant past. Dinosaurs were long extinct, and dragons existed only in ancient myths—or so everyone thought, until science announced the discovery of a real-life dragon of sorts, one that was every bit as mighty and monstrous in appearance as its mythological namesakes. Just eastwards of Java and Bali is a collection of much smaller Indonesian islands known as the Lesser Sundas. One of these is Komodo, composed of volcanic rock and covered in dense tropical forests. A very small island, less than 25 miles long and no more than 240 square miles in total area, Komodo may not be everyone's initial concept of a likely home for a population of giant reptiles—which only goes to show how wrong first impressions can be!

Originally uninhabited by man, in the early 19th century Komodo had become of great use to the sultan of the neighbouring island of Sumbawa, who used it as a 'Devil's Island' on which to maroon convicts and political opponents from his own land. It was also visited voluntarily from time to time, by hunters and pearl-seekers from Sumbawa, Flores, and other parts of Indonesia, and once back home again they often spoke of what they had learnt about Komodo from the island's enforced residents, and of their own experiences there.

In particular, they told of huge, frightening monsters, of a type supposedly somewhat akin to a crocodile in appearance, but which lived on the land, and thought nothing of devouring pigs, deer—and humans. This giant form became known as the *boeaja darat* ('land crocodile'), reports of which soon attracted so much attention that even the scientific world, usually so sceptical of native tales and testimony, began to take notice.[12-13]

Goliath frog. (Dr. Jordi Sabater Pi)

One authority who found the Komodo 'land crocodile' rumours decidedly intriguing was Major P.A. Ouwens, director of Java's Botanical Gardens at Buitenzorg, near Batavia (now Jakarta). In 1910, in an effort to learn something conclusive regarding this matter, he contacted J.K.H. van Steyn van Hensbroek—at that time Governor of the island of Flores, Komodo's much larger, eastward neighbour. By good fortune, the governor was a keen amateur naturalist, who promised to find out whatever he could for Ouwens when he next visited Komodo. Meanwhile, Ouwens's interest was heightened by a report received from an airman, who had recently made a forced landing on this island, only to discover to his very considerable alarm that it was populated by 'horrible dragons'.[12-13]

In 1912, van Steyn van Hensbroek finally arrived on Komodo, and he soon met up with two Dutch pearl-seekers, named Kock and Aldegon (members of a Dutch pearl fleet stationed there), who recounted to him many a hair-raising tale of their encounters with the terrifying *boeaja darat*, and declared that it could attain a length of up to 23 ft! Disentangling fact from fiction whenever possible, examining the skin of an unusually large lizard discovered in one of the resident's huts, and following the tracks of what was said by some Sumbaya exiles to be a genuine *boeaja darat* (but failing to observe the tracks' originator), the Flores governor ascertained that (although still very impressive) the maximum length reached by Komodo's 'land crocodile' was appreciably less than the figure claimed by his pearl-seeking companions. Moreover, he concluded that it was not a crocodile at all, but was most probably an extra-large member of the monitor lizard (varanid) family. Eventually, he succeeded in shooting a fair-sized specimen, measuring 7 ft 4 in long, and sent its skin, plus a photo of the corpse, to Ouwens in Java.[12-14]

It was indeed a varanid, much bulkier than the familiar monitors known from Africa, Asia, and Australasia, but otherwise quite comparable. Although some monitor species are small, others can attain lengths of several feet, but almost all are characterised by their elongate, mobile necks, extremely long tongues and tails, and principally carnivorous lifestyles.

Not long afterwards, the Buitenzorg Botanical Gardens sent a collector to Komodo to procure some living specimens of this major new lizard. Four were brought back to Java, including an adult that measured no less than 9.5 ft.[14] More recently, specimens up to 10 ft 2 in long have been recorded, thereby making this giant monitor the *largest* of all lizard species—but *not* (as many books mistakenly allege) the *longest*.

Salvadori's monitor *Varanus salvadorii* of New Guinea quite often exceeds 10 ft in length (Komodo equivalents are much rarer). However, whereas the tail of this and all other species of large monitor constitutes at least *two thirds* of their total length, that of the Komodo species only constitutes about *half* of its total length, so it is far sturdier and heavier than other monitors of comparable length, and thus a much more formidable predator.[15-16]

Later in 1912 Ouwens published a scientific description of this new, giant lizard, which he named *V. komodoensis*. As for its common name—its huge size and impressive appearance, coupled with the fire-spurting illusion created by its long, bright-yellow tongue's flickering, flame-like movements, were sufficient inspiration for it to be referred to thereafter as the Komodo dragon. This name was first committed to print towards the end of the 1920s, by dragon-stalking American naturalist W. Douglas Burden.[17]

Van Steyn van Hensbroek learnt much from the pearl-fisher Aldegon and others on the island regarding the Komodo dragon's behaviour and lifestyle (functioning both as an active predator and, especially, as a prodigious devourer of carrion), and passed on this information to Ouwens to be incorporated within his formal description of this species. The accuracy of most of that information has since been confirmed by herpetological researchers, including the following excerpts, in which Ouwens noted that the dragons live:[14]

> . . . on land, where they make great holes under the stones and rocks, in which they always remain at night. Their feet are fairly long, and in spite of their awkward build, they can move with great rapidity. . . . The neck is rather long and extraordinarily mobile. The animal can move its

Komodo dragons.

London Zoo's first Komodo dragon, photographed during the late 1920s.

head in every direction, and so it can see everything: this is of great use to the creature, as it seems to be remarkably deaf. Mr. Aldegon says, that, if only care is taken, that the animal does not see the hunter, the latter may make as much noise as he pleases, without the animal being aware of his presence [more recent studies, however, have challenged this statement, revealing that these lizards can hear shouting voices]. . . . They live either singly or in troops. Their food is exclusively of animal nature. If Mr. Aldegon shot wild pigs or birds and left them on the ground, they were eaten by the *Boeaja darat*, which sometimes fought desperately for the prey.

The widespread interest initially generated by the Komodo dragon's discovery was inevitably eclipsed in the west by the all-consuming events of World War I. But it was revived in 1926 by W. Douglas Burden, when he succeeded in bringing to New York's Bronx Zoo the very first living specimens of Komodo dragon ever seen outside Asia. Others have since been exhibited at other zoos, including London Zoo and Chester Zoo in Britain, and Indonesia's own Jakarta Zoo.

Our knowledge of its recorded range in the wild state has also increased, because it has since been found not just on Komodo but also on the even smaller neighbouring islands of Rintja and Padar, as well as in the southeastern coastal areas of Flores.[18]

Almost a century has passed since the Komodo dragon made its scientific debut in Ouwens's paper, but there are still many mysteries relating to it that await a satisfactory solution. One concerns its present-day location. As pointed out by Floridian herpetological researcher Dr. James M. Kern, fossil evidence suggests that its ancestors evolved neither on Komodo nor on other members of the Lesser Sundas, but instead in Australia, many millions of years *before* the Lesser Sundas rose up out of the sea. Yet Australia is separated from these islands by at least 500 miles of notoriously turbulent waters, and the Komodo dragon is far from being a masterly swimmer. So how did it reach Komodo?[18]

Equally perplexing is why it should have attained such colossal proportions. Physiologist Dr. Jared Diamond from California University has noted that in earlier days Flores and certain other southeast Asian islands housed two species of small stegodont elephant, and he suggested that the dragon evolved as a specific predator of these pygmy pachyderms. However, MacQuarie University palaeontologist Dr. P.B. Mitchell challenged this hypothesis by recalling that the stegodonts were not the only medium-sized herbivores available as prey, so the concept of a predator evolving to feed *exclusively* on these two species is unlikely.[19]

Finally, it is worth mentioning that as recently as the early 1980s an unconfirmed report was recorded that suggests that Komodo dragons may also exist on the island of Sumbaya.[20] Moreover, in 2006, the Komodo dragon was found to be able to reproduce by parthenogenesis (i.e. virgin birth, without any male input). And in 2009, it was revealed that this species has a pair of venom glands in its lower jaw that secrete venom containing several toxic proteins. Clearly Komodo's spectacular monitor lizard has quite a few surprises still in store.[20a]

TAIL-WAGGING FROGS WITHOUT TAILS

The most primitive species of modern-day frog constitute the trio belonging to the genus *Leiopelma* (originally spelt '*Liopelma*'), confined exclusively to New Zealand—home of so many other 'living fossils' on account of its many millions of years of isolation from all

Do Komodo dragons also exist, still-undiscovered by science, on Sumbaya?

other land masses. Together with their only close relative—an American species, *Ascaphus truei*—their vertebrae have two concave faces (the amphicoelous condition), like those of most fishes but unlike those of any other frog. They lack eardrums and vocal sacs, but possess tail-twitching muscles—even though they do not possess a tail! (Ancestral frogs were tailed, but the tail disappeared during subsequent evolution.)[4,21]

Prior to the 20th century, the only known *Leiopelma* frog was Hochstetter's frog *L. hochstetteri*, inhabiting mountainous regions of North Island. Then in September 1916, Harold Hamilton of Wellington's Dominion Museum collected some specimens of a similar but hitherto unrecorded frog on Stephens Island, a tiny dot of land in the Cook Strait, close to the northernmost tip of South Island. This was described in 1919, and given the name *L. hamiltoni* in honour of its discoverer. Although Hamilton's frog was later believed to have become extinct, it was rediscovered in January 1950, and some years afterwards a population was also found on nearby Maud Island (deemed by some researchers to constitute a separate species, which was dubbed *L. pakeka*, the Maud Island frog, in 1998). Even so, it nowadays exists in such small numbers that it lays claim to the dubious privilege of being New Zealand's rarest species of frog.[22]

In 1942, a third *Leiopelma* species was described. Christened *L. archeyi*, Archey's frog, its scientific recognition was somewhat belated. S.P. Smith had discovered some small, greenish-golden frogs now known to be of this 'new' species as long ago as June 1862, high in the Coromandel Peninsula's Tokatea Ridge, North Island, but these had simply been considered to be a variety of Hochstetter's frog, and thus had attracted little attention.[23]

THE EXTRAORDINARY CASE OF QUEIMADA GRANDE'S ULTRA-VENOMOUS VIPERS

The Queimada Grande pit viper is one of the world's most venomous species of snake, and was officially described in 1921. Also known as the golden lancehead, and formally christened *Lachesis* [now *Bothrops*] *insularis* by Dr. Afranio do Amaral, its discoverer, it is 3-4 ft long, predominantly pale golden-brown in colour with widely-spaced transverse stripes, and restricted solely to a miniscule uninhabited islet called Queimada Grande. This rocky chunk of steep slopes interspersed with scattered patches of tropical foliage is located on the coast of Brazil's São Paulo State, roughly 40 miles southwest of the Bay of Santos.[24]

On such a tiny outcrop of rock as Queimada Grande (with under 0.75 mile^2 of surface area above sea level), suitable prey for this species is, hardly surprisingly, something of a rarity. Only one notable source exists—birds, flitting back and forth from the mainland. Thus, for the viper to survive it has to be inordinately adept at obtaining prey.

Compelled to subsist upon such elusive, highly-mobile creatures as birds—which could only too readily fly safely away from the island and leave all serpentine foes behind, wholly unable to follow—the Queimada Grande viper cannot afford to make mistakes and allow a potential victim to escape. The means that it has evolved to reduce such mistakes to a minimum are two-fold. Firstly, unlike its larger, principally ground-living relatives on the mainland, this viper is markedly arboreal, with a partially prehensile tail, thereby allowing it to pursue birds perching on the islet's foliage as well as upon the ground. Secondly, and most devastatingly, its venom is extraordinarily potent—far more so than that

Queimada Grande viper. (Mark O'Shea)

of other pit vipers—and capable of paralysing prey almost instantaneously. This ensures that once a viper actually catches a bird, its meal is guaranteed, because its venom's singularly rapid action prevents the bird from flying away before the poison takes effect.[4,24-5]

It sounds like a very efficient, successful arrangement for the viper; and as there appeared to be 3000-4000 specimens on Queimada Grande at the beginning of the 1920s, there seemed little chance that it would ever become an endangered species. However, in 1930, when a scientific survey took a closer look at these snakes on their islet home, it found that whereas 50% of the total number were males, only 10% were females. All of the remaining 40% were intersexual specimens, predominantly female in anatomy but with male copulatory organs—and hence more than likely to be infertile.[4,25-6]

The production of intersexes is not an uncommon occurrence within highly inbred populations in many animal species—and it would be difficult under natural conditions to obtain a more inbred population than from a species wholly confined to an island less than a square mile in area. However, the longer an inbred population inbreeds, the greater the probability that its offspring will be intersexual. There were far fewer specimens on Queimada Grande by 1955, and only 3% were females, with 70% of the remainder consisting of intersexes. It seemed as if this could only end in one way; there would ultimately come a time when every viper born would be an intersex, after which—bearing in mind that intersexes are usually infertile—the species would become extinct. Also hastening the species' decline was a series of bush fires, some started deliberately by mainland locals anxious to eradicate the vipers from Queimada Grande so that it could then be established as a site for banana cultivation purposes.[4,25-6]

Moreover, during the 1940s and 1950s, several hundred specimens were captured and taken off the island for laboratory studies. Although this reduced the wild population even further, it did mean that the species would not become entirely extinct—as long as at least some of these relocated specimens bred in captivity, that is.[4,25-6]

No vipers were found on a two-day visit to Queimada Grande in 1965, but a year later Butantan herpetologist R. Hoge captured seven of them.[4,25] For a long time afterwards, however, nothing more was heard of this species. Hence, as I noted in this book's first edition, *The Lost Ark* (1993), it was widely assumed that—in what must surely be one of the most macabre of all paradoxes—on its native islet the Queimada Grande viper had bred itself into extinction!

In fact, this is now known not to be the case. In recent years, Brazilian biologist Dr. Marcelo Duarte has paid regular visits here, revealing that the viper is still very much alive—and well. Remarkably, although there are indeed many intersexes, which have substantially replaced female specimens, these are actually fertile and serve as functional females, thereby permitting this species to reproduce after all. Indeed, although the total number of vipers on their islet remains a mystery, Duarte has logged several hundred specimens on just the one trail that he has been using. Recent visits to Queimada Grande which confirmed that the species is indeed still thriving here include one in February 1999 by British herpetologist Mark O'Shea (which was filmed as part of one of his TV series on reptiles), and one by Robin Eveleigh, also in 1999. Nevertheless, it is presently categorised as critically endangered by the IUCN.[26]

REDISCOVERING THE TAIPAN—AUSTRALIA'S LARGEST VENOMOUS SNAKE

Sometimes more than 10 ft long, the taipan *Oxyuranus scutellatus* is indeed Australia's largest venomous snake, but following its type specimen's procurement in 1867 near Cooktown, Queensland, this slender, brown-scaled species nevertheless succeeded in eluding science for 56 years—until two more were collected in 1923, on Queensland's Cape York Peninsula. By that time, however,

Taipan. (Mark O'Shea)

Fierce snake. (Mark O'Shea)

the Cooktown specimen had been forgotten, so these two were thought to represent a hitherto-undescribed species, which was promptly christened *Oxyuranus maclennani* and referred to as the giant brown snake. The interest generated by this 'new' species inspired the search for other specimens of it, and eventually quite a number were obtained—whereupon comparisons duly disclosed that these, the two from Cape York, and the Cooktown example all belonged to the same species.[27]

Until quite recently, the taipan was also deemed to be Australia's *most* venomous snake (probably exceeded worldwide only by the sea snakes, king cobra, and black mamba), and its noticeably large poison fangs are capable of inflicting a fatal bite. Fortunately, human fatalities are rare—due to its relatively restricted range along Queensland's coasts (as well as southern New Guinea's), and generally elusive nature. In 1976, however, Drs. J. Covacevich and J. Wombey revealed that its western, inland contingent actually constituted a wholly distinct species—now known variously as the fierce snake, small-scaled snake, or inland taipan *O.* [formerly *Parademansia*] *microlepidotus*—whose venom is believed to be four times as potent as that of the taipan (nowadays referred to as the coastal taipan)[27]

Moreover, in 2007 a third species of taipan was recognised and described—*O. temporalis*, the Central Ranges taipan. Its species was originally based upon a single immature female specimen that had been captured alive by South Australian Museum herpetologist Dr. Mark Hutchinson, but in May 2010 a second specimen, this time an adult female, was captured in Western Australia's Great Victoria Desert.[27]

CROCODILE LIZARD

Known to the local Chinese people as 'the lizard of great sleepiness' as a result of its tendency to fall soundly asleep while sunning itself on boughs overhanging mountain streams and rivers in Kwangsi Province, the Chinese crocodile lizard *Shinisaurus crocodilurus* is an unusual species. It was first made known to science in 1928, when specimens were collected during an expedition to Canton by researchers from Sun Yat-sen University.[4,28]

Named after the crocodile-like double row of horny scales that it bears on the upper surface of its tail (with a single row running down the middle of its back), this pale-brown species is the sole member of its genus. It seems to be most closely related to an equally obscure group of lizards termed xenosaurs or knob-scaled lizards (see p. 172), even though they live many thousands of miles further east—in Mexico and Guatemala. Indeed, some herpetologists nowadays call it the Chinese xenosaur. Unlike the dorsoventrally flattened skulls of the New World species, however, that of *Shinisaurus* is laterally compressed, yielding a high, narrow outline.[4,28]

. . . AND THE GENUINE ARTICLES

In 1929, eminent herpetologist Dr. Karl Schmidt, from Chicago's Field Museum, described and named a new species of crocodile—*Crocodylus novaeguineae*—from the Sulu Archipelago, various other Philippine islands, and New Guinea itself. Measuring 10 ft in total length, it is characterised by a noticeably long snout.[29]

Chinese crocodile lizard.

Morelet's crocodile.

The following year, Schmidt received three specimens of a small crocodile form from the Philippine island of Mindanao. He expected them to belong either to the saltwater crocodile *C. porosus* (that island's most common crocodile) or to his newly-described New Guinea species. Instead, they proved to be distinct from both, evidently representing another new species, which Schmidt described on 15 May 1935, naming it the Philippine crocodile *C. mindorensis*. Although most closely related to the New Guinea crocodile, it was readily distinguished by several morphological features, such as a proportionately heavier, broader skull with a greater degree of pitting, more pronounced ridges around the eyesockets, and a larger eustachian tube (the internal connection between the ear and the back of the throat). The Philippine crocodile is one of the world's most threatened species of reptile, and is presently categorised as critically endangered by the IUCN.[30]

Just a few years earlier, in 1923, a team from the Chicago Natural History Museum had found Morelet's crocodile *C. moreletii* alive and well in a swamp inland from British Honduras (now Belize). Until that find, it had been known solely from some skins brought to the Paris Museum in 1851 by French naturalist P.M.A. Morelet, and had not been recognised as a separate species in its own right. This relatively small but very broad-snouted species is also known to exist in Mexico and Guatemala.[31]

LAFRENTZ CAECILIAN—A DISCOVERY IN DUNG

Caecilians are legless, worm-like amphibians of the African, Asian, and American tropics and subtropics, whose discovery is often due more to luck than to intention. This was certainly true for the Lafrentz caecilian *Dermophis oaxacae* from Mexico. Scientifically described in 1930 (when it was initially classed merely as a subspecies of *Gymnophis multiplicata*), it had been discovered by Dr. K. Lafrentz on a coffee plantation in Oaxaca State. Dr. Lafrentz had been digging through a pile of donkey dung, when suddenly a throng of blue-black caecilians emerged from it. Measuring 1-1.5 ft in total length, this creature was well-known to the local Indians, who referred to it as *metlapil* (and were convinced, mistakenly, that it was poisonous), but constituted a species previously unknown to science.[4,32]

GOLDEN FROGS AND BLACK TOADS

The existence of one of Panama's most popular, if unusual, tourist attractions only gained zoological attention at the relatively late date of 1933—for that was when E.R. Dunn formally described the so-called golden frog of Panama's El Valle de Anton (though in reality it is a toad, not a frog). He named it *Atelopus varius zeteki* (after James Zetek, who had collected its type specimen in 1929), but nowadays it is categorised as a species in its own right, as *A. zeteki*.[33]

Not normally more than 2.25 in long, these engaging amphibians have black-dappled, conspicuously-vivid orange skin. Over the years, scores of tourists have visited their mountain-encircled, oval-shaped valley, eager to espy its most famous inhabitants reposing like tiny droplets of molten gold upon the rocks and lush green foliage fringing the streams of El Valle de Anton. Inevitably, however, human nature being what it is, harmless observations of these delightful little creatures has not been sufficient to satisfy some tourists, who have been unable to resist the temptation to collect some of the frogs and attempt (unsuccessfully) to bring them back home alive as exotic, living souvenirs. In addition, an invasive fungal pathogen called chytridiomycosis reached El Valle de Anton in 2006.[34]

As a result, the total population size of the Panamanian golden frog has fallen calamitously, to such an extent in fact that today this frog is virtually extinct, and is classed as critically endangered by the IUCN. None has been seen in the wild since 2007, when the last few specimens recorded were all taken into captivity for breeding and protection purposes. Destroyed through

The golden frog of Panama is actually a toad.

greed and lack of consideration just like the goose that laid the golden eggs in Aesop's fable, El Valle de Anton's unique golden gathering may now be gone, forever.

Only a year after the description of *A. zeteki* was published, herpetologists learnt of a second species of endangered, distinctly-coloured, valley-confined batrachian new to science. In September 1934, zoologist Dr. Carl L. Hubbs visited Deep Springs Valley—an isolated depression within the desert mountains of northeastern Inyo County, California—in search of some cyprinodont-like fishes seen there by Prof. G.F. Ferris of the Stanford Natural History Museum, in case they were new to science. He was unable to find any of these, but he did obtain some unusual black toads, whose dark skins were decorated with delicate white spots and fine tracings. In 1936 he showed these amphibians to Dr. George S. Myers at the University of Michigan, who recognised that they were distinct from all known toads.[35]

In March 1937, Prof. Ferris brought back five living black toads from Deep Springs Valley. Their species was officially described by Myers in 1942, naming it *Bufo exsul* and allying it with the northern toad *B. boreas*.[35]

Endemic to this single locality, the black toad is naturally vulnerable to outside interference, and when various draining and irrigation work was later carried out in the valley a great number of juvenile toads were killed, endangering the species' longterm survival.[34] It is to be hoped, therefore, that future work undertaken here will recognise the importance of preserving this notable member of the area's fauna, and act accordingly.

ADDITIONAL ANACONDAS

It is widely known that there are two species of anaconda—the common green anaconda *Eunectes murinus* from Brazil and much of South America's upper half; and the smaller, yellow or Paraguayan anaconda *E. notaeus* from this continent's central-southern portion. What is far less well known, however, is that there is also a third species of anaconda—and, in spite of a fairly recent reclassification, a fourth too.[35a]

In 1936, Drs. Emmett R. Dunn and Roger Conant described two new species of anaconda. The first of these was dubbed *Eunectes barbouri*, Barbour's anaconda, whose type was a living specimen given to Philadelphia Zoo on 31 July 1924 by Rudolphe Meyer de Schauensee, a renowned researcher of Neotropical fauna. Its provenance was not verified, but Dunn and Conant deemed it likely that this specimen had been collected on the extremely sizeable island of Marajó, situated at the mighty mouth of the Amazon River. Although very similar to the green anaconda, it was differentiated by its dorsal markings, comprising a double row of eye-like spots, rather than large solid black spots as sported by the green anaconda.[35a]

Dunn and Conant's second new species was de Schauensee's anaconda *E. deschauenseei*, named in honour of its discoverer, who had collected its type on 6 October 1924, again most probably on Marajó, and had donated this specimen to Philadelphia Zoo too. Although it closely resembled the yellow anaconda, this specimen sported a single row of large dark spots along its back, whereas the yellow anaconda's back bears a single row of dark bars.[35a]

Since then, both of these new species have been largely forgotten outside specialised herpetological circles. Within those circles, conversely, much discussion has occurred regarding their validity, or otherwise. Although most researchers supported the claim to taxonomic distinction of de Schauensee's anaconda, which was found to have a localised distribution in northeastern Brazil and coastal French Guiana, Barbour's anaconda received short shrift—even, ultimately, from one of its original co-describers, Dr. Roger Conant. In 1997, four finally became three, when a paper co-authored by Dr. Peter Strimple, Conant, and three other researchers presented a detailed set of morphological data which very persuasively demonstrated that those characteristics purportedly distinguishing Barbour's anaconda from the green anaconda actually fell well within the range recorded for the latter species.[35a]

In contrast, de Schauensee's anaconda is still recognised as a valid third (albeit scarcely-known) species. Interestingly, this anaconda and the yellow anaconda have indistinguishable scalation—despite the remarkable zoogeographical fact that these two species, which both inhabit flooded savannah terrain, are separated from one another by more than 1000 miles of Amazonian rainforest. On account of this, British snake expert Mark O'Shea considers it possible that they are sister species: ". . . surviving from a time when the Amazon forests were more reduced and continuous savanna[h] covered a greater proportion of Amazonia".[35a]

Moreover, in 2002 a new species of anaconda was discovered in the flood plains of northern Bolivia's Pando province. Similar in overall appearance to the green anaconda but currently known only from a few specimens, it was duly dubbed the Bolivian anaconda *E. beniensis*, thereby restoring the total number of known anaconda species to four.[35a]

GEORGIA BLIND SALAMANDER— A SUBTERRANEAN SURPRISE

North America is home to several species of cave-dwelling salamander, but its most extraordinary example remained unknown to science until 19 May 1939. This was the day on which an air-lift pump operating in a 200-ft-deep Artesian well at Albany, in Georgia's Dougherty County, brought to the surface a pallid, shovel-headed, eel-like creature that was unlike any species seen before. Measuring 3 in long and with bright pink plume-like gills, but also possessing two pairs of spindly limbs that betrayed its allegiance to the amphibian (rather than to the anguilline) lineage, it was sent to arachnologist Dr. Howard W. Wallace, who in turn submitted it for identification and study to U.S. herpetologist Dr. A.F. Carr.[36]

In superficial outward appearance it recalled the olm *Proteus anguinus* (a bizarre Balkan relative of the American mudpuppies that was once thought to be the larva of a mountain-dwelling dragon!). However, x-ray examinations of the living animal revealed that it was in reality a morphologically degenerate member of a quite different family, consisting of the plethodontid salamanders. Of these, the species that it most closely resembled was the Texas blind salamander *Typhlomolge rathbuni*, but whereas the latter had a prominently pointed snout, the entire head of the new species was strikingly spatulate. Moreover, its skull was simpler in construction, and it lacked external eyes, intimating that this salamander was of even greater spelaean (cave-dwelling) persuasion than *Typhlomolge*.[36]

In July 1939, Carr formally named it *Haideotriton wallacei*, the Georgia blind salamander, designating it as the sole member of a new genus (though in recent times, some researchers have re-assigned it to the genus *Eurycea*). An aquatic species with a finned, laterally-compressed tail to augment its propulsion through underground water sources via fish-like undulations of its slender body, its lack of pigmentation bestows upon it an opalescent, semi-transparent appearance. This enabled Carr to perceive that its type specimen was a gravid female—her eggs could be readily discerned through her body's wall. *Haideotriton* is highly adapted for life in total darkness, bearing along the sides of its head and lower jaw a very pronounced lateral line system, consisting of numerous finger-like projections that detect water currents and thereby reveal the presence of potential prey swimming nearby.[36]

By 1954, *Haideotriton wallacei* was still known only from its type specimen (though others have since been recorded), but during that year another new species of plethodontid cave salamander came to notice, when Dr. Edward McCrady formally described *Gyrinophilus palleucus*, the Tennessee cave salamander. Once again, it was virtually unpigmented, but unlike *Haideotriton* it had well-developed limbs and readily visible eyes. *G. palleucus* was a major herpetological find, as it proved to be America's largest species of cave-dwelling salamander, with a total length sometimes slightly exceeding 6 in. Its type specimen, an adult female, was obtained in 1944 from Sinking Cove Cave in Tennessee's Franklin County, and others were later procured from four more limestone caves in this same county.[37]

PALESTINIAN PAINTED FROG— A VICTIM OF HUMAN CONFLICT

The year 1940 saw the discovery of the Palestinian painted frog *Discoglossus nigriventer*, an attractive ochre-coloured species bearing red blotches and spots on its limbs and upperparts, and with dark underparts. Today, it is considered to be among the world's rarest amphibians—assuming that it still survives at all (it has been classed as extinct by the IUCN since 1996). The first specimens collected were two tadpoles (later lost) and two sub-adults (the larger of which subsequently ate the smaller while in captivity), from the swamps of Hula Lake's eastern shore, situated directly upon the Israeli-Syrian border—the scene of intense fighting for many years. Needless to say, the continuing existence of any species living in such a disrupted location is so uncertain a matter that few would care to speculate optimistically regarding it.[8,38]

In 1955, one other specimen was obtained, this time an adult female, which lived for quite a time in a well-cared-for terrarium owned by its collector, M. Costa. He recorded that it was only active at night, spending most of the daytime buried in the sand with just its head protruding above the water—indicating, as noted by David Day in his own coverage of this species, a specialised existence within shallow swamps. No additional specimen has been recorded (an unconfirmed sighting of a single frog in Lebanon's Aammiq marshes was reported during 2003). And as the swamps have since been drained and have now disappeared, it is unlikely that this important frog (the only *Discoglossus* species native to the Mediterranean's eastern side) will ever be seen again—an innocent, little-highlighted victim of human conflict.[39]

XENOSAURS—REPTILIAN RECLUSES FROM THE NEW WORLD

The xenosaurs ('strange lizards') or knob-scaled lizards are inconspicuous, brown-scaled species from Mexico and Guatemala, fairly small (less than 1 ft long) but powerfully built, and sufficiently different from all other lizards to merit their own genus—*Xenosaurus*. Reclusive insectivorous forest-dwellers, remaining hidden during daylight hours beneath tree roots or within shadowy crevices, until the 1940s only one species—*X. grandis*—was known. Then in 1941, Rackham's xenosaur *X. rackhami*, with brilliant red eye-irises, was officially described from Finca Volcán, east of Cobán in Guatemala. This was followed in 1949 by Newman's xenosaur *X. newmanorum*, from the Xilitla region of Mexico's San Luis Potosi State. Another species, the flat-headed xenosaur *X. platyceps*, from Tamaulipas, Mexico, was described in 1968; the pallid xenosaur *X. rectocollaris* from Puebla, Mexico, followed in 1993; and further xenosaurs have been described in more recent times.[40]

Most closely allied to the Chinese crocodile lizard (see p. 168) and the limbless lizards known as anguids (e.g. the slow worm *Anguis fragilis*), morphological characteristics of the xenosaurs include their cylindrical teeth (small in size but numerous in number), non-linked osteoderms (in most reptiles possessing them, these hard bony plates beneath the body's epidermal scales are closely connected together), and flattened skulls.[4,40] In view of their secretive lifestyle, there may well be species of xenosaur still awaiting scientific detection.

BLOMBERG'S GIANT TOAD

In 1951, 45 years after the world's longest species of frog had been officially described, the world's longest species of toad hopped onto the scientific stage.

Its story began a year earlier, while Dr. John Funkhouser from Stanford University's Natural History Museum was visiting his explorer-photographer friend Rolf Blomberg in Quito, Ecuador. He learnt from Blomberg that a giant form of toad supposedly existed in the area of Nachao, in southwestern Colombia's Nariño Province. Spurred by Funkhouser's interest, Blomberg set out to discover more concerning this mysterious creature, and in August 1950 he travelled to Nachao in search of specimens.[41]

A month later, he returned to Quito with the exciting news that this titan of the toad world was not a myth but a genuine creature—verified by the spectacular example that he had captured on 11 September and had brought back with him. This became the type specimen of *Bufo blombergi*, described in 1951 and named after its discoverer. Its head-and-body length was 8.28 in, and it weighed 2.2 lb when captured.[41]

Blomberg's giant toad. (Paignton Zoological & Botanical Gardens)

Moreover, Blomberg had been informed by Nachao locals that even larger specimens existed, and that the species was most plentiful during the area's wet season (he had arrived there during the height of its dry season). Accordingly, in May 1951, during the rainy period, Blomberg returned to Nachao, and succeeded in capturing three living specimens, which were maintained thereafter at New York's Bronx Zoo.[41] These were indeed bigger than the species' type, but probably the biggest of all captive specimens on record is one that was exhibited more recently at Germany's Ruhr Zoo, with a head-and-body length of 10 in, and a weight of 2.475 lb.[16]

Not only is this species notable for its great size, it is also very handsome in colour, with a thick black stripe separating the copper glow of its upperparts from the brownish-purple hue of its underparts. Although the world's *longest* toad species, Blomberg's giant toad is not its *largest*. That title is claimed by the cane toad *B. marinus*—for even though the latter is slightly shorter than *B. blombergi*, it is more massively-built and somewhat heavier. Interestingly, some time prior to 1971 a male cane toad and a female Blomberg's giant toad mated while in captivity together at Stuttgart's Wilhelma Zoo; the resulting hybrid offspring closely resembled their paternal species externally, but their growth was unusually strong.[16,41]

Due to its impressive appearance, Blomberg's giant toad soon became very much in demand as a zoo exhibit and in particular among private collectors of herpetological creatures—to such an extent that it is now listed as near threatened by the IUCN. Yet, disturbingly, it can still be purchased from herpetological dealers, with the continuing collection of specimens from the wild threatening its longterm future.[41]

THE SNAKE-NECK WITHOUT A SNAKE-NECK!

The snake-necked tortoises earn their name from their long, slender neck, which, when extending out of the shell, can appear rather like an emerging snake. This makes Australia's western snake-neck *Pseudemydura umbrina* something of an anomaly, because its neck is only very short, not snake-like at all. Its type specimen was obtained as long ago as 1839, by Ludwig Preiss, and its species was formally described and named in 1901, but no other example was discovered until 1907. After that, this odd species disappeared again, but in 1953 a third specimen was located, in swampland near Perth. Initially, this was mistakenly assumed to represent a new species, which in 1954 was christened *Emydura inspectata*, but its true identity as a specimen of *P. umbrina* was recognised in 1958.[42]

Since then, others have been found here (where they undergo a form of summertime hibernation termed aestivation—as do certain lungfishes). Nevertheless, it is sufficiently uncommon to have been classed at one time as the world's rarest species of tortoise, and is currently categorised as critically endangered by the IUCN. By the late 1970s, less than 100 were known to exist in the wild, but a small population is maintained at Perth Zoo, which may assist in sustaining numbers of this unusual reptile.[16]

BOLSON TORTOISE— OVERLOOKED FOR 71 YEARS

It may seem difficult to overlook the largest species of terrestrial reptile in North America, but for 71 years this is precisely what happened with the Bolson tortoise. Although it was first scientifically documented in 1888, by A. Duges, it was not formally described, and for much of the time thereafter was thought to be one and the same as the Florida gopher tortoise *Gopherus polyphemus*. Only when specimens were found in north-central Mexico during the late 1950s did the Bolson tortoise regain scientific interest—whereupon it was recognised to be a wholly distinct species. In 1959, it received its much-belated formal description, penned by J.M. Legler, who christened this rediscovered reptile *Gopherus flavomarginatus*.[43-4]

Fossil finds have since shown that the Bolson tortoise was once distributed as far north as Oklahoma, but it is nowadays confined exclusively to the Bolsón de Mapimí in Mexico's famous Chihuahua desert. Tragically, despite its current IUCN status as a vulnerable species, it is still hunted by the local populace, especially in the northeastern portion of its range. If stringent protection measures are not enforced, in the near future this significant species could truly become extinct, with no second comeback.[44]

AGAK (CARN-PNAG)—A FORMIDABLE FROG FROM NEW GUINEA

In December 1960, news emerged concerning the discovery of what was said to be an exceptionally large form of frog in New Guinea. Known locally as the agak or carn-pnag, it supposedly measured up to 15 in from snout to vent (anal opening), thereby rivalling the West African goliath (see p. 162) for the title of the world's longest frog, and allegedly weighed more than 6 lb. Three years later, however, this would-be usurper of the goliath frog's record was exposed as a charlatan, when it was officially described by Prof. Michael Tyler. Naming it *Rana jimiensis* (later renamed *Hylarana jimiensis* by some workers), Tyler revealed that its precise provenance was Manjim, on the Jimi River Valley's Ganza River in New Guinea's Western Highlands province, and that the first reports concerning its size had been exaggerated. The maximum authenticated length on record for the agak constituted a rather more modest 160 mm (just under 6.5 in). Even so, this still makes it the second largest species of frog in New Guinea—exceeded only by the Arfak Mountains frog *R. arfaki*, whose females can attain a snout-vent length of 8 in.[45]

ONE-TOED AMPHIUMA—AN 'EEL' WITH LEGS

Also known as Congo eels, amphiumas are neither eels nor from the Congo (and should not be confused with conger eels either!). Admittedly, their long slender bodies *are* distinctly anguilline, especially as their vestigial, non-functional limbs are rarely noticed on first glance. Zoologically, however, these highly unusual creatures are amphibians (large aquatic salamanders, to be precise), which inhabit swamps in the southeastern U.S.A. Prior to the early 1960s, only two species had been documented, whose major external differences from one another centred upon their number of toes: *Amphiuma means* has two toes on each foot, *A. tridactylum* has three. However, this situation was soon to change, thanks to an accidental find that had occurred more than a decade earlier, on the evening of 8 September 1950.[46]

This was when herpetologist Dr. Wilfred T. Neill had collected two stranger-than-usual amphiumas in an area between Otter Creek and Cedar Keys in Florida's Levy County. Amphiumas normally bury themselves in the murky soil at the bottom of their swamps, but this area had recently been flooded by torrential rain, so the pond and stream beds had been washed away, exposing their reclusive fauna, including the amphiumas.[46]

Following subsequent laboratory examination of his captures, Neill had recognised that they were conspicuously different from both of the known species—and when, 13 years later, he discovered three more specimens of this new type, he prepared a formal description of it, published in 1964, in which he named it *Amphiuma pholeter*.[46]

The most pronounced distinguishing features of *A. pholeter* were the presence of just one toe on each foot, limbs even tinier than those of the other two species, reduced eyes, and a shortened, simple-shaped head. Also, its body exhibited an unusual colour scheme, whereby its upperparts were lighter than its underparts (the reverse is true with other amphiumas), and it measured no more than 271 mm (just under 11 in) in total length—much smaller than either *A. tridactylum* or *A. means*, which can attain lengths in excess of 3 ft.[46]

Evidently, *A. pholeter* is a dwarf species, and is also the most morphologically degenerate of the three known amphiumas, seemingly adapted for an extensively secretive, fossorial (burrowing) lifestyle.[46]

THE ANOMALOUS ATRETOCHOANA— A LUNGLESS ENIGMA

Staying with worm-like amphibians: even more vermiform than amphiumas are the caecilians (see also p. 169), a taxonomic order of entirely limbless species, many of which greatly resemble earthworms. Most are predominantly subterranean, fossorial species (although there are also some aquatic river-dwelling forms), and are rarely seen above-ground unless flushed out during rainfall. They inhabit wet tropical localities in Africa, Asia, and South America, feed principally upon insects and worms, and most are relatively small, but a few extra-large species also exist, growing up to 4 ft long and superficially similar in outward appearance to snakes.

One feature that all species share is a pair of lungs—or at least they did until 1968, or, to be exact, 1995; as will be seen, the history of the following ground-breaking species is nothing if not complicated. It began in 1968, when a notably large preserved specimen of caecilian originating from an unspecified location in South America became the type (and for a long time thereafter the only) specimen of a new species, which was formally christened *Typhlonectes eiselti* by Kansas University herpetologist Dr. Edward H. Taylor in a major monograph entitled *The Caecilians of the World*.[46a]

Measuring 27.5 in long, this new species' unique specimen was an adult female housed in the collections of Vienna's Naturhistoriches Museum, but Taylor's official description of it had been unusually brief, and concentrated primarily upon its external morphology. Not until the early 1990s did it attract any further scientific attention, even though it was on full public display inside a glass exhibition case at the museum during the intervening years. When it was finally re-examined, however, by zoologist Dr. Ronald A. Nussbaum from Michigan University and Bristol University caecilian specialist Dr. Mark Wilkinson, they closely investigated its internal form, which duly revealed a very remarkable and hitherto-unsuspected surprise—this large, enigmatic, but previously neglected caecilian was lungless.[46a]

Indeed, their discovery meant that *T. eiselti* was now the largest species of lungless tetrapod known to science, and was sufficiently removed from all other caecilians to require them to publish a major redescription of it, which appeared in 1995. Moreover, in their paper, they also created a new genus for it, renaming this radically disparate species *Atretochoana eiselti*, which highlighted the fact that even its choanae (internal nostrils) were sealed. It also totally lacked any pulmonary blood vessels. Clearly, therefore, this caecilian respired entirely through its skin (cutaneous respiration is also carried out by other caecilians, but only as a supplement to normal pulmonary respiration). Its cranium exhibited fundamental differences from that of other caecilians too. If only there were other examples of this extraordinary species to examine, but where to look? The provenance within South America of its one and only specimen was unknown.[46a]

Happily, in 1998, after 30 long years, a second specimen was finally recorded—documented by Wilkinson and three co-researchers—which was confirmed to have originated from a river in Brazil, but its precise location was unknown. Measuring 31.7 in (80.5 cm), it was longer than the first one but otherwise very similar in form. And that is where the remarkable history of *A. eiselti* comes to an end, at least for now, as no further specimens have been discovered.[46a]

What *has* been discovered, meanwhile, is a second species of lungless caecilian, and whereas *A. eiselti* is aquatic, this new one is land-based. Found in the scrub of Guyana's Iwokrama Forest and described online in late 2009 (in hard-copy format in 2010), when it was formally named *Caecilita iwokramae*, it is the only member of a new genus and is much smaller than *A. eiselti*, its holotype only measuring 4.4 in long.[46b]

ABINGDON ISLAND TORTOISE— ADEPT AT CONCEALMENT, DESPITE ITS SIZE

The South American Galapagos archipelago is named after its giant tortoise *Chelonoidis nigra* (formerly known as *Geochelone elephantopus*); 'galapagar' is Spanish for 'a place where tortoises thrive'. An imposing sight, weighing in at a hefty 330-440 lb, and with a burly carapace (shell) at least 3.5 ft long, it once existed on no fewer than 11 of the islands, and occurred in so great a variety of shell shapes that it was once split into at least

An Abingdon Island giant tortoise at London Zoo in 1914.

15 different species. On some islands, its carapace was domed (as in smaller tortoises), on others it was flattened like a saddle. The largest island, Albemarle (also called Isabella), had five species, and ten other islands each had one; but nowadays these are all treated merely as distinctive subspecies of a single species.[34,44]

Regardless of their shell shapes, however, all of the islands' giant tortoises were united by at least one shared feature—a feature that proved to be their undoing. Their flesh was extremely tasty—prompting their slaughter *en masse* during the early 1800s by visiting sailors, whalers, and other seafarers, until several subspecies were exterminated.[34,44]

One of the most distinctive was the saddle-shelled form on Abingdon (Pinta) Island, *C. n. abingdoni*, whose carapace was unusually thin. By the 20th century's opening years, its population had virtually disappeared, and during scientific expeditions to Abingdon in the 1930s and 1950s not a single specimen was observed (though it is now known that local fishermen found and slaughtered some for meat in the early 1950s). To make matters even worse, goats were introduced onto the island from 1954 onwards, whose insatiable appetites soon converted its all-too-small covering of foliage into an arid wilderness. Not surprisingly, the Abingdon Island giant tortoise was written off as extinct, but in 1964 no fewer than 28 dead specimens were discovered there. They appeared to have died about five years before, which meant that they must have been alive, but concealed, during the earlier searches. Even so, 28 dead specimens could hardly resurrect their subspecies from extinction.[34,39,44,47]

Nevertheless, it was resurrected in March 1972, for this was when—to the astonishment of herpetologists everywhere—a *living* specimen was encountered on Abingdon. Furthermore, tracks indicating the presence of others were also sighted. The live tortoise, a male (later christened Lonesome George), was swiftly transferred to the Charles Darwin Research Station on Santa Cruz for security. Still alive today (August 2011), George is believed to be 60-90 years old and in good health, but attempts to mate him with females of other Galapagos subspecies in order to preserve his genes in future generations of hybrids has so far proven unsuccessful.[44,47]

Meanwhile, no other living pure-bred Abingdon giant tortoise has been found on Abingdon since the discovery there of George. Nor have any been confirmed existing in any zoo or private collection (although as first publicised in 2006, there is a possible pure-bred adult male, Tony, living at a Prague zoo, whose taxonomic identity is now under close investigation). However, in view of the tracks observed, it seems remotely feasible that specimens do still survive on Abingdon—a possibility reinforced in 1981 by the discovery on this island of some tortoise faecal droppings that appeared to be no more than a few years old. To encourage searches, a

Abingdon Island giant tortoises.
(Fortean Picture Library)

reward of $10,000 is being offered by the Charles Darwin Research Station to anyone who successfully discovers a living female Abingdon Island giant tortoise.[44,47]

Even more mysterious than the current status of Abingdon's subspecies is that of *C. n. phantastica*—the Narborough Island giant tortoise. It is presently known only from a single specimen, obtained there in 1906 by Rollo H. Beck, leader of an expedition from the California Academy of Sciences. Yet in 1964, several fresh faecal droppings were found by J. Hendrickson on Narborough's southern slopes, implying that this is another subspecies skilled in the art of eluding scientific searches.[34,44]

GOLDEN DEATH FROM THE FORESTS OF COLOMBIA

The brightly-coloured arrow-poison frogs of tropical Central and South America possess granular glands in their skin that secrete highly toxic compounds. The common name of these frogs refers to the utilisation of their secretions by Indian tribes to coat the tips of arrows or blow-darts and thereby ensure hunting success. When shot into the body of an animal, the projectile's deadly coating immediately paralyses it, with death ensuing rapidly.

Already ranked among the world's most poisonous animals, the arrow-poison frogs gained even further toxicological prestige in 1973, when a previously unknown species, later named *Phyllobates terribilis*, the golden arrow-poison frog was discovered in the rainforests of Colombia. For this species, whose secretions are used by the Embera Indians to coat the darts of their blowpipes, proved to be not only the largest but also the most toxic arrow-poison frog ever recorded by science![48]

So deadly is the alkaloid poison (known as batrachotoxin) secreted by its skin that chickens and dogs are known to have died merely from contact with a paper towel on which a golden arrow-poison frog had walked, and deaths of humans who have directly touched one of these frogs have also been confirmed. Yet its appearance belies its deadly status, as it is a rather small, attractive form, most famously golden-yellow in colour, though there is also a green morph native to Colombia's La Brea region that is commonly seen in captivity, as well as a rarer orange morph.[48]

GASTRIC-BROODING FROGS—THE FEMALE'S EGGS HATCH IN HER STOMACH!

The year 1973 also marked the scientific debut of one of the most bizarre frogs known to mankind. Discovered in a mountain creek in southeastern Queensland, Australia, but resembling a snub-nosed form of African clawed toad *Xenopus*, it became known as the platypus frog on account of its large flat feet ('platypus' actually means 'flat foot'), and in its formal description it was christened *Rheobatrachus silus*. As the first wholly aquatic frog to be recorded from Australia it was already a significant species, but just a few months later it manifested the extraordinary nature of its brooding behaviour—which ever afterwards assured it of herpetological immortality. On 23 November 1973, 19 days after having been collected from the wild, a captive laboratory specimen suddenly opened its mouth and disgorged into its aquarium a handful of fully-formed young![49]

Later studies uncovered its species' spectacular secret. After laying her eggs, the female swallows them. These enter her stomach, where they hatch into tadpoles that live off their own yolk and undergo their entire 6-7-week development within this very unusual form of nursery—after which they make a most unceremonious debut into the outside world, by being vomited up and spat out by their mother. By then, she must be very hungry, for the tadpoles release a substance that inhibits the secretion of her gastric juices, thus preventing her from digesting her offspring—but also preventing her from digesting any food.[49]

One of the greatest zoological tragedies of recent times was staged just seven years later—when, in 1980, searches for more specimens of this singular species in its native homeland met with total failure. Thus it seemed that, less than a decade after its scientific discovery, the world's only known gastric-brooding frog had died out in the wild. Even worse, in November 1983 the last *captive* specimen died—*R. silus* was lost, forever.[49]

By an exceptional fluke, however, the phenomenon itself was not lost, because less than two months later a second species of gastric-brooding frog was discovered—at Eungella, Queensland—and although larger and darker, was clearly closely-related to the first one. Named *R. vitellinus*, the northern platypus frog (with *R. silus* thereafter becoming the southern platypus frog), it became the subject of renewed studies relative to gastric brooding (and also to possible applications of

Southern platypus frog. (Prof. Michael Tyler)

A female southern platypus frog regurgitating her young into the outside world from the security of her stomach, where she had been brooding them since swallowing them as eggs. (Prof. Michael Tyler)

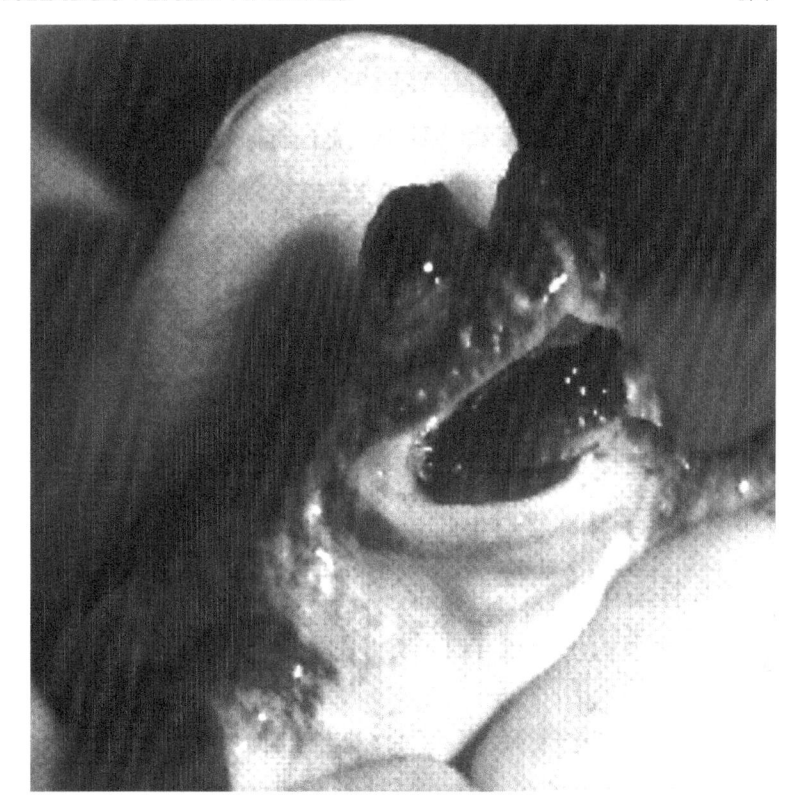

its underlying principles in the treatment of human stomach ulcers). But as its known distribution range is extremely small, it is imperative that this species is not over-collected. Yet these words may already be too late—no wild specimen has been found since 1985.[50]

The platypus frogs might have been saved from extinction if they had been discovered by science at an earlier date, when they were presumably still common. Ironically, there is now evidence to suggest that the southern species had indeed become known to science earlier, but had promptly been forgotten again. In 1991, Dr. Glen Ingram of the Queensland Museum disclosed that as long ago as 1915 the museum had received a specimen of *R. silus*, obtained at Montville by Heber A. Longman. Sadly, however, it had not been recognised to be a new species, and had been overlooked until Ingram brought its existence to attention 76 years later![51]

GRAY'S MONITOR—THE LOST IS FOUND, ON LUZON

Until 1976, one of the world's largest lizards, Gray's monitor *Varanus olivaceus* [=*grayi*] (related to the Komodo dragon and now known to attain a length of around 6.5 ft in some adults), was represented by just three museum specimens, and was dismissed by many zoologists as extinct. Formally described in 1845, its type specimen was a juvenile collected by Hugh Cuming from an unknown locality somewhere in the Philippines, and housed in the British Museum (Natural History). The other two representatives consisted of a skull from the Philippine island of Luzon, discovered in 1942 by noted herpetologist Dr. Robert Mertens in the Berlin Museum, and a stuffed adult found in the U.S. National Museum by Walter Auffenberg in 1976.[52]

This last specimen was of particular significance, because it carried a data tag recording a precise provenance—Pasacao, Luzon. Following up that vital clue, Auffenberg travelled there to ascertain whether this large-sized but little-known species still survived. To his surprise, he found not only that it *was* still in existence but also that it was widely distributed, especially in the forested areas of southern Luzon's Bicol region, where it is known locally as the *butaan*. Of interest is that it feeds *regularly* on fruit and other plant material (unlike most other monitors), though it does also take birds, birds' eggs, and rodents.[52]

A BRONZEBACK COMEBACK

In 1978, a deceptively serpentine species of legless lizard called the bronzeback *Ophidiocephalus taeniatus* was rediscovered. Ten were collected in soil and leaf litter during early January by herpetologists Dean Metcalf and Harold Ehrmann at Charlotte Waters, near Alice Springs, South Australia. Until then, this species had been known only from its type specimen, caught in the 1890s by P.M. Byrne at Charlotte Waters and described in 1897.[53]

BRACHYLOPHUS AND BRACHYASPIS— EACH GAINED A NEW SPECIES IN 1979

In January 1979, Dr. John Gibbons, of the University of the South Pacific, found a new and very beautiful iguana on the Fijian island of Yadua Taba. Its bulky, pale green body striped with white bands, the crest of spines along its back, and its conspicuous dewlap all distinguished it from the more slender, darker, crestless, and dewlap-lacking Fijian banded iguana *Brachylophus fasciatus*. Until that time, the latter species was the only Fijian iguana known, and was already recorded from Yadua Taba as well as from the principal Fijian islands.[16,54]

Shortly after its discovery, this new crested iguana entered the record books, though in a quite unusual way—by becoming the first species to be bred in captivity before even receiving a scientific name. This occurred when a female specimen, given to Ivy Watkins of the Fijian Cultural Centre on Orchid Island, promptly laid three eggs, which were successfully incubated at Mrs Watkins's home. The species itself was eventually named *B. vitiensis*, the Fijian crested iguana; and to secure its protection on Yadua Taba, in 1980 the World Wildlife Fund helped to bring about the island's designation as the first Fijian wildlife reserve.[16,54]

More recently, detailed genetic and morphological analyses of a selection of 61 specimens of *B. fasciatus* collected from 13 different Fijian islands revealed that those originating from the larger central and northwestern islands were sufficiently distinct from the others to warrant separate specific status, and in 2008 this hitherto-cryptic new species was formally christened *B. bulabula* ('bula' is Fijian for 'hello'), the central Fijian banded iguana.[54a]

Australia has the somewhat dubious distinction (at least as far as Antipodean ophiophobes are concerned!) of being the only continent to house more species of venomous than non-venomous snake. This unique reptilian ratio was further emphasised on 6 October 1979, when two specimens belonging to a species of venomous snake not previously known to science were collected by P. Griffin and G. Barron in open eucalyptus woodland at Lake Cronin, Western Australia. Both specimens measured less than 18 long, and had a striking black head, noticeably large eyes, a narrow neck, and a slender body dark brown in colour above, reddish-brown below. Their species was described in 1980 by herpetologist Dr. G.M. Storr from the Western Australian Museum, who named it *Brachyaspis* [now *Echiopsis*] *atriceps*, the Lake Cronin snake.[55]

It belongs to the elapid family—one of the largest snake families, which includes such (in)famous, highly venomous types as the mambas, cobras, kraits, coral snakes, and about 80% of *all* Australian snake species.[55]

MALLORCAN MIDWIFE TOAD—
FIRST KNOWN TO SCIENCE AS A FOSSIL
The discovery in 1979 of the Mallorcan midwife toad *Alytes muletensis* was a big surprise for two quite separate reasons. Firstly, it is a rare event for a new species of amphibian to be found in Europe—even on islands, such as the Balearics. As it turned out, however, the Mallorcan midwife toad was not exactly new. Its second surprise was that it had originally become known to science two years earlier, in 1977—as a fossil, which had been dubbed *Baleaphryne muletensis*. Dating from the Upper Pleistocene, the age of those remains had suggested that their species was long extinct, but the living specimens recorded on Mallorca (=Majorca) in 1979 succinctly disproved this. Instead, studies disclosed that this small black-spotted toad's success in avoiding earlier scientific discovery lay in its chosen habitat—it lives in colonies hidden from sight within inaccessible crevices in the island's steep limestone cliffs.[56]

INDIAN CAVE TURTLE—ITS
REDISCOVERY WAS A GIFT, LITERALLY
During a visit in mid-1982 to the Anaimalai Hills in India's Cochin region, Indian zoologist Jaganath Vijaya

Young male Fijian crested iguana, from Matacawa Levu Island, Fiji. (John Gibbons Memorial Trust, courtesy of Dr. William Kenchington).

was presented with an unusual gift that proved to be of substantial scientific importance. Given to her by a native from a village sited on the upper Chalakudi River, it was a living specimen of the Indian cave turtle *Heosemys silvatica*. This was a species known until then only from its type, obtained in 1911 from some Kadar natives who captured it on the Anaimalai Hills' northwestern slopes, near Kavalai, and from one additional specimen collected in the same locality shortly afterwards. Subsequent searches for it had always been in vain.[57]

Although a relatively small, unimpressive-looking chelonian, this species is noteworthy for its terrestrial lifestyle, contrasting markedly with the primarily aquatic mode of existence practised by most of its closest relatives. Following Vijaya's unexpected present, a short field survey seeking more evidence for its current survival took place between 26 October and 5 November 1982 by one of her colleagues, Dr. B. Groombridge, who succeeded in obtaining a number of specimens at various localities in the Chalakudi valley. This implies that it is not as rare as traditionally believed—its apparent scarcity more a manifestation of its capability for concealment than a genuine reflection of its numbers—although it is currently categorised as endangered by the IUCN. In recent years, in recognition of its distinctive nature, this species has been rehoused in a genus of its own, *Vijayachelys* (honouring its rediscoverer), so it is now known as *Vijayachelys silvatica*.[57]

COURESSE—REDISCOVERING
AN EXTINCT SPECIES THE EASY WAY
Until fairly recently, the couresse or St. Lucia racer *Liophis* (=*Dromicus*) *ornatus* was held to be the world's rarest species of snake. Indeed, for a number of years it was feared to be extinct. Limited to the larger of two tiny specks of land called the Maria Islands, just off-shore from the West Indian island of St. Lucia, this non-venomous species was rediscovered in April 1984 by Dr. David Corke, an ecology lecturer at the North East London Polytechnic, who was visiting the Marias in search of this secretive serpent and also the islands' other reptilian speciality, the Maria Island ground lizard *Cnemidophorus vanzoi*.[58]

As it happened, Corke did not have to do very much to rediscover the couresse—one morning, a surprisingly bold specimen glided through a forest's leaf litter directly towards him! Startled, but prepared, he promptly captured it, measured, weighed, and photographed it, and then called in some other scientists on the island to verify the record—after which he released his welcome visitor back into the forest.[58]

When the couresse was first discovered, in the 1800s, it also inhabited St. Lucia, along with the highly venomous fer-de-lance. In an attempt to exterminate the latter snake on this island, man subsequently introduced mongooses—which for the most part duly ignored the fer-de-lances, and killed all of the couresses instead![58]

TWO GIANT GECKOS—IS ONE A LIVING LEGEND?
Geckos are well known to tourists in southern Europe (especially Greece), and further eastwards into Asia, as those small lizards with suction-pad toes that emerge at night to stalk insects across the ceilings of hotel bedrooms. In 1984, however, the zoological world learnt of two species of gecko that are unlikely ever to be seen engaging in activities of this nature—due to their notably large size.

In July 1984, during the Iran-Iraq war, a shell landed on Fakke, Khouzistan, in southwestern Iran. As it exploded, it disturbed a large lizard, which was spotted by Iranian zoologist Mohammed Reza Ensaf, who was serving there as a corporal and medical orderly. He

Delcourt's giant gecko, ventral (top) and dorsal (bottom) view—known only from this single museum specimen, which for over a century had been on public display without anyone realising that it represented a significant new species wholly unknown to science. (Photo from N.Z. J. Zool. (13: 141, Fig. 4), reproduction courtesy of SIR Publishing and the Royal Society of New Zealand)

Are Delcourt's giant gecko and the legendary kawekaweau one and the same? (Markus Bühler)

succeeded in capturing it, after which it was passed on to Tehran's Faculty of Sciences, where M. Baloutch considered it to represent a new species of gecko. In 1986, this species was formally described by Baloutch and French scientist Michel Thireau (from Paris's National Museum of Natural History), who named it *Eublepharis ensafi*, in honour of its discoverer (tragically, Ensaf was killed during the Iran-Iraq war).[59]

E. ensafi is richly patterned with dark stripes and spots, but its status as a valid species was challenged in 1989—when herpetologist Dr. L. Lee Grismer from San Diego Natural History Museum put forward a detailed case for believing its type specimen to be merely an unusually large specimen of the Iraqi eyelid gecko *E. angramainyu* (described by Drs. S.C. Anderson and A.L. Leviton in 1966[59a]). Furthermore, although frequently alleged to measure 16 in (even by Baloutch and Thireau), in reality this specimen measures just under 11.5 in, thereby diminishing its claim as a giant gecko.[60]

By a remarkable coincidence, within weeks of this 'giant' gecko's discovery, science finally became aware of another, even larger species—one whose history was even stranger, and longer, because it had begun sometime between 1833 and 1869.

This was the period during which France's Marseilles Natural History Museum had received a specimen of a very unusual lizard from an unrecorded locality. As a mounted taxiderm exhibit, it was subsequently put on open display at the museum—where, for many years, it remained in full view of countless numbers of visitors, not to mention generations of museum scientists and many others who passed through from elsewhere. Yet, unbelievably, never once in all that time did anyone realise, or even suspect, that it belonged to a dramatically new species—one that had never been recorded by science![61]

The decades rolled by, but still the ignored lizard's true identity remained undisclosed and uninvestigated—until as recently as 1979, when this strange specimen attracted the curiosity of the museum's herpetology curator, Alain Delcourt. Eager to learn more about it, Delcourt took some photographs, and along with the specimen's measurements he sent them for identification to a number of reptile experts around the world.[61]

They ultimately reached Canadian biologist Dr. Anthony P. Russell, who in turn showed them to Villanova University herpetologist Dr. Aaron M. Bauer. Russell and Bauer recognised that the specimen was clearly a gecko, but of grotesquely gigantic proportions, measuring fractionally over 2 ft in total length—twice as long as the newly-discovered *E. ensafi*. It was a short-headed, bulky-bodied creature, with sturdy legs and a long pointed tail, and was handsomely marked along its back with dark reddish-brown, longitudinal stripes overlying its yellowish-brown background colouration. In overall appearance, it compared fairly closely with geckos of the genus *Hoplodactylus*—except, once again, for its huge size.[61]

The existence of this enigmatic lizard finally became known to the world at large in 1984, when Bauer's investigations of its possible origin led him to New Zealand. And in 1986 its species was formally described by Bauer and Russell, who named it *Hoplodactylus delcourti*—in recognition of Delcourt's laudable action in rescuing this long-neglected form from more than a century's worth of zoological obscurity.[61]

Its identification as a *Hoplodactylus* species had provided an important indication to its likely origin, because this genus's species are mostly limited to New Zealand, thus implying very strongly that this was also the home of the giant *H. delcourti*. Extra support for this conclusion came from Bauer's investigations here, because he learnt that Maori legends spoke of a strange New Zealand creature called the *kawekaweau* or *kaweau*. No-one had previously succeeded in identifying this mysterious animal with any known species inhabiting New Zealand, but various reports from the 19th century described alleged encounters with such creatures. One of the most detailed of these accounts, documented in 1873 by W. Mair, reported the killing of a *kawekaweau* three years earlier in North Island's Waimana Valley by a Urewera Maori chief. He had informed Mair that it was a large forest-dwelling lizard about 2 ft long, as thick as a man's wrist, and brown in colour with red longitudinal stripes. This description is a near-perfect match with that of Delcourt's giant gecko.[62]

Bauer and Russell thus believe that the *kawekaweau* and *H. delcourti* may indeed be one and the same. Sadly, however, there seems little hope that this can ever be conclusively tested, because it is almost certain that *H. delcourti* has been extinct for many years. How ironic, that a species as striking as this one should vanish into oblivion while a specimen was actually on public display for many years at a renowned natural history museum.

Or *has* it really vanished? Wellington's *Dominion* newspaper reported on 11 September 1984 that Welling-ton resident Dave Smith allegedly saw one on the western portion of North Island in the 1960s! Also, following a New Zealand radio programme broadcast on 23 March 1990 in which this species' remarkable history was recounted by James Mack, assistant curator of New Zealand's National Museum, the museum was contacted by several people who claimed to have spied *living* specimens of *H. delcourti* in recent times. The eyewitness accounts included three independent, reliable sightings all made at the same locality near Gisborne, on North Island's eastern coast. These, and various other reports, were followed up by herpetologist Anthony Whittaker and government scientist Bruce Thomas, but without success. Nevertheless, Whittaker believes that this species might still survive in the remote East Cape Forests.[63] Perhaps, after all, there will come a time when Delcourt's giant gecko will be known from more than just a single, long-forgotten taxiderm exhibit.

Yet another gecko milestone from the 1980s was the rediscovery in 1986 of a beautiful species known as the golden gecko *Calodactylodes aureus*. Discovered in India in 1870 but never seen again, its continuing existence 116 years later was confirmed by J.C. Daniel, a member of the Fauna and Flora Preservation Society, who encountered it in residence on the Tirumalai Hills, one of southern India's most sacred Hindu sites.[64]

TALE OF A TV MONITOR—
BUT OF THE REPTILIAN VARIETY

In February 1987, Dr. Wolfgang Böhme and two colleagues documented a new monitor lizard, native to what was then North Yemen. Officially described and christened *Varanus yemenensis* two years later, it seems to be

Yemen monitor. (Mark K. Bayless)

most closely related to *V. albigularis*, a South African species.[65]

Its existence had first become known to science in a rather unusual manner. One evening in 1985, Dr. Böhme was at home watching a television documentary about North Yemen when the programme screened a piece of film featuring a large tree-climbing monitor apparently indigenous to that area. This was of great interest to Böhme, because he knew that there was no *known* species of monitor native to North Yemen. Moreover, as he was unable to identify the monitor with *any* known species, he concluded that it must be new to science, and he urged herpetologists planning to visit Yemen in the future to look out for it.[65]

In October 1986, eight of these mysterious monitors were caught in Yemen's As Sokhna region during a herpetological field-trip from Zurich's Zoological Museum. Six were given to Zurich Zoo, the other two were sent to Böhme for observation in the living state. Ironically, it was later discovered that some specimens of this 'new' species had actually been collected as far back as the late 1800s, and had been held at the British Museum (Natural History) since the early 1900s, but researchers back then had presumed (wrongly) that these specimens' Yemeni provenance labels were incorrect and that they must have originated from Africa instead.[65]

ST. VINCENT WHIPSNAKE—
OVERLOOKED BY SCIENCE FOR 93 YEARS

In November 1987, a snake of slaty black colouration received by Wisconsin's Milwaukee Public Museum from St. Vincent's Department of Forestry proved to be a St. Vincent whipsnake *Chironius vincenti*—related to the familiar grass snake *Natrix natrix*, but wholly confined to this small West Indian island, where it had last been reported as long ago as 1894. The specimen sent to Milwaukee had been captured earlier in 1987, within a somewhat inaccessible forest, which may explain its species' success in eluding science for almost a century. Since its rediscovery, enquiries have ascertained that this species is familiar to the local people, who call it the black snake.[66]

A TUATARA TURNABOUT—AT LEAST FOR A TIME

One of the world's most remarkable reptiles is New Zealand's famous tuatara *Sphenodon punctatus*—for although outwardly lizard-like, it is in reality the only modern-day representative of a once-notable lineage of dinosaur contemporaries called sphenodontids (formerly classified as rhynchocephalians). During most of the 20th century, only a single living species of tuatara had been recognised. Yet back in 1877, the eminent New Zealand naturalist Sir Walter Buller had described a second species—almost a third smaller in size and of brighter colouration than *S. punctatus*, but confined solely to two tiny Cook Island islets (North Brother Island and East Island). Buller named this species *S. guntheri*, Gunther's tuatara,[44,67] but other researchers subsequently discounted this classification, deeming it instead to be nothing more than a morphologically distinctive representative of *S. punctatus* (and even denying it separate subspecific status).

In December 1989, however, biologist Dr. Charles Daugherty and colleagues from Wellington's Victoria Museum, working in conjunction with Dr. M.B. Thompson from Sydney University, publicly released the first details concerning the results obtained from their recent morphological and genetic comparisons of several separate, isolated tuatara populations. These comparisons were published in full a year later, and revealed that the North Brother tuataras (those of East Island have died out) seemed to be sufficiently different from the others to warrant classification as a completely distinct species after all. Thus, in their paper, Daugherty and his colleagues reinstated *S. guntheri* as a valid, second species of tuatara.[68]

Tuatara.

Having said that, however, in 2009 an online paper (published in hard copy in 2010) reexamining the genetic bases used to distinguish these two species of tuatara concluded that they only represent geographic variants, and that only one species should therefore be recognised.[68a]

In May 1995, New Zealand's Department of Conservation publicised the highly unexpected discovery by two medical students of a living tuatara at Tutukaka, on the northeast coast of North Island. Although tuataras did indeed once exist on New Zealand's mainland, pressures of competition with European animals imported by man supposedly caused their extinction here long ago, with representatives persisting only on a few tiny offshore islets. Having said that, it is remotely conceivable that a small mainland colony has somehow escaped detection. However, it seems more likely that this mystifying specimen, a large healthy male, had escaped from black-market animal smugglers, for whom tuataras are very lucrative. In autumn 2009, a tuatara hatchling was born at the Karori Sanctuary on North Island, and is thought to be the first tuatara born on the New Zealand mainland for over 200 years outside captive-breeding facilities.[68b]

PINK SNAKES, HARLEQUIN FROGS, AND JAMAICAN IGUANAS

Madagascar, already renowned as a treasure trove of endemics scarcely studied by science, may have quite a few more secretive species still awaiting formal recognition. During a three-month expedition to this island 'mini-continent' led by Dr. Chris Raxworthy in early 1990, at least four herpetological forms were found that could not be readily identified with known species. Two were *Boophis* tree frogs—one with white skin ornamented by brown harlequinesque blotches, the other a bright emerald species with yellow eyes and red feet. The two remaining forms were a species of skink (a short-limbed lizard); and a nocturnal, dark-headed snake, whose pink-scaled body bore a series of longitudinal dark-brown stripes. The snake ultimately proved merely to be a rare colour variant of a species already documented, but the other three are indeed new to science.[69]

And in June 1990, a specimen of the Jamaican iguana *Cyclura collei* was obtained by a hunter in southern Jamaica's Hellshire Hills. Until then, the last specimen on record was one that had been killed by a hunter's dog in 1969. Up to 17 in long, this greenish-blue lizard is Jamaica's largest species of native land animal, but is categorised as critically endangered by the IUCN.[70]

ELECTRIC FROGS AND SKUNK FROGS

In 1975, while working for the Queensland Forest Service, Chris Corben encountered a tree frog at west-central Queensland's Polygammon Creek that closely resembled a species already known to science, *Litoria rubella*—until it opened its mouth to emit its mating call. For the sound that emerged was truly extraordinary—a wavering, buzzing cry reminiscent of the sound produced by a high-voltage, long-duration electric arc! When the frog was captured, studies showed that it did differ a little from *L. rubella* externally too, via brown and yellow blotches on its thighs, and chocolate-brown bands on its upperparts.[71]

Other specimens were obtained in 1981, by Queensland Museum herpetologist Dr. Glen Ingram near Cloncurry (again in west-central Queensland). In 1990, Ingram and Corben described this species, and christened it the buzzing tree frog *L. electrica*—a fitting tribute to the unique nature of its vocal outpourings.[71]

No less extraordinary, albeit for a very different reason, was the yellow-patterned, green-skinned species of arrow-poison frog discovered in 1981 by Venezuelan biologist Dr. Alfredo Paolillo O, within the Andean cloudforests of the extreme northeastern corner of Trujillo, northwestern Venezuela. To begin with, and in stark contrast to other such species, this particular frog was nocturnal rather than diurnal, principally aquatic rather than terrestrial, and unexpectedly large for an arrow-poison frog. Its most distinctive characteristic, however, which has earned it its English name, was revealed when scientists attempted to handle it— for it swiftly registered its disapproval by secreting a liquid whose smell was not only foul, but actually resembled the infamous odour of the anal liquid ejected by threatened skunks.[72]

So far removed taxonomically from other arrow-poison frogs that it was housed in a new genus and a new taxonomic family, the Venezuelan skunk frog was described in March 1991, and as a reminder of two of its most characteristic features it was named *Aromobates nocturnes*. Worryingly, however, this remarkable species has not been seen again since its discovery, despite being searched for, is classed as critically endangered by the IUCN, and may even be extinct.[72]

PYGMY BLUETONGUE— REGURGITATING A REDISCOVERY

When herpetologist Graham Armstrong opened up the gut of a squashed snake (killed by a car) spotted by him on a road near Burra, South Australia, in late October 1992, he found inside it the corpse of a pygmy bluetongue skink *Tiliqua adelaidensis*—a species of lizard hitherto believed extinct since 1959. The area where Armstrong found the snake was duly searched, and a colony of pygmy bluetongues was uncovered, from which some were then sent to Adelaide Zoo for captive breeding purposes.[73]

CRYPTOPHIDION—THE SECRET SERPENT OF ANNAM

The volume of the International Society of Cryptozoology's journal *Cryptozoology* for 1992 marked a notable first in the history of zoological discovery. Contained within its pages was a paper presenting the formal description of a hitherto unknown snake—the very first time that a new species had been officially described and named within the world's only refereed, scientific journal devoted to cryptozoology.[74]

Even so, all was not quite as straightforward as it initially seemed. Sole occupant of a new genus too, this species had been dubbed the Vietnamese sharp-nosed snake *Cryptophidion annamense* by snake anatomist Dr. Van Wallach and vertebrate taxonomist Dr. Gwilym S.

Jones from Boston's Northeastern University. Its type (and only) specimen, obtained west of Da Nang, in Annam, central Vietnam, during 1968, was one of a series of zoological specimens collected by Jones and various other colleagues while serving as U.S. naval officers during the Vietnam War. These were destined for the Smithsonian Institution's National Museum of Natural History, but somehow the type of *Cryptophidion* was lost before it could be sent to the Smithsonian, and has never been seen since. All that remain to confirm its erstwhile existence are three poorly-focused colour photographic slides, deposited in the slide collection of Harvard University's Museum of Comparative Zoology (Department of Herpetology), and it is upon these that its description by Wallach and Jones was based.[74]

Judging from the morphological details of its type specimen (probably an adult female) depicted in the slides, *Cryptophidion* seems to be a snake adapted for a fossorial (burrowing) existence. However, it possesses a number of very distinctive features that collectively render it quite unlike any other southeast Asian snake. These include a depressed snout with a laterally pointed rostrum, greatly reduced concave nasal shields (scales) separated from the rostrum, an enlarged preorbital shield instead of a loreal and preocular shield, large temporal-like postocular shields, a temporolabial shield, and a short tail. Indeed, in the opinion of Wallach and Jones *Cryptophidion* appears to be so distinct from all other serpents that they are not even entirely certain as to which taxonomic family it rightfully belongs. On account of its distinct neck, almost flat belly, fairly slim tail, and typical anterior ventral scales, however, they consider that it most probably belongs to the colubrids.[74]

Not everyone, conversely, shares their view regarding the dramatically discrete taxonomic status of *Cryptophidion*. On the contrary: in response to the Wallach and Jones paper, herpetologists Dr. Olivier Pauwels from France's National Museum of Natural History in Paris and Dr. Danny Meirte from the Royal Museum for Central Africa in Tervuren, Belgium, presented a detailed morphological comparison of *Cryptophidion* with another somewhat enigmatic, rare burrowing serpent from Asia known as the sunbeam snake *Xenopeltis unicolor*, and they concluded that the two species are one and the same.[75]

This identification was in turn discounted by vertebrate zoologist Dr. James D. Lazell who acted as one of the three editorial reviewers of the Wallach and Jones paper. Lazell readily dismissed any close degree of similarity between what he considered to be an undeniably colubrine *Cryptophidion* and the neckless, thick-tailed, boa-like *Xenopeltis*. This opinion was also reiterated in a response by Wallach and Jones themselves.[76]

Without any further specimens of *Cryptophidion* having been procured so far to resolve the problematical issue of its identity, this is where the saga of Annam's hidden snake draws to a close, at least for the present time—caught up amid ongoing discussion and not a little dissension as to whether it truly constitutes a major herpetological discovery, or is merely an unfortunate artefact of misidentification.

In 1994, another new genus and species of unusual burrowing snake was described from Asia. Its type specimen was collected on 12 July 1988 by Louis Deharveng at Malawa in southeastern Sulawesi, during the Pyrenean Speleological Association's 'Maros 88' expedition. Named *Cyclotyphlops deharvengi*, Deharveng's blind snake, what makes it so unusual is that its large central circular head shield, around which smaller scales radiate, lies above what could be a third, parietal eye—which has been recorded in lizards but has never before been found in snakes.[77]

Still on the subject of new burrowing snakes: until recently, the genus *Typhlophis* contained only a single species, *T. squamosus*. In 1998, however, a second was recognised. Dubbed *T. ayarzaguenai*, specimens were unearthed during excavations for bauxite deposits at Serranía de Los Pijiguaos, in Venezuela's Bolivar State. Lately, however, some researchers have claimed that this new species is in reality conspecific with *T. squamosus*.[78]

NEW FROGS, NOVEL VOICES

Frogs are no stranger to water, but during the 1990s a couple of very curious new frogs were discovered whose voices were also emphatically—indeed, exclusively—hydrophilic.

The first of these was a new species of leopard frog, which was first collected in 1988 at Ramsey Canyon in the Huachuca Mountains of southeastern Arizona's Cochise County by herpetologist Dr. James E. Platz from Nebraska's Creighton University. In 1993, Platz formally named it *Rana subaquavocalis*, the Ramsey Canyon leopard frog, for good reason, because unlike all other leopard frogs the male of this species gives voice to its mating call entirely underwater, and from a depth of more than 3 ft—thus making it completely inaudible in the air. The obvious advantage of this highly unusual behavioural trait is that female specimens of *R. subaquavocalis* know that any calls heard by them underwater will be from males of their own species, thus avoiding any interspecific competition. Gravely threatened by habitat loss, this recently-revealed species is categorised as critically endangered by the IUCN.[79]

In 1997, a new species of frog was described from Bolivia's cloudforests that only calls when the weather is rainy. Named *Eleutherodactylus* [=*Pristimantis*] *pluvicanorus* and less than 2 in long, this interesting little amphibian remains hidden under stones and logs during the day, but if rain or fog occurs it comes out at night and voices its distinctive call—a soft whistle.[80]

THE TORTOISE THAT TIME FORGOT

One of the most memorable reptilian revelations of late must be the Gulf snapping turtle, for this Antipodean anachronism is a bona fide prehistoric survivor. Its remarkable history began with a partial carapace and associated plastron from a previously undocumented species of Australian freshwater tortoise (but colloquially termed a turtle) unearthed at the Riversleigh fossil deposits site in northwestern Queensland. Dating from the Pleistocene, in 1994 these remains were officially described as a new species by Drs. Arthur White and Michael Archer, who named it *Emydura lavarackorum*. This species belonged to the short-necked chelid group of freshwater tortoises, and was believed to have died out 20,000-50,000 years ago.[81]

In 1996, however, Australian zoologists Drs. Arthur Georges and M. Adams documented the discovery of a living form of tortoise in the Nicholson River drainage of the Lawn Hill National Park, in northwestern Queensland's Gulf region, that looked amazingly similar to the 'extinct' *E. lavarackorum*, yet which had been hitherto overlooked by zoologists. Dubbed the Gulf snapping turtle, it had first come to scientific attention in 1994, at much the same time as the discovery of its fossil counterpart, when Queensland government conservation official Dr. Col Limpus had spotted a strange-looking tortoise in photos of wildlife hunted during a recent land rights campaign that had taken place in the Lawn Hill National Park.[82]

The modern-day tortoise's skeleton and the unusual arrangement of its carapace's plates subsequently confirmed that it was a veritable 'living fossil', because it did indeed prove to be one and the same as *E. lavarackorum*. Moreover, studies on the modern-day tortoise suggested that its fossil equivalent had been misclassified, and that this recently-resurrected species is actually more closely aligned with the genus *Elseya* than *Emydura*. Consequently, in 1997 Canberra researcher Dr. Scott A. Thomson together with White and Georges assigned the Gulf snapping turtle (present and past) to *Elseya*, renaming it *Elseya lavarackorum*.[82]

MARY RIVER TORTOISE—THE AUSTRALIAN PET TRADE'S BEST-KEPT SECRET?

No less notable Down Under is the Mary River tortoise *Elusor macrurus*, readily differentiated from other short-necked freshwater tortoises by its distinctively long, sturdy, laterally-flattened tail, which in the adult male can be 70 per cent as long as the creature's carapace and almost as thick as a small human wrist. Despite being one of Australia's largest species of tortoise, it was only scientifically described and named as recently as 1994, and constituted a totally new genus. Yet this intriguing species had been familiar to the public since the early 1960s, because hatchlings were often sold in Australian pet shops! It even had its own particular names in the pet trade—being variously referred to as the saw-shelled snapper, the penny turtle, and (aptly) the pet shop turtle.[83]

Yet scientists had never found where these enigmatic pet shop specimens had originated, because none of the shop owners would disclose their sources (some of which included illegal egg-collectors). After more than 30 years of investigations, however, Australian herpetologist John Cann and Prof. John M. Legler from Utah University—who saw their first specimens in 1961 while looking around some pet shops in Victoria—finally achieved success. During the early 1990s, after being taken there by a poacher, Cann uncovered this secretive species' hideaway—the Mary River, in southeastern Queensland. Here he found four adult individuals, which were the first wild specimens ever knowingly recorded by a scientist.[83]

RACING TO HELP A RACER

Mongooses were introduced onto Antigua just over a century ago to eradicate the rats that were devastating this Caribbean island's sugar plantations. Once here, however, they also displayed their famous talent for serpent-killing, rapidly decimating an inoffensive native species of brown and black snake—the Antiguan racer *Alsophis antiguae*. Within a short time, this non-venomous lizard-eating reptile seemed to have been entirely wiped out.[84]

Nothing more was heard of it for several decades, but since the 1960s occasional reports emerged concerning the Antiguan racer's possible survival on a small, rocky, densely-forested, uninhabited isle called Great Bird Island, just a mile off Antigua. During the early 1990s, these reports were confirmed by a team of West Indian conservationists. However, their joy was tempered by horror when, assisted by British experts from Fauna and Flora International, they found that the few racers on Great Bird Island (estimated at below 50) were under imminent threat of annihilation by the teeming population of black rats thriving here. Most of the racers were old and bore scars of bite wounds inflicted by the rats; some had actually been castrated by rats biting their tails. And there were few if any young racers—for the simple reason that the rats were eating the eggs laid by the adult snakes.[84]

The Antiguan racer, only lately rediscovered after a century of supposed extinction, now faced genuine extinction. Happily, however, the team was able to avert this by carrying out an effective programme of poisoning that removed the rats while leaving the racers unaffected. By spring 1998, freed from rodent persecution, the racer population had risen to 130, and a captive breeding population was established at Jersey Zoo. Although it is categorised as critically endangered by the IUCN, numbers continue to increase steady, and there are currently some 500 individuals in existence.[84]

Also of note from the snake world: a new, aptly-named genus of snake, *Xenophidion* ('strange snake'), was described in 1995, with two species, the spinejaw snake *X. acanthognathus* and Schaefer's spinejaw snake *X. schaeferi*, both hailing from Malaysia. Echoing the situation with *Cryptophidion annamense* (see p. 183), however, *Xenophidion* is so anomalous that it cannot presently be satisfactorily assigned to any existing taxonomic family. Its discoverers tentatively placed it with the colubrids, but in 1998 snake expert Drs. Van Wallach (who had earlier co-described *Cryptophidion*) and Reiner Günther created an entirely new family for it—Xenophidiidae—reflecting its anatomical idiosyncrasies.[85]

ARNOLD'S GIANT TORTOISE AND THE HOLOLISSA OF SEYCHELLES— LONG FORGOTTEN, BUT NO LONGER GONE

The Galapagos giant tortoises were once rivalled for size by several huge species inhabiting the Indian Ocean's granitic Seychelles group, the coral atoll of Aldabra, the Mascarene islands, and Madagascar. Of these, only the Aldabra giant tortoise *Dipsochelys* [=*Aldabrachelys*] *gigantea* [=*dussumieri*, =*elephantina*] is traditionally thought to have survived into the present day, the remainder having been killed for their meat during the 1700s and 1800s—or so it was thought, until Arnold's giant tortoise and the hololissa unexpectedly reappeared.

There has been much debate concerning the precise number of giant tortoise species native to the Seychelles. Four are currently recognised, one of which was formally described in September 1982, by Dr. Roger Bour from France's National Museum of Natural History. He based his description upon three old taxiderm specimens (two at the above museum, the third at the British Museum). They possessed various skeletal modifications that seemed to be adaptations to browsing, and originated from the granitic Seychelles islands. Bour named their species *Dipsochelys arnoldi*, but as there did not seem to be any giant tortoises (other than Aldabra's) in the Seychelles today, he naturally assumed that it was extinct—belatedly recognised as a distinct species, yet irretrievably deceased.[86]

Imagine, then, his surprise when, while still preparing his paper, Bour was shown some photos by film producer Claud Pavard (who had taken them in August 1981) depicting two *living* giant tortoises that seemed to belong to his supposedly extinct species *D. arnoldi*. Nor was this the only surprise. The tortoises, males and very old, were living in semi-captivity at a sugar estate, but not in the Seychelles—instead, on Mauritius! Naturally, Bour hoped to visit Mauritius, to ascertain conclusively these potentially significant specimens' identity.[86]

And that is where this most promising saga seemed to come to an abrupt end. During my preparation of this book's first edition, *The Lost Ark* (1993), I was unable to locate a single publication carrying any further news regarding these tortoises, and none of my zoological colleagues and correspondents had any details (sadly, I never succeeded in eliciting a reply from Dr. Bour himself), though they were all as intrigued by it as I was. Happily, however, the mystery was finally solved in May 1992, when I learnt from British Museum herpetologist Dr. Nick Arnold (after whom *D. arnoldi* had been named) that Dr. Bour had indeed visited the two Mauritius specimens, but had found that they were not representatives of *D. arnoldi* after all.[87]

Hololissa. (Alan Pringle)

Moreover, Dr. Ian Swingland, Founding and Research Director of the Durrell Institute of Conservation and Ecology (DICE), informed me that giant tortoises reared in captivity sometimes have shells that have become distorted in shape, due to the way in which these animals have been fed. In some cases, therefore, it is possible that they may even resemble the shells of quite unrelated species, and this is presumably what had happened in the case of the two Mauritius specimens, which were probably individuals originating from Aldabra. Captivity-induced distortion of shell shape can cause problems for tortoise taxonomists too, especially if they are dealing with specimens whose life histories are unknown (and which, therefore, may have been reared in captivity).[87]

Of course, one objection that could immediately have been raised in relation to this entire episode is the fact that supposed specimens of *D. arnoldi* were discovered not in the Seychelles, but instead in Mauritius. As it happens, however, this objection can be effectively countered—because a number of giant tortoises from the Seychelles are known to have been introduced there after that island's own indigenous species had been exterminated during the 1700s. In particular, the French explorer Marion de Fresne transported five such specimens in 1776 from the Seychelles to his military barracks on Mauritius. What was assumed to be the last of this quintet died there in 1918,[86] but there may have been others too, whose records have failed to survive to the present day.

In any event, what did seem clear at the time of writing this book's first edition was that none of the long-lost species of Seychelles giant tortoise had been resurrected after all. During 1995, however, another discovery was made—one that added a new and much more dramatic chapter to this long-running saga of mistaken and incognito identities.

In January of that year, the Nature Protection Trust of Seychelles (NPTS) learnt of two very large, and very old, male tortoises living in the garden of a Seychelles hotel. When examined by Dr. Justin Gerlach and K. Laura Canning, chief scientists with the NPTS, they were found to exhibit pronounced flaring, flattening, and scalloping of the carapace, especially over the hind legs—characteristics that distinguished them from the Aldabra giant tortoise but corresponded closely with those of Arnold's supposedly long-extinct species.[88]

Enquiries revealed that these and one other male specimen had been purchased in 1994 from an old local man, in whose family they had been throughout living memory. The third had died in December 1994, but its skeleton was preserved and donated to the NPTS's scientific collections. Cranial studies subsequently determined that it was indeed distinct from the Aldabra species. Genetic studies were also set in motion, to bypass any possible misclassification based solely upon morphological characteristics—which can, as already ably demonstrated with the earlier episode of the Mauritius specimens, be very deceptive.[88]

By early 1997, several additional specimens of unusual giant tortoise had been discovered in various Seychelles localities and examined by Gerlach. Moreover, whereas some of these resembled Arnold's giant tortoise, eight others closely recalled a second supposedly long-vanished species—the hololissa *Dipsochelys hololissa*. Previously, this latter species had been known only from two shells found in 1810, described in 1877, and destroyed in the 1940s by German bombing raids during the London Blitz. It formerly inhabited various granitic islands of the Seychelles, where it grazed vegetation on the edges of streams and marshes, but had vanished in the wild by 1840.[89]

In March 1997, Dr. Les Noble conducted genetic tests at Aberdeen University on blood samples taken by Gerlach from a large selection of live Seychelles giant tortoises, including the controversial ones. These tests showed that three distinct groups could be identified, revealing that eight of the specimens were hololissas, two were Arnold's giant tortoises, and the remainder were Aldabra giants. But this was still not the end of the story.[89]

A year later, Blackpool Zoo in England announced that Darwin, the Aldabra giant tortoise that had been living there for the past 25 years, was not a member of the Aldabra species after all. While closely scrutinising photos of recently-discovered living specimens of the hololissa, staff at the zoo were astonished to discover that they looked just like Darwin. Anxious to learn more, they duly contacted Gerlach, who visited the zoo, examined Darwin, and confirmed that he was indeed a living hololissa. This presumably explains why he has never successfully mated with Beagle, the female Aldabra giant tortoise that accompanied him when he arrived at Blackpool in 1972—because they belong to separate species.[90]

I subsequently learnt from Alan Pringle of the Cotswold Wildlife Park that their supposed Aldabra giant tortoise had also been unmasked as a hololissa. Moreover, by the end of 1999 at least 12 living hololissa individuals and 18 living individuals of Arnold's giant tortoise had been revealed in various locations around the world. All of which invites speculation as to how many other incognito specimens of hololissa and Arnold's giant tortoise may still be awaiting identification elsewhere. By the end of 1997, the NPTS had introduced several specimens of hololissa and Arnold's giant tortoise to Silhouette Island (third largest of the central Seychelles islands) in order to initiate captive breeding programmes for both of these recently-revived species and thus ensure their continuing survival, and it continues to search for more possible examples in captive collections worldwide.[89-91]

Officially known as the NPTS Seychelles Giant Tortoise Conservation Project, its patron is veteran wildlife film maker and broadcaster David Attenborough. Its longterm goal is to increase the numbers of both species in order to permit reintroduction to secure reserve sites within the Seychelles group—thereby restoring in viable form two remarkable endemics to their native island homeland after more than 150 years of 'official' non-existence.[91]

COSTA RICA'S PHANTOM FROG—AND A CRYPTIC FROG FROM RWANDA

The year 1997 saw the formal description of a new species of frog characterised by its skin's distinctive spectral sheen. Currently known only from a single specimen, a female, discovered several years earlier on the rocky banks of a forest stream in the mountain forests of southern Costa Rica by Dr. Karen R. Lips from New York's Saint Lawrence University and Dr. Jay M. Savage from Miami University, it has defied all attempts by scientists since then to uncover more specimens. This memorable species was fittingly christened *Eleutherodactylus phasma*, the phantom frog, on account of the ghostly pallor of its almost pure white skin, but was subsequently reassigned to the genus *Craugastor*, becoming *Craugastor phasma*. Nearly 2 in long, it was initially thought to be a freak albino specimen of a previously described *Eleutherodactylus* species, but this was later disproved.[92]

Speaking of mystifying white frogs: I also have on file a tantalising snippet of text mentioning what may be a still-undiscovered species in Rwanda, eastern Africa. It featured in *Among Pygmies and Gorillas* (1923), a book by Prince William of Sweden, which detailed his explorations of Rwanda. The passage in question reads as follows:[93]

> The white, or blind, frog which was said to live in Mutanda was also conspicuous by its absence. We hired the natives to collect amphibians from the whole of the lake, and they came regularly once a day with big baskets absolutely full of frogs. In hundreds and thousands, in cubic yards. The whole camp was alive with frogs. In vain! There were no white ones among them, and every one had two big staring eyes.

Could this mysterious blind white frog have been a specialised cave-dwelling species? If so, it would compare with various spelaean salamanders and freshwater fishes, which also lack eyes and skin pigment, as these are superfluous when living in a permanently lightless world.

TURNING TURTLE IN HANOI

During the past five centuries, Hoan Kiem Lake, a small algae-choked but seemingly well-oxygenated expanse of freshwater situated in the centre of modern-day Hanoi, Vietnam, has been associated with legends and stories of giant turtles (i.e. freshwater tortoises), and for many years at least three such creatures have been regularly reported here. On 24 March 1998, however, events escalated when a passing cameraman succeeded in filming three of them with his videocamera while they surfaced to gulp air.[94]

After the film had been shown on television, Hanoi National University biologist Professor Ha Dinh Duc announced that he had been studying these outsized chelonians since 1991, and believed that they constitute a new species. Conversely, a number of other herpetologists consider them to be conspecific with *Pelochelys bibroni*, but this identification was discounted by Dr. Peter C.H. Pitchard, co-chairman of the Tortoise and Freshwater Turtle Specialist Group within the Survival Specialist Commission (SSC), after he had visited Hanoi to view them himself. Instead, he announced that in his opinion, "these turtles are either an outlying population of the Chinese species *Trionyx* [=*Rafetus*] *swinhoei* or that they are a new species. Certainly they are not *Pelochelys bibroni*". This conclusion was also favoured by Dr. Patrick P. McCord, a leading U.S. expert on soft-shelled freshwater tortoises, of which *R. swinhoei* is one such species. Accordingly, Prof. Duc later dubbed the Hoan Kiem turtles 'Rafetus hoankiemensis', thereby classing them as a distinct species within the existing genus *Rafetus*, which in 2000 he formally named *R. leloii*.[94]

A stuffed specimen, preserved almost 40 years ago, is currently on exhibition at a small temple on an island in Hoan Kiem Lake. According to Duc, *R. leloii* is the world's largest freshwater tortoise, attaining lengths of up to 6.5 ft and weights of up to 440 lb, with a greenish-brown

Turquoise monitor. (Mark K. Bayless)

Moluccan yellow monitor—first made known to science in 1997 from a photograph shown at a varanid conference. (Al Baldogo)

carapace, pink belly, a green and yellow football-sized head, downcast mouth, oval shell, and peeling skin.[94]

However, there have already been claims that the narrow-headed soft-shelled turtle *Chitra indica* (another Asian freshwater tortoise) can attain lengths of 6 ft. In any event, the existence of what could be a new species of very sizeable reptile remaining 'undetected' by science while existing in full view of downtown urban Hanoi's teeming populace is sufficiently noteworthy in itself to encourage further investigation of these remarkable animals—especially as they may be the only representatives of their kind. Having said that, in 2003 herpetologists Drs. B. Farkas and R.G. Webb published a paper in which they denounced *R. leloii* as an invalid species, reclassifying its trio of representatives as specimens of *R. swinhoei*.[94]

MONITORING THE MONITORS

One of the most attractive reptiles to be formally described and named during the 1990s, the turquoise monitor lizard is in fact a decidedly belated arrival on the official scene. Native to the small Indonesian island of Halmahera in the Moluccas group, it was christened *Varanus caerulivirens* by Drs. Thomas Ziegler, Wolfgang Böhme, and Kai Philipp in 1999. However, it had first been procured by a Mr. Kukenthal way back in 1895, and was also reported by Berlin Museum herpetologist Dr. Robert Mertens in 1942, yet remained undescribed. Attaining a total length of possibly 4 ft, this species is certainly very eyecatching, with a turquoise-gold head, face, and back, and a turquoise-blue tail.[95-6]

The 1990s were a good decade for finding and refinding monitor lizards (varanids). The late Mark K. Bayless, a notable varanid expert, prepared an extensive review of those new and rediscovered monitors[96], which he very kindly permitted me to utilise here. The most notable species are as follows.

Described in 1991 by varanid researcher Dr. Robert G. Sprackland from a specimen collected back in 1978 by Dr. Gregory Czechura at Buthen Buthen, Queensland, and residing ever since in a bottle of alcohol at the Queensland Museum, the canopy goanna *V. teriae* was first photographed in the living state in 1992 by varanid seeker Theo Tasoulis. Five years later, thanks to the diligence of Australian television wildlife presenter Steve Irwin, who successfully captured and brought back some specimens to Australia Zoo where he was director prior to his untimely death in 2006, this little-known species bred in captivity for the first time.[96-7]

A second new varanid described in 1991 by Sprackland was the Rossel Island monitor *V. telenestes*. It is currently known only from its type specimen, deposited at the Queensland Museum, which had been collected on Rossel Island in Milne Bay, Papua New Guinea, and had earlier been classed as a specimen of Bogert's monitor *V. bogerti*.[96-7]

In 1994, D. Yang and Wanzhao Liu described and named a new species, *V. vietnamensis*, on the basis of a specimen discovered at a market in Hekou, Yunnan, on

Peacock monitor. (Al Baldogo)

15 November 1992. Four specimens were imported into the U.S.A. in 1995 and sold to various varanid keepers. Many herpetologists nowadays deem the Vietnam monitor to be nothing more than a colour morph of *V. nebulosus*, but not everyone agrees with this classification.[96,98]

The Moluccan yellow monitor first came to scientific attention when a photograph of this then-unknown species was shown by zoologist Rafe Brown to colleagues while attending a varanid conference in Bonn during 1997. Brown had snapped the picture during a recent visit to the Moluccas islands. Christened *V. melinus* later in 1997 by Böhme and Ziegler, its survival in the wild is greatly threatened because of logging on its island homes, persecution by the loggers (who casually behead any hapless monitors that they encounter while working), and mass exporting of specimens by the international pet trade.[96,99]

A year later saw the description and naming of another Moluccan speciality, the tricoloured or black-backed monitor *V. yuwonoi*. This species had been photographed in 1994 by Dr. Hans J. Jacobs from Germany while holidaying on Halmahera, and the first specimen to be collected was procured by Frank Yuwono (after whom the species is named) from Jakarta, Java. Almost 6 ft long, it is distinguished by its dark anterior dorsum, nine blackish bands on its posterior dorsum, and a blue tail.[96,100]

Also native to Halmahera is a pygmy mangrove monitor, which has already appeared in pet trade listings even though it is apparently still undescribed by science. So too may be the Moluccan blue monitor, inhabiting Buru, and possibly the other Moluccan islands as well. In addition, Mark K. Bayless owned a varanid of a species yet to be described for many years. Even the very striking peacock monitor *V. auffenbergi* of Roti Island in the Timor group was only formally recognised and named as recently as 1999, by Sprackland. Until then, this handsome blue-scaled tree varanid had been lumped together taxonomically with the common Timor monitor *V. timorensis*. Clearly, there seems good reason to believe that further species remain to be discovered by enterprising varanid researchers.[96,101-2]

El Hierro giant lizard.

GIANT LIZARDS IN THE CANARY ISLANDS

Staying with large lizards: in 1974, the El Hierro giant lizard *Gallotia simonyi*, a moderately sizeable lacertid up to 3 ft long and indigenous to the Canary Islands, was rediscovered by local goatherd Juan Machin on the Tibataje crags—an expanse of volcanic cliffs at the isle of El Hierro's northwestern end. Until then, this species had last been recorded alive (it is also known in subfossil form) in 1935, when some specimens were collected on the Roque Chico de Salmor, a tiny islet just off El Hierro. But its reappearance was only the beginning of this story.[103]

In 1985, herpetologist Rainer Hutterer described a new but supposedly long-extinct subspecies, *Gallotia simonyi gomerana*, from subfossil bones in an archaeological site dated at about BP 500 (c.500 years ago), on the island of La Gomera in the Canaries. A decade later, during June-October 1999, a team from La Laguna University conducted an intensive search of this island, and trapped several specimens of a large, unfamiliar lacertid lizard with a dark back and blue spots on its flanks. Its body size corresponded with the subfossil remains of *G. s. gomerana*, and some (but not all) herpetologist eventually concluded that it was indeed one and the same as La Gomera's 'extinct' giant lizard; it was later renamed *G. bravoana*. A hitherto-undescribed species, the Tenerife speckled lizard *G. intermedia*, was discovered as recently in 1996, on northwestern Tenerife's coastal cliffs, with a second population of at least 15 specimens being uncovered in April 2003 on this same island's southernmost cliffs.[103]

La Gomera giant lizard.

THE FISHES

Coelacanths, Megamouths, and More

I stood as if stricken to stone. Yes, there was not a shadow of doubt, scale by scale, bone by bone, fin by fin, it was a true Coelacanth. It could have been one of those creatures of 200 million years ago come alive again. I forgot everything else and just looked, and then almost fearfully went close up and touched and stroked . . . it was true, it was unquestionably a Coelacanth. Not even I could doubt any more.

PROF. J.L.B. SMITH—*OLD FOURLEGS: THE STORY OF THE COELACANTH*

The discovery of megamouth does one thing. It reaffirms science's suspicion that there are still all kinds of things—very large things—living in our oceans that we still don't know about. And that's very exciting.

DR. LEIGHTON R. TAYLOR—*WAIKIKI BEACH PRESS*

The lobe-finned coelacanth, resurrecting its ancient lineage from many millions of years of supposed extinction, and the megamouth, adding an entirely new taxonomic family to the world catalogue of modern-day sharks, are certainly the most celebrated ichthyological finds of the 20th century, but they are very far from being the *only* such finds of note since 1900. As will be revealed here, others include: the formal recognition of the Siamese fighting fish as a species in its own right, and the discovery of another species so pugnacious that it was named after a famous boxer; a six-eyed spookfish, and a blood-sucking vampire fish; a fish with fur, and a fish with teeth on top of its head; the first deepwater stingray, and a land-living catfish that actually dislikes water(!); the world's largest species of freshwater fish, and one of the world's most popular species of tropical aquarium fish; plus the identification of a goblin unicorn from the deep, the finding of a living fish-trap with built-in illumination on the sea bottom, the enigma of Denmark's bottled sea serpent, and a second species of living coelacanth discovered thousands of miles beyond the first species' distribution range.

GOBLIN SHARK—A NIGHTMARISH UNICORN FROM THE OCEAN DEPTHS

The beginning of the 20th century was a time of great excitement in ichthyological circles—due to the debut of one of the most grotesque fishes ever seen by man.

Up to 14 ft long and generally living at depths of about 1500 ft (but occasionally in more shallow localities too, such as Tokyo's Sagami Bay), the aptly-named goblin shark was first found in Japanese waters, but would not look out of place in a painting by Salvador Dali or Hieronymus Bosch! True, its fairly slender, grey-pink body is rather nondescript, bearing two rounded dorsal fins, a pair of similarly-rounded pectorals, plus a pair of larger, straight-edged pelvics, and a single anal fin just in front of the tail. In contrast, the tail fin is very odd, with a long but extremely low ventral half, and little more than a terminal bob for the dorsal half. Most bizarre of all, however, is its head.[1-2]

Extending forward from the top of the head like a fixed, horizontally-held spear is a long, shovel-shaped horn, which overhangs the fish's jaws and provides this surrealistic shark with a disturbingly menacing, malevolent expression. Interestingly, this peculiar protuberance is totally separate from its mouth; by comparison, the

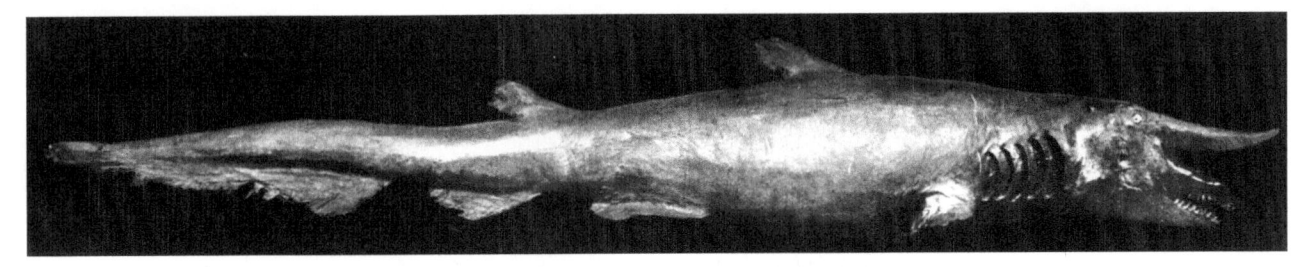

Rare early photograph of a preserved goblin shark.

'sword' of swordfishes and marlins, the 'saw' of sawfishes and sawsharks, and the corresponding structures of other fishes similarly endowed all take the form of a greatly enlarged upper jaw. As for the goblin shark's mouth, this is rather odd too—terminally-sited (in most sharks it is placed ventrally), highly protrusible (in most sharks it is fixed), and crammed with diagnostically-shaped teeth that occupy a central role in this species' curious history.[1-2]

During the mid-19th century, fossil shark teeth of a distinctive, thorn-like shape with smooth edges were found in Lower Cretaceous rocks (i.e. dating back 140 million years) in several widely separated localities, including Syria, New Zealand, England, and India. They did not correspond with any living shark known at that time—nor did their owner, a remarkable creature bearing a long, horizontal, shovel-like projection on its forehead. This strange shark was thus deemed to be a once widely-distributed species that had ultimately died out entirely, leaving no modern-day descendant. In 1889, this seemingly defunct experiment in shark evolution was formally named *Scapanorhynchus* ('shovel-beak'), and was afterwards largely forgotten—until 1897, which heralded its astonishing resurrection.[1-2]

This was when some Japanese fishermen, working in deep waters near Yokohama, hauled up an extraordinary-looking shark—the very first goblin shark ever recorded by science, whose weird, nightmarish appearance instantly distinguished it from all other modern-day species of known fish. Fortunately, after having been rescued by Yokohama-based naturalist Alan Owston, it swiftly came to the attention of Japanese ichthyologist Prof. Kakichi Mitsukari. After examining it, Mitsukari was so impressed by its form that in 1898 he transported it personally to the U.S.A., to be studied by Leland Stamford University researcher Dr. David Storr Jordan, an internationally-renowned expert on Japanese fishes. Later that year, Jordan officially described Mitsukari's mystery shark, which, in honour of the two men who had brought it to scientific attention, he named *Mitsukarina owstoni*.[1-2]

The advent of the 20th century witnessed Jordan's paper attracting very appreciable zoological attention worldwide. Yet whereas students of present-day fishes were primarily intrigued with the goblin shark's most readily visible morphological attributes, palaeontologists were becoming equally interested in certain less conspicuous features—namely, its teeth. This was because they had recognised that these were identical in form to those of the supposedly long-extinct shark genus *Scapanorhynchus*.[1-2]

Accordingly, the newly-discovered, modern-day goblin shark was rechristened *Scapanorhynchus owstoni*, but nowadays it is considered to be sufficiently different from its fossil counterpart to require its own genus, so '*Mitsukarina*' has been restored to use.

Sadly, despite its unique appearance and history the goblin shark is almost as mysterious today as when first brought to science's attention a century ago. Even the function of its characteristic shovel-horn is unknown. However, some researchers have suggested that this species is a sea-bottom feeder, so perhaps it uses its shovel to stir up the sea-bottom's ooze and sediment in order to dislodge the creatures living there, which it can then engulf within its extensible jaws. Its distribution range is uncertain too; specimens have been caught off Japan, South Africa, and possibly Portugal.[1-2]

Thanks once again to its unmistakable teeth, moreover, there also is evidence to indicate its presence in the Indian Ocean. At a depth of 4500 ft in this ocean, a submarine cable unexpectedly malfunctioned, so it was brought back to the surface for examination. This revealed the presence of a broken shark's tooth embedded between the cable's strands of wire—implying that a shark had been feeding on organisms that had settled upon the wire. As for the tooth itself, its smooth-edged, thorn-shaped form was immediately recognisable—it was the tooth of a goblin shark! As with everything else associated with this enigmatic creature, a significant new piece of information concerning it had been obtained totally by accident.[2]

MACRISTIUM CHAVESI AND OTHER INFANTS INCOGNITO

In 1903, an extremely strange fish, caught severely injured off the Azores, was described by British Museum ichthyologist Dr. Charles Tate Regan, and the hitherto unknown species that it seemed to represent was dubbed

Macristium chavesi.[3] Totally unlike anything ever recorded before, it eventually found itself housed within a completely new taxonomic order—Ctenothrissiformes—as the only modern-day member amongst a series of fossil species.

Measuring 11 cm (about 4.5 in) from head to tail base, its slender body was grey in colour and somewhat herring-like in overall shape, but was laterally flattened and scaleless. Its most striking features, however, were its very large, prolonged body fins. The dorsal fin, for example, had a very broad base, occupying almost half the length of the fish's back, and contained several extremely long rays. The pectoral fins were also long, and just behind these were the very elongate pelvic fins, which extended back towards the tail fin. Also present was a single, median anal fin, containing 12 rays and positioned just behind and beneath the rear end of the dorsal fin (but on the fish's undersurface).[3-5]

The head of *Macristium* was rather small, and it only had a weak dentition, consisting of a set of small sharp teeth on its lower jaw and tongue, as well as on a nasal-related bone termed the vomer, but its diet was unknown. In fact, apart from its deepsea origin, nothing whatsoever was on record concerning this strange species.[3-5]

Even its type specimen remained unique for several decades—not until 1961 was another *M. chavesi* specimen recorded, this time from the Bay of Biscay. Obtained by N.B. Marshall, it only measured 3.3 cm (less than 1.5 in), and appeared to be a younger individual than Regan's specimen. In 1967, Drs. F.H. Berry and C.R. Robins described a second modern-day member of this order, a short-finned species that they named *Macristiella perlucens*, based upon a young specimen procured in the Gulf of Mexico.[5]

By now, the long-neglected ctenothrissids, living and fossil, were engaging the interest of several ichthyologists; but from their studies (especially those of Dr. D.E. Rosen), evidence emerged that the *Macristium* and *Macristiella* specimens were nothing more than immature individuals of some already-known deepsea fishes (genus *Bathysaurus*), known as deepwater lizardfishes, rather than separate species in their own right. This identification is nowadays accepted by most experts, with Ctenothrissiformes looked upon as a wholly extinct order (perhaps not even a genuine order either), and its former modern-day representatives demoted to incongruous infants.[5]

This is not an isolated case. Also likely to be an infant incognito is *Korsogaster nanus*, a peculiar fish whose body is covered with small prickles instead of scales. The sole member of its family, it has long been known only from two specimens. Its type was caught near the Bahamas on 26 February 1927, and a second was captured in the central equatorial Pacific by the Scripps Institution of Oceanography's Styx Expedition in August 1968. Researchers now believe that these are merely juveniles of a fish known, somewhat unappealingly, as a slimehead (genus *Hoplostethus*), which is related to the squirrelfishes.[6]

As for *Rosaura rotunda*, a tiny 8.4-mm-long fish known only from its type specimen, caught in the Atlantic off northeastern Brazil during the 'Rosaura' expedition and described in 1954, this is now thought by some to be an aberrant immature giganturid, specifically *Gigantura indica*. These are grotesque deepsea fishes, distantly related to salmon, with huge bulging eyes.[7]

PUGNACIOUS PISCEANS—
THE SIAMESE FIGHTING FISH

Notorious for the aggressive behaviour of males in close proximity to one another, the fighting fishes (gouramis belonging to the genus *Betta*—'warrior') are among the most familiar stars of tropical freshwater aquaria. This makes it all the more ironic that their most famous representative was among the last to be recognised by science.

The first to attract ichthyological attention were the Sumatran fighting fish *B. brederi*, the Javan *B. picta*, and the Malayan *B. pugnax*, all of which had been described by the mid-1800s. A somewhat rarer species, *B. bellica*, had also been described by the 19th century's close, but it was not until 1909, following Dr. Charles Tate Regan's studies of these flamboyant fishes, that another three species were distinguished. Two of those, the banded fighting fish *B. fasciata* (another rare form) and the Bornean *B. taeniata*, received little attention afterwards. The third, conversely, was not only one of its country's most well-known animals, but had already gained fame among European aquarists too—and all long before its separate specific status had been recognised by science. For this species, which gained from Regan the formal name *Betta splendens*, was none other than the Siamese fighting fish.[4,8]

Introduced into France as early as 1893, followed three years later by its German debut, Thailand's *B. splendens* is undoubtedly the most beautiful of all fighting fishes, and occurs in a varied range of vivid colours in the wild state, with northern males tending towards greenish hues and southern males exhibiting the brilliant red shades that are so popular in aquaria. Additionally, careful selective breeding has created an even greater spectrum of colours, including deep blues and mauves, as well as spectacularly transforming the wild strain's short dorsal and anal fins into greatly enlarged, diaphanous veils that recall the ostentatiously-elaborated sleeves and lace-exaggerated cuffs of the Victorian dandy.[4]

Based upon outward appearances, therefore, the male Siamese fighting fish may seem to have been emasculated,

Siamese fighting fish.

rendered harmless—but this illusion can be quickly dispelled. All that we need to do is place a mirror inside a tank containing one of these piscean popinjays—and the fop instantly becomes the fighter. Opening wide its mouth and gill covers, and spreading its fins as far apart as possible, it prepares to do battle with the equally formidable, belligerent rival that in reality is nothing more than its own reflection in the mirror![4]

...AND A FISH CALLED JACK DEMPSEY

During the same decade that he unmasked the Siamese fighting fish, Dr. Charles Tate Regan also described another pugnacious species nowadays very popular in aquaria[9]—a species that could very accurately be referred to as an Amazonian boxing fish! Indeed, so close are this species' behavioural links with the much-revered sport of pugilism that it was soon named after a famous world heavyweight boxing champion—Jack Dempsey.

Jack Dempsey.

Originally christened *Heros octofasciatus*, then later becoming known as *Cichlasoma biocellatum*, but now renamed *Rocio octofasciata*, the Jack Dempsey is a tropical freshwater cichlid from Central America, inhabiting bogs and other warm, slow-moving swampy water. When adult, it is an outstandingly beautiful fish, whose dark flanks, fins, and tail bear an iridescent mosaic of brilliant green spots. Almost nothing is known of its habits or lifestyle, however, except for its famous territorial behaviour, readily observed in aquaria. If one male enters the territory of another during the breeding season, the two circle one another and raise their fins to make themselves appear bigger and more ferocious. But if neither fish gives ground, then they begin to 'spar', by butting each other with their jaws in a manner that has all the appearance of a genuine boxing match! Nonetheless, it is a largely ritualistic 'bout', as serious injuries are rarely inflicted upon one another, and usually ends with the territory's owner chasing away the invader—rather than with a straight knockout![4]

TSURANAGAKOBITOZAME— JAPAN'S DIMINUTIVE SHARK WITH THE LONG FACE (AND NAME!)

In 1912, the world's smallest species of shark was described—a cigar-shaped pygmy known as the spined (or spiny-finned) pygmy shark *Squaliolus laticaudus*, and referred to by the Japanese as *tsuranagakobitozame*—'dwarf shark with long face'. Its type specimen, a slender, mouse-sized male measuring a mere 5.75 in, had been caught along with a comparably-sized female on 8 June 1908, at a depth of around 600 ft in Batangas Bay, Luzon, during the 1907-1910 Philippine Expedition of the U.S. Fish Commission Steamer *Albatross*. Inky black with white fins, and numerous light-emitting photophores on its undersurface (which may assist in camouflaging this species from predators), for the next 49 years these remained the only known specimens of this Lilliputian shark on record—until 2 June 1961, when some Japanese fishermen caught five more, in Honshu's Suruga Bay. It is now known to be widely distributed around the world.[2,10]

GREGORYINA AND INDOSTOMUS— FISHES THAT DEFY CLASSIFICATION

We owe our entire knowledge (such that it is) of an extraordinary little fish called *Gregoryina gygis* to the fortuitous find of a Hawaiian seabird. In 1923, the Pacific islands west of Hawaii were visited by the Tanager Expedition, and during their voyage its scientists landed on Laysan, one of the Hawaiian Islands' northwestern members, where they encountered the nest of a white tern that contained a small but most unusual fish, which the bird had recently captured to eat. Measuring just 2 in long, with a small mouth, a keeled undersurface, and a very distinctive complement of rays and spines in its dorsal and anal fins, this fish seemed wholly different from any species recorded by science, and became the type specimen of *Gregoryina gygis*, formally described the following year. For quite some time afterwards, many ichthyologists placed it within an entire family of

Spined pygmy shark.

its own, most closely allied to the flagfishes and morwongs. In 1957, however, ichthyologist J.R. Norman showed that this mysterious fish was in reality a late postlarval specimen of the Hawaiian morwong *Cheilodactylus* [=*Goniistius*] *vittatus*.[11-12]

Still very problematical, conversely, is *Indostomus paradoxus*, discovered in 1926, and known variously as the paradox fish, armoured stickleback, and Burma stichling. The first specimens were caught in Lake Indawgyi, a shallow expanse of water in Upper Burma's Myitkyina District, but 30 years later some were also recorded from a stream running to the lake from neighbouring Thailand. A relatively inactive creature, and only just over 1 in long, with a very slender, spindle-shaped body, and the unusual ability (for a fish) to move its head up and down, this enigmatic species has often been outwardly likened to a composite of pipefish (syngnathid) and stickleback.[11,13]

Certainly its large tail fin, overall shape of head, and the external plates encircling its body are reminiscent of the former type of fish, whereas its general body shape and the series of isolated spines running along its back are characteristic of the latter type. Due to its mixed morphology, fish taxonomists have proposed many different classifications for *Indostomus*, aligning it variously with sticklebacks, pipefishes, and cling fishes (gobiescoids).[11,13]

Anatomically, however, it possesses a number of features peculiar to its genus, including a near-absence of muscles in its posterior body region (even the tail fin is operated by a collection of extra-long tendons originating in the body's anterior half), and a lower jaw whose total length is *twenty times* that of the upper one, stretching back a considerable distance along the fish's underside. Due to its unique overall anatomy, in 1970 Dr. K.E. Banister placed *Indostomus* in its very own taxonomic order—Indostomiformes—a classification still adhered to by some ichthyologists, for the sake of convenience if nothing else, whereas others house it tentatively in the stickleback order Gasterosteiforms. In short, its precise relationship to other fishes remains a subject very much open to speculation.[11,13]

In 1999, two additional *Indostomus* species were described—the Thailand paradox fish *I. crocodilus*, and the Mekong River paradox fish *I. spinosus* (which is also found in the Mekong's tributaries and adjoining swamps).[13a]

PA BEUK—THE WORLD'S LARGEST FRESHWATER FISH

Also called the Mekong giant catfish and the pla buk, and inhabiting the Mekong River's deep waters (flowing through Laos, Thailand, Cambodia, Vietnam, and China), the pa beuk is currently deemed to be the largest species of fish that spends its entire life in freshwater. Yet, surprisingly, it was not described until 1930, whereupon it was formally named *Pangasianodon gigas*. Up to 10.5 ft long and 660 lb in weight, its huge size sets it well apart from other catfishes, but so too does its diet—whereas the great majority of catfishes are carnivorous, *P. gigas* is exclusively vegetarian (indeed, older specimens are often toothless). Sadly, this impressive creature is nowadays a critically endangered species, because for many years it has been caught in great numbers during

Pa Beuk—despite being the world's largest species of freshwater fish, this enormous catfish remained undescribed by science until 1930. (Neg. #121812/Courtesy Department Library Services, American Museum of Natural History)

the rainy season, when it migrates upstream for spawning in China's Lake Tali. Today it occurs in appreciable numbers only in Thailand's Chiang Khong.[4,14-15]

BATHYSPHAERA—
AN 'UNTOUCHABLE' DEEPSEA RIDDLE

One of the 20th century's most mysterious ichthyological discoveries was a remarkable deepsea fish that was officially described and named in 1932, yet which has never been examined in the flesh and has no known representative in any of the world's museums.[16]

On 22 November 1932, Bermuda-based zoologist Dr. William Beebe was 2100 ft beneath the surface of the sea in a bathysphere, sited 5 miles southeast of Bermuda's Nonsuch Island. While he was observing the denizens of the deep passing by the bathysphere's windows, two very unusual fishes became illuminated in the craft's electric beam of light as they twice swam past it, no more than 8 ft away. Their long slender bodies, each of which was at least 6 ft long with strongly undershot jaws housing numerous teeth, reminded Beebe of barracudas, but running along either side of each fish was a single, laterally-sited, horizontal row of luminous organs (photophores), little short of twenty in total, and every one emitting a powerful pale blue light.[16]

Equally striking were the two twitching, tentacle-like structures that hung down beneath each fish—one arising from its lower jaw, the other from the beginning of its short anal fin. Once again, each of these structures emitted light, by virtue of a pair of organs at its tip; the organ attached directly to the tentacle shone red, the other one (attached to the red organ) shone blue. Also noteworthy was their vertical dorsal fin, positioned well back towards the tail-end of the body. Beebe was unable to discern any pectoral fins or pelvic fins.[16]

From these fishes' general morphology, Beebe concluded that their species was most probably allied to the melanostomiatids, popularly known as the scaleless black dragon fishes. However, its single line of lateral photophores, not to mention its pair of ventral tentacles with light-emitting terminal organs, unequivocally distinguished it from any known species within that family. As a result, Beebe christened his mystifying discovery *Bathysphaera intacta* ('untouchable bathysphere fish'), sole member of a new genus.[16]

Bathysphaera was not the only hitherto unknown species of deepsea fish that Beebe discovered and named, but was unable to capture, during his Bermudan bathysphere observations in the early 1930s. He spied a mysterious, 2-ft-long, torpedo-shaped fish at depths of 1500 ft and 2500 ft, and named this grey-coloured species the pallid sailfin *Bathyembrix istiophasma*. He also described the three-starred anglerfish *Bathyceratias trilychnus*, a 6-in-long species sighted at 2470 ft, bearing three 'fishing

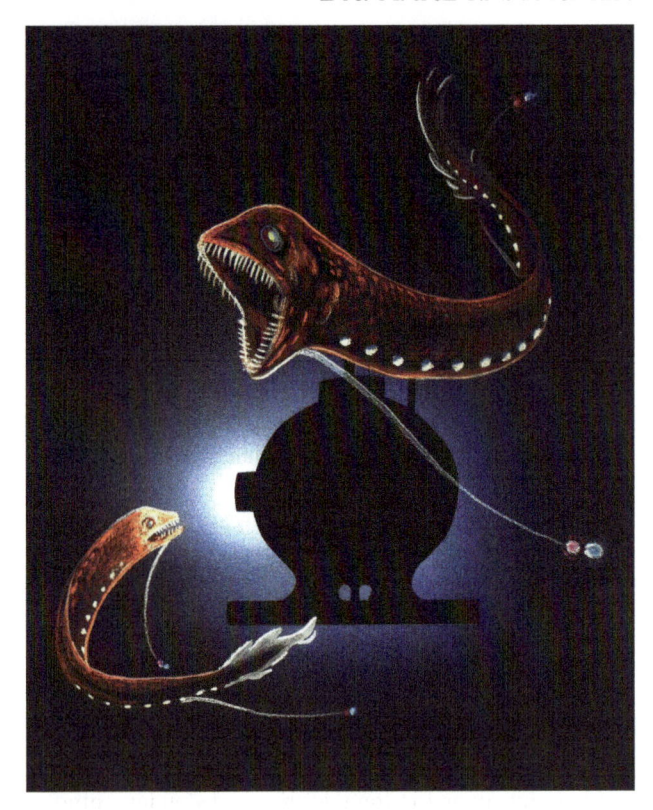

Bathysphere fish. (William M. Rebsamen)

rod' structures (illicia) on its head, and clearly allied to the deepsea anglerfishes or ceratioids (see p. 201); as well as the five-lined constellation fish *Bathysidus pentagrammus*, spotted at a depth of 1900 ft and resembling a *Chaetodon* butterfly fish or an *Acanthurus* surgeon fish, but exquisitely decorated with five glowing lines of yellow and purple photophores on each side of its roundish body.[16]

Just as beautiful was a long-beaked multicoloured fish with scarlet head, blue body, and yellow tail that Beebe informally named the abyssal rainbow gar—four of which he observed swimming together in a stiff, almost upright posture at a depth of 2500 ft.[16]

No specimens of any of these 'untouchable' species have so far been obtained. Consequently, as with *Bathysphaera*, and assuming that Beebe's testimony was truthful, they are secrets of the sea that were revealed to humanity only briefly before disappearing back into its depths' dark and alien anonymity.[16]

A TRILOGY OF TETRAS

In 1936, just six years after the reappearance of the golden hamster (see p. 49), another future favourite of western pet-owners hit the headlines—and completely by accident.

Canoeing in the Rio Putumayo, one of the Amazon's countless tributaries, A. Rabaut, a French animal collector, noticed a shoal of tiny but exquisitely-coloured fishes, swimming nearby, so he scooped up a handful to

Neon tetras.

examine them more closely. Measuring just under 2 in long, each had a greyish back, colourless fins, and silver belly; the front half of each flank's lower portion was also silvery, whereas the rear half of each was red. The feature that attracted most attention, however, was the horizontal iridescent stripe running backwards along each flank's *upper* portion, from just behind the fish's eye to a point a little in front of its tail's base, and separating its grey back from its silver-and-red lower flanks. This stripe continually flickered from brilliant blue to glittering green, according to the angle at which it was viewed, and closely resembled a fluorescent strip of neon lighting.[17]

The species represented by these beautiful little fishes was unknown to science, so Rabaut collected a selection of live specimens, which were passed on to Chicago's John G. Shield Aquarium. The species belonged to the characin family—comprising tiny, carp-like forms—and later in 1936 it was formally described.

Neon tetras (top, and bottom left) and cardinal tetras (centre). (Burkhard Kahl)

It was originally christened *Hyphessobrycon innesi*, in honour of William T. Innes, a leading American aquarist;[18] more recently, its generic name has been changed to *Paracheirodon*. Its English name, conversely, has remained the same, and is known to fish-fanciers worldwide—the neon tetra.[17]

It attracted so much attention in Chicago, with aquarists clamouring for specimens and journalists praising it as the most beautiful aquarium fish known to man, that others were swiftly sought in its native Amazon homeland. One German collector brought back at least two thousand, captured at the mouth of the River Nanay just inside Peru, with the assistance of the Javaro Indians. Specimens were also obtained from the Rio Putumayo again.[17]

Mexican cave tetra. (Dr. Jürgen Schramme, courtesy of Dr. Thomas Teyke)

Unfortunately, the neon tetra initially proved very difficult to maintain and breed in captivity, but by the end of World War II attempts were proving more successful, thanks to the discovery that soft, slightly acidic water at a temperature of up to 24°C and the presence of water plants in the tank were favoured by it for spawning. Nowadays, the neon tetra is more commonly bred and is one of the most widely available of all tropical freshwater fishes, known to every fish-lover from the lowliest schoolboy to the most learned ichthyologist.[17]

In 1956, a closely related species was discovered, adorned with a somewhat wider 'electric' stripe on each flank, and further distinguished from the neon tetra by its entirely red lower flanks (in contrast to the neon tetra's half-silver, half-red counterparts). This species, first called *H. cardinalis* but later renamed *Cheirodon axelrodi* (after noted fish collector-discoverer Dr. Harold Axelrod), is popularly known as the cardinal tetra, and was first found within forest pools in the Rio Negro's upper waters, but is also known from the Orinoco's tributaries. It too soon became extremely popular as an aquarium species.[19]

The same is true of the Mexican blind cave tetra, a curious 3-in-long species discovered in 1936 and shortly afterwards named *Anoptichthys jordani* ('Jordan's eyeless fish'), in honour of its finder, Basil Jordan, who recorded it from Mexico's Cueva Chica. Although related to the neon tetra, *A. jordani* is radically different from it in appearance, thanks to certain morphological specialisations symptomatic of a cave-dwelling (spelaeic) existence.[20]

As it is never exposed to sunlight and therefore has no need of body pigmentation to protect it from ultraviolet radiation, *A. jordani* has none. Thus its body is translucent, with the red colour of its blood, coursing through the vessels beneath its skin's surface, affording it a pale pink hue. And as it never encounters light, it does not need eyes either; hence they are covered by a layer of skin. Nevertheless, it *is* still capable of detecting short wavelengths in the visible spectrum (i.e. violet, indigo, and blue light), so it is not totally blind.[20-1]

Although its unpigmented body seems similar to those of freak albino specimens of fishes, the Mexican blind cave tetra is not incapable of producing pigment. Individuals kept in aquaria exposed to sunlight will gradually darken. Consequently, its normal lack of pigmentation is not an irreversible, genetically-induced phenomenon like that of genuine albinos, but is merely an environmentally-determined feature capable of reversion under correct conditions.[21]

In more recent years, its scientific name has changed to *Astyanax jordani*; and due to its strange, somewhat eerie appearance, this species has become very popular among fish fanciers, especially in America, to the extent that today there are probably more specimens in pet aquaria than in its native Mexican cave streams.[21]

NORMANICHTHYS—A SCALY SCALELESS FISH

The bullheads and sculpins comprise a mostly marine family (Cottidae) housing a diverse range of species, but all were traditionally united by at least one important characteristic—they were all scaleless. In 1937, however, this orderly classification was disrupted with the description of *Normanichthys crockeri*, the mote sculpin, native to fairly shallow waters off the coasts of Chile, and discovered during the Templeton-Crocker Expedition of 1934-35. Measuring 4.5 in long, although it seems to be most closely allied to the bullheads and sculpins it is set apart from all of them by its shining array of scales, covering not only its entire body but much of its head too. Equally distinctive is the many-spined composition of its first dorsal fin (it has two). Due to these discrepancies from the basic cottid configuration, *Normanichthys* was placed in a taxonomic family of its own, of which it remains the only member.[22]

'OLD FOURLEGS': THE INCREDIBLE COELACANTH—PART 1: A LIVING FOSSIL CALLED LATIMERIA

If the okapi constitutes the most significant new species of terrestrial animal to be discovered during the 20th century, then its aquatic counterpart must surely be the coelacanth.

Towards mid-day on 22 December 1938, Marjorie Courtenay-Latimer, the curator of the then small and little-known East London Museum at South Africa's southernmost tip, received a phone call from the manager of a local commercial fishing company called Irwin and Johnson, informing her that their trawler *Nerine* had brought in a pile of fishes that she might like to examine. This was nothing new—anxious to acquire specimens of local wildlife for the museum, Miss Courtenay-Latimer had encouraged everyone connected in some way with wildlife to contact her if any specimens came to hand that may be of importance to the museum. As a result, Irwin and Johnson had kindly permitted Captain Hendrik Goosen, skipper of the *Nerine*, to collect fishes for the museum during trawling excursions, and had informed her on many previous occasions about the arrival of fishes, which she had always inspected for anything suitable for exhibition and preservation at the museum. So, although this latest phone call was, as ever, most welcome, she did not attach any significance to it, and simply made her way to the wharf with an assistant to examine the fishes, just as she had done so many times before. Little did she realise, it was probably the most important phone call that she would ever receive in the whole of her life![23-4]

The type specimen of Latimeria chalumnae.

When she arrived at the wharf and saw the pile of fishes waiting for her, she recognised that they were principally sharks—but then she noticed something very different, almost hidden beneath them. Upon her request, the fish that she had spotted was disentangled from the others and stretched out for her to observe it properly. It was the most extraordinary specimen that she had ever seen, and although she was not a trained ichthyologist she realised straight away that this could well be something important. Mauve-blue in colour, faintly flecked with white spots, it was 5 ft long, weighed 127 lb, and had thick bony scales that gave it an armour-plated appearance; but what attracted her especial attention was its set of very remarkable fins. It possessed quite a number—two separate dorsal fins, a pair of pectorals and a pair of pelvics, a single anal, and a unique type of tail fin.[23-4]

Like the body fins of most other modern-day fishes, the first dorsal fin was borne directly upon the fish's body, and contained rays that all arose from the very base of the fin, thereby yielding a fan-like structure. Conversely, each of its remaining body fins was borne upon a muscular lobe (surrounding a complex skeleton), from which the rays arose, and which gave these fins the appearance of stumpy legs![23-4]

If its body fins were remarkable, this curious creature's tail fin was even more so. In most fishes, the backbone ends at the base of the tail fin, which in turn consists of two distinct lobes—the upper one directed upwards and backwards, the lower one directed downwards and backwards. In contrast, the tail fin of Courtenay-Latimer's fish was pointed, with the fish's backbone running right through it into its very tip, thereby completely separating the fin's upper lobe from its lower one. This compares with the structure of the tail fin in eels and lungfishes—but unlike theirs (and indeed, unlike that of any other modern-day fish known to science), the tail fin of Courtenay-Latimer's fish had a *third* lobe, sandwiched between its upper and lower lobes, and arising from the very tip of its backbone, thus yielding a tripartite or three-lobed tail.[23-5]

Also of note was this fish's excessive oiliness—later found to be caused by a layer of oil-secreting cells below the surface of its skin.[24]

Questioning one of the trawlermen concerning this strange creature, Courtenay-Latimer learnt that it had been caught at the mouth of the River Chalumna, and that it had snapped at Captain Goosen's hand when it had been hauled aboard. None of the men had ever seen a fish like it before, even though some had been fishing there for the past 30 years.[23-5]

Naturally, she was acutely aware of the importance of preserving this unique specimen for formal identification by a fish specialist, so with the help of her assistant and a very reluctant taxi driver (who needed a lot of persuading before finally agreeing to transport in his vehicle a large, oil-secreting fish whose smell he felt sure would be both pronounced and persistent!), she succeeded in bringing it back to the museum, where she made arrangements for its body to be prepared by a local taxidermist, requesting him to retain for her all of the soft parts that he would be removing during its preparation.[23-5]

The next day, she wrote to a longstanding correspondent, Prof. J.L.B. Smith—a world-renowned ichthyological researcher at Rhodes University College in Grahamstown. She enclosed with her letter a sketch of the fish, noting its salient features, sizes of fins, body, etc, and enquired whether he could offer her any opinion regarding its likely identity.[23-5]

She posted her letter later that day—but due to ill-health, Prof. Smith had left the university a few days earlier, to recuperate at home over the Christmas holiday period. Consequently, coupling the time required for rerouting his post to his home with the inevitable Christmas postal delays, he did not receive her letter until 3 January.[23-5]

He opened it, read her note, looked at her sketch of the fish—and then suddenly:[24]

> ... a bomb seemed to burst in my brain, and beyond that sketch and the paper of the letter I was looking at a series of fishy creatures ... fishes no longer here, fishes that had lived in dim past ages gone, and of which only often fragmentary remains in rock are known.

Almost reeling with incredulity, Prof. Smith had recognised the significance of the fish's combination of extraordinary features—its thick, armour-like scales, limb-like lobed fins, and tri-lobed tail. Although no previously recorded *modern-day* fish shared them, these features characterised certain fishes of the distant past—a peculiar group belonging to the extinct order Crossopterygii—known as coelacanths. These fishes had flourished 200 million years ago, but were supposedly long extinct; the most recent fossils were more than 60 million years old. Yet here in his hand was a sketch of a *modern-day* fish, captured *alive* only a couple of weeks earlier, that compared precisely with these long-deceased creatures—creatures that had been contemporary with the dinosaurs! Somehow—unbelievably, inconceivably—a living coelacanth had been caught![23-4]

To cut a long and intricate story short: although university commitments prevented him from examining the fish immediately, by 16 February Smith had arrived at the East London Museum and finally, allaying weeks of fear and doubt regarding the identity of Courtenay-Latimer's astounding specimen, he saw with his own eyes the coelacanth—for that is indeed what it was. As expressed in this section's opening quotation (p. 191), from Smith's book *Old Fourlegs* (the title referring to the coelacanth's limb-like pectoral and pelvic fins), despite all of his mental preparations beforehand, the sight of a creature lying before him that science had firmly believed to be as deceased as the dinosaurs, plesiosaurs, pterosaurs, and so many other celebrities of the prehistoric world was one that physically immobilised him, rendering all other thoughts and speech momentarily impossible. Not until he hesitantly touched it, to convince himself that it was real and not illusion, was the spell broken. It really *was* there, and it really *was* a coelacanth![24-5] But what was to be done now?

First and foremost, a formal description of the new species that this specimen represented was needed—for although it was unmistakably a coelacanth, more than 60 million years of evolution (taking place after the date from which the most recent fossil forms were known) had produced a modern-day species with various prominent differences from its long-extinct ancestors (see p. 206). To produce such a description, it is best to have available as complete a specimen as possible—but this requirement caused considerable problems for Smith. In between Courtenay-Latimer's sending her letter to him and Smith's receipt of it, the fish's precious soft parts (removed by the taxidermist during its preservation as a museum exhibit) had deteriorated to such an extent that they had been discarded.[24-5]

Nevertheless, the specimen was still of such tremendous value, with so many features still present that readily confirmed its identity as a 20th-century coelacanth, that the publication of an official description of its species was fully warranted. On 18 March and 6 May 1939, preliminary notes introducing the coelacanth's discovery, and containing some brief notes regarding its morphology plus a photograph of it as a taxiderm specimen, were published in the British scientific journal *Nature*.[26] A longer account devoted to its morphological characteristics appeared in the Royal Society of South Africa's *Transactions*.[26-7]

The modern-day coelacanth was acknowledged throughout the world as one of the greatest zoological finds ever made. Thus, in recognition of its provenance and, especially, the person who made its discovery possible, Smith named its species *Latimeria chalumnae*. Marjorie Courtenay-Latimer was thereby granted ichthyological immortality by an anachronistic anomaly that was very nearly lost to the world beneath a pile of scuppered sharks![26-7]

Now that one *Latimeria* specimen was on record, Smith was keen to seek more, so that the species' internal structure could finally be recorded. But how would he obtain another coelacanth? He decided to advertise for one—by printing thousands of leaflets illustrating the coelacanth, offering a reward of £100 for the first two specimens obtained, and giving firm instructions to the finders of such fishes *not* to clean or cut them in any way.[24]

The leaflets were distributed *en masse* throughout southeast Africa, and to ensure that language did not pose any problem every copy carried Smith's message in Portuguese, English, and French. Nevertheless, it took almost 14 years for Smith's advertisement to receive a possible answer, and when it did come it was from a wholly unexpected locality with a stranger-than-normal specimen as its subject (as revealed on p. 203).[24]

AN EYE-OPENER OF A BLINDCAT

The year 1947 saw the official description of one of the world's most extraordinary catfishes, the wide-mouthed blindcat. Its story began in 1938, when biologist Dr. Carl L. Hubbs was visiting the Withe Memorial Museum at San Antonio, Texas, during a zoological collecting trip to the area. At the museum, its director, Ellen S. Quillin, made available to him for study a most peculiar catfish specimen. An immature male measuring 68.7 mm (just under 3 in) long, it had been collected from an Artesian well 1250 ft deep, in the Edwards Plateau region near San Antonio, and had been donated to the museum by William Kempin.[28]

A long thin fish with a flattened head, it had no pigmentation, its ghostly white colouration relieved only by the pale pink tinge of blood vessels beneath its skin. Moreover, it was eyeless, and its nasal apertures were extremely reduced. In contrast, its mouth was notably large and wide, its body bore an extremely prominent adipose fin, and its lateral-line sensory system (for detecting water currents) was very distinct and highly developed.[28]

Hubbs and ichthyologist R.M. Bailey made a detailed study of this unique specimen, which represented a radically new species of ameiurid freshwater catfish highly specialised for life in pitch-dark underground streams, wholly devoid of sunlight. And in their paper of 28 April 1947, introducing the wide-mouthed blindcat to the scientific world, Hubbs and Bailey provided it with a singularly distinctive scientific name to match its uncommon appearance and lifestyle. Its broad mouth earned it the specific name *eurystomus* ('wide-mouthed'), and its lifestyle as an inhabitant of eternal lightlessness, a piscean prince of darkness, was the inspiration for its somewhat startling generic name—*Satan*, of which it remains the only member species.[28]

SMITH'S BLIND WHITE FISHES IN PERSIA

Qanats are underground water channels that criss-cross Iran (formerly Persia); and in 1949, Anthony Smith, a zoology undergraduate at Oxford University, read in a travel book that these channels reputedly contain blind white fishes that are good to eat. Smith was very intrigued by this, because at that time not a single species of blind subterranean fish had been scientifically documented anywhere in Eurasia.[29]

Eager to discover the identity of this mysterious qanat-contained curiosity, Smith organised a four-man expedition to southern Iran, setting off in June 1950 and returning home to England in October. During their visit, they saw many interesting sights—enabling Smith to write an entire book based upon their experiences. Published in 1953, it became a bestseller, and was entitled *Blind White Fish in Persia*—which was a little ironic, because despite being the principal inspiration for the expedition, this unidentified species was, tragically, the one important sight that they did *not* succeed in observing while there![29]

Even more ironic, however, was the news that they had received not long after their return home from Iran. A newly-published paper by Danish ichthyologists Drs. Anton Bruun (of *Dana* and *Galathea* fame) and E.W. Kaiser revealed that Iran's blind white fish was *not* unknown to science. On the contrary, it had been discovered in 1937, within the Zagros Mountains' Ab-i-Serum valley, and was actually a sightless *pink* species of carp, which Bruun and Kaiser formally described in their paper, naming it *Iranocypris typhlops*, sole member of a new genus.[29-30]

Nevertheless, Smith's expedition was still able to make a valuable contribution, thanks to his book. Following its publication, Smith received a letter from an amateur pot-holer called A.G. Widdowson, working in Iraq. Widdowson had recently captured some blind white fishes in a 300-ft-deep pot-hole at Haditha, and, having read Smith's book, decided to contact him in case this information may be of interest (the fishes themselves had died shortly after capture). Smith forwarded Widdowson's letter to the British Museum (Natural History), who requested Widdowson to send to it any further specimens that he may succeed in obtaining. Not long afterwards, a series was duly delivered to the museum, and were found to represent another new species, which in 1955 was christened *Typhlogarra widdowsoni*, the only member of another entirely new genus.[29-30]

In 1976, the third episode in this long-running saga of subterranean fishes took place, when Smith visited Abi-i-Serum, the type locality of *Iranocypris*, and succeeded in procuring a trio of specimens. One of these was donated to a helpful assistant, but Smith retained the other two, which he transported alive back to England, taking them shortly afterwards to the British Museum—where he received a great surprise. After examining them, ichthyologist Dr. P. Humphrey Greenwood informed him that whereas one was certainly an *Iranocypris*, the other was a totally different species—one that was undescribed by science! It was a cave-dwelling loach—the first such loach recorded anywhere in the world—and was officially dubbed *Nemacheilus* [now *Paracobitis*] *smithi*, Smith's cave loach, a much-deserved honour for the zoologist who had been so unlucky in his previous attempt to discover a new species of subterranean fish.[29,31]

Several other species of subterranean fish have since been discovered in the Middle East; and in 1978 a second species of cave loach was found, but this time in the Far East, when eight specimens were captured at Bajianjing, near Qiafang, Gejiu, in China's Yunnan Province. Lacking eyes (even its eye sockets were vestigial) and body pigmentation, it was closely related to Smith's cave loach from Iran, and was dubbed *Nemacheilus gejiuensis*. As it differed from Smith's species via its shorter body, and absence of a deep adipose ridge between its dorsal and tail fins, however, it was later transferred to a different genus and thus renamed *Triplophysa geijuensis*.[32]

GALATHEATHAUMA AND COMPANY—
FISHES THAT GO FISHING

For two very different reasons, the ceratioids or deep-sea anglerfishes inhabiting the ocean's inky-black abyssal depths are among the most bizarre fishes known to science.

Firstly: the female of many of these species possesses an extraordinary structure called the illicium—the highly-modified first ray of the dorsal fin, which has parted company with the remainder of the fin and is now sited on top of the fish's head (or thereabouts). A long pole-like apparatus, it resembles a human angler's fishing rod, and serves precisely the same purpose. Just as the angler attaches bait to the end of his rod, so at the end of the ceratioid's illicium is a luminous organ called the esca that the fish uses as a lure, tempting potential prey to swim within range of its cavernous jaws.[4,11]

Secondly: the male ceratioid is a radically stunted, morphologically reduced form, a veritable dwarf that actually parasitises the much larger, fully-formed female of its species when mature, clamping itself to her body with its pincer-like jaws. In some species, moreover, it remains permanently attached to her, becoming so intimately united that its circulatory system fuses with hers, enabling it to derive all of its own nutrients directly from her bloodstream in an outstanding physiological parallel to the placenta-mediated association between a foetus and its mother. In such cases, the male's only function appears to be the fertilisation of the female's eggs; in many ways it can be looked upon as simply an organ of the female—which in turn can therefore be thought of as a functional hermaphrodite![4,11]

The ceratioids come in a diverse range of grotesque forms, and some were first discovered long before the 20th century. Of particular interest, however, is that the most extraordinary and extreme examples (and hence the very ones that are most frequently illustrated in books) remained unknown to science until more recent times. During the 1920s, for example, the oceanographic explorations of Danish research vessel *Dana* obtained many new ceratioids. In 1925 alone, British Museum ichthyologist Dr. Charles Tate Regan described 8 new genera and 24 new species, which included *Lasiognathus saccostoma* and *Linophryne arborifera*—two of the most celebrated species so far recorded.[33]

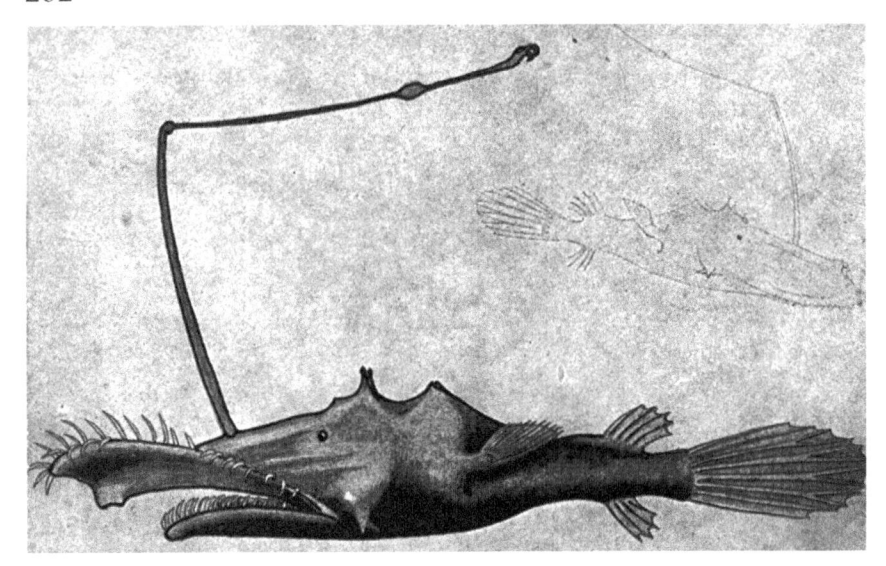

Lasiognathus saccostoma anglerfish—its type specimen was described in 1925, and remained its only specimen until 1962.

Lasiognathus saccostoma was described from a single 3-in-long female obtained in the Caribbean, which remained its only specimen until 1962, when two more were procured, this time from Madeira (they were initially classed as a new, separate species, which was dubbed *L. ancistrophorus*). A black, dorsoventrally flattened species with bizarre facial flaps that greatly extend its upper jaws laterally so that its upper teeth point outwards rather than downwards, *Lasiognathus* possesses the most elaborate illicium currently documented from any ceratioid. Borne upon an extremely long and mobile basal bone called the pterygiophore (four-fifths of the fish's entire length!), it carries at its tip a luminous esca, which in turn bears three hooks at the end of a long line. Thus equipped with an effective fishing rod, line, float, lighted bait, and hooks, it is little wonder that *Lasiognathus* has been hailed as, in ceratioid terms, 'the compleat angler'![4,11,33-4]

Black in colour with a stout, rounded body (and renamed *Linophryne brevibarbata* in 1932 by William Beebe, but this name is rarely used), *Linophryne arborifera* has a much shorter, less imposing illicium than *Lasiognathus*, arising from its snout and somewhat swollen in shape, with a luminous subdivided esca at its tip. Instead, what makes this species so spectacular in appearance is the intricately branched, tree-like barbel that hangs down from its chin. The purpose of this remarkable structure is currently uncertain. In some specimens it has been found to be faintly luminescent; thus it may act as an accessory lure. It may also detect water-borne movements of potential prey nearby. This species' type specimen was a 3-in-long female that was captured in the North Atlantic.[4,11,33-4]

In 1930, Regan described yet another truly bizarre ceratioid—the hairy anglerfish *Caulophryne polynema*, first made known to science from a female and male caught off Madeira. However, their species attracted little if any public attention until 2001, when it starred in a television programme that exposed some of the extraordinary creatures inhabiting the abyssal region of the sea where sunlight cannot penetrate, and which featured in the BBC documentary series 'The Blue Planet'. About 18 in long, red in colour, and filmed at a depth of around 13,000 ft, the hairy anglerfish looked more like a gooseberry than a fish—and for good reason. Its entire body is covered with long fine hair-like sensory filaments, which are extremely responsive to water movements, thus enabling it to detect the approach of a suitable prey victim even in this inky, lightless deepsea realm.[34a]

As already noted (see p. 196), during his Bermudan bathysphere observations in the early 1930s William Beebe sighted a still-uncaptured species of ceratioid that possesses *three* illicia; he named this exceptionally endowed fish *Bathyceratias trilynchus*.[16] However, the most sophisticated ceratioid presently on record was not revealed until the early 1950s.

On 15 October 1950, the Danish oceanographic research vessel *Galathea* began a major exploration of the oceans' deepwater trenches, during which it discovered some major new species (see p. 237 for its greatest triumph). One of these was hauled up from the sea bottom off Central America's western coast on 6 May 1952, at a depth of over 11,500 ft. It proved to be an 18.5-in-long female ceratioid (relatively large for this suborder of fishes), black in colour with a broad head and a huge mouth, whose upper jaw was fringed with curved, pointed teeth. It clearly represented a hitherto unknown species, but regrettably it seemed to be a damaged specimen because it lacked the ceratioids' diagnostic feature—an illicium. Presumably, therefore, this delicate structure had been torn off during the fish's capture and subsequent haul to the surface—or so the ichthyologists examining it initially assumed. Fortunately, however, one of these researchers decided to take a look inside its mouth—and, in so doing, exposed the astonishing secret of this cryptic creature.[34-6]

It was not damaged at all; it was merely different—very different. Instead of hovering in midwater, seeking to coax prey within range of its mouth via a fishing rod bearing a luminous lure on the end, this newly-found species had taken the ceratioid concept one notable step further. The luminous lure, forked in shape, was actually present *inside* its mouth, suspended from the roof just behind its upper jaw's 'comb' of teeth. All that this fish needed to do, therefore, was to remain stationary on the sea bottom with its jaws agape, knowing that its lure would entice unsuspecting prey to swim directly *into* its mouth![34-6]

In 1953, the *Galathea*'s leader, Dr. Anton Bruun, named this remarkable species *Galatheathauma axeli*—commemorating its discovery by the *Galathea* expedition, and honouring the chairman of the expedition's committee, Prince Axel of Denmark. Subsequent re-examination of this species, however, revealed that it should be reclassified as a member of the pre-existing genus *Thaumatichthys*, so it is now known as *Thaumatichthys axeli*—'Axel's wonder fish'.[36]

Other strange ceratioids may still await discovery—after all, we have yet to obtain any specimen of Beebe's *Bathyceratias*. Meanwhile, *T. axeli* remains one of the most fascinating and enigmatic species on record—a living fish-trap with built-in illumination, content to repose lazily upon the sea floor, and to emulate Lewis Carroll's famous little crocodile as, with the assistance of its remarkable lure, it 'welcomes little fishes in with gently smiling jaws!'

'OLD FOURLEGS': THE INCREDIBLE COELACANTH—PART 2: A COMEBACK IN THE COMOROS

As documented on p. 198, the discovery off South Africa in December 1938 of *Latimeria chalumnae*, the first modern-day coelacanth species known to science, was not

Linophryne arborifera anglerfish—the purpose of its elaborate tree-like barbel has never been satisfactorily ascertained.

Prof. J.L.B. Smith (foreground) with the 'Malania' coelacanth. (J.L.B. Smith Institute of Ichthyology)

just an international sensation in zoological circles. It had also captured the imagination of the general public worldwide, so that everyone was clamouring for more news of this archaic fish that had ostensibly risen like a piscean phoenix from the ashes of the past.

No-one was more eager for news—and especially a new specimen—than the species' leading researcher, Prof. J.L.B. Smith, but the 1940s came and went, without bringing either. It seemed the inevitable doom of *Latimeria* to be known only from a single, incompletely-preserved representative.[24,37]

It was now December 1952, only two days away from the fourteenth anniversary of the first specimen's capture. Ahmed Hussein, a fisherman from Anjouan Island in the Grand Comoro archipelago northwest of Madagascar, had been fishing offshore from the southeastern Anjouan port of Comoni when he captured a very strange-looking fish measuring about 5 ft in length, hauled up from a relatively shallow depth of around 120 ft. At the time, he paid little attention to the fish's extremely odd appearance, and the next morning he took it to the local market to sell it, just as he did with all of his catches—but he did not go alone. Destiny and good fortune evidently accompanied him, because just before his fish was about to be cut up a fellow Anjouan mentioned that it very closely resembled the valuable fish depicted on some leaflets that had recently been distributed on the Comoro Islands.[24,37]

The leaflets were of course those of Prof. Smith, advertising his reward for the capture of a coelacanth, and as they requested anyone who caught such a fish to take it at once to someone responsible, who would know what to do with it, Hussein and a group of other natives carried it some 25 miles to one of Smith's friends, Captain Eric Hunt, who was based at the opposite end of Anjouan. Hunt had supervised the distribution of Smith's leaflets there, and he identified the fish immediately as a bona fide coelacanth. Hunt instructed the natives to cut it only for preservation (by salting), which they did (albeit more enthusiastically than scientifically!), after which he also injected formalin into it, supplied by a local doctor. Hunt then contacted Smith in South Africa with the historic news that his long-sought-after second coelacanth had finally surfaced, and was awaiting collection.[24,37]

Sadly, however, many obstacles prevented Smith from travelling at once to take charge of the specimen that he so desperately needed, and he feared that in the Comoros' extremely hot climate the precious fish would rot before he could even see it, let alone study it.[24,37]

Happily, fate took a hand, in the shape of no less a personage than the prime minister of South Africa, Dr. Daniel F. Malan, who recognised the extreme urgency of the situation and very generously loaned Smith a military aeroplane in which to fly directly to Anjouan. On 28 December, Smith landed on the island, Hunt lifted the shroud of cotton wool covering the specimen in its box, and revealed the armour-plated steely-blue fish with lobed fins that Smith knew only too well to be a coelacanth.[24,37]

Even though the heat of the Comoros' climate had induced from the fish a singularly powerful odour, Hunt's makeshift preservatives had functioned satisfactorily, so that the precious internal organs and tissues were available for study. As it happened, however, the specimen's *external* morphology was also of considerable interest—due to two features that instantly differentiated it from the first specimen.[24,37]

Whereas the first coelacanth had two dorsal fins, this second specimen only had one—the posterior, lobed version; the anterior dorsal fin, arising directly from the body of the first specimen (and possessing rays arranged in a fan-like manner) was absent. Also absent from the second coelacanth was the remarkable third lobe of the tail fin, positioned at the tip of the first specimen's backbone. As a result, Smith initially looked upon Specimen #2 as a representative of a new, second species of modern-day coelacanth, which he christened *Malania anjouanae*, acknowledging its provenance and the person who had been so instrumental in enabling him to reach it.[24,37]

Yet tragically, his tribute to Malan was short-lived, because the detailed examination of *Malania* that Smith subsequently carried out convinced him that it was really nothing more than a deformed *Latimeria chalumnae*, and not a new species after all.

Nevertheless, that disappointment was more than tempered by this episode's crowning triumph—it had unveiled the coelacanth's true home, because during the next few years an impressive series of specimens was obtained off the Comoros. By contrast, no further example would be recorded for many years off South Africa's East London area, implying that the type specimen of *Latimeria* was a straggler, a wanderer far from its native home and, as such, an exceptionally lucky find. Also, as most of the later specimens have been captured in deep water, the relatively shallow depth at which the 'Malania' individual had been caught was again atypical for the species.[24,37]

Upon questioning the Comoro natives, Smith learnt that some of them knew the coelacanth well, referring to it as *kombessa*. No doubt with a mixture of incredulity and horror (in view of the professional awe that the coelacanth received from scientists worldwide), he also learnt that the tough scales of this species, which in the 1950s would have been priceless exhibits in any western museum, were traditionally used by the natives as sandpaper, for the humble purpose of roughening bicycle tyres when mending a puncture![24,37]

As can be expected, a great deal has happened in relation to the coelacanth since the early 1950s, so that only the very briefest of summaries can be included here. On 12 November 1954, once more off Anjouan, the eighth coelacanth specimen known to science was captured, and proved to be the first female recorded. As the Comoros are French-owned, coelacanth research since the capture of the 'Malania' individual has been carried out predominantly by French workers, but other countries have also contributed from time to time.[24,37] In March 1972, for example, it was a joint British-American-French expedition, backed by the Royal Society, that was privileged to succeed in observing and filming for the very first time a *living* coelacanth. It had been captured by Comoros fisherman Madi Youssouf Kaar, who was fortunate enough to recognise what it was and bring it to the attention of the scientific team while it was still alive. It was a catch that he will never forget, because his reward was £10,000—roughly equivalent to *a century's worth* of wages for a Comoros fisherman![38]

Still to be achieved, however, was the ultimate goal of any coelacanth film-maker—to film it underwater in its natural habitat, but in January 1987, using a two-man submersible, this seemingly impossible objective was successfully accomplished—by Prof. Hans Fricke, a marine biologist and documentary film-maker from Germany's Max Planck Institute for Animal Behaviour. He discovered much about coelacanth lifestyle. For example: it frequented crevices in lava rock at depths of around 650 ft; it could swim backwards as well as forwards, sometimes even drifting upside-down; and contrary to expectations, it did not *walk* on its limb-like lobed fins across the sea bottom, although it did brace itself with them when resting on it. When swimming, its synchronisation of fin movements yielded the same gait as that of a trotting horse; and it was responsive to weak electrical fields. Most unexpected of all was its pièce de resistance—for no apparent reason, a coelacanth would suddenly drift nose downwards

When swimming, the coelacanth's fin movements mirror the gait of a trotting horse. (William M. Rebsamen)

with the current and stand on its head (for up to two minutes at a time!). This extraordinary behaviour may be some sort of response to the presence of electrical fields nearby, but this has not been conclusively established.[39]

The considerable studies of the coelacanth's morphology and physiology initiated by Prof. Smith and continued to this day by other ichthyologists have also uncovered some major surprises—so much so that even its formal classification is nowadays a subject for controversy. There are two modern-day classes of jawed fishes. One is Osteichthyes (also called Pisces), the bony fishes—so-called because one of their principal characteristics is a skeleton composed wholly or predominantly of bone. The other class is Chondrichthyes (also called Selachii), the cartilaginous fishes—consisting of the sharks and rays, and so-called because their skeletons are composed wholly of cartilage (gristle). Fossil species of coelacanths had bony skeletons, and because many of their other features also allied them with Osteichthyes this was the class in which they had been traditionally housed.[24,37]

Studies on the *modern-day* coelacanth, conversely, have exposed various features that make the above classification of the coelacanths as a group somewhat unsatisfactory. Some of the most important of these features as are follows:[24,37]

1) Its backbone is not made of bone but of cartilage, and bears spines that are not bony and solid but are instead cartilaginous and hollow (this is the feature from which the term *coelacanth*—meaning 'hollow spine'—is derived).
2) Its gills are bony and hard, and bear teeth—very unlike the soft, cartilaginous, untoothed gills of other Osteichthyes species.
3) Its blood contains a high level of urea, similar to that of cartilaginous fishes but unlike that of bony ones.
4) Its intestine contains a spiral-shaped valve, again like that of cartilaginous fishes but unlike that of bony ones.
5) Like those of its fossil ancestors, its skull is a two-part structure with a hinge, unlike the single unit of bony fishes.

In certain ways, it even differs from its fossil relatives. For instance, it is about five times as long as most of them; and it is a marine form, whereas most fossil coelacanths were freshwater species. There are also several basic anatomical differences.[24,37]

Thus, there are several conflicting schools of thought regarding the coelacanth group's correct taxonomic position. There are those who still look upon them as bony fishes (albeit within their own subclass, Sarcopterygii, also containing the lungfishes), notwithstanding *Latimeria*'s varied morphological contradictions. Others prefer to cite the latter's similarities to cartilaginous fishes as evidence that these and the coelacanths shared a common ancestor. And some feel that the coelacanths merit a class of their own (also housing the lungfishes), and of equal status to the bony fishes and to the cartilaginous fishes.[24,37]

Finally: based solely upon fossil evidence, coelacanths were initially classed as the fish group most nearly allied to the land vertebrates (tetrapods). However, findings from later anatomical and biochemical studies with *Latimeria* seemed to refute that idea—until recently. From comparisons of the amino acid sequence in *Latimeria*'s haemoglobin with that of various other fishes, and of some amphibians, in May 1991 a research team presented persuasive new molecular evidence for reinstating the coelacanths as the tetrapods' closest piscean relatives.[37] With ideas turning full circle, it is clear that there is still much to resolve regarding the lobe-finned lineage of 'Old Fourlegs'.

It has traditionally been assumed that *Latimeria*'s type specimen, the only example ever captured in South African waters, was a unique straggler, having wandered far from its species' typical Comoros provenance. However, there may have been at least one precedent. When, on 18 March 1939, Prof. J.L.B. Smith documented the above specimen's existence, he noted that a responsible citizen-angler of East London had claimed that about five years earlier he had seen just such a fish himself, washed ashore on a lonely part of the East London coastline. He had left it there in order to fetch some assistance in transporting it (as it was apparently somewhat larger than *Latimeria*'s type specimen), but by the time that he and his helpers had returned, the tide had carried the carcase back out to sea.[26]

In August 1991 a female was captured off Quelimane, Mozambique. However, results from Fricke's recent DNA fingerprinting studies with the Mozambique example and five Grand Comoro specimens revealed no significant differences between them, thus implying that the Mozambique and South African coelacanths were not derived from some undiscovered coastal African population, but were, after all, merely wanderers originating from the Comoros.[39a]

On 5 August 1995 the first coelacanth recorded from Madagascan waters was caught, in a deep-set net off the southwestern coastal village of Anakaó at a depth of 420-450 ft. Weighing 70 lb and in excellent condition, it was frozen and taken to Toliara University's Institut Halieutique et des Sciences Marines, where it was preserved in formalin.[39b]

In October 2000, while making a deep dive 320 ft off Sodwana, a bay off KwaZulu-Natal's northeast coast, divers Pieter Venter and Peter Timm spied three coelacanths in the St. Lucia Marine Protected Area. These were the first coelacanths recorded off South Africa since the historic type specimen of *Latimeria chalumnae* was caught and revealed to Marjorie Courtenay-Latimer back in December 1938. Returning as the 'SA Coelacanth Expedition 2000', on 27 November Venter and eight colleagues successfully filmed a trio of coelacanths at a depth of 350 ft, the biggest measuring 4.8-6 ft long. However, this extremely significant zoological discovery was overshadowed by tragedy, as one of the members of Venter's diving team, Dennis Harding, ascended too rapidly afterwards, and died from a cerebral embolism. This was the first time that coelacanths had been spied by divers outside a submersible, and was also the shallowest coelacanth encounter on record at that time.[39c]

And in early 2001, a coelacanth was captured in the nets of a commercial trawler off the Kenyan coastal resort of Milindi—the first time that a specimen of this famous 'living fossil' fish has been caught in East African waters. Its discovery, however, was not publicly revealed until October 2001, when the 170-lb specimen in question, after remaining for several months in the Mombasa-based fishery company's cold storage depot, was delivered to the Kenyan National Museum in Nairobi.[39d]

As revealed on p. 219, moreover, in 1997 *Latimeria* would be sensationally discovered in a location far beyond the reaches of Africa, Madagascar, and the Comoros—a discovery that would, in best coelacanthian tradition, send shock waves throughout the zoological world.

MIRAPINNA—A FURRY FISH FROM THE AZORES
The year 1956 saw the description of a 2.5-in-long hump-backed fish so grotesque in appearance that it required a wholly new family to accommodate it. Caught many years earlier (in June 1911) at the surface of the mid-Atlantic about 550 miles north of the Azores, this extraordinary fish seemed to be covered in hair! Closer examination, however, disclosed that its 'fur' was really a mass of living body outgrowths (hair, conversely, is composed of dead cells) containing secretory cells. Their function is unknown, though they may produce distasteful compounds to deter would-be predators.[11,40]

No less extraordinary than its 'hair' were its fins. The pelvic fins were large and wide, their wing-like appearance additionally enhanced by their long rays, diverging outwards from a broad base, and directed upwards rather than downwards. Equally odd was its pair of very small, reduced pectoral fins, placed much higher up on its body's flanks than those of most other fishes. Most peculiar of all, however, was its tail fin, because the rays in the lower half of its upper lobe uniquely overlapped those in the upper half of its lower lobe, and like the rays of the pelvic fins they were extremely long and spine-like. Little wonder that this bizarre species, commonly termed the hairy fish, was given the generic name *Mirapinna*—'wonderful fins'; its full scientific name is *Mirapinna esau*.[11,40]

Until very recently, the hairy fish, along with two previously described genera of much slimmer fishes (*Eutaeniophorus* and *Parataeniophorus*, collectively containing five species known as tape-tails), was placed by many authorities within a distinct order, Lampridiformes, set apart from all others. In January 2009, however, an international team of ichthyologists published a revelatory paper in which they announced that their morphological and genetic studies of specimens of these fishes had shown them to be nothing more than juvenile forms of certain deepsea species known as whalefishes, housed in the taxonomic family Cetomimidae. After more than half a century, therefore, the hairy fish with the wonder fins was a wonder no longer, and the family Mirapinnidae is now obsolete.[11,40]

BATHYLYCHNOPS EXILIS—A FISH WITH SIX EYES

Discovered in 1958, inhabiting depths of 300-3000 ft within the northeastern Pacific Ocean, and belonging to the spookfish (opisthoproctid) family, *Bathylychnops exilis* is a slender, 18-in-long, pike-like species with noticeably large eyes. These provide efficient prey detection in the dim light of its deep undersea world, but that is not all. Housed within the lower half of each eye is a second, smaller eye, pointing downwards and termed the secondary globe, which comes complete with its own retina and lens. This extraordinary arrangement probably increases the species' sensitivity to light within its shadowy surroundings. Yet, as if its possession of four eyes were not sufficiently strange, behind each of the two secondary globes is a third, even tinier eye-like organ, though these do not have retinas, serving instead merely to direct incoming light into the fish's large, principal eyes. Intriguingly, its pair of secondary globes were once thought to be light-producing organs—hence its name, *Bathylychnops* ('deep lamp-eye').[41] In popular parlance, it is called the six-eyed or javelin spookfish.

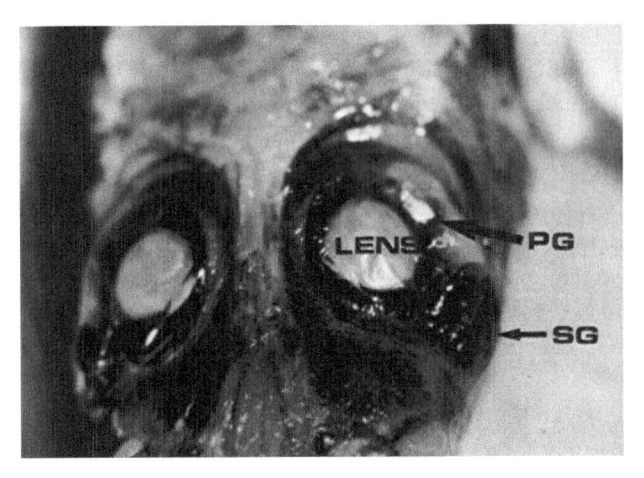

Six-eyed spookfish's head. (Dr. David Stein)
[PG = primary globes (i.e. primary eyes)]

DENTICLE HERRING—NIGERIAN ANOMALY WITH TEETH ON ITS HEAD!

In 1959, a little silver-coloured fish with a notably large anal fin was discovered that proved so peculiar that the creation of a wholly new family and suborder (within Clupeiformes, the herring order) were needed, in order to define adequately its emphatic taxonomic independence from even its closest relatives.[42]

Recorded only from the medium-sized alkaline streams of certain rivers in Benin, Cameroon, and Nigeria, this 2-in-long, shoal-forming species was named *Denticeps clupeoides* ('herring-like tooth-head'), and is commonly called the denticle herring. Both names not only allude to its herring-like form but also stress its most bizarre characteristic. Its head and its body's most anterior region are covered in small teeth! Known as denticles, they are so densely packed on the underside of its head that they make it appear almost furry on first sight.[11,42]

Coincidentally, at about the same time that this modern-day species was discovered, a very similar fossil form was also reported, its remains having been found in Tanzanian rocks dating back to the Miocene or Oligocene epochs (15-45 million years ago). As a token of its comparability to the denticle herring, it was dubbed *Palaeodenticeps tanganyikae*.[42]

LEPIDOGALAXIAS AND GRASSEICHTHYS— A PAIR OF VERY PUZZLING PISCEANS

The pikes are a northern-based family of fishes related to another northern-based family, the salmon. The counterparts of salmon in the southern hemisphere constitute a family known as the galaxiids, but the pikes did not seem to have any southern equivalent—and then along swam *Lepidogalaxias salamandroides*.

An inhabitant of acidic pools all over southern Australia, this small, slender species was described as recently as 1961, and is commonly known as the salamanderfish or Shannon mudminnow. It bears a distinctive black band on each side of its body, running from its gills along the entire length of its flank to the base of its tail. It looks like a galaxiid on first sight, but its dorsal and anal fins are much closer to the front of its body than are those of true galaxiids, and the structure of the adult male's anal fin is very specialised. Most distinctive of all, and setting it completely apart from galaxiids, is its well-scaled body (hence *Lepidogalaxias*—'scaled galaxiid'); galaxiids are scaleless. Based upon detailed anatomical studies, some ichthyologists suspect that this strange fish is more closely related to the pikes than to the galaxiids, but its precise taxonomy remains controversial. Since 1991, its family has been housed within the order Osmeriformes (containing the galaxiids, as well as the smelts and noodlefishes among others), which is allied to both the pike and the salmon orders.[43]

On 29 October 1964, an equally strange fish was obtained from the Ivindo Basin in Gabon, West Africa. In honour of eminent French zoologist Prof. Pierre Grassé, its species was christened *Grasseichthys gabonensis*. Although only 1 in long, it is very distinctive, as it lacks scales, teeth, and a lateral-line sensory system; even its gill slits are restricted. Although its overall anatomy reveals that it is an offshoot of the salmon lineage of fishes, there has been much dispute concerning its precise classification within this taxonomic order. Dr. Jacques Géry, who described it in 1964, placed it in a family of its own, but it is currently housed within Kneriidae, a family of gonorhynchid fishes (carp relatives) containing the knerias and the shellears.[4,44]

SIPHONOPHORE FISH—APPEARANCES CAN DECEIVE (IN MORE WAYS THAN ONE)

In 1965, an astonishing fish called *Kasidoron edom* was described.[45] Sole member of a totally new family (until a second, similar species, *K. latifrons*, was recorded in 1969[46]), it became known as the siphonophore fish—due to its wonderful pelvic fins. These were greatly modified, having transformed into a long, multi-branched tree-like organ hanging underneath its body, and closely resembling the tentacular appendages of those jellyfish-like composite creatures the siphonophores (exemplified by the famous Portuguese man-o'-war *Physalia*).[45]

Known only from waters of around 6-165 ft depth, about 150 miles east of Florida's Cape Canaveral and northeast of Bermuda, this 1.25-in-long velvet-black fish attracted appreciable interest, on account of its conjoined pelvic fins' unique, extraordinary structure. This was assumed to be a device for warding off predators, as they would be likely to mistake its harmless form for the deadly stinging tentacles of genuine siphonophores.[45]

After a time, however, the remainder of this fish's anatomy began to receive attention too, and researches ultimately disclosed that in spite of its distinctive appearance the siphonophore fish was not a new species at all.[47] On the contrary, it was unmasked as the hitherto-unknown juvenile form of an odd little species called the gibber fish *Gibberichthyes pumilus*, described in 1933.[48] Previously known only from four specimens, this deepwater denizen attains a total length of 4.5 in, and inhabits the western North Atlantic, as well as the South Pacific waters close to the Samoan Islands. With a very large head, a deep, laterally flattened body, and perfectly normal fins lacking any vestige of its juvenile's astounding tentacle-impersonating structure, the gibber fish was placed within a family of its own, most akin to the squirrelfishes and slimeheads.[48]

THE CURIOUS CASE OF THE BOTTLED SEA SERPENT

The 1970s began with a twist to one of ichthyology's most enigmatic episodes—the curious (and confusing) case of the bottled sea serpent. This had attracted particular attention in 1965, when sea monsters enjoyed a renaissance in scientific respectability—thanks to the publication that year of a now-classic tome by cryptozoologist Dr. Bernard Heuvelmans, entitled *Le Grand Serpent-de-Mer* (a somewhat different English version, *In the Wake of the Sea-Serpents*, appeared in 1968, also incorporating a greatly condensed version of another of his books, dealing with the giant squid and alleged giant octopuses).[49]

Within his book, Heuvelmans proffered evidence for believing that 'the great sea serpent', one of cryptozoology's most celebrated creatures, might actually be a non-existent composite—i.e. it had been 'created' via the erroneous lumping together (by previous investigators) of eyewitness reports that in reality feature a number of totally different types of animal. In short, there was no *single*, morphologically heterogeneous species *wholly* responsible for all sea serpent reports on record; instead, there were *several* well-defined, separate species *collectively* responsible for those reports.[49]

Some of them, according to Heuvelmans, were species still unknown to science, and included various unusual seals and whales, a giant turtle, a marine crocodile-like reptile, and a giant-sized 'super eel'. If his hypothesis

Dana giant leptocephalus—for many years this was mistakenly believed to be the larva of an unknown species of gigantic eel, estimated to measure up to 180 ft long. (Prof. Jørgen Nielsen, courtesy of Lars Thomas)

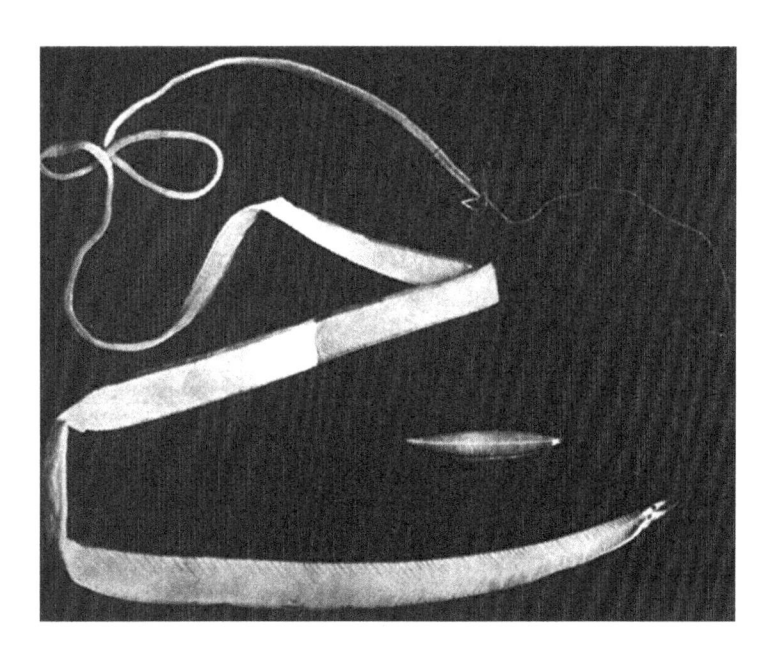

was correct, this would have profound ichthyological implications; for as a result of a chance discovery made over 30 years earlier, it meant that at least one bona fide sea serpent had already been captured—a sea serpent whose remains, moreover, were preserved, bottled, and available for scientific scrutiny![49]

On 31 January 1930, the Danish research vessel *Dana* unexpectedly captured an exceptionally long eel larva (leptocephalus) at a depth of about 900 ft, west of the Agulhas Bank and south of the Cape of Good Hope, South Africa. Whereas leptocephali of the common eel *Anguilla anguilla* measure a mere 3 in long at most, and even those of the formidable conger eel *Conger conger* only reach 4 in, the *Dana*'s remarkable specimen was a colossal 6 ft 1.5 in! This in itself was quite staggering, but its implications were even more astounding.[50]

During their metamorphosis from leptocephalus to adult, true eels (anguillids) greatly increase their total length—the precise index of growth varying between species. In the common eel, the increase is generally eighteen-fold, producing adults measuring around 4.5 ft; in the conger, it can be as much as thirty-fold, yielding adults up to 10 ft. Consequently, as conceded by *Dana* ichthyologist Dr. Anton Bruun, in the case of the *Dana* leptocephalus, which was already *6 ft* long, there existed the incredible possibility that this would have metamorphosed into a monstrous adult measuring anything between 108-180 ft, with a length of 50 ft seemingly the very *minimum* (less than a nine-fold increase) that even the most prudent estimator might expect of such a larva! Needless to say, any species of eel attaining such stupendous lengths as these would make an excellent candidate for those sea serpents grouped within Heuvelmans's 'super eel' category.[49]

After its capture, the *Dana* leptocephalus was preserved in alcohol and has since resided in a specimen bottle within the collections of Copenhagen University's Zoological Museum. Periodically, it has been taken out of its bottle to be examined, and as a result it has gradually shrunk, but it remained a notable riddle in need of an answer—especially when, as the years progressed, a few other inordinately long leptocephali were obtained.[50]

In 1959, an anatomically similar but somewhat shorter specimen, just under 3 ft long and procured in New Zealand waters, was described as a new species—dubbed *Leptocephalus giganteus*—to which the *Dana* specimen was later assigned. Interestingly, the Danish research vessel *Galathea* supposedly obtained a 6 ft leptocephalus during its voyages in the early 1950s, but no formal record of this (let alone the specimen itself) appears to exist.[50]

Even so, two specimens of *L. giganteus* certainly did exist, and the reality of the infamously elusive sea serpent, or at least one of its constituent members, seemed at last to have been fully endorsed. Inevitably, however, the truth proved very different. In 1966, two much smaller specimens of *L. giganteus* were documented. Measuring just under 4 in and 11 in respectively, they had been sifted from the stomach contents of an *Alepisaurus* lancet fish captured in the western Atlantic. Except for their modest lengths, they corresponded very closely to the New Zealand example, and were carefully studied by Miami University ichthyologist Dr. David G. Smith, in a bid to pinpoint conclusively the taxonomic affinities of *L. giganteus* in relation to the many other species of eel known to science.[52]

In March 1970, he exploded the sea serpent scenario for *L. giganteus*—by asserting that its two smaller specimens were the larvae of a notacanthid (spiny eel), *not* of an anguillid (true eel). This spelled doom for their species' claim to fame as (in its adult phase) a genuine sea serpent—because in stark contrast to the leptocephali of true eels, those of notacanthids do *not* greatly increase their length during metamorphosis from larva to adult. Thus, predictions that mature specimens of *L. giganteus* would measure over 100 ft were totally unfounded. Instead, when the unknown adult phase of this species was finally collected, it would be very little longer than the leptocephalus, i.e. a mere 6 ft or so.[52]

More recently, however, this reclassification of *L. giganteus* as a notacanthid has itself been overturned, so that nowadays it is classed as a species of short-tailed eel or worm eel instead, within the family Colocongridae, housed in turn within Anguilliformes, the order of true eels, and it has accordingly been renamed *Coloconger giganteus*. Like the notacanthids, however, the short-tailed eels do not display a sizeable increase of length during larva-to-adult transformation, so its identity as a sea serpent remains null and void.[52]

Of course, there may indeed be eels of gigantic length still eluding scientific detection in the vastness of the oceans—giant anguillids, for example, that are compatible with Heuvelmans's concept of the 'super eel' category of sea serpent—but unlike *C. giganteus*, these have yet to be captured, preserved, and bottled.

MEGAMOUTH SHARK—
THE 'BIG ONE' THAT DIDN'T GET AWAY!

Most fishermen have a cherished tale or two about 'the big one that got away', but none can surely compete with the following version—in which, just for a change, the whopper in question did *not* get away, much to the delight of marine biologists throughout the world.

On 15 November 1976, a team of researchers from the Hawaii Laboratory of the Naval Undersea Center (now known as the Naval Ocean Systems Center) was aboard the research vessel *AFB-14*, sited about 26 miles northeast of Kahuku Point, Oahu, in the Hawaiian Islands.

During the course of their work, two large parachutes employed as sea anchors were dropped overboard, and lowered to a depth of 500 ft. Later that day, when the boat was ready to leave for home, the researchers hauled the parachutes back up—and found that one of them had drawn up the greatest ichthyological discovery since the coelacanth! Entangled in the parachute was a gigantic shark, measuring 14.5 ft in total length, weighing 1653 lb, and differing radically in appearance from all other sharks on record.[53-5]

Recognising its worth, the team hauled its mighty body aboard on rollers, and sent it at once to the Naval Undersea Center's Kaneohe Laboratory, where biologist Lieut. Linda Hubbell lost no time in contacting the University of Hawaii. Next morning, it was examined by Dr. Leighton R. Taylor, director of the university's Waikiki Aquarium, after which its body was quick-frozen at a firm of tuna packers, and retained there until, on 29 November, it was transported (still frozen) to a specially-constructed preservation tank at the National Maritime Fisheries Service's Kewalo dock site. It was then thawed and injected with formalin, procedures that marked the commencement of what was to be an intensive period of study in relation to this unique specimen—swiftly recognised to represent a dramatically new species never before brought to the attention of science. The study lasted almost seven years, and was undertaken jointly by Dr. Taylor, Dr. Paul Struhsaker of the Fisheries Service, and shark specialist Dr. Leonard Compagno from San Francisco State University. Preserved, the specimen is now held at Honolulu's Bernice P. Bishop Museum.[53-5]

The head of this strange shark was very large, long, and broad, but not pointed like that of more typical sharks, whereas its lengthy, cylindrical body tapered markedly from the broad neck to the slender heterocercal tail (i.e. the tail's upper lobe was much longer than its lower lobe). Its pectoral fins were also long and slender, but its pelvic fins and anal fin were very small—smaller than the first of its two dorsal fins. Identifying it straight away as a male, its pelvic fins bore a pair of elongate claspers (a male shark's copulatory organs).[54-5]

The specimen's huge size made its species, on average, the sixth largest species of modern-day shark known to science, but even more striking than its overall bulk was its mouth. Relative to the rest of its body, its mouth was exceedingly large and wide—a feature that soon earned it in newspaper reports a very fitting soubriquet—'megamouth', which became accepted by science as this species' official English name. In addition to its size, the megamouth's mighty orifice was distinguished by its thick lips, more than 400 tiny teeth arranged in 236 rows, a very unusual anatomy which meant that its jaws did not lower at the bottom like those of most sharks but flapped open at the top instead, and—most startling of all—a silvery mouth lining that glowed in the dark![54-5]

Despite initial speculation that this unexpected last-mentioned feature was due to light-emitting structures comparable to the bioluminescent organs of many deepsea fishes and other benthic life, insufficient evidence was obtained from the study to verify this. Even so, when taken together with the megamouth's immense size but only tiny, relatively useless teeth, various other anatomical attributes, plus the great depth at which it was captured, its glowing jaws indicated that this mysterious marine form was itself a deepsea denizen, whose lifestyle probably consisted of slow cruises through the

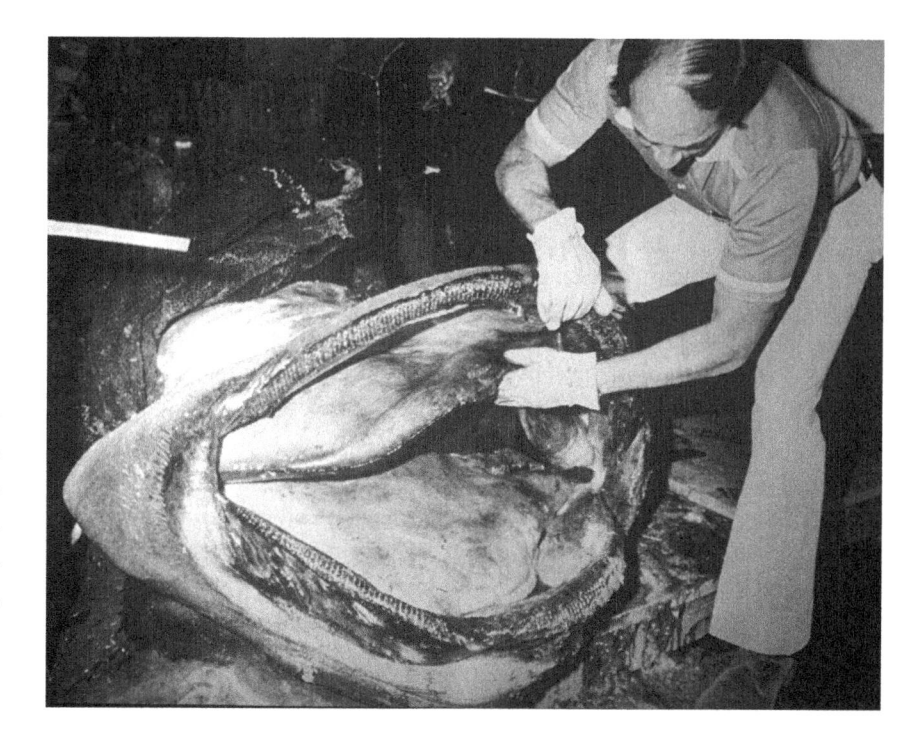

The first scientifically-recorded specimen of the aptly-named megamouth shark—this species remained unknown to science until as recently as 1976. (Charles Okamura/Honolulu Advertiser)

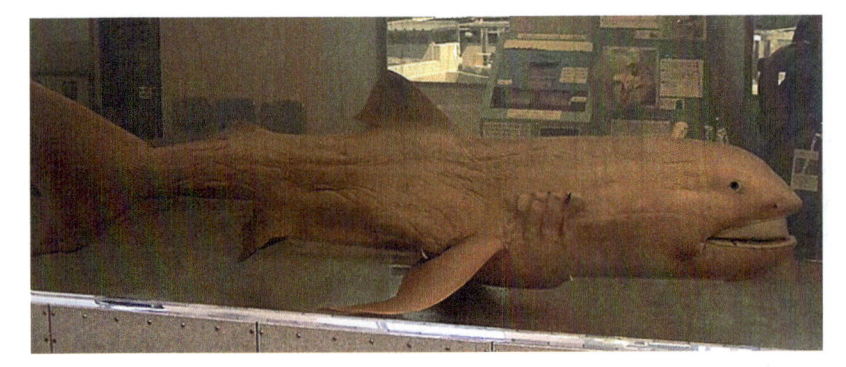

Preserved megamouth specimen in Japan.

inky darkness of the sea's depths with its huge, glowing jaws held open, to entice inside great numbers of tiny marine organisms. Thus, the megamouth was a harmless plankton feeder, a gentle giant.[54-5]

All of this and much more was recorded in the paper prepared by Taylor, Struhsaker, and Compagno, constituting the megamouth's formal scientific description and published on 6 July 1983.[55] Their study had revealed this mighty creature to be so unlike all other sharks that they had not merely classed it as a new species, they had also placed it in an entire genus *and* family all to itself. Approving of 'megamouth' as its common name, Taylor and colleagues made it the basis of this species' scientific name too, christening it *Megachasma pelagios* ('great yawning mouth of the open sea')—sole member of the family Megachasmidae, but most closely allied to the basking shark *Cetorhinus maximus*, another plankton feeder.[55]

Attempts to catch a second megamouth for comparison purposes proved unsuccessful until November 1984, when another megamouth *was* caught—but, once again, completely by accident. This time, a commercial fishing vessel named *Helga* took the honours, snaring it unknowingly within a gill net at a depth of only 125 ft, while based close to California's Santa Catalina Island, near Los Angeles. Needless to say, this priceless specimen was carefully brought ashore, and was sent at once to the Los Angeles County Museum of Natural History. Tissue samples were taken and stomach contents removed, after which its 14-ft-long body was stored in a frozen state within a temporary case until work upon a specially-prepared fibreglass display unit was completed, whereupon the new megamouth was preserved and retained thereafter within its 500 gallons of 70 per cent ethanol.[56]

The megamouth's known distribution range expanded dramatically with the third specimen's discovery. On 18 August 1988, an adult male almost 17 ft long was found washed up on a beach near Mandurah, a holiday resort south of Perth, Western Australia. When news of its appearance reached the Western Australian Museum, ichthyologist Dr. Tim Berra (visiting from Ohio State University) and a team of fish researchers swiftly travelled to the beach to salvage the shark's body. This was just as well, because some of the resort's residents, not realising its immense scientific significance, had been attempting (albeit unsuccessfully) to push it back into the sea![57-8]

The scientists were delighted to find that this latest megamouth was still in good condition, and it was ultimately preserved and housed in a fibreglass display tank like that of the Los Angeles specimen. During the tank's construction, it was retained in a frozen state, enabling the museum's taxidermist to prepare a plaster cast of its body for exhibition.[57-8]

On 23 January 1989, a fourth megamouth appeared, stranded dead on the sandy beach of Hamamatsu City in Japan's Shizuoka Prefecture, yielding the first record of this species from the western Pacific. An adult male, estimated at over 13 ft in total length, it attracted the notice of a photographer who took some good pictures of it that demonstrated beyond any doubt that it really was a megamouth—all of which was very fortunate, because shortly afterwards, before there was time to rescue it, this scientifically invaluable specimen was washed back out to sea and lost. The photos, however, were sent to Dr. Kazuhiro Nakaya, who published them in a short *Japanese Journal of Ichthyology* report.[59] Less than six months after this specimen's brief appearance, a second Japanese megamouth made the headlines, when on 12 June a *living* specimen was caught in a net in Suruga Bay. Photographs confirming its identity as a megamouth were taken, after which it was released unharmed.[57]

The next episode in the megamouth saga however, was truly spectacular. On 21 October 1990, a sixth specimen turned up, measuring 16 ft 3 in and snared in a drift net off Dana Point, in California. It was towed to shore by the net's vessel, and was found to be still alive. Marine biologist Dr. Dennis Kelly, from the Orange Coast College, gently examined the huge fish, and decided that although it would not survive in captivity, it would probably live if released back into the sea. And so, very carefully, it was set free, and was filmed underwater as it swam slowly down into the depths from which it had earlier arisen. Moreover, capitalising upon this unique opportunity to discover a little more about its species' lifestyle, a radio transmitter was attached to its body. This enabled researchers to track it in the sea for the next three days (after which time the transmitter's batteries ran out), and revealed that it exhibited vertical migration—moving to the ocean surface only at night, and descending back into the depths at dawn—which explains how this extremely large and striking species had escaped scientific detection for so long.[57]

Almost exactly 18 years after the first one was hauled up by a research vessel off Oahu, a seventh megamouth appeared. This proved to be the first female specimen on record, and was washed up in Hakata Bay, Kyushu, on 29 November 1994. The third Japanese megamouth, its body measured 15.5 ft, weighed 0.8 ton, and was transported to Fukuoka's Marine World Museum, where it was deep-frozen, prior to permanent preservation.[60]

A hitherto unsuspected portion of this species' distribution range was revealed on 4 May 1995, when the first megamouth to be recorded from the Atlantic Ocean was captured by a French tuna fishing vessel, *Le Bougainville*, in its purse seine, roughly 40 miles off Dakar, Senegal. This eighth megamouth was a young male, measuring only 6 ft or so in total length. Regrettably, however, its body was not preserved.[61]

Megamouth #9 extended its species' known distribution even further, for this specimen, another young male, approximately 6 ft 3 in long, was procured off southern Brazil, on 18 September 1995. Its body was retained by the Instituto de Pesca, in São Paulo, Brazil.[61]

It was the fourth time for Japan when the megamouth made its next confirmed appearance, courtesy of only the second known female specimen turning up on 1 May 1997 near Toba. More than 16 ft long, its carcase was taken to Toba Aquarium.[62]

On the evening of 20 February 1998, yet another specimen (#11) of this maritime megastar surfaced—caught by three Filipino fishermen in Macajalar Bay, Cagayan de Oro, in the Philippines, and estimated to measure around 16 ft. The first on record from this island group, its taxonomic identity was confirmed on 21 March 1998 by Dr. Leonard Compagno. Unfortunately, its body was hacked to pieces after it had been landed and photographed. Moreover, a female megamouth captured at Atawa in Mie, Japan, on 23 April 1998 was subsequently discarded.[63]

Megamouth #14, another female specimen and measuring approximately 17 ft long, was captured in a drift gillnet roughly 30 miles west of San Diego, California, on 1 October 1999. The third megamouth to be caught off southern California, after being photographed it was released again, still in good health.[64]

In addition, Genoa Aquarium worker Pietro Pecchioni claimed in an internet shark discussion group that he saw and photographed what may have been a living megamouth, being harassed by three sperm whales near the island of Nain, off northern Sulawesi, Indonesia, on 30 August 1998. The shark measured 15-18 ft long, and Pecchioni spied it while in the company of a group of people participating in a WWF whale-watching programme. When the whales saw the watchers, they came towards them, then swam away, so the shark survived. At the time of his claim (4 September 1998), Pecchioni's photos had not been developed, but when they were, the shark's identity as a megamouth was confirmed (making it megamouth specimen #13); and the encounter was formally documented by Pecchioni and Milan University zoologist Dr. Carla Benoldi in 1999 on the website of the Florida Museum of Natural History's ichthyology department.[65]

On 19 October 2001, megamouth #15, a male specimen roughly 18 ft long, was caught alive in a drift gillnet by a commercial swordfish vessel sited about 42 miles northwest of San Diego. After a United States National Marine Fisheries Service observer aboard the vessel had photographed its unexpected catch and had also taken a tissue biopsy from it, this megamouth was also released in good condition.[65a]

Another notable specimen was megamouth #23, which was washed up on 13 March 2004, onto Gapang Beach in northernmost Sumatra. Only relatively small, measuring just over 3 ft in length, it was subsequently frozen at the Lumba Lumba Dive Centre, and following formal examination by scientists it was deposited at Cibinong Museum. Interestingly, because of marked differences in shape between this megamouth's dorsal fins and those of all previously recorded specimens (and also between its anal fin and those of previous specimens), when formally documenting it later that year a team of researchers suggested that these differences may indicate the existence of a second species of megamouth, but no further evidence for such a situation has been presented since then.[65b]

Megamouth #26 was discovered on 4 November 2004, stranded but still alive at Namocon Beach, in Tigbauan, Iloilo City, in the Philippines. An adult female measuring approximately 16.5 ft long and weighing roughly a ton, this was the third megamouth to have been recorded in the Philippines, and bore a wound that may have been a spear wound, or possibly a bite from the cookie cutter shark *Isistius brasiliensis*. The megamouth was formally identified the day after its discovery by an official from the Southeast Asia Fisheries Development Center (SEAFDEC), after which 16 men carried it to a SEAFDEC aquarium where it lived for a day. It was then preserved in 10 per cent formalin within a 1-ton fibreglass tank.[65c]

As of August 2011, 53 specimens of the megamouth shark have been obtained or conclusively sighted. The three most recent ones are: #51, a specimen of unknown sex caught off eastern Taiwan on 19 June 2010, but later cut up for meat that was sold at a local market, with only a jaw retained; #52, a dead juvenile male specimen captured by fishermen close to the western Baja California peninsula of Mexico on 12 June 2011; and #53, an individual of unspecified sex but measuring approximately 10 ft long that was recorded from Japan's Kanagawa Prefecture on 1 July 2011. A comprehensive listing of these, together with pertinent details of their respective discoveries, plus a detailed bibliography of sources, can be accessed online in Wikipedia.[65d]

Finally: an intriguing footnote (fin-note?) to the megamouth history is that this species' own discovery set the scene for a remarkable parasitological parallel. During the study of the very first megamouth specimen, an extremely strange form of tapeworm was found inside its intestine. When closely examined, this peculiar parasite proved to be not just a new species (later named *Mixodigma leptaleum*), but one so different from all others that it required a completely new genus and family—exactly like its megamouth host![66]

HEXATRYGON—WORLD'S ONLY KNOWN SPECIES OF DEEPWATER STINGRAY

On 5 July 1980, a 41-in-long fish found washed ashore on a beach at Port Elizabeth, South Africa, by former angling journalist Dave Bickell proved to be a species so drastically different from all others that a completely new suborder was created to receive it. On first sight, it resembled an ordinary stingray, but closer inspection exposed some fundamental differences. First and foremost, it had six pairs of gill openings—in stark contrast to every other modern-day ray or skate, which only have five pairs. This is the diagnostic feature of the new fish that inspired its common and scientific names—when it was officially described by Drs. P.C. Heemstra and M.M. Smith from the J.L.B. Smith Institute of Ichthyology, they named it *Hexatrygon bickelli*, the sixgill stingray.[67]

Like other rays and skates, just behind its eyes was a pair of reduced gill openings called spiracles, but *Hexatrygon* was uniquely able to open and close them, thanks to a pair of large, mobile flaps of skin. Equally distinctive was its very long, thin, flaccid snout, which contained a transparent, jelly-like substance that rendered it translucent. The underside of this snout, moreover, was well supplied with remarkable little organs known as the ampullae of Lorenzini, which can actually detect the weak electrical currents running through the nervous systems of organisms lying in the thick sludge on the sea bottom.[67]

The possession of such organs by *Hexatrygon* suggests that it is a deepwater species, using them to detect potential prey buried in this ooze. Other aspects of its anatomy and morphology indicating this mode of existence are its tiny eyes, unusually small brain, and thin black dorsal skin—all features typical of deepwater fishes—so that ichthyologists concluded that in life it dwells at depths of around 1300-3300 ft. That was subsequently confirmed by the collection of additional specimens, and from areas as widely dispersed as Japan, Queensland, and Hawaii, as well as more from South Africa. This adds a further distinction to an already radically distinct creature, because it constitutes the world's first-known deepwater stingray—all other species inhabit fairly shallow zones. In later years, four subsequent *Hexatrygon* species were described, based upon specimens obtained from Hawaii, South Africa, and Taiwan, but these are now believed merely to represent intraspecific variations within the single original species, *H. bickelli*.[67]

A VERITABLE FISH OUT OF WATER!

An ostentatiously out-of-place creature—one that could be described not only figuratively but also perfectly literally as a veritable fish out of water—was the extraordinary 1-in-long entity that ichthyologist Dr. Peter Henderson spotted wriggling amongst the leaf-litter on a bank at least 2 ft *above* an Amazonian blackwater stream called the Taruma-Mirim, near Manaus, Brazil, in 1984. Bright red, eyeless, worm-like, with scaleless skin and whisker-like projections around its face, it presented a formidable challenge to anyone merely wishing to assign it to any known *phylum* of animals, let alone any known *species*! Closer observation, however, exposed its true nature—it was, of all things, a fish![68]

Later researches uncovered that it was a radically new species of trichomycterid (also termed pygidiid) catfish—freshwater species that are normally parasitic—highly-modified for survival on land. Indeed, this bizarre little

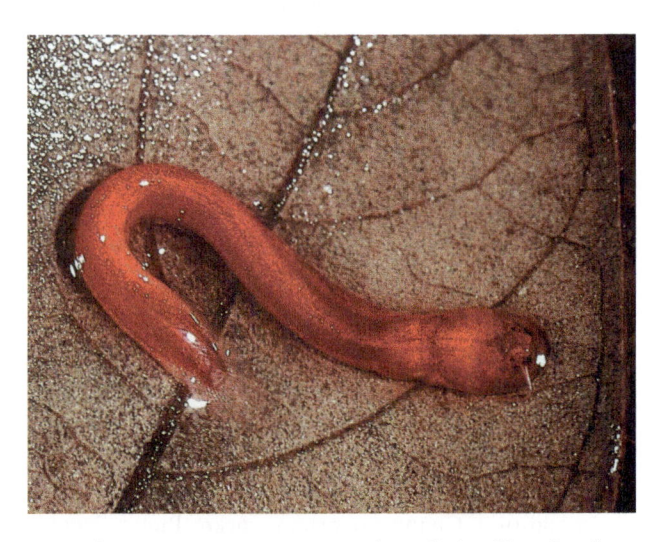

A bizarre semi-terrestrial catfish. (Dr. Paul Sterry/Nature Photographers Ltd, Basingstoke)

catfish, since dubbed the land catfish, seems actively to prefer a terrestrial existence to an aquatic one—when some captive specimens were taken from their leaf-litter and placed into water, they promptly climbed back out! Moreover, as its gills are greatly reduced structurally, it appears to obtain oxygen by absorbing it directly through its skin—which is extremely vascularised (explaining its intensely red colouration). Its food apparently consists of small insects and other organisms sharing its leaf litter habitat.[68]

While preparing this book's first edition, I was informed by Dr. Henderson that he was planning to name this new catfish 'Pheatobius walkeri', but almost two decades have since passed by, during which time the anticipated paper containing its scientific description and name has never appeared. When preparing this book's second edition (*The New Zoo*, 2002), I contacted Henderson again, and learnt from him that he had agreed to allow some Brazilian ichthyologists to describe it but they had never done so. He had therefore decided to prepare its description himself, but this has still not appeared either, and 'Phreatobius walkeri' is now a nomen nudum, a name unlinked to any formal scientific description.[68]

THE MONSTER FISHES OF CHINA'S LAKE HANAS/KANASI

Officially, the largest specimen of freshwater fish on record is a 15-ft European catfish *Silurus glanis*, caught in Russia's Dniepper River sometime prior to the mid-1800s (though this species *as a whole* is generally shorter than the pa beuk, officially deemed to be the world's largest freshwater fish—see p. 195). As a consequence, the lake-dwelling fishes reported in July 1985 by no less an authority than China's eminent biologist Prof. Xiang Lihao (also transliterated as Yuan Guoying in some reports but indisputably one and the same person), from Xinjiang University, attracted appreciable scientific interest.[69]

In July, the professor and a party of students arrived at a large but remote body of water known variously as Lake Hanas (in Kazakh) or Lake Kanasi (in Tuvan), situated in northwestern China's Xinjiang Autonomous Region, in order to examine its potential as the site of a future nature reserve. On 24 July, one of the students observing the lake from a watchtower built two years earlier noticed several huge reddish-coloured objects moving at the water's surface. When the professor and students scrutinised them closely through binoculars, they discovered to their astonishment that they were enormous salmon-like fishes, whose heads, tails, and spiny dorsal rays could all be clearly discerned. Just *how* enormous they were, however, was not revealed until the next day.[69]

That morning, while again being observed through binoculars by Xiang Lihao, one of the fishes very obligingly aligned itself in parallel with a stretch of the bank extending between two trees. Armed not only with binoculars this time but also with a camera, the professor took some photos, then measured the distance between the trees. Using this measurement, he was able to calculate from the photos that the fish was at least 33 ft long![69]

A large salmon known as the taimen *Hucho taimen* is indeed known from several rivers in northern China, but this species' maximum recorded length is a mere 6.5 ft—far short of the Lake Hanas monsters. Worth noting is that giant red fishes in this lake have been reported for decades by local villagers, but as the lake had not previously attracted scientific attention such reports had not been widely circulated. Now, with an eyewitness of Prof. Xiang Lihao's scientific standing, there should be no question concerning their existence or authenticity as giant fishes. So unless they are abnormally huge taimen, the Lake Hanas fishes must surely constitute a spectacular new species, requiring formal description and study.[69]

In mid-August 2005, an expedition was launched by Xinjiang Ecological Association to investigate Lake Hanas's gargantuan inhabitants, but nothing conclusive resulted from the search. Two years later, in July 2007, a tourist claimed to have videoed these mystery beasts. Aired on Chinese television and also online, the video appeared to show several very large animate objects moving rapidly across the lake just beneath the water surface.[69a]

HEUVELMANS'S MEDITERRANEAN MIDGET

Blennies are mostly small, predominantly marine fishes of worldwide distribution in temperate and tropical seas. There are many different species, but they are characterised by their scaleless skin, as well as their numerous small teeth, packed closely together.

In February 1986, a new species of blenny was described by French ichthyologist Dr. François Charousset, after it had been collected by him in Mediterranean waters off Croatia's Istrian coast. A small, yellow-headed blenny, commonly known as the midget, the discovery of such a fish would not normally be considered particularly remarkable or significant. As it happens, however, the midget is especially noteworthy, because it epitomises the definition of a cryptozoological animal. Namely, a species (or subspecies) known to the local people sharing its domain, but whose existence long remained unconfirmed by science. So it was with the midget, reported by the locals for several decades but eluding scientific detection until now.[70]

Accordingly, Charousset very appropriately christened this new blenny *Lipophrys heuvelmansi*—in honour of French zoologist Dr. Bernard Heuvelmans, popularly called 'the father of cryptozoology' due to his seminal work in this emerging investigative science.[70]

A WHIP-TAILED WHOPPER FROM THAILAND

In 1983, a series of reports appeared in various Thailand newspapers revealing the existence of an enormous form of freshwater stingray (reputedly weighing at least 660 lb) in the Chao Phraya and Mekong Rivers, and the capture of two such fishes, one from each river. Unfortunately, they were not preserved for scientific examination, but from the description and photos in the reports it seemed clear that they represented a species undescribed by science, so ichthyologists eagerly awaited news of further captures.[71]

On 9 November 1987, their period of waiting came to an end, with the procurement of a male Chao Phraya specimen at Pichit, roughly 286 miles upriver from the Gulf of Thailand. With a total length of almost 11 ft and a body disc measuring 3.5 ft in width, it proved to be a hitherto unknown species of whip-tailed stingray—with a noticeably projecting snout, small eyes, a very narrow tail base, and a broad but thin, oval body disc, brownish-grey on top, white underneath, and black around its underside's margins. On 14 June 1988, an even larger specimen was captured, again in the Chao Phraya but this time at Chumsang (about 37 miles upriver of the Gulf of Thailand). A female, it measured just under 15.5 ft in total length, with a body disc width of almost 6.5 ft. A third Chao Phraya specimen, a female just over 7 ft long with a body disc 2 ft 7 in wide, was caught on 31 May 1989 at Ayutthaya (about 62 miles upriver from the Chao Phraya's mouth), and was designated the holotype of this impressive new species—which in 1990 was named *Himantura chaophraya*, the giant freshwater stingray.[71]

According to fishermen along the Chao Phraya and Mekong rivers, the largest specimens that they have captured so far have weighed as much as 1100-1320 lb. If this is true, the giant freshwater stingray is one of the largest new species of fish revealed in recent years.[71]

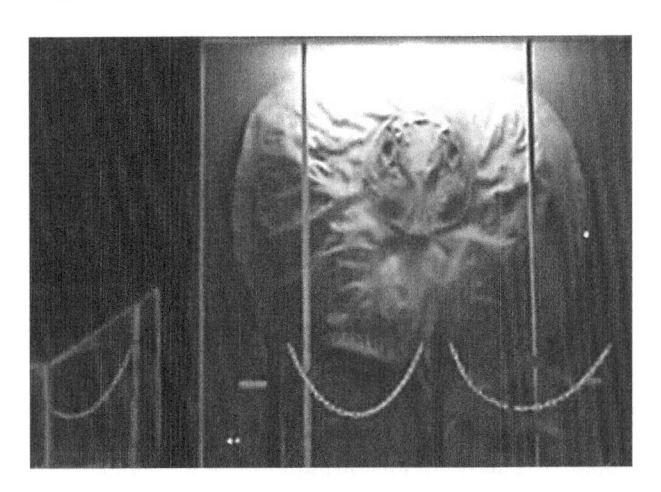

Chao Phraya's giant freshwater stingray. (Richard Freeman)

OPAL ALLOTOCA—WHEN HOME IS A POLLUTED CATTLE TROUGH!

Even the least glamorous of tasks can occasionally prove rewarding—as ichthyologist Michael L. Smith from the American Museum of Natural History found out in 1990. In a remote valley tucked away amid the mountains of west-central Mexico, he and three colleagues had been undertaking the distinctly unpleasant task of wading neck-deep in a scum-covered, sulphur-polluted, artificial pond used for watering cattle and covering about three acres—when, after spending several hours scooping up water from its murky, stench-exuding depths, their long-handled nets drew up a mass of small, iridescent fishes that instantly made all their discomfiture well worthwhile.[72]

The survival of anything in such a seemingly hostile environment as this extensively-polluted pond was in itself quite remarkable, but far more so was the identity of these fishes. They were opal allotocas *Allotoca maculata*—a species hitherto believed extinct for 20 years! Relatively nondescript in shape, and only 2 in long, its unassuming form is compensated for by the shimmering scales on its flanks, which glisten like the precious stone after which it is named. Closely related to the killifishes (popular in aquaria), the opal allotoca belongs to the goodeid family, which has many species in this region of Mexico.[72]

Following their highly unexpected discovery of this 'extinct' fish, Smith and his colleagues brought back ten living specimens, and successfully established a breeding population of opal allotocas within the luxury of the New York Aquarium—a far cry indeed from the sulphur and scum offered by their previous accommodation![72]

RED-FINNED BLUE-EYE—A PRIZE IN A PUDDLE

No less surprising than the opal allotoca's reappearance was the discovery of the red-finned blue-eye *Scaturiginichthys vermeilipinnis*. Sole member of a new genus *and* subfamily, this small but brilliantly-coloured species made its scientific debut in a tiny 3-in-deep puddle, for this is where Peter Unmac of the Australia-Papua New Guinea Fishes Association found its type specimen in December 1990. It is currently known from just four springs, all situated on a now-protected grazing property called Edgbaston Station, near Longreach, in central Queensland, but is currently categorised as critically endangered by the IUCN, as is the opal allotoca.[73]

BEWARE OF THE BRAZILIAN VAMPIRE FISH!

The skin on the researcher's hand was twitching and writhing, as if something small but very much alive was wriggling just beneath its surface—which is not surprising, because that is precisely what was happening. If we were talking about a parasitic worm, this scenario may seem relatively mundane (albeit unpleasant!). However, the creature involved here was not a worm, but a fish! Moreover, it was heading directly for a vein, from which it hoped to imbibe its species' sole source of sustenance—blood. For this tiny but tenacious terror, which had surreptitiously slipped into a cut on the researcher's hand, was a bona fide vampire![74]

That chilling little vignette was drawn not from reverie but from reality, for the aptly-dubbed vampire fish is a recently-revealed species that thrives exclusively upon a diet of blood. On account of its minuscule size (less than half an inch long) and transparent body, however, it can be quite a problem to study, as discovered by the researcher whose macabre experience was outlined above.[74]

Hailing from the Araguaia River in the Amazon basin's southeastern section, the vampire fish was brought to scientific attention in 1994 by Brazilian biologist Dr. Wilson Costa, from the Federal University of Rio de Janeiro. He revealed that it spends its evenings sucking blood from the flesh of other fishes—remaining attached to their gills via a pair of special hook-shaped teeth until it is swollen with ingested blood like an engorged leech. Its closest relative is the very slightly larger but equally unsavoury candiru *Vandellia cirrhosa*, a tiny Amazonian river catfish infamous for swimming up inside the urethra of any bather urinating in the water. Once inside, the candiru cannot be removed because of its erectile spines, and can cause agonising pain, even death, to the hapless human penetrated by it.[74]

MORE NEW SPECIES FROM VIETNAM

Vietnam proved to be a veritable wonderland during the 1990s for seekers of significant new species of animal, as already revealed in this book's previous chapters, dealing with mammals, birds, and reptiles and amphibians. Moreover, ichthyology also recorded some memorable Vietnamese discoveries.

In 1996, while participating in a three-week WWF-sponsored field study within the now-famous Vu Quang Nature Reserve, Vietnamese zoologist Dr. Nguyen Thai Tu discovered a hitherto-undescribed species of barb. Belonging to the genus *Crossocheilus*, measuring 8-10 in long, and weighing 3.3 lb, this new cyprinid was readily differentiated from other *Crossocheilus* species by its scales, which were larger and more rounded. Equally distinctive was the golden stripe running along its back, and the silver stripe on its belly. Named *C. vuha* in 1999 (but subsequently demoted by some ichthyologists to a subspecies of *Neolissochilus benasi*), although hitherto new to science this fish is well known to the locals, who call it the *co* and catch it for food.[75]

This is not the first new species of cyprinid discovered in Vietnam by Nguyen. In 1992, he found a new species of river carp, also in Vu Quang, which in 1995 he dubbed *Parazacco vuquangensis*, and in 1987 he discovered two new species of Vietnamese *Opsariichthys* carp—*O. bea* and *O. hieni*. Four years earlier, he found a new Vietnamese loach, *Cobitis yeni*.[75-6]

SOME SHOCKING TAILS FROM THE AMAZON'S ABYSS

In contrast to those inhabiting its surface regions, the fishes dwelling in the mighty Amazon River's dark, deep-water zones have remained largely unknown—until the 1990s, that is. Beginning in 1992, Arizona University ichthyologist Dr. John G. Lundberg has led an ongoing, pioneering exploration of this unexplored underwater territory using deepwater nets. By February 1997, his team had sampled the piscean fauna from 2500 nautical miles of river and had caught 125,000 fishes, comprising an ever-increasing species count standing at 240 by the end of the 20th Century—of which at least 30 are totally new to science.[77]

One of the most bizarre of these lately-revealed fishes is *Magosternarchus duccis*—a new predatory species (and genus) of electricity-generating gymnotid. Pinkish-white in colour, it is characterised by its greatly enlarged teeth and jaws, especially its lower jaw, which projects strongly, including and often extending dorsal to its upper jaw and snout. This distinctive fish's principal diet is also memorable, for it consists of the tails of other electric fishes, which it bites off—but because its victims soon regenerate their tails, its food source is undiminished. Its type specimen was collected in the Rio Branco, Roraima State, Brazil, on 8 December 1993 by Lundberg and colleagues, 2-7 miles upriver from the Rio Branco's confluence with the Rio Negro, during a river-bottom trawl at a depth of 19-22.5 ft.[78]

On 29 November 1993, the type specimen of *M. raptor*—a close relative of *M. duccis* and sharing its curious dietary proclivity for tail-devouring, but distinguished by its uniquely enlarged premaxillary bones—had been collected by Lundberg and company in the Rio Solimões, Amazonas State, Brazil, 10.5 miles downriver from the confluence of the Rio Purus, during a bottom trawl at a depth of 13-19 ft. Both species were formally described and named by Lundberg and three co-workers in a *Copeia* paper published in 1996.[78]

Another notable fish discovered during this lengthy exploration of the Amazon's abyss was *Micromyzon akamai*, a new genus and species of banjo (aspredinid) catfish—lacking eyes, with reduced melanic pigmentation and extremely reduced premaxillae, but displaying hypertrophied development of cranial roofing elements and body armour. Its type was collected on 20 November 1994 by Lundberg and colleagues in the Rio Tocantins, Pará State, Brazil, above the confluence with the Rio Pará, at a depth of 32-45 ft.[79]

A memorable rediscovery was also achieved, when specimens of the anteater gymnotid *Orthosternarchus tamandua* were trawled up from the Amazon's inky depths. Named after its long, slender anteater-like snout, this almost blind species of electric fish with tiny eyes had previously been known only from two museum-preserved examples.[77]

BORNEO'S ONCE AND FUTURE RIVER SHARK

As almost all shark species are exclusively marine, the finding of any freshwater shark is a notable ichthyological event, but the discovery made in March 1997 by scientists working in the Kinabatangan River of Sabah, a Malaysian state in northern Borneo, was momentous even by these standards. It was a young female shark, 32 in long, with black beady eyes, and a blunt snout, which had been captured in the river by some local fishermen. When it was shown to the scientists, from the Elasmobranch Biodiversity, Conservation and Management project being conducted by the IUCN Shark Specialist Group in Sabah, they were amazed to see that it instantly recalled a near-legendary species—the Borneo river shark.[80]

More than 100 years ago, the sole recorded specimen of this shark, known even in taxonomic parlance only as '*Glyphis* species B', was obtained from an unknown river in Borneo, after which it had been preserved and sent to a museum in Vienna where it still resides. Now, however, over a century later, there was at last a second specimen, and, ironically, there might have been even more. Apparently, the fishermen responsible for rediscovering this long-lost species had caught not one but several specimens, as confirmed by photos taken of them while they were still in the net. Tragically for science, however, the fishermen had only retained one, discarding all of the others, but several additional young specimens were subsequently caught and preserved for study. In 2010, this species finally received a formal scientific description and name—*Glyphis fowlerae*.[80]

Equally notable was the capture in 1996 of an adult female specimen of the Ganges river shark *Glyphis gangeticus*, and the finding of two fresh jaws from this species. Until then, it had been known only from three museum specimens obtained more than a century ago.[80]

SPOTTING THE LEOPARD CHIMAERA

Named after a monstrous composite beast from classical Greek mythology, the chimaeras are renowned for their phantasmagorical forms, and also for bewilderingly combining various anatomical features drawn from both of the major taxonomic classes of fishes—Chondrichthyes (cartilaginous fishes) and Osteichthyes (bony fishes). For whereas they possess a skeleton composed of cartilage, they also possess a number of characteristics normally typifying bony fishes, most notably an operculum over their gills.

Although many chimaeras are decidedly distinctive, one of the most eyecatching species remained undescribed by science until the late 1990s. In 1998, it was dubbed *Chimaera panthera*, the leopard chimaera, after the bold brown markings dappling its body and fins, and measures 3 ft in length. This zoological newcomer's first-known specimens were a couple that were collected in deep waters off northern New Zealand by a team of Japanese scientists aboard the *Fukuyosi-maru* in 1990, but they failed to attract much attention.[81]

In 1995, however, a third specimen was caught, this time by a commercial fishing vessel from New Zealand itself, and was handed over to its National Museum—just when chimaera expert Dr. Dominique A. Didier Dagit happened to be paying a visit to this country from the Academy of Natural Sciences in the U.S.A. After examining the specimen, she realised that it belonged to a hitherto undescribed species, and not long afterwards she uncovered the two earlier specimens—which had been residing unpublicised at a Tokyo museum. The first chimaera species found in New Zealand waters, it is the sixth species within the genus *Chimaera*. Since its discovery, however, an estimated nine new species of chimaera have been found off New Zealand, which now await formal description by Didier.[81]

Moreover, in July 2011, a new chimaera, the opal ghost-shark *C. opalescens*, was described from Ireland's West Coast—making it the first new species of fish to have been discovered in Irish waters for over a century. It owes its scientific unveiling to 31 specimens spotted in a French fish market after having been caught by a trawler fishing in the northeast Atlantic Ocean off Ireland.[81a]

HE SAW A SAWBELLY

Another significant ichthyological discovery from the late 1990s that was brought about by an extraordinary coincidence involving the right person being in the right place at precisely the right time was the giant sawbelly. While sorting through his catch one day in 1998, trawlerman Tim Parsons from Port Adelaide, South Australia, noticed a large, rather unusual fish, which he set aside. It was very fortunate that he did so, because it just so happened that only a few minutes earlier on that

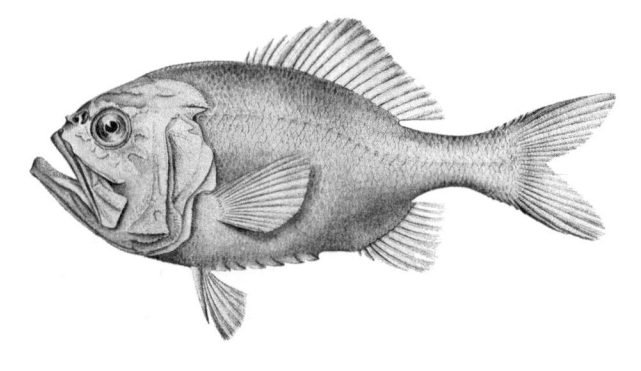

Giant sawbelly.

very same day, Australian fish taxonomist Dr. Peter Last had given to Parson's manager, Craig McDowall, a description of a long-lost species of fish—the giant sawbelly or giant roughy *Hoplostethus gigas*—that Last was very anxious to track down. Pink in colour with a deep, laterally-compressed body, a large multi-ridged head, and spine-bearing fins, this marine species earns its name from the rough spiny scales on its belly. It had not been seen by scientists since its original discovery back in 1914 by Dr. Harald Dannevig, during a fisheries survey in the Great Australian Bight, but Last hoped that it might still exist, overlooked but surviving.[82]

And sure enough, when Parsons's odd fish was compared with the latter species' description, they were found to correspond exactly. After more than 80 years, the giant sawbelly had finally resurfaced, caught at a depth of 580-1130 ft in a location close to where Dannevig's original specimen had been trawled in 1914. Scientists believe that it may have avoided previous trawling by living near or close to the sea bottom in this region.[82]

ADAPTIVE RADIATION IN ANTARCTICA

Adaptive radiation is the evolutionary diversification of descendants from an original single species to yield a wide range of morphologically different species that can collectively occupy a wide range of different ecological niches. Darwin's finches constitute the most famous example, and there are others known among birds, mammals, insects, and certain additional major animal groups. Ohio University anatomist Dr. Joseph Eastman, however, recently claimed a significant first in this evolutionary field—by revealing the first known example of adaptive radiation among marine fishes.[83]

The fishes in question are nototheniids, which are blenny-like Antarctic species characterised by a dorsal fin divided into two parts, and having the pelvic fins in the jugular position. In March 1999, Eastman announced that his studies aboard the *Nathaniel B. Palmer*, an ice-breaker of the National Science Foundation's polar research fleet, had revealed that all habitats in the relatively shallow waters of the Antarctic shelf are dominated by nototheniids, numbering 95 different species at that time, rather than by a diversity of unrelated species of fish. Moreover, at least four of these nototheniids were new to science, and included the first new species of the nototheniid genus *Artedidraco* to have been identified for 80 years.[83]

Geologically speaking, this example of adaptive radiation is of fairly recent origin. Fossil evidence from this region shows that in past ages, when Antarctic waters were warmer, its ecological niches were occupied by species belonging to many different taxonomic groups of fish. Today, conversely, these have all been replaced by nototheniid counterparts.[83]

'OLD FOURLEGS':
THE INCREDIBLE COELACANTH—
PART 3: AN INDONESIAN REVELATION

It isn't everyone who makes an astonishing zoological discovery while on their honeymoon, but that is precisely what happened to naturalist Arnaz Mehta Erdmann.[84-5]

On September 18 1997, Arnaz was honeymooning in Indonesia with her newly-married husband, biologist Dr. Mark V. Erdmann from the University of California, Berkeley, when she spied a large, extraordinary-looking fish being wheeled along in a cart across the fish market in Manado, on Sulawesi. What was so extraordinary about it was that it looked precisely like a coelacanth! Yet modern-day, living coelacanths were known only from marine waters off South Africa, Mozambique, Madagascar, and (principally) the Comoros.[84-5]

A closer look by the Erdmanns, however, swiftly dispelled any doubts. Zoogeographically astounding though it may be, this was indeed a coelacanth, only recently demised, and seemingly conspecific with the world's only known living species, *Latimeria chalumnae*—first brought to scientific attention in 1938 when a specimen was captured off South Africa.[84-5]

As they were due to fly back to California the next day, the Erdmanns would have been unable to bring the fish's body back with them, but the fish was sold anyway while they were standing observing it. Fortunately, however, they had been able to take a series of excellent, close-up photos—photos that readily confirmed its amazing identity.[84-5]

Inspired by their chance but zoologically-remarkable discovery, the Erdmanns later returned to Sulawesi, and, with the support of the National Geographic Society, spent time painstakingly interviewing local fishermen there to discover all that they could regarding the possible existence of more coelacanths in Indonesian waters. To their delight, they learnt that the Manado specimen was far from unique. On the contrary: although uncommon, the coelacanth as a species was well known to the Sulawesi fishermen, was caught by them using deep gill-nets, and even had its own local name—*raja laut* ('king of the sea'). Eager to obtain a specimen for formal study, the Erdmanns requested the men to contact them as soon as another coelacanth was captured.[84-5]

This notable event took place on 30 July 1998, when fisherman Om Lameh Sonathan and his crew hauled up a living coelacanth from waters off the northern Sulawesi islet of Manado Tua at a depth of 328-492 ft, using a shark gill-net. Measuring just over 4 ft long and weighing 64 lb, this specimen's most distinctive feature was its brown colouration. African coelacanths are brown when dead, but when alive they are steely blue. Moreover, although it shared the African specimens' characteristic white mottling pattern, Sonathan's Indonesian coelacanth was further patterned with numerous eye-catching flecks of gold dorsally—seemingly a prismatic effect caused by light reflecting off its scales' denticles.[84-5]

Sonathan's specimen was transferred to a large tank, where it lived for over three hours, and was filmed during that time by Dr. Erdmann before it died. Tissue samples were taken for molecular analysis, to determine its taxonomic status in relation to African coelacanths. Afterwards, its carcase was deep-frozen for preservation purposes, and presented to Indonesian scientists for safekeeping.[84-5]

In September 1998, Erdmann and two colleagues published a preliminary account of their significant discovery within the British science journal *Nature*. One of these colleagues, integrative biologist Professor Roy L. Caldwell, also from Berkeley, expressed an interest in taking a submarine journey with Erdmann to explore the sea at a site where they consider coelacanths may occur. In 1999, such a journey was accomplished by veteran coelacanth film-maker Prof. Hans Fricke, who succeeded in obtaining the first film of a living Indonesian coelacanth in its native habitat.[85-6]

The presence of a population of coelacanths in Indonesia, over 6200 miles from the Comoros, yields a remarkably discontinuous distribution for this fish—so remarkable, in fact, that it seems more probable that coelacanths have a much wider geographical distribution than originally realised. Perhaps, like the megamouth shark (another deepwater species initially known from just a few widely-dispersed specimens), further representatives will eventually turn up elsewhere, in the Indian Ocean and possibly even in the Pacific.[84-5]

This would not be wholly unexpected. As I noted in this book's first edition and also in my *In Search of Prehistoric Survivors* (1995), coelacanth-like scales have been reported from Florida and even Australia.[87]

But was the Indonesian population truly conspecific with the Comoros/African one, or did they actually constitute two distinct species? In spring 1999, French genetics expert Dr. Laurent Pouyaud and a team of Indonesian researchers drew the latter conclusion, and in a *Comptes Rendus* paper they formally christened the Indonesian coelacanth *Latimeria menadoensis*. They arrived at their decision to classify it as a separate species after conducting morphological comparisons between the specimen caught in 1998 and specimens of the Comoros species, *L. chalumnae*. These studies revealed differences between the two geographically isolated groups that in the opinion of Pouyaud and his team were significant. They also sequenced mitochondrial DNA from cytochrome b and 12S rDNA genes using the 1998 Indonesian specimen and some Comoro specimens. These again exposed differences between the two geographical groups that were deemed notable by the Pouyaud team.[88]

The findings of a subsequent, independent study, conducted by Texas University molecular biologist Dr. Mark T. Holder, Erdmann, Caldwell, and two other colleagues, were published in *Evolution* in October 1999. Based upon the degree of mitochondrial DNA divergence recorded between the two populations, these findings similarly indicated that the two populations were taxonomically discrete (albeit still closely-related) species.[89]

The Indonesian coelacanth's own discovery may have had a noteworthy precedent. In 1995, fish specialist Georges Serre, working for ORSTOM (Office de la Recherche Scientifique et Technique Outre-Mer) in southwestern Java, was fishing for spiny lobsters one night when he reputedly caught a 25-lb coelacanth. He took some photographs of his remarkable catch, before sending it to an oceanographic institute in Jakarta, but it was reputedly lost in transit. Moreover, Serre claimed that he was later robbed, and all of his photos were lost.[90] In 2000, however, one of these Java coelacanth photos was allegedly rediscovered, and was submitted by Serre, Pouyaud, and French ichthyologist Dr. Bernard Séret to *Nature*—whose staff recognised that the coelacanth depicted in it was remarkably similar in appearance to a Sulawesi specimen portrayed in an earlier, published photo by Erdmann, thus resulting in claims being made that the Java photo was a hoax.[91]

Notwithstanding all of this, the coelacanth is apparently familiar to fishermen in this region of Indonesia, separated from northern Sulawesi by approximately 1000 miles, who refer to it as *ikan fomar*. All of which emphasises that *Latimeria* may well occupy a much more expansive territory than science has hitherto suspected.[90]

The discovery of *L. menadoensis* is every bit as significant as that of *L. chalumnae* 59 years earlier. To quote Dr. Mark Erdmann, speaking in September 1998:[84]

> Even if the Sulawesi population is the only other area where the coelacanth is found, the fact that it could escape detection from the scientific community in an area well-studied by ichthyologists for over 100 years is wonderful. It is a humbling and exciting reminder that humans have by no means conquered the oceans, and provides fodder for our imagination about other, as-yet-undiscovered 'sea monsters' and oddities from the deep. And it underscores the importance of protecting our oceans, lest we lose things forever which we have not yet even discovered!

Let us hope now that the newly-revealed Indonesian domain of 'Old Fourlegs' will be rigorously protected, to ensure that this cryptozoological celebrity is not threatened by over-fishing, and that its reign here, as king of the sea, is a long and trouble-free one.

THE INVERTEBRATES

Cooloola Monsters and Wonders of the Vent World

Shortly after arriving at my new post as Curator of Orthoptera, CSIRO, Canberra in 1977 I was presented with a small parcel from Mr. E.C. Dahms, Curator of Insects, Queensland Museum, Brisbane with a note 'Here's something to introduce you to the Australian fauna'. After some amusement at the technical excellence of the apparently manufactured monster, it was determined that it was a genuine complete cricket-like insect.

Dr. David C.F. Rentz (describing the Cooloola monster's discovery)
—Memoirs of the Queensland Museum (1980).

Writers and adventurers have often speculated about a "lost world," where supposedly extinct or unknown animals thrive in their own isolated environment, unknown to humans and unaffected by human activities. Not only has such a "lost world" been found, but even more astonishing, it does not depend on the sun for energy, a situation believed almost unique in the Earth's natural history. The "lost world" in question is located at a depth of 8,500 feet in the Pacific Ocean, about 150 miles south of Baja California, where scientists aboard the U.S. Navy's three-person submarine Alvin have been studying it. The focus of interest is a number of volcanic vents with slowly seeping lava which support enormous worms, snails, crabs, clams, jellyfish, and other invertebrates, all of which are new to marine science.

J. Richard Greenwell—'Mini-"Lost World" at 8500 Feet' (*ISC Newsletter*, winter 1982).

Although they may not incite as much public interest as do the more showy and familiar mammals, birds, reptiles, amphibians, and fishes, in terms of scientific significance the invertebrates (animals without a backbone) include among their multifarious assemblage an appreciable number of the 20th century's most important new and rediscovered animals.

Indeed, as will be seen in this section, four such species have each been responsible for the creation of an entirely new animal phylum (the highest echelon within the hierarchy of taxonomic categories employed in animal classification). Others of distinguished scientific standing documented here include: a giant bee and a super flea; sea daisies and spaghetti worms; an ostensibly innocuous insect ultimately unmasked as a bona fide vampire moth, and a devastatingly dangerous sea creature capable of inflicting the most excruciating pain known to mankind; an assortment of marine curiosities collectively representing four totally new subclasses of crustacean, plus a forgotten little crustacean belonging to a group previously believed extinct for at least 50 million years; methane worms and mimic octopuses; social shrimps and carnivorous sponges; a monstrous grasshopper, and a monster-sized earwig; as well as an uncommonly long-tongued hawk moth whose existence was predicted by Charles Darwin several decades before it was actually discovered. Most spectacular of all: while humanity was seeking new life forms in Outer Space, an entire new world was awaiting discovery back on Planet Earth—a world teeming with extraordinary animals never before seen!

JEWELLED SQUID—LIVING RAINBOW FROM THE SEA

Despite its close affinity to the often less-than-handsome squids and octopuses, *Lycoteuthis diadema* (=*lorigera*) is unquestionably one of the world's most exquisite sea creatures, and is popularly called the jewelled squid—for very good reason. Its body is transparent, thereby enabling observers to discern clearly its internal organs—of which no fewer than 22 are special light-emitting (bioluminescent) structures known as photophores, whose combined visual effect is a surrealistic, psychedelic panorama of glistening rainbow colours. Two of these remarkable organs are present on each member of the squid's pair of extra-long prey-capturing tentacles, a further five along each eye's lower rim, and the remaining eight on its body's undersurface.[1]

Jewelled squid—equipped with photophores.

In 1899, the first two specimens of this marvellous species to attract scientific attention were examined while still alive by German marine biologist Dr. Carl Chun. He formally described their species in 1900; and in 1903, documenting its photophores, he wrote:[1]

> Among all the marvels of colouration which the animals of the deep sea exhibited to us nothing can be even distantly compared with the hues of these organs. One would think that the body was adorned with a diadem of brilliant gems. The middle organs of the eyes shone with ultramarine blue, the lateral ones with a pearly sheen. Those towards the front of the lower surface of the body gave out a ruby-red light, while those behind were snow-white or pearly, except the median one, which was sky-blue. It was indeed a glorious spectacle.

Inhabiting the Atlantic's open waters, the jewelled squid is relatively small, no more than 5 in long, but its unparalleled display of multicoloured, glittering bioluminescence certainly compensates for any deficiencies in dimensions.[1]

A HAWK MOTH FORETOLD

Charles Darwin and Alfred Russell Wallace are immortalised in the annals of science as the originators of the Theory of Evolution. Less well known is that they accurately predicted the existence of a very unusual moth—more than 40 years *before* it was actually discovered.

One of Madagascar's native orchids is *Angraecum sesquipedale*, whose nectar-producing organs (nectaries) are almost 12 in deep (much deeper than those of other orchids), but with only the lowermost 1.5 in containing nectar. In 1862, Darwin recognised that for this species to be insect-pollinated like other orchids, Madagascar must house a butterfly or moth with an extraordinarily long proboscis (tongue), measuring at least 10-11 in, in order to be able to reach the nectaries. (In so doing, the insect would inevitably brush against the orchid's pollen-producing organs, causing pollen grains to stick to its body and thus be passed to the next *A. sesquipedale* orchid that the insect visited, thereby effecting pollination.)[2]

Madagascan long-tongued hawk moth—Charles Darwin predicted its existence 41 years before its scientific discovery. (Marcel Lecoufle)

Indian stick insect.

Yet no such insect was known from Madagascar, hence Darwin's prediction was scorned by many scientists; but the existence of *Angraecum sesquipedale* was sufficient for him to remain convinced that one would certainly be discovered there eventually. In 1870, Wallace suggested that such a species might be closely related to Morgan's hawk moth *Xanthopan* (=*Macrosilia*) *morgani* of tropical Africa, which has an 8-in-long proboscis. And in 1903, their predictions were confirmed, by the documentation of a new, Madagascan representative of *X. morgani*—with an 11-in-long proboscis! Initially considered to be a separate subspecies, it was aptly christened *X. m. praedicta*, the predicted hawk moth, but is nowadays deemed to be taxonomically identical to its mainland African counterpart. Nevertheless, its inordinately lengthy proboscis remains an undeniable testament to Darwin's accuracy in predicting this moth's existence in Madagascar long before it was actually discovered there.[2-3]

In 1992, Ohio entomologist Dr. Gene Kritsky pointed out that another Madagascan orchid, *A. longicalcar*, could be pollinated only by a hawk moth with a *15-in-long* proboscis. Hence he has predicted that such a moth, as yet unknown to science, must still await discovery here.[4]

Speaking of extraordinary insects: the Indian stick insect *Carausius morosus* is today one of the world's most familiar captivity-maintained invertebrates—with millions in laboratories and schools as study specimens, and in countless homes as unusual children's pets. Remarkably, however, it remained undescribed by science until 1901. As this species reproduces readily via parthenogenesis (females producing eggs without having mated with males), especially in laboratory cultures, males rarely occur.[5]

ENORMOUS BUTTERFLIES AND GIANT SEA URCHINS

With a wingspan of up to 11 in recorded from some female specimens, Queen Alexandra's birdwing—a rainforest species mostly restricted nowadays to Papua New Guinea's Popondetta Plain in Oro Province—is undoubtedly the world's biggest butterfly. This makes all the more surprising, at least upon first consideration, the fact that it eluded scientific documentation until 1907. Like many other birdwings, however, it flies only at heights well beyond the reach of even the most ardent collector, generally remaining 50-100 ft above the ground. Consequently, the first specimen to come to scientific attention, in 1906, was not obtained by traditional means (i.e. via a capacious net), but instead with the aid of a somewhat less orthodox piece of butterfly-acquiring apparatus—a shot gun![6-7]

That specimen, a large female with handsome chocolate-brown wings edged with cream scalloping, procured by animal collector Albert Meek, became the type specimen of this previously undescribed species, which Meek's employer, Lord Walter Rothschild, officially documented a year later, naming it *Ornithoptera alexandrae* in honour of Queen Alexandra, consort to King Edward VII of the UK. Male specimens were later obtained, and these, although smaller in size, proved to be much more showy than the females, displaying delicate shades of shimmering green and blue upon a jet-black background. In both sexes, the head and thorax are black, the abdomen yellow.[6-7]

Another invertebrate record-breaker first described in 1907 was *Sperosoma giganteum*, the world's largest sea urchin. A Japanese species, found off Omai Saki Light in 1906 by the U.S. Fish Commission Steamer *Albatross*, it attains a diameter of just over 1 ft.[8]

PROTURANS—WORLD'S MOST PRIMITIVE INSECTS(?)

An entirely new insect order, Protura, was created in 1907 by Italian entomologist Dr. Filippo Silvestri, to house some

Queen Alexandra's birdwing, male —females of this species are the world's largest butterflies.

minute creatures recently discovered at Genoa that were totally unlike any previously known to science, and which were subsequently deemed by many researchers to be the world's most primitive insects. Under 0.2 in long, these eyeless, wingless soil-dwellers have slender bodies tapering to a pointed tip at the rear, and peculiar mouthparts that look like a collection of tiny needles. Indeed, proturans (also called myrientomatans or coneheads) have so strange a general appearance that they seem to constitute a major exception to several fundamental zoological rules hitherto used in defining just what makes an insect an insect.[9]

For example: whereas all other insects have 11 abdominal segments when adult, proturans have 12; and while all other insects hatch from the egg with all 11 of these segments already present, proturans hatch with only nine, adding the remainder by successive subdivision of the original ones when moulting. This process of segment multiplication *after* hatching is termed anamorphosis, and is unique to proturans among insect but prevalent among various other arthropods (jointed-limbed animals), notably millipedes.[9]

Juvenile proturans (Sinentomon sp.). (Prof. G. Imadate, courtesy of Prof. Yin Wen-Ying)

Equally important is that whereas all other insects bear a pair of antennae or feelers on their head, the proturans have none. Instead, they hold up their first pair of legs and use these as antennae. Indeed, this behaviour is deemed by some to be an indication of how true antennae evolved in higher insects—by specialisation of a front pair of legs. Moreover, although most insects bear legs only upon the segments of their thorax, the proturans have leg-like appendages upon their abdominal segments too. (This is also true of the springtails, but these latter creatures are in any case so peculiar that many entomologists nowadays remove them from the insect class altogether and treat them as a class all to themselves, Collembola—see also below.) Also of note is that the proturans have all manner of mysterious structures, projections, and pits of unknown function on their body surface (cuticle); these may serve as types of sense organ, but are not present on any other form of insect.[9]

As for their lack of wings, this is due to the fact that proturans have descended from an early line of insects that lived *before* wing evolution took place. Thus, unlike fleas and lice (whose present-day winglessness is due to their ancestors' gradual loss of wings during successive generations of adaptation for an efficient parasitic existence), proturans have no wings simply because their ancestors never had any either.[8] The first modern-day proturan to be formally recognised was *Acerentomon doderoi*, from Genoa. Since then, over 200 species have been described, from all over the world, but unfortunately their lifestyle and biology remain almost as obscure today as they were at the time of Silvestri in 1907.[9]

In recent times, some entomologists have removed the proturans from the insect class altogether and, in company with the afore-mentioned springtails and another enigmatic group, the diplurans, have placed them within their own separate taxonomic class, Entognatha.[9]

TITANUS GIGANTEUS—EXTRAORDINARY REAPPEARANCE OF AN INSECT GIANT

The world's longest species of beetle (if jaws and horns are not taken into account; if they are, then the title goes to the giant sawyer beetle *Macrodontia cervicornis*) and also one of the most legendary is the fittingly-named *Titanus giganteus*, the titan longhorn beetle—a shiny black colossus of a coleopteran that can attain a total length of 8 in (in the male). A member of the cerambycid or longhorn family, it is indisputably the premier prize of any beetle collector, but few can boast a specimen—on account of its very *un*accountable rarity. Even its scientific debut was shrouded in mystery.[10]

It was formally named in 1771, by Linnaeus, but there was no type specimen. Instead, Linnaeus based his description of the species upon a coloured plate (illustrating it alongside various Brazilian birds), contained in a work published over a decade earlier by French zoologist Louis J.M. Daubenton. During the next century, collectors scoured Amazonia for this very impressive insect, but mostly in vain. Those that *were* obtained consisted of dead specimens found floating in the Rio Negro, near to Manaus, Brazil. As *T. giganteus* is not an aquatic species, they must have fallen (or been blown) into the river during long, exhausting flights from southern Guyana's relatively unexplored mountainlands.[10]

By 1900, only a handful of specimens were on record, and the species was in such demand among the wealthier collectors that during this same year one specimen was sold at a London auction for what was then the very princely sum of £150. Ten years later, however, a strange event occurred that revealed a regular, if unexpected, means of procuring further specimens of this elusive insect. One evening in 1910, while setting up his tent near to Brazil's Rio Branco, German orchid collector Jacob Wörner watched one of the local Indians spear a big fish, open it up, and take something large out of its gut. The Indian was just about to throw the object into the river when Wörner intervened, taking it from him and examining it closely. It proved to be a complete specimen of *T. giganteus*! This remarkable incident came to the attention of a German entomologist called Dr. Bossmann, who thereafter paid the Indians to fish for more specimens in the river, which he then sold to European collectors. Consequently, the presence of *T. giganteus* in collections gradually increased.[10]

Moreover, in late 1957 a number of *living* specimens were finally discovered—captured during the rainy season (when they fly), after the *National Geographic Magazine*'s consultant zoologist, Dr. Paul Zahl, had placed 'reward' posters throughout a mining camp in Brazil's Territory of Amapa. Since then, others have made their presence known in a memorable way—attracted by the bright illumination, they have been spied crawling like diminutive tanks beneath various towns' street lamps![10]

Intriguingly, the larval form of *T. giganteus* is still undescribed by science. Judging from the size and conformation of bore-holes in wood thought to have been made by this elusive creature, however, it is probably as much as 12 in long and over 2 in wide—another veritable titan in every sense![10]

Titan longhorn. (Matthias Forst/Cologne Zoo)

SCLEROSPONGES—FORMERLY IN A CLASS OF THEIR OWN

The year 1911 saw the official description of *Ceratoporella nicholsoni*—a perplexing sea creature that has baffled zoologists for over a century. It first came to scientific attention in the 1870s, when it was dredged up off the coast of Cuba at a depth of just over 65 ft by the U.S. Coast Survey Steamer *Blake*, and in 1878 eminent biologist Prof. Louis Agassiz classed it as a living representative of a hitherto extinct genus of tabulate coral. Ten years later, however, he changed his mind, reassigning it to a completely different phylum—by labelling it as a bryozoan (moss animal). In 1911, its identity changed again, when S.J. Hickman reclassified it as a coral—but this time as a species of hard coral (octocoral).[11-12]

Thus it remained for several decades, until its correct identification was at last achieved—whereupon it proved to be neither coral nor bryozoan. Instead, it was a sponge, but of a type most comparable to the extinct stromatoporoids, and hence markedly distinct from all three previously-recorded classes of modern-day sponge. Its skeleton contained three different constituents—organic fibres, spicules of silicon, and crystalline aragonite (calcium carbonate).[12] In contrast, that of the calcareous sponges (class Calcarea) is composed solely of calcium carbonate spicules; the skeleton of the glass sponges (Hexactinellida) consists solely of siliceous spicules; and that of the demosponges (Demospongiae) comprises siliceous spicules and/or fibres of spongin (a collagen-like protein).

During the taxonomic transference of *Ceratoporella* from one group of animals to another over the years, some new species were being discovered that were more readily recognisable as sponges but which nonetheless shared its characteristic tripartite skeleton. In 1900, for example, *Astrosclera willeyana* was described, hailing from the Pacific islands of Funafuti and Lifu. Interestingly, scientists initially believed that its siliceous spicules were not an intrinsic part of its skeleton but had instead been incorporated artificially by some means.[13] In 1910, however, R. Kirkpatrick of the British Museum (Natural History) revealed that they were a genuine constituent, secreted by the sponge itself. Just a year earlier, Kirkpatrick had described a related species, *Merlia normani*, from Madeira's Porto Santo Island.[14]

In 1970, these various intriguing species received discrete taxonomic classification when marine biologists Drs. W.D. Hartman and T.F. Goreau grouped them all together within their very own class, Sclerospongiae, and to which they added five new species that they had discovered in the Jamaican fore-reef slope environment. One of these was a new *Merlia* species; the other four collectively yielded three new genera, consisting of *Stromatospongia vermicola*, *S. norae*, *Hispidopetra miniana*, and *Goreauiella auriculata*.[12] Additionally, in 1975 a totally new order of sclerosponges was created by Hartman and Goreau, to house another newly-found species, *Acanthochaetetes wellsi*.[15]

On account of their coral-like appearance (exemplified by *Ceratoporella*), the sclerosponges are often referred to as the coralline sponges. In 1985, however, sponge specialist Dr. Jean Vacelet revealed that Sclerospongiae was not a natural, monophyletic (single-origin) taxonomic class, because he had discovered that sclerosponge species occur in other classes of sponge too. In other words, the ability to build a sclerosponge-type skeleton had evolved independently in various totally-unrelated species of sponge, and at different times through the history of Earth too, so Sclerospongiae was duly abandoned thereafter as a valid taxonomic class.[15]

ZORAPTERANS—TO BE WINGED, OR NOT TO BE WINGED? THAT IS THE QUESTION!

Zorapterans (also known as angel wings) constitute an order of insects that were entirely unknown to science until 1913, when Dr. Filippo Silvestri (discoverer of the first proturan—p. 223) described the first species;[16] there are now more than 30 modern-day species on record, all belonging to the single genus *Zorotypus*. They are tiny creatures, none exceeding 0.1 in long, which live in colonies concealed under moss, leaves, or tree bark, or in soil, decaying wood, and sometimes even in termites' nests. Their most notable external characteristics include their distinctive nine-segmented antennae (each resembling a series of beads strung together), and the unusual construction of their thorax—its three segments are quite separate from one another (in other insects, at least two—if not all three—of these segments are intimately fused together).[17-18]

With regard to certain other features, zorapterans fall into two well-delineated types, both of which can occur within a single species. Winged (alate) individuals have two pairs of slender, sparsely-veined wings of which the front pair is largest, as well as compound eyes and simple light-sensitive cell clusters termed ocelli, plus a tough, shield-like structure covering the thorax. Wingless ('morph') individuals lack all of those features, and their unprotected thorax has only a thin, delicate cuticle. Some species do not produce winged individuals at all, whereas in certain others the latter lose their wings after attaining maturity, like termites.[17-18]

Zorapterans have a widespread distribution, occurring in the U.S.A., South America, the West Indies, the Hawaiian Islands, southeast Asia, and western Africa. They appear to be most closely related to an order of equally inconspicuous insects known as embiopterans or webspinners, although they also exhibit certain affinities with cockroaches and termites.[17-18] Again like termites, at least one zorapteran species, *Z. gurneyi* from Panama, has been found to display social behaviour, living in colonies of up to several hundred individuals.[18a]

GRYLLOBLATTIDS—INSECTS OF ICE AND FIRE

Scarcely before the ink had dried on Silvestri's description of the first zorapteran species, news emerged of another miniscule insect of great significance—differing so radically from all others that a second new taxonomic order was required. Inciting this latest entomological revolution was a tiny wingless form found under some stones in the Rocky Mountains of Alberta, Canada, by entomologist Dr. E.M. Walker, who described it in 1914.[19]

Morphologically, it seemed to constitute an anomalous amalgam of at least three completely different orders of insects. Its ovipositor (egg-laying tube) was made up

Grylloblattid. (Paul Pesson, courtesy of Harrap Publishing Group)

of six components, joined together in a manner that invited comparison with the ovipositors of certain crickets (of the genus *Gryllus*); whereas its five-segmented legs and pair of long segmented filaments (cerci) at the posterior tip of its abdomen were features shared by some cockroaches (genus *Blattus*). Yet its delicate body cuticle, blindness, and absence of wings superficially allied it with *Campodea*, a primitive insect known as a dipluran. In recognition of its composite construction, Walker named his minute discovery *Grylloblatta campodeiformis*.[19]

In truth, however, its resemblance to *Campodea* was due merely to convergent evolution (unrelated animals evolving a similar morphology due to their existence in the same type of habitat) rather than to any close taxonomic kinship. Equally, its correspondences with cockroaches were present simply because they were basic insect features retained by *all* primitive insect groups. Nor did its cricket-like ovipositor provide sufficient reason on its own to include the new insect within Orthoptera—the cricket and grasshopper order. And so a brand new order came into being—Grylloblattodea, which is considered by some entomologists to represent a primitive level of evolutionary development comparable to that attained by the ancestors of orthopterans.[17-18]

Since 1914, a few other species of grylloblattid (also known as ice crawlers or ice bugs) have been found—in the U.S.A., Japan, and the former U.S.S.R. Measuring 0.5-1.5 in long, with soft, pale brown or yellow bodies, they have unusually long lifespans (relative to those of most insects). Their entire development, from egg, through various larval stages, to adulthood and eventual death, can take up to eight years. And as a further claim to fame, the Canadian *G. campodeiformis* is renowned for thriving at temperatures near 0°C in the icy portions of its Rocky Mountain haunts.[17-18]

There is at least one species with the unhappy status of an endangered species. This is the Mount St. Helens ice crawler *G. chirurgica*, described in 1961.[20] Confined exclusively to a single lava flow on Washington's volcano, Mount St. Helens, its survival is threatened by its home's unpredictable tendency for violent eruption, and also by human interference.[6]

In recent years, some entomologists have united Grylloblattodea with another enigmatic insect order, Mantophasmatodea (p. 303), to yield the single order, Notoptera, within which Grylloblattodea and Mantophasmatodea have been duly demoted to the level of suborders.[6]

A VERY ODD OCTOPUS

The large, prominent eyes of cephalopods (octopuses, squids, cuttlefishes, nautiluses) are noted for their remarkably complex, highly-evolved structure. Consequently, *Cirrothauma murrayi*—a reddish-pink species discovered during the *Michael Sars* North Atlantic Deep-Sea Expedition of 1910, first described in 1914 by Dr. Carl Chun, and not recorded again until 1967—is little short of heretical, because this highly aberrant species of deepsea cirrate octopus is blind.[21]

Even though this condition is quite common among deepsea creatures, all of the squids existing in such habitats have well-formed eyes, so why should their octopod counterpart be sightless? In fact, it does possess eyes, but as they have neither retina nor lens, and are embedded beneath the surface of its gelatinous tissues, they are non-functional.[18,21]

The rest of its morphology is equally strange. Whereas the body of more familiar octopuses is globular and bulky, that of *C. murrayi* is gelatinous and fragile, like a jellyfish's. This is an adaptation for living at depths of 6650-10,000 ft, where the surrounding water pressure is immense (around 1 ton per square inch!). Similarly, whereas the tentacles of more familiar octopuses are separate from one another, those of *C. murrayi* (and other cirrate octopuses) are interconnected by a membranous web that reaches almost to their tips, giving the animal itself the appearance of a semi-transparent, eight-spoked umbrella.[18,21]

And whereas most cephalopods move by ejecting water in best jet-propulsion style through a tube termed the funnel, in *C. murrayi* this latter structure is greatly reduced. Its locomotory function has been superseded by the tentacles' interconnecting web—for *C. murrayi*

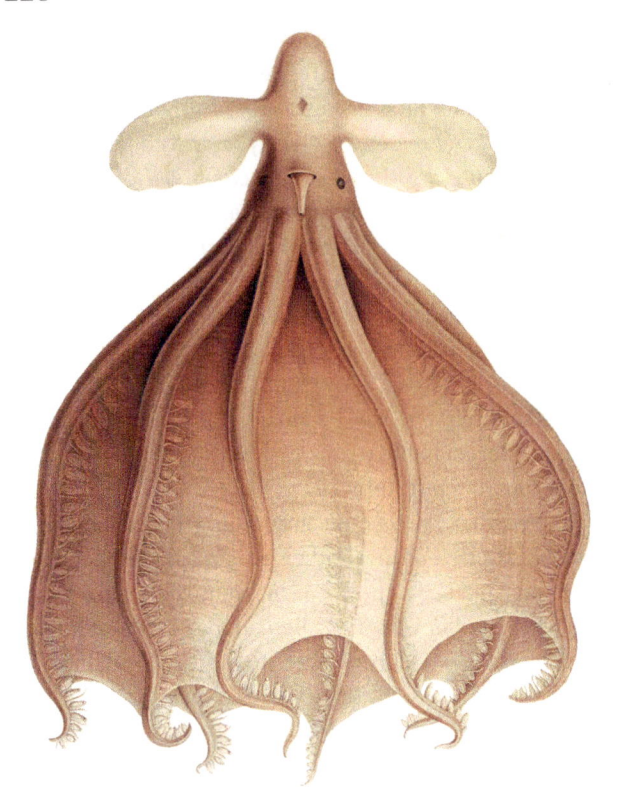

Cirrothauma murrayi—a blind cirrate octopus.

moves by slowly opening and closing this web, once more in faithful imitation of an animated parasol. Like other cirrate ('hairy') octopuses, in addition to suckers each of its tentacles has two rows of delicate hair-like filaments (cirri), which are probably used for trapping minute particles of food. Its only other noticeable external feature is a single pair of paddle-like fins, possibly of auxiliary assistance in locomotion.[18,21]

SUPER FLEA!

In 1921, publication of the formal scientific description of *Hystrichopsylla schefferi* introduced zoologists to the world's largest species of flea. When its original name, *H. mammoth*, was disallowed on a nomenclatural technicality, it was renamed *H. schefferi*, after its discoverer, agricultural researcher Theophilus Scheffer, from the U.S.A.'s Bureau of Biological Survey. He had collected the type specimen whilst in Washington State, finding it inside a nest belonging to the world's most primitive species of rodent—*Aplodontia rufa*, the sewellel, though also referred to popularly but very inappropriately as the mountain beaver (it is neither a mountain-dweller nor a beaver!). Other specimens have been collected since, some of which are more than 8 mm (0.31 in) long.[22]

Subsequently nicknamed 'Super Flea', *H. schefferi* appears to be a specific parasite of the sewellel; most specimens have been obtained from individuals of this rodent, or from their nests. A few have also been taken from the fur of mink and spotted skunks, but as these are carnivores that are known to prey upon sewellels it is likely that they received their over-sized parasites directly from their prey or, once again, from its nests. Little is known either about the natural history of 'Super Flea' or about that of its host, so the mystery of why the world's most primitive rodent should be exclusively parasitised by the world's largest flea has yet to be solved.[22]

OPHIOCANOPS—WORLD'S MOST PRIMITIVE BRITTLE STAR

Brittle stars look like starfishes with unusually long and slender, mobile arms that are sufficiently serpentine in appearance and movement for brittle stars to be known scientifically as ophiuroids—'snake tails'. Unlike starfishes, which rely upon the tiny suckered tube-feet underneath their arms for locomotion, brittle stars move by direct motion of the arms themselves. They also differ from starfishes via certain key anatomical traits.

Despite such differences as these, however, starfishes and brittle stars are quite closely related, and are believed to have diverged from a common ancestor many millions of years ago. An inkling of what this common ancestor may have been like was obtained in 1922, when the world's most primitive living species of brittle star was discovered. Obtained off Indonesia, and named *Ophiocanops fugiens*, it uniquely bridges the gap between

Sewellel or mountain beaver—exclusive host of Scheffer's 'Super Flea'.

brittle stars and starfishes, because although it is evidently a brittle star, it possesses certain features normally characterising starfishes. Most notable of these is the presence within each of its arms of extensions from the gut, the coelom (body cavity), and the reproductive organs (other brittle stars lack these extensions). Also, whereas the longitudinal (ambulacral) groove underneath each arm in other brittle stars is covered by calcareous plates called ossicles, in *Ophiocanops* each such groove is covered with nothing more than a very thin layer of skin—thereby corresponding more closely with the uncovered grooves of starfishes.[23]

Underlining the singular nature of its anatomy, *Ophiocanops* is set apart from all other brittle stars in its own taxonomic order—Oegophiurida.

THERMOSBAENACEANS—SOME LIKE IT HOT

In 1923, French zoologist Dr. L-G. Seurat collected some small but very peculiar crustaceans inhabiting the shallow hot spring of Hel Hamma, close to the ruins of an ancient Roman bath near Tunis in Tunisia. Little more than an eighth of an inch long, lacking any vestige of eye-stalks or eyes, and with a short body composed of a series of near-identical segments bearing unspecialised limbs, these tiny, cylindrical organisms were at first thought to be the larval stage of some currently unknown crustacean, but careful examination of the specimens finally convinced scientists that in reality they were adults, but nonetheless of a wholly new species.[24-5]

The following year, another French zoologist, Dr. Theodore Monod, described this species, naming it *Thermosbaena mirabilis*—emphasising its hot spring home and novel form. It seemed most closely related to the peracaridan crustaceans (woodlice, sandhoppers, and many shrimp-like species—all characterised by the possession of a brood pouch formed from special plates on the limbs of their thorax). Yet in conspicuous contrast to these, whose brood pouch is situated on their body's underside, in *Thermosbaena* it was located on its back. Together with a number of other unique features, its overall morphology was distinct enough to convince Monod that *Thermosbaena* required special taxonomic treatment, so he created an entirely new peracarid order especially for it, which he named Thermosbaenacea.[24]

Resulting from the interest aroused in zoological circles by the discovery of *Thermosbaena*, other bodies and sources of water were examined for the presence of any related species, and several were eventually found. Between 1949 and 1953, three close relatives were obtained—one from a freshwater Tuscan cave, the others from the slightly brackish waters within an Italian Adriatic cave and a Croatian cave at Dubrovnik. These species were all assigned to a new genus, *Monodella*, but shared the diagnostic dorsal pouch of *Thermosbaena*.

In 1976, two more new genera were described—*Limnosbaena* and *Halosbaena*, from Yugoslavia and the West Indies respectively. Over 30 species in seven genera are currently known (though *T. mirabilis* remains the only member of the genus *Thermosbaena*), including a single mainland North American species—*Tethysbaena texana* from Texas. However, whereas *Thermosbaena* is a hot spring dweller, capable of surviving in water temperatures of 36.5-47.5°C, the others are cool-water species, at temperatures of 12.5-15.5°C.[25]

COLOSSAL SQUID—DISCOVERING THE WORLD'S LARGEST INVERTEBRATE

The omens did not look good. On 1 April 2003, reports appeared in the media of an enormous, monstrous squid that had attempted to attack an Antarctic fishing vessel in the Ross Sea. Nor was this creature just big. According to the reports, it was armed with swivelling razor-sharp hooks on its tentacles, had eyes the size of dinner plates, made the formidable giant squid *Architeuthis dux* look positively tame in comparison, was itself referred to as the colossal squid, and had the jaw-breaking scientific name *Mesonychoteuthis hamiltoni*.[25a] Not too surprisingly, it was widely dismissed as just another fraudulent beast engendered by the imagination of some reporter hoping for the biggest April Fool's hoax of 2003. But then again....

Whereas most such reports are readily exposed by the suspiciously-odd names of hitherto unknown 'experts' and non-existent institutes, those cited in the colossal squid accounts were all genuine, including among them the name of, and quoted statements attributed directly to, Dr. Steve O'Shea, one of the world's leading experts on the giant squid. So out of curiosity, I looked up '*Mesonychoteuthis hamiltoni*', and was intrigued to learn that such a species did indeed exist. Having said that, it was barely known to science.

The only member of its genus, which was created in 1925 to accommodate it on account of its unique combination of arm, tentacle, and radula characteristics, the colossal squid (this name seems to be of much more recent origin) first came to scientific attention during the winter of 1924-25. That was when a Mr. E. Hamilton, while conducting investigations for the government of the Falkland Islands on the whales of the South Shetland region, obtained from the stomach of a sperm whale some fragments of two large squids, which he presented to the British Museum's Zoological Museum for formal examination and description.[25b]

Other incomplete remains were subsequently retrieved from other sperm whale stomachs, and eventually some juvenile specimens were procured in this way too, but until the 1 April 2003 specimen, a half-grown adult female, was reported, only one complete post-juvenile

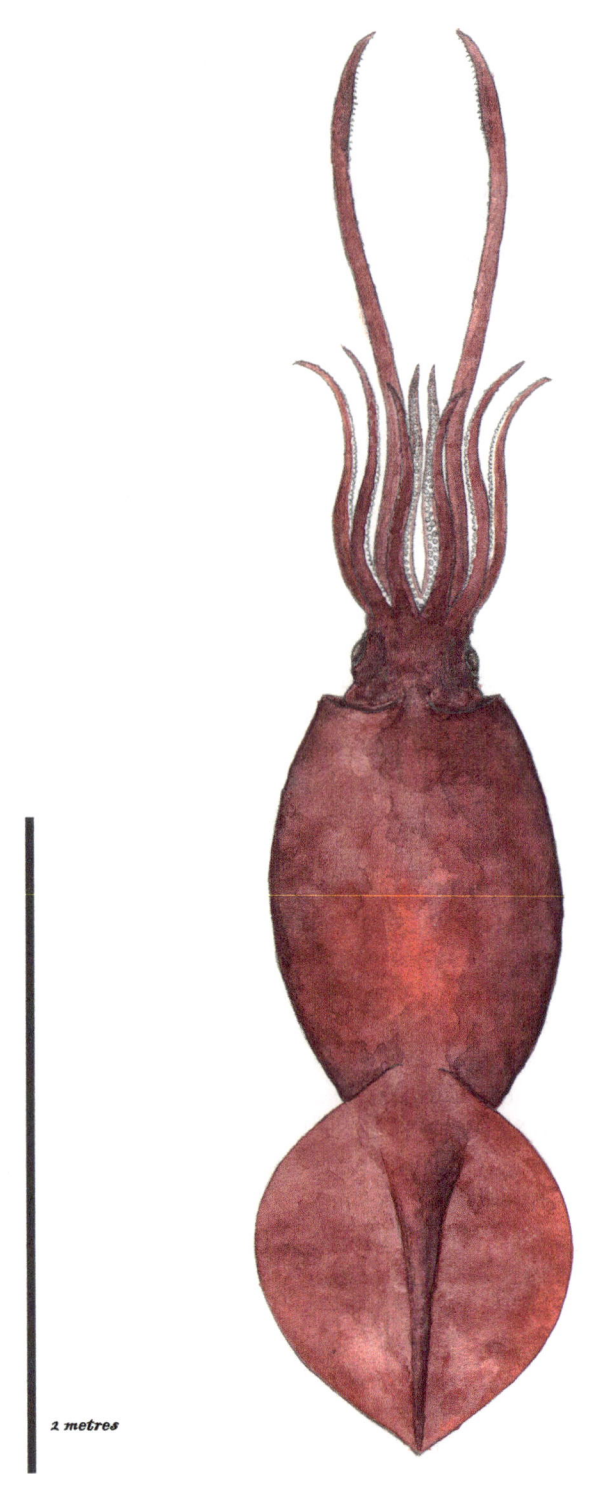

Colossal squid; scale bar = 2 m (6.56 ft). (Markus Bühler)

20 ft long, but could have grown to 40 ft, weighed 330 lb but might have attained a full-grown weight of 2000 lb, and was feeding on Patagonian toothfishes caught on longlines set by the fishermen. In addition, its species was said to be larger and meaner than the giant squid, a phenomenal predator, and highly dangerous to humans, with Dr. O'Shea quoted as saying: "It would kill you if you fell in".[25a]

Not everyone was happy with these media reports, however, and on the online cephalopod discussion list, CEPH-LIST, a number of discrepancies between media speculation and zoological reality were highlighted. To begin with, based upon behaviour exhibited by other members of the taxonomic family, Cranchiidae, to which the colossal squid belongs, it is more likely that this species ambushes its prey by hanging motionless in the water, rather than being forcefully aggressive. Coupled with this, its bizarre-sounding tentacle hooks are not as extraordinary as suggested by the tone of the media reports, because several other squid species, even including very small ones, also possess them, modified from tentacle suckers. So there should be no inference drawn that these are designed for attacking humans, but merely that they are modifications for capturing certain types of prey. In addition, giant squid author Richard Ellis personally discounted the possibility that the colossal squid is any more of a monster than the giant squid, or any more dangerous to humans.

Even the claim that the Ross Sea specimen was only half grown and might therefore have attained a much greater size had it lived is unsubstantiated. As noted in the cephalopod group's discussions, during their later stages of development squid mature extremely rapidly, with almost all of their energy being channelled into sperm and egg production, not into body size increase. So the Ross Sea individual could have become fully mature, mated, spawned, and died without gaining any additional size at all.

In late February 2007, however, the largest specimen of the colossal squid ever recorded was not only seen but also captured, when it was snared on a line by a New Zealand fishing boat's crew in the Antarctic's Ross Sea after it attempted to eat a hooked Patagonian toothfish. This spectacular specimen took two hours to land, after which it was frozen, and transported to the Museum of New Zealand's Te Papa Tongarewa, where it was thawed out in a bath of saltwater, and carefully dissected. It was initially estimated to measure 35-39 ft long (though it had shrunk to only 14 ft after thawing out), and it weighed 495 kg (1091 lb). This made it around 430 lb heavier than the next biggest colossal squid specimen known (an intact subadult female found floating on the surface of the Ross Sea in March 2003[25c]). It was therefore the heaviest specimen of any species of

specimen had been recorded—caught at a depth of 2500 ft in the Antarctic, off Dronning Maud Land, by the Soviet trawler *Evrica* (aka *Eureka*) in 1981 and photographed by Alexander Remeslo. It had a mantle length of 8 ft, and an estimated entire length (i.e. from end of tentacles to end of tail) of 16.7 ft.[25c]

According to the plethora of reports that swiftly followed the 1 April ones, the new specimen was almost squid ever captured (giant squids can be longer, but are far less bulky and hence much lighter, than colossal squids).[25d]

Following examination and selected dissection, this preserved colossal squid was placed on display at the museum in an exhibition that opened on 13 December 2008. With a collapsed eye diameter of 10.63 in, and an estimated non-collapsed diameter of 12-16 in, its eyes were the largest ever recorded from any animal. Yet its beak was considerably smaller than some colossal squid beaks that have been found inside sperm whale stomachs, thus suggesting that there may be colossal squid specimens out there that are much bigger even than this monstrously large individual![25d]

EXPOSURE OF A VAMPIRE MOTH!
In 1926, entomologist G.F. Hampson described a small species of noctuid moth with brown forewings and black-edged orange hindwings, related to Britain's familiar red, yellow, and orange underwing moths, but native to Malaysia and Sri Lanka. Now known as *Calyptra eustrigata*, it attracted little attention—until its extraordinary, sinister secret was uncovered in April 1967 by Swiss entomologist Dr. Hans Bänziger, while working in Kuala Lumpur's zoological gardens.[26-7]

There are a number of moth species that idly lap up animal fluids—tears, sweat, urine, blood seeping from an already-open wound—but following his observations of a *C. eustrigata* specimen discovered on the flank of one of the zoo's tapirs, Bänziger suspected that this particular species had taken that process one crucial step further, involving behaviour hitherto unknown among moths. There was only one way to obtain a conclusive answer—so he caught a specimen of *C. eustrigata*, made a small incision in the back of one of his own fingers, allowed the blood to ooze gently from the wound, and then offered his finger to the moth, to observe its reactions. Would it be content merely to lap up the blood flowing from the wound, or would it proceed beyond this?[27]

Losing no time in answering Bänziger's question, the moth totally ignored the blood. Instead, it immediately stabbed its unusually short, sturdy, non-tapering proboscis directly through the skin exposed by the wound, and began actively sucking up the blood from beneath the skin's surface! Bänziger's suspicions were confirmed. This was no innocent lapper of wound-leaking blood. On the contrary, preferring to make its own wound and purposefully suck the blood from it, *C. eustrigata* was, quite literally, a vampire moth![27]

This unique, skin-piercing, blood-sucking species probably descended from fruit-piercing forms (which have similar types of proboscis), subsequently transferring its attention from fruit juices to animal blood. Ironically, like so many notable zoological discoveries, the sanguinivorous secret of *C. eustrigata* might have been exposed much earlier if science had heeded the words of local people. Malaysia's Jakun natives, who inhabit part of this species' wide distribution range, have long known that certain moths are capable of 'biting'.[27-8]

PLANCTOSPHAERA PELAGICA—ITS ADULT FORM IS STILL UNKNOWN
In 1932, the *Michael Sars* Deep-Sea Expedition collected two specimens of a very strange organism from the plankton of the North Atlantic's Bay of Biscay, at a depth of 830 ft. Each resembled a totally transparent, gooseberry-sized sphere, containing a glassy jelly through which its almost U-shaped gut and other internal organs could be clearly perceived. On its external surface it bore a highly-branched tract of hair-like structures called cilia.[29]

Later in 1932, this peculiar organism was formally named *Planctosphaera pelagica*, and not long afterwards it was classed as the larval form of some member of the phylum Hemichordata, but sufficiently distinct from any species within it to warrant the creation of a new hemichordate class—Planctosphaeroidea. In time, however, detailed studies of the fine structure of *Planctosphaera* convinced researchers that it was most similar to the larval type (known as a tornaria) produced by the acorn worms or enteropneusts—the most well-known class of hemichordate. Even so, it was much larger than the tornaria of any *known* acorn worm, and its ciliary tract was much more complex in form. Researchers thus assumed that it was the larva of some still-undescribed acorn worm, probably an abyssal species.[29]

Since its initial discovery, only a handful of *Planctosphaera* larvae have been obtained. Moreover, until the 1970s it was still known only from the North Atlantic. But in May 1974, and again in September 1977 and 1982, four specimens (including the largest so far recorded) were trawled up from the North Pacific, at depths of 250-1660 ft, thereby expanding this enigmatic larva's known range by roughly 5000 miles.[29-30] Nevertheless, science is still no nearer to solving the fundamental mystery concerning *Planctosphaera*—what does its adult form look like? For no adult specimen of this species has ever been knowingly obtained, and its likely shape and size cannot even be predicted, because metamorphosis from the tornaria to the adult in acorn worms is a very drastic process—the spherical tornaria bears no resemblance to the slender, vermiform adult.[30]

One intriguing suggestion, by Hawaiian researchers Drs. M.G. Hadfield and Richard E. Young, is that *Planctosphaera* larvae do not belong to any distinct species at all; instead, they are merely abnormally-enlarged (hypertrophied) tornariae of some already-known species, their enlargement an outcome of failure to find suitable substrata for settlement (and thence metamorphosis into the adult), so that they continue to drift through the sea, growing larger and larger. According to Hadfield and Young, their specific morphology implies that they may be swollen tornariae of some species from a group of acorn worms termed ptychoderids.[30]

Ultimately, however, there would seem to be only one conclusive way of determining the identity of the *Planctosphaera* larvae, and that is to maintain them in aquaria, in order to observe their metamorphosis and thus reveal the precise nature of the resulting adults. Yet sadly, as these larvae are so very rarely encountered, it is likely to be a long time before this vital and quite fascinating experiment can be accomplished.

SEA SPIDERS—'ALL LEGS'!

Without a doubt, sea spiders are among the strangest of all animals known to man. Referred to scientifically as pycnogonids, on first glance they appear to be composed entirely of legs, hence they are also known as pantopods—'all legs'! Only on closer observation can they be seen to possess a small head and body too. Despite their common name, sea spiders are only distantly related to true spiders. Indeed, these bewildering, exclusively marine creatures, whose species range in size from well below an inch to a little over 2 ft across, do not seem to be very closely allied to any other group of animals.

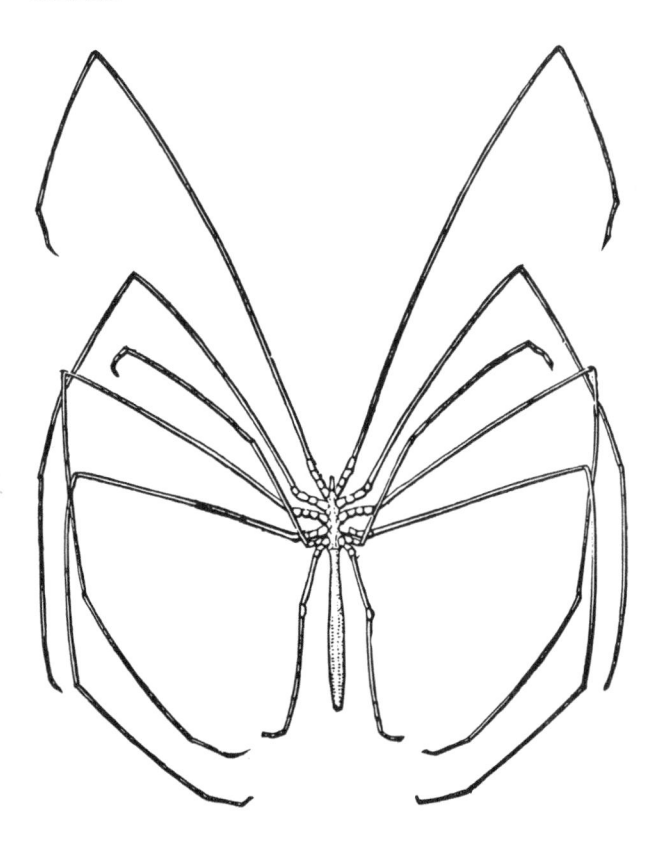

Twelve-legged sea spider.

Most species of sea spider have four pairs of legs, but a few have five. In 1933, a new precedent was set when Dr. W.T. Calman, Keeper of Zoology at the British Museum (Natural History), announced that during the recent British, Australian, and New Zealand Expedition, Sir Douglas Mawson had captured a sea spider in the Antarctic Ocean that possessed *six* pairs of legs. This visibly distinct species was duly dubbed *Dodecolopoda mawsoni* ('Mawson's twelve-legs'). At first, various authorities suspected it to be simply a freak, extra-limbed individual of some already-known species, but later finds of other *D. mawsoni* specimens (as well as specimens of some new 12-legged species) refuted this theory.[31]

RICINULEIDS—WORLD'S MOST ELUSIVE ORDER OF ANIMALS?

Moving from one group of strange spider-like creatures to another, ricinuleids (also called podogonids or hooded tickspiders) are notoriously mysterious arachnids, inhabiting damp caves and lurking beneath leaf mould in various regions of West Africa and tropical South America. Superficially they resemble tiny eyeless spiders, and were originally classed as harvestmen (opilionids), but in 1904, closer examination revealed them to be sufficiently different from all arachnids to require a taxonomic order of their own, and today it is generally considered that their closest relatives are the mites.[32]

Notable among their many curious features is a peculiar hood-like structure (cucullus) capable of being lowered over their mouthparts, plus an abdomen divided dorsally into 12 separate regions. Equally odd is their extraordinarily thick body cuticle, which is amazingly resistant to even the sharpest of scalpels and other cutting tools—as used by researchers on dead specimens when attempting to investigate these strange animals' internal design.[32]

Originally known only from a late Carboniferous fossil specimen discovered in 1837 and wrongly described by English geologist William Buckland as a beetle, the first recorded example of a modern-day ricinuleid—a round-bodied individual measuring all of 8 mm in total length—was collected in 1838 near the coast of Guinea by Swedish naturalist C. Westermann, and was named *Cryptostemma* [now *Ricinoides*] *westermanni* after him by Félix Édouard Guérin-Méneville. Although it was subsequently lost, other specimens, belonging to several additional species, were later obtained—but only infrequently, and in very small numbers. By 1900, only 19 specimens were known (ultimately shown to represent eight different species), and zoologists despaired of ever obtaining enough examples to discover anything substantial about these most obscure arachnids.[32] And then along came Ivan T. Sanderson.

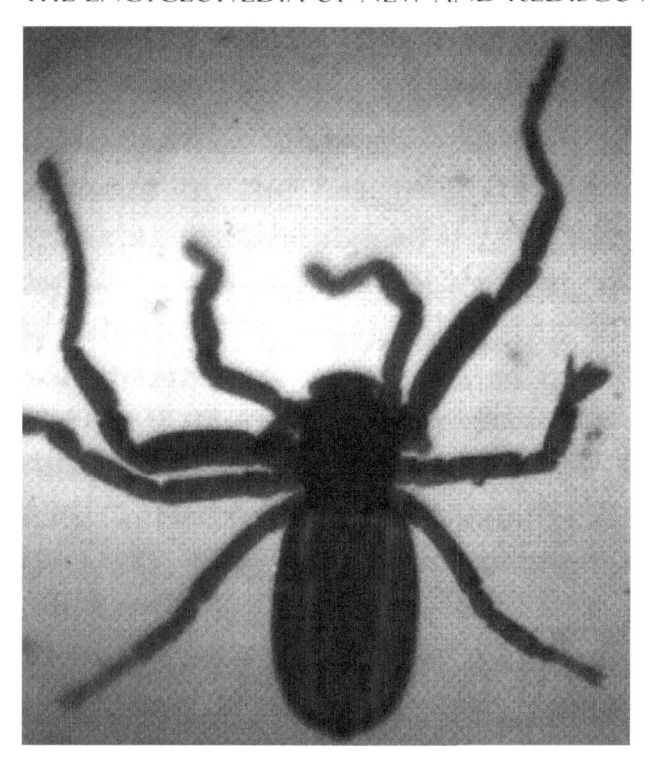

Ricinuleid. (Prof. John L. Cloudsley-Thompson)

Meant only as a light-hearted joke (being only too aware of their famous rarity), in 1933 zoologists from the British Museum (Natural History) playfully beseeched Sanderson—a noted zoologist and leader of the forthcoming Percy Sladen Trust Expedition to Nigeria and Cameroon—to be sure to come back with plenty of ricinuleids. To their amazement, they learnt upon his return that Sanderson had taken them at their word—for he had succeeded in bringing back from Cameroon no less than 317 ricinuleids! The number of ricinuleids on record had instantaneously soared upwards to yield a very respectable three-figure total.

Sanderson's collection all belonged to the species *C.* [now *R.*] *sjostedti*, and about 20 of them actually lived for a year in an incubator at the British Museum, giving scientists a unique opportunity to observe their activity—though as they refused to eat anything, their feeding behaviour and preferences remained a mystery.[32-3]

Although further specimens and species have since been uncovered, and a little more is known of their biology, it would be difficult to match the outstanding record for ricinuleid detection set by the resourceful Mr. Sanderson.

NOTHOMYRMECIA—AN
ARCHAIC ANT FROM AUSTRALIA

Australia is home to *Nothomyrmecia macrops*—the world's most primitive species of ant, on account of which it is commonly referred to as the dinosaur ant. Officially described in 1934, from two worker specimens housed in the zoological collections of Victoria's National Museum, and first collected in 1931 on a track near Western Australia's Mount Ragged, *N. macrops* is distinguished by its noticeably large eyes, unusually short wings (which in queens are incapable of sustaining flight), the location of its stridulatory organ (produces chirping sounds) underneath its abdomen rather than on top of it, and by the close morphological similarity of its queens to its ordinary workers.[34]

After the 1930s, *N. macrops* was not reported again for many years; but in 1977 it was formally rediscovered, near Poochera on South Australia's Eyre Peninsula—when on 22 October a CSIRO field party collected some workers and dealate queens (i.e. queens whose wings had been shed, following fertilisation) while camping there overnight. It is currently categorised as critically endangered by the IUCN.[34]

THE MYSTERIOUS MYSTACOCARIDS—
CRYPTIC CRUSTACEANS FROM CAPE COD

Measuring in total length no more than 0.02 in, the 60 or so specimens of an unusual but tiny, colourless, cylindrical crustacean collected from intertidal sand at Cape Cod, Massachusetts, by ecology researchers Drs. Robert W. Pennak and Donald J. Zinn in 1942 could have been easily overlooked. Thankfully, as Pennak and Zin specialised in studying the microfauna of these habitats, those odd creatures attracted their fair share of attention—which was just as well. At first, they seemed to be nothing more than copepods—a well-known subclass of crustaceans with the ubiquitous freshwater *Cyclops* and the planktonic *Calanus* among its most familiar members. Closer examination, conversely, exposed significant differences.[25,35]

In contrast to previously known copepods, for example, the tiny Cape Cod crustacean had a distinct carapace ('shell'), its thoracic limbs were much simpler in structure, both pairs of antennae were relatively long and well-developed (in copepods, the second pair is extremely short), the front portion of its head was divided from the remainder by a distinct transverse groove, and its genital opening was located on the first thoracic body segment (most crustaceans bear it close to the body's rear end). As a result, this highly aberrant species, named *Derocheilocaris typicus* by Pennak and Zinn in 1943,[35] became the sole member of a new crustacean subclass, Mystacocarida, the first of four erected during the 20th century.[25]

Since then, however, other species have been found—from as far afield as New England, South Africa, the French, Italian, and Spanish Rivieras, and the coasts of South America; but only a dozen are currently known, housed in just two genera. Inhabiting the meiobenthos, these interstitial crustaceans feed on minute particles of detritus; and on account of being so very small themselves, they do not possess a blood-circulatory system—oxygen simply diffuses directly into their body across its external surface.[25]

POGONOPHORANS—BIZARRE BEARD-BEARERS THAT ONCE CONSTITUTED AN ENTIRELY NEW ANIMAL PHYLUM!

Scientific sensations come in all shapes and sizes. In this particular instance, it took the form of a long, bedraggled worm-like creature—represented by specimens dredged up from the sea bottom's sludge off the coast of Indonesia in 1900 by the Dutch research vessel *Siboga*. More closely resembling slender lengths of mangled string than any type of animal, these strange entities seemed totally unclassifiable, and so after being bottled they were duly ignored for the next decade.[36]

In 1914, however, their pickled remains attracted the attention of French zoologist Dr. Maurice Caullery, who became so amazed and fascinated by them that he spent the next 30 years painstakingly studying every minute aspect of their morphology and anatomical structure, in a determined bid to extract the abtruse secret of these enigmatic animals' identity. He named their representative species *Siboglinum weberi* (after the *Siboga* expedition and its leader, M. Weber), and could see that it was a singularly novel creature, seemingly not allied to any other major (or minor) animal group. Very tentatively, he placed it within the phylum Hemichordata—housing the pterobranchs and the acorn worms (see p. 262).[37]

Meanwhile, in 1932 Soviet scientist P.V. Uschakow had discovered an odd-looking creature in the Sea of Okhotsk. The following year, he named it *Lamellisabella zachsi*, and classed it as an aberrant species of sabellid fanworm, belonging to the polychaete class of segmented (annelid) worms.[38]

This classification of *Lamellisabella*, however, did not satisfy Swedish researcher Dr. K.E. Johansson. His studies, undertaken during the late 1930s, assured him that it was not a polychaete at all, and that it deserved a class completely to itself, which he named Pogonophora ('beard bearer')[39]—but in which phylum should it be placed? It did not seem to have any affinity either with Annelida or with any of the others!

In 1944, zoologist Dr. V.N. Beklemishev proposed that there was only one satisfactory solution for the problematical *Lamellisabella*—to elevate Pogonophora from a class to a phylum.[40] As the scientific world agreed with this new status for Pogonophora, it became the first new animal phylum to be created and accepted during the 20th century.

So far, no-one had suggested that *Lamellisabella* might be related to the equally odd *Siboglinum*—both were perplexing, but that was all. By careful comparison, however, a third Soviet scientist, Dr. A.V. Ivanov, demonstrated beyond any doubt that the two were indeed closely allied to one another. Accordingly, in 1951 *Siboglinum* was formally removed from Hemichordata and reassigned to Beklemishev's new phylum, containing *Lamellisabella*.[36,41]

Since then, many new species have been discovered and described. More than 200 are currently recognised, and have been collected from a variety of different geographical locations in the Pacific, Indian, Atlantic, and Antarctic Oceans—and not just from marine, deepwater localities. Some specimens, for example, have been obtained from the Norwegian fjords in relatively shallow water.[42] Most ironic of all, it is now known that innumerable specimens had been scooped up from the Pacific long *before* their scientific significance had been recognised, but as they had simply looked like fibrous rubbish, they had been unceremoniously tipped back into the sea again![43]

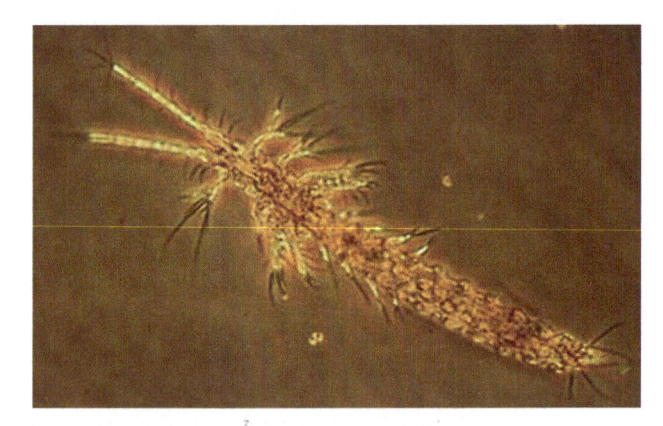

Ctenocheilocaris sp., a mystacocarid.
(Dr. M.R. Lee)

Moreover, Dr. Eve Southward's re-examination of the *Siboglinum weberi* specimens studied by Caullery revealed in 1961 that this 'single species' actually consisted of at least 16 different species! Caullery's failure to detect this had been due to the specimens' poor state of preservation, and the lack of other pogonophorans for comparison at that time.

And so, these once-discarded denizens of the deep (and fjords) are now unequivocally recognised to be one of the 20th century's most remarkable zoological discoveries. Having said that, in more recent times the common zoological consensus has been to reclassify the pogonophorans as merely a taxonomic family, Siboglinidae, of highly-specialised polychaete worms within the phylum Annelida, and thus abandon Pogonophora as a separate phylum. Taxonomic considerations aside, however, what makes a pogonophoran a pogonophoran, and why are they still so very special? Numbering approximately 100 different species that range in length from 2-12 in (depending upon the species concerned) and inhabit somewhat longer tubes, they exhibit a unique combination of morphological features.

When removed from its tube, the body of a pogonophoran is seen to consist of three sections. The short, front section or forepart houses the brain, and a special region called the cephalic lobe bears a crown of long tentacles (the 'beard'), which may number up to 250 (and even exceed 2000 in one rather special species, discussed later). Behind the forepart is the long, slender, worm-like trunk. This is the principal body section, yet is so extraordinarily slender that its width can measure less than 1/500th of the animal's length! Nevertheless, within this fragile, gossamer-like body beats a well-formed heart supplied by a complex system of blood vessels. It also houses paired reproductive organs (the sexes are separate in pogonophorans), and a sturdy nerve cord running down its middle from the brain into the third, rearmost body section. This latter section is a very short segmented portion called the opisthosoma, which is so delicate that it is readily detached and lost during the collection of specimens. Indeed, its very existence was unknown until 1970, because it had always broken off from specimens obtained prior to then.[18,36]

So far, a pogonophoran would seem to have (in some form or another) most of the major body constituents present in animals of all types—but there is one fundamental omission. Pogonophorans lack any form of digestive system—they have neither a mouth nor a gut! Instead, it is believed that they feed by absorbing nutrient micro-particles from their all-embracing seabottom ooze directly through their body surface—analogous to the way in which the endoparasitic tapeworms obtain their food (absorbing it in their instance from the surrounding supply ingested by their unfortunate host). As for breathing, it seems likely that pogonophorans respire by exchanging gases across the surface of their tentacles.[18,36]

All in all, a remarkable combination of characteristics that seemed unlikely ever to be duplicated—an assumption which, as it happened, proved to be somewhat premature. In 1966, some very odd worm-like creatures were collected at a depth of 3750 ft off the coast of California, near San Diego. Their species was described three years later by Dr. Michael Webb, from South Africa's Stellenbosch University, who named it *Lamellibrachia barhami*.[44] A little over 3 ft long but less than 0.5 in thick, and with more than 2000 tentacles, it proved to be a pogonophoran, but one so essentially different from all others that Webb placed it in a brand new pogonophoran class—Afrenulata (now housed within the pogonophoran family of polychaetes).[44]

Notable among the distinguishing features of *L. barhami* was the vestimentum—a special body section present between its crown of tentacles and its trunk. *L. barhami* thus became known as a vestimentiferan. In 1970, a second species of vestimentiferan was found, at a depth of 1665 ft from Guyana's continental slope, and christened *L. luymesi*. Moreover, 6-ft-long *Lamellibrachia* specimens found at a depth of 1800 ft in the Gulf of Mexico have since been estimated to live for up to 250 years—longer than any other species of non-colonial marine invertebrate.[44]

A pogonophoran, Oligobrachia ivanovi. (Drs. Alan and Eve Southward)

Yet despite their distinction agr... pogonophorans, the two *Lamellibrachia* species pale to insignificance when compared with the vestimentiferans found in 1977—because these latter proved so exceptional that they are recognised today as being among the most bizarre, visually spectacular creatures ever witnessed by mankind, as revealed on p. 250.

VAMPIRE SQUID—UNMASKING A NIGHTMARE FROM THE DEEP

In 1946, the published exposé of a certain grotesque marine creature's true identity sent shock waves through zoological circles worldwide.

First collected during the *Valdivia* deepsea expedition of 1903, and introduced that same year to the scientific world by noted marine biologist Dr. Carl Chun,[45] it was immediately a source of wonder and puzzlement, for whereas the most famous attribute of any octopus species hitherto described had been its possession of *eight* arms, this new, superficially octopus-like creature had *ten*, like squids. The rest of its external morphology was equally strange. Its eyes were distinctly sinister in appearance—huge, deep-red orbs that glittered eerily. Even more macabre, however, was the purplish-black web present between its eight longest arms, connecting each to the next like a dark, membranous bat wing. Hardly surprisingly, the resulting combination of glittering crimson eyes, dark nightmarish web, and squid-like complement of ten arms inspired Chun when officially describing this

Vampire squid—portrayed here in an illustration from 1911, it resembles an octopus, is referred to as a squid, but is neither.

dramatically new creature to name it *Vampyroteuthis infernalis*—'hellish vampire squid'.[45]

Yet despite its ten arms and corresponding squid appellation, in overall external form *Vampyroteuthis* seemed similar enough to octopuses to be classed with them. As for its arms: eight were long, and on the underside of each of these was a central row of suckers, running down its length of its distal section, and bordered on each side by a row of pointed, protruding structures called papillae or cirri. The remaining pair of arms was much shorter, each of its two members resembling a coiled, reduced version of the longer arms, and capable of being withdrawn into a pouch. Unlike the eight main arms, these two smaller retractile ones, known as velar filaments, were not linked to the membranous web, and appeared to act as tactile (touch) organs or feelers.[18,45-6]

The vampire squid's body was fat and bulbous, as with typical octopuses, but its ink sac was very degenerate, unlike the prominent structure sported by octopuses. Its body bore a short pair of paddle-like dorsal fins, resembling those of squids, and certain internal features also suggested a squid affinity. Clearly, its classification as an aberrant octopus was unsatisfactory, but it remained as such for the next four decades.[18,45-6]

Yet for much of that time it was not a solitary species, because between its discovery in 1903 and its identity's denouement in the mid-1940s no less than eleven fresh species of vampire squid (assigned to eight different genera) were brought to scientific attention, and were distinguished from one another by reference to certain external idiosyncrasies.[18,46]

The most sensational surprise that *Vampyroteuthis infernalis* provided zoologists with, however, was still to come. During the early 1940s, it had been meticulously studied by American biologist Dr. Grace Pickford—who revealed two remarkable findings when she published her researches in 1946. Firstly, the other species of vampire squid were based upon nothing more than poorly-preserved specimens of *V. infernalis*; the chemicals used in preserving them had distorted their appearance to give the impression that they were quite different from the single original species. Secondly, and most startling of all, the vampire squid was neither an octopus nor a squid! Instead, it was a last survivor of an ancient group of cephalopods related to (but separate from) both of those taxonomic orders yet hitherto known only from various fossil species; in particular, its horny but uncalcified endoskeletal shield (gladius) resembled those of long-extinct species called loligosepiids.[46]

As a result, the vampire squid required the creation of a modern-day taxonomic order all to itself, and once again this species' weird appearance inspired the choice of name—the newly-established order was duly christened Vampyromorpha ('vampire-shaped'), later changing to Vampyromorphida. A totally new order of molluscs had been added to the catalogue of present-day animals.[46]

Although the last member of an ancient lineage, as a species the vampire squid is by no means uncommon. Since its discovery, well over a hundred specimens have been obtained, the larger ones almost 1 ft long, with a worldwide distribution in subtropical and tropical oceans, inhabiting the cold, lightless, abyssal zone present at depths of about 5000-8350 ft. In a world without sunlight, its huge eyes might seem somewhat incongruous and superfluous, but they probably serve to detect the many types of bioluminescent life forms that occur at these depths, and upon which it presumably preys. Indeed, the vampire squid itself has a very impressive array of external bioluminescent organs (photophores) almost entirely covering its body's surface, their lights flickering like a veritable firmament of coloured stars amongst the dark folds of its arms' interconnecting wing-like veil.[18]

NEOPILINA—ENDING 350 MILLION YEARS OF EXTINCTION!

One of the most spectacular zoological comebacks of all time occurred in 1952, restoring to life an archaic lineage previously believed to have vanished over 350 million years ago.

On 6 May, trawling off Mexico's western coast at a depth of almost 12,000 ft in dark, muddy clay, the Danish research ship *Galathea* hauled up ten complete specimens and three empty shells of a small, seemingly unremarkable mollusc superficially resembling a limpet.[47]

Closer observation, conversely, disclosed that they were not limpets. On the contrary, they constituted a hitherto undescribed species fundamentally distinct from all other living molluscs known to science. Yet they were not wholly unfamiliar. The zoological world was astounded to learn that they were unquestionably akin to a class of molluscs previously known only from fossils—whose most recent representatives had died out over 350 million years ago, during the Carboniferous Period, when the dominant species were ammonites and giant fishes. For these newly-discovered limpet-resembling animals were monoplacophorans, and as such were now the most primitive modern-day molluscs known to man.[47]

Of especial note was the close similarity in outward appearance between this 20th-century monoplacophoran species and its primeval fossil ancestors, suggesting that monoplacophorans had occupied much the same ecological niche throughout their lineage's many millions of years of existence. Indeed, because of this species' resemblance to one particular long-extinct genus, *Pilina*, when officially described in February 1957 by Danish zoologist Dr. Henning Lemche it was dubbed *Neopilina galatheae* ('similar to *Pilina*, obtained by *Galathea*').[47]

Spoon-shaped and no more than 1.5 in long, all of the *Neopilina* shells, inhabited and uninhabited, were very fragile, semi-transparent, and extremely thin—with a shimmering mother-of-pearl inner surface, a yellowish-white outer shading, and a slightly slanted peak. The mollusc itself, still present within ten of the shells, had a circular fleshy foot, similar to but smaller than that of a limpet. Yet unlike a limpet, or any other living species of mollusc known at that time, its body was divided internally into segments—a condition reminiscent of that in earthworms, ragworms, and leeches (collectively termed segmented worms or annelids). Zoologists considered this to be extremely exciting, for the following reason.[3,47]

The fertilised eggs of annelids and those of molluscs undergo the same basic pattern of division (known as spiral cleavage), and ultimately transform into almost identical larvae (of a type called a trochophore). These similarities have been interpreted by many zoologists as indicators of a close evolutionary relationship between these two animal phyla. As a consequence, the news that *Neopilina* was a primitive mollusc with annelid-like segmentation fuelled widespread speculation that monoplacophorans constituted a 'missing link' between other molluscs and the annelids, and that molluscs had descended from annelids.[3,47]

Neopilina galatheae—its discovery in 1952 sensationally resurrected the monoplacophoran molluscs from over 350 million years of presumed extinction. (Geert Brovad)

In later years, however, this theory lost favour to a quite different one, which proposes that molluscs and annelids arose *independently* of one another, from a common ancestor. According to this, therefore, the presence of segmentation in annelids and in monoplacophorans is also of independent origin, not evolving until *after* the separate emergence of these two phyla from their shared ancestor.[3,47]

If the latter theory is correct, this reduces the significance of the segmentation exhibited by *Neopilina*, but its identity as a living monoplacophoran was still more than enough to make it one of the foremost zoological discoveries of the 20th century.[3,47-8]

It was not long before *Neopilina galatheae* was joined by a number of other present-day monoplacophorans. In December 1958, four specimens of a second species, christened *N. ewingi*, were dredged up from the Peru-Chile Trench off northern Peru by the research vessel *Vema* (owned by the Lamont Geological Observatory, of New York's Columbia University). This was followed in 1960 by the discovery of *N. valeronis*, hauled up from the Cedros Trench of Baja California, again by *Vema*. Seven years later, Peru's Trench offered up *N. bacescui* and *N. bruuni*[18,48]. All of these were deepwater species, brought up from 3000-6500 m (10,000-21,660 ft), but in 1976 this trend was broken by Dr. J.H. McLean, who obtained a new, tiny species from depths of as little as 174 m (580 ft) on the Cortes Ridge, off southern California. Similar finds were made in 1977 by Prof. Heinz A. Lowenstam of the California Institute of Technology.[49] These finds show that *N. galatheae* is far from being the lone survivor of an otherwise long-vanished line that it was initially assumed to be. Moreover, recent molecular studies have suggested that of all monoplacophorans, *Neopilina* is actually closest taxonomically to a separate but closely-related molluscan class, Polyplacophora, housing the multi-shelled chitons.

HUTCHINSONIELLA—AN ODDITY FROM THE OOZE

In 1955, a wholly new crustacean subclass—dubbed Cephalocarida—was created by Dr. Howard L. Sanders in order to accommodate some diminutive but quite extraordinary animals discovered in the mud and ooze at the bottom of New York's Long Island Sound. Nowadays popularly known as horseshoe shrimps, they were tiny, colourless, shrimp-like creatures no more than 4 mm (1.5 in) long; with a horseshoe-shaped head bearing two pairs of short antennae but lacking eyes; an elongate body composed of 19 segments (but not assembled into a thorax and an abdomen) of which the first nine each bore a pair of leaf-like limbs; and a terminal anal segment that bore two pairs of tail-like appendages (one pair of which was extremely long). The undescribed species to which these curious specimens belonged was christened *Hutchinsoniella macracantha*, and caused a sensation in the zoological world—all because of the shape of its limbs.[50]

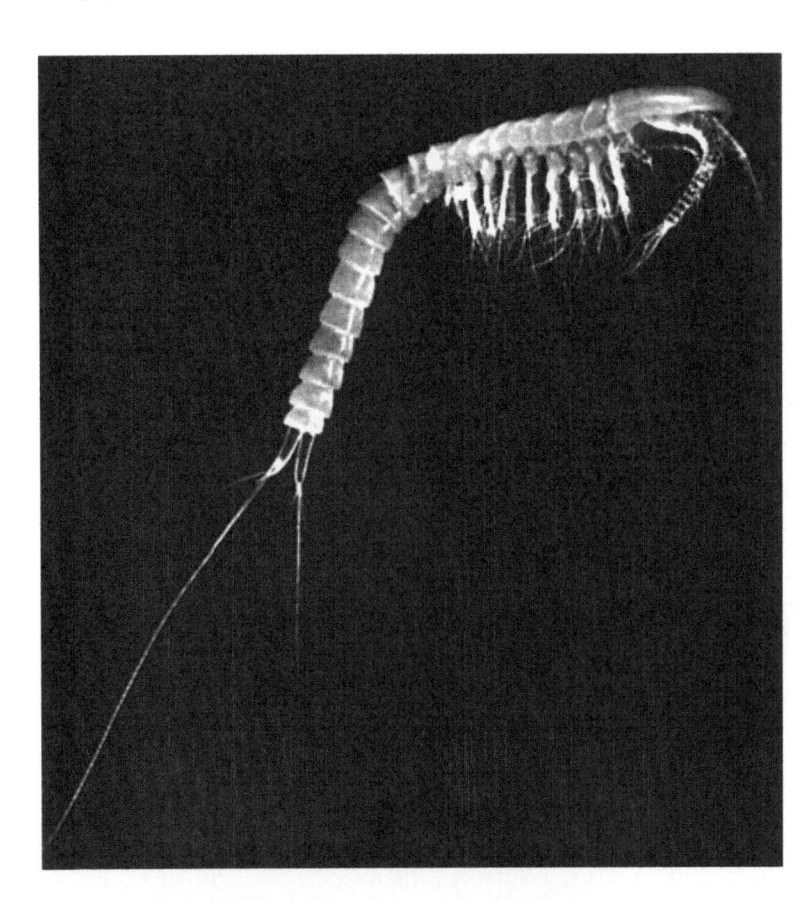

Hutchinsoniella macracantha—possibly the world's most primitive crustacean, it was first documented in 1955. (Smithsonian Institution)

With every previously known species of present-day crustacean, the body's limbs consisted of two or more morphological types, each type having become specialised to fulfil a different function. With *Hutchinsoniella*, however, all nine pairs were virtually identical to one another, a condition comparable in the opinion of many researchers to that of the earliest ancestral crustaceans. And that was not all—for whereas the base of the limb in previously-known modern-day crustaceans often gave rise to two terminal appendages (such limbs are termed biramous), in *Hutchinsoniella* each limb's base gave rise to three such appendages (i.e. triramous), comparable to the limb structure of the ancient trilobites. All of this implies that *Hutchinsoniella* may be the most primitive crustacean alive today.[25,50]

A second cephalocarid species, lacking a pair of limbs on its eighth body segment, and obtained from San Francisco Bay in 1957, was formally documented in 1961. Named *Lightiella serendipita*, it was the first of a series of species belonging to this genus that turned up during the early 1960s. *Sandersiella acuminata* was described in 1965, its generic name commemorating the cephalocarids' discoverer, and at least two more *Sandersiella* species have been uncovered since then. In 1977, a new cephalocarid from New Zealand was *Chiltoniella elongata*; and in 2000, a new Brazilian species, *Hampsonellus brasiliensis*, was described, yielding a fifth genus. A total of twelve species in five genera are currently recognised, of which *Lightiella* is sometimes separated from the others within a family of its own.[25, 50]

Little is known of cephalocarid biology, but they appear to feed upon the tiny organic-rich particulate matter or detritus present within their habitat's thick mud. Unlike most other non-parasitic crustaceans, cephalocarids are hermaphroditic (developing both male and female sexual systems), rather than having separate sexes.[25]

SPELAEOGRIPHACEANS—CAVE-DWELLING CRUSTACEANS FROM CAPE TOWN

Sanders's description of the new crustacean subclass, Cephalocarida, had no sooner appeared in print before another major new crustacean taxon was created, this time to house a small, blind, transparent species from South Africa. In 1956, a series of specimens representing this hitherto unknown crustacean was obtained by members of the South African Spelaeological Association from a pool at a depth of 110 ft, located inside a cavern known as Bat Cave, on Cape Town's Table Mountain. The specimens ultimately reached crustacean specialist Dr. Isabella Gordon of the British Museum (Natural History), who recognised that the new species clearly belonged to the subclass Malacostraca (containing the familiar lobsters, crabs, shrimps, and prawns), but was sufficiently dissimilar from all of its other members to require a new taxonomic order.[25,51] In 1957, Gordon named this species *Spelaeogriphus lepidops* ('scale-eyed cave riddle'); its order was dubbed Spelaeogriphacea.[51]

Measuring no more than a few millimetres in length, the subterranean *Spelaeogriphus* has a long cylindrical body, superficially shrimp-like in appearance, bearing more than ten pairs of limbs that comprise several different morphological types, and an eyeless head equipped with two pairs of antennae (the posterior pair of which is well developed). Instead of eyes, it has only a pair of scale-like structures called ocular lobes, which lack visual components or pigments, so that the animal is undoubtedly blind.[25,51]

Curiously, this cryptic little crustacean combines features from several separate malacostracan groups, including the woodlouse-allied tanaidaceans, the mountain shrimps or anaspidaceans, and the thermosbaenaceans (see p. 229). Three more species (requiring two new genera) were later discovered. Two of these species, *Mangkurtu mityula* and *M. kutjarra*, are from Australia; and the third, *Potiicoara brasiliensis*, is from Brazil. In 1974, moreover, the fossil crustacean *Acadiocaris novascotica*, housed until then within the order Syncarida, was recognised by crustacean expert Dr. W. Frederick Schram to be a spelaeogriphacean.[25,51]

FLECKER'S SEA-WASP—BEWARE THE TENTACLES OF DEATH!

Except for its rounded apex, the 3-4-in-high bell of the Australian sea-wasp jellyfish *Chiropsalmus quadrigatus* is virtually cuboidal in shape. This is a characteristic of most of the species sharing its taxonomic group (nowadays widely treated as a separate class, distinct from the true jellyfishes or scyphozoans), so they are collectively termed cubomedusans or box jellies. For many years, there have been cases reported from northern Australian waters involving swimmers and bathers stung very severely (sometimes fatally) by jellyfishes that leave weals in their victims' flesh and inflict such intense agony that death has sometimes ensued within minutes. Investigations suggested that a cubomedusan species was the most likely culprit, but as these are almost transparent it proved very difficult to obtain a conclusive identification. However, one such species—*C. quadrigatus*—was definitely known to frequent these waters. So in the absence of precise data, science assumed that this was probably the creature responsible, until a case occurred that exposed the *real* offender.[52-3]

On 20 January 1955, at 9.35 a.m., a five-year-old boy was swimming in the sea at Cardwell, north Queensland, when he was stung by a jellyfish and died shortly afterwards. The manner of stinging was of the type hitherto assumed to be due to *C. quadrigatus*, but on this occasion there was actually some physical evidence to consider too. Tentacles from his attacker were found in the boy's hair; moreover, his mother had observed the jellyfish responsible. The following day, zoologist Dr. Hugo Flecker instigated a series of nettings in the expanse of sea where the boy had been attacked, in the hope that its assailant would still be there. The nettings resulted in the capture of several box jellies, which were closely examined. Their tentacles appeared to be the same as those found in the boy's hair, and his mother stated that the jellies themselves were of the same type as her son's attacker. All of this was obviously very important, but most important of all was that the species to which these jellies belonged was *not* the familiar *C. quadrigatus*. Instead, they proved to be of a species wholly unknown to science until then, superficially like *C. quadrigatus* but differing from it significantly with regard to certain specific anatomical details.[52-3]

In 1956, Dr. R.V. Southcott, a jellyfish researcher who had examined the specimens after receiving them from Dr. Flecker, christened this newly-exposed species *Chironex fleckeri*—known popularly as Flecker's sea-wasp, and the sole member (for many years thereafter) of a new genus.[52]

With a bell up to 7 in tall at maturity, but otherwise resembling *C. quadrigatus* externally, the virtually transparent *Chironex* has several unique anatomical features, including its lobe-shaped gonads (reproductive organs); those of *C. quadrigatus* and other cubomedusans are leaf-shaped. At each lower corner of its cuboidal bell is a bunch of tentacles, faintly tinged with mauve; each tentacle can measure up to 1 ft long *prior* to any active extension induced by the box jelly itself, and is encircled by rings of devastatingly effective stinging cells called nematocysts. Analyses of these cells and the poison that they release upon contact, not to mention the harrowing cases on record involving the stinging of humans by this species, confirm that Flecker's sea-wasp is unquestionably the world's most dangerous jellyfish.[52-4]

Its venom is cardiotoxic, acting upon the victim's heart, and is so potent that according to one Australian scientist, "only a bullet kills faster than a sea wasp". Worst of all, however, is the degree of pain that it inflicts. Wielding an armoury of almost *40 million* nematocysts, *Chironex* is capable of eliciting the most horrific, excruciating pain known to mankind—so terrible that even before its victim has died (in cases of extensive stinging), he/she has been driven to the point of raving insanity by the indescribable agony of the stings.[52-4]

Fortunately, a toxoid for active immunisation is now available; but there is also a simple (if unusual) way of avoiding the stings of *Chironex*. Despite their formidable effect, its nematocysts do not have great powers of

Flecker's sea wasp. (Tim Morris)

penetration, so swimmers can actually withstand brushing against its tentacles—if they are wearing an all-embracing, loosely-fitting swimsuit made from ladies' panty hose! This odd costume has thus become the unlikely 'uniform' of some lifeguards working in offshore waters frequented by *Chironex*, where, needless to say, it is acknowledged that an unusual swimsuit is infinitely preferable to an agonising death.[54]

In 2009, a second *Chironex* species was formally described. Native to waters around Japan, where it is known locally as the habu-kurage, it was originally thought to be conspecific with *C. fleckeri*, but when its separate taxonomic identity was recognised it was christened *C. yamaguchii*.[54a]

GNATHOSTOMULIDS—AN OVERLOOKED PHYLUM OF ANIMALS

Zoologists with a specialised knowledge of the flatworm phylum, Platyhelminthes (housing such notorious species as tapeworms, liver flukes, and many other parasites), had been aware since the 1920s of a small group of ostensibly simple creatures known as gnathostomulids ('jaw mouths'). Rather than parasitising other animals, however, these tiny, worm-like creatures lead innocuous, inconspicuous lives buried in the thick black ooze at the bottom of the sea. Indeed, as they seem perfectly able to spend their entire lives completely submerged within this slimy sludge, it is likely that they respire anaerobically (without oxygen), an impressive feat that many bacteria but few animals can accomplish.[55]

Even so, with transparent bodies no more than 1 mm long, the gnathostomulids remained undescribed by science until as recently as 1956, when Göttingen zoologist Dr. Peter Ax published the findings of his detailed studies concerning them—with far-reaching results.[56]

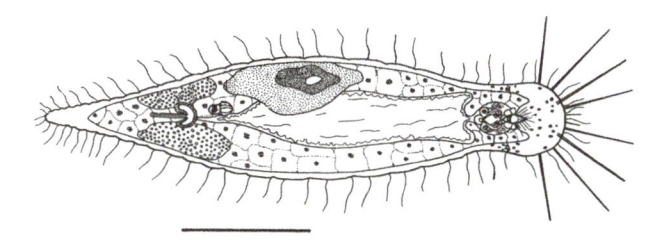

A gnathostomulid, Valvognathia pogonostoma. (Dr. Reinhardt Kristensen)

His investigations drew comparisons between the gnathostomulids and the turbellarian flatworms—the only platyhelminth class of predominantly non-parasitic species—and likened them to certain turbellarians lacking an internal body cavity (i.e. acoelomate), as the bodies of gnathostomulids are totally solid throughout too. Overall, however, they seemed sufficiently discrete to warrant separate categorisation, and so in 1960 he formally assigned them to a new, specially-created taxonomic class within the flatworm phylum.[57]

Less than a decade later, moreover, they acquired even greater taxonomic significance. The last time that an entirely new phylum of animals had been established was in 1944, to house the pogonophorans (see p. 234); this had raised the total number of modern-day animal phyla to 33. Resulting from his examinations of gnathostomulids, however, in 1969 Dr. R. Riedl announced that they constituted a valid, independent phylum too. This was because he had discovered that these onetime overlooked and understudied creatures embodied an exceptional combination of characteristics drawn from two totally separate levels of morphological complexity within the animal kingdom.[58]

On the one hand, their acoelomate body, lack of an anus, and hermaphroditic reproductive system drew comparisons with various turbellarians—animals fairly low on the scale of structural, evolutionary development. On the other hand, they also possessed sensory bristle-like cilia, a well-delineated head and jaws in most species, and a range of specific cellular details that were characteristic of some much more advanced animal groups,

such as 'wheel'-bearing creatures called rotifers, and related organisms called gastrotrichs. Consequently, Riedl's recommendation was accepted by science, and a thirty-fourth phylum of modern-day animals, Gnathostomulida, was duly added to the zoological register.[55,58]

GLORY OF THE SEA

Throughout the history of shell collecting, the most sought-after species has always been the glory of the sea *Conus gloriamaris*, one of the many species of cone shell, but totally without peers as the most coveted prize of all—conchology's *crème de la crème*.

The reason for its near-legendary status is twofold. Firstly, it is an exquisite species. Its slender shell can attain a total length of up to 16 in, which includes an extremely graceful spire, composed of several almost straight whorls and elevated much higher than the spire of most other cone species. Equally elegant is its shell's patterning, for its entire surface, of pale background colour, bears a fine filigree of tiny triangular reticulations, overlain by four or five darker spiral bands of varying degrees of visibility. Once seen, the glory of the sea cone is quite definitely never forgotten. Secondly, it has always been famed for its great rarity, with fewer than a dozen specimens known prior to the 20th century, owned only by various museums and wealthy private collectors, and each one worth thousands of pounds.[59-60]

Glory of the sea cone

Indeed, 1896 saw the discovery of what seemed destined to gain immortality as the very last specimen of the glory of the sea cone ever collected, because in spite of numerous searches and enquiries within the conchological fraternity during the years that followed, no additional find was made—until 1957, when a specimen was discovered off Corregidor Island in the Philippines. The glory of the sea had finally resurfaced, quelling widespread fears that it had become extinct. Yet nothing could have been further from the truth, for since then quite a number have been obtained, from the waters east of New Guinea, with its habitat being confirmed in 1969. So although it is no longer a rare species (on the contrary, specimens can often be purchased nowadays from online shell dealers and retailers for less than £100), it will forever be associated with the aura of conchological romance and desirability.[59-60]

In 1959, another of conchology's celebrated rarities reappeared, when several specimens of the impressive bull conch *Strombus taurus* were collected by R.C. Willis in the Marshall Islands. This was the first time that the species had been reported since its description back in 1875.[60]

STYGIOMEDUSA—EXTRAORDINARY ENTITY FROM THE STYGIAN DEPTHS

On 18 October 1959, the research vessel *Sarsia* was cruising off Santander in northern Spain's Bay of Biscay, collecting deepsea animals on behalf of the Development Commission's Fisheries Research Biomedical Unit, when their Isaac Kidd pelagic net brought to the surface a spectacular new denizen of the deep. Dredged up from a depth of at least 10,000 ft, it was a large jellyfish, deep brownish-red and plum in colour, with a bell (umbrella) of firm composition and a diameter measuring roughly 20 in.[61]

Following its capture, it was closely examined by Dr. F.S. Russell of Plymouth's Marine Biological Association, who recognised that it was a member of the semaeostome order of jellyfishes, but appeared to be sufficiently different from all known semaeostomes to warrant the creation of a new genus. Semaeostomes are named after the four lobe-like structures termed oral arms that extend down from their mouth (manubrium). In the case of the *Sarsia* specimen, these arms were enormously elongate, superficially resembling tentacles and attaining a length of around 5.5 ft. In contrast, whereas most semaeostomes also have a fringe of long, true tentacles encircling their bell's rim, the *Sarsia* individual had none.[61]

Most interesting of all: whereas sexual reproduction in most semaeostomes involves the shedding of eggs and sperm into the sea for external fertilisation to occur, the new jellyfish was viviparous, i.e. its young develop internally and are born alive (rather than encased inside eggs). This was revealed by the presence of fully-developed young jellyfishes inside each of the adult's four specialised brood chambers that protrude into its stomach cavity.[61]

Later in 1959, Dr. Russell prepared a short paper in which he named this brood-caring, live-bearing surprise *Stygiomedusa fabulosa*, likening its lightless, deepsea home to the inky blackness of Greek mythology's fabled River Styx, and emphasising the extraordinary nature of the jellyfish itself. This was followed in 1960 by a detailed morphological description, prepared jointly by Russell and Dr. W.J. Rees of the British Museum (Natural History), where the *Sarsia* example now resides as the type specimen of *S. fabulosa*.[61]

On 3 May 1962, a specimen of what appeared to be a new species of *Stygiomedusa* was caught at a depth of 3160-3260 ft in the Gulf of Guinea, about 100 miles northwest of the Congo Estuary. The most important feature that appeared to differentiate it from *S. fabulosa* was the layout of its radial canals. These canals are actually fine tubes that radiate outwards to the edge of the jellyfish's bell from its centrally-located stomach (gastric cavity). In *S. fabulosa*, these canals did not combine with one another over their entire length, and were readily discerned. By contrast, those of the Guinea *Stygiomedusa* specimen were totally obscured. Thus in 1967 French marine biologist Dr. R. Repelin designated the latter as the type specimen of a new species—*S. stauchi*.[62]

Also in 1962, a second undeniable specimen of *S. fabulosa* was recorded, this time from Greenland, but sadly it was not preserved. Nine years later, however, on 24 June 1971, the research vessel *Sarsia* obtained another *S. fabulosa*; and, like the first, this one *was* preserved. It had been caught in a region of the Bay of Biscay near to where the species' type specimen had been captured, but this time at a depth of only 4000-4660 ft.[63]

On 1 July 1972, yet another *Stygiomedusa* was procured. It had been captured by the research vessel *Chain*, about 745 miles north of the Azores in the north Atlantic, at a depth of 2500 ft, and proved to be of special significance. With a bell diameter of 56 in, it was the largest *Stygiomedusa* on record; but of much greater importance, it was undeniably intermediate in appearance between specimens of *S. fabulosa* and the single, type specimen of *S. stauchi*. Consequently, zoologists now consider *S. fabulosa* to be more variable morphologically than previously suspected, discounting *S. stauchi* as a separate species and reclassifying its type as merely a distinctive individual of *S. fabulosa*.[64]

More recently, an interesting revelation occurred, showing that far from being a new species discovered in 1958, *S. fabulosa* had first been described in 1910 (by Edward T. Browne), when it was formally christened *Diplulmaris gigantea*, and was known from material collected in Antarctic waters from as far back as 1899. However, it is still deemed sufficiently distinct to warrant its own genus, and so today this giant deepsea jellyfish's official scientific name is *Stygiomedusa gigantea*. In 2008, a huge specimen, with oral arms of around 20 ft long, was filmed in the Gulf of Mexico by a remote-operated vehicle during the Serpent project—an international collaboration between marine scientists and energy companies. A short video of this specimen can be viewed on YouTube.[64]

WHITE-TOOTHED COWRY—REDISCOVERING THE WORLD'S MOST VALUABLE SEASHELL

In the opinion of many shell collectors, the most valuable shell is that of the white-toothed cowry, and in view of its remarkable history this distinction is well-deserved. The type specimen of this species, of unknown provenance, was purchased in May 1828 from G.B. Sowerby by naturalist W.P. Broderip, who formally described it later that year. He named it *Cypraea leucodon*—after the white colouration of the tooth-like projections fringing its shell's aperture, and by which it could be readily distinguished from other cowry species. In 1837, with no additional specimen of *C. leucodon* reported in the meantime, Broderip's entire shell collection was purchased by the British Museum's trustees. In so doing, the museum gained a shell that was destined to become one of the most famous in conchological history, because in spite of every effort made by shell collectors throughout the world to locate more *C. leucodon* specimens, not a single one was obtained.[60,65]

Over a century passed by in this way, until some experts began to wonder if *C. leucodon* really was a genuine species, or, alternatively, just a freakish form of some more familiar cowry. But in 1960, conchologists were startled to learn that a second specimen had been in existence for many years, in the collections of the Boston Society of Natural History, but with its true identity hitherto unrecognised. This find supported the status of *C. leucodon* as a valid species, but as the Boston specimen's origin was unknown, its species' provenance remained a mystery—until 1965. In that year, shell expert S. Peter Dance examined some photos of two shells that had been taken from the stomach of a fish caught in the Philippines' Sulu Sea—and realised that one of the shells was a perfect *C. leucodon*! After 137 years, the riddle of the white-toothed cowry had finally been solved. Moreover, by the year 2000 it was being found on a sufficiently regular, abundant basis even for perfect ('gem') specimens to be available for under £1500 from shell dealers.[60]

This history recalls that of another rare cowry. In 1903, Fulton's cowry *Barycypraea fultoni* was described from a specimen taken out of the stomach of a fish caught off South Africa's Natal coast. Remarkably, many specimens obtained since have also been found inside fishes, though it is now well-recorded from natural habitats in the Indian Ocean along the coasts of South and also East Africa.[60,66]

PLATASTERIAS LATIRADIATA—
RISE AND FALL OF A LIVING FOSSIL
The most familiar members of the phylum Echinodermata are the starfishes, which have traditionally been divided into two subclasses—Euasteroidea (the true starfishes), housing all of the modern-day species; and Somasteroidea, known only from fossil species dating back to the early Ordovician Period, some 488-478 million years ago. (More recently, Somasteroidea has been elevated from a sub-class to a class in its own right.) During the 1960s, however, Somasteroidea experienced a most unexpected (albeit only temporary) renaissance.

The seeds for this had been sown some years earlier, when Prof. H. Barraclough Fell, from New Zealand's Victoria University, had become very interested in the obscure species *Platasterias latiradiata* from Mexico's Pacific coast, described in 1871 but still virtually unknown. After studying the meagre amount of published data concerning its structural anatomy, he began to wonder whether it really was a true starfish after all, because it seemed to exhibit certain features more comparable to those of somasteroids than euasteroids—including petal-shaped arms whose skeletal components resembled rod-like structures called virgalia (possessed by somasteroids but not by euasteroids).[67]

Yet to satisfy himself totally, Fell needed to examine a specimen of this enigmatic echinoderm—which immediately posed a problem, as earlier enquiries to museums in Mexico had failed to elicit one. Undaunted, he contacted Alisa M. Clark, Curator of Echinoderms at the British Museum (Natural History), and promptly received a portion of an arm from a preserved *Platasterias*. From his study of this vital material, he felt sure that the arm's ventral skeleton was indeed constructed from virgalia-like rods, and other anatomical features likewise seemed to substantiate a somasteroid identity for it.[67]

Thus, in December 1961 Fell announced that *Platasterias* was a living somasteroid, thereby resurrecting an entire subclass of echinoderms from 400 million years of extinction. In 1962, he published details of his structural analysis of *Platasterias* and his conclusions in several scientific journals, which attracted appreciable zoological interest.[67]

By 1966, however, doubts regarding the somasteroid affinities of *Platasterias* had begun to be voiced. In particular, Dr. F. Jensenius Madsen, of Copenhagen's Zoological Museum, opined that the virgalia of somasteroids were not so significant as previously thought. The skeleton of euasteroids is composed of numerous bone-like knobs called ossicles, and Madsen considered virgalia to be nothing more than ventrolaterally-sited versions of these, thereby reducing their taxonomic value. Leading on from this, he postulated that *Platasterias* was merely a somewhat aberrant member of the starfish genus *Luidia*, and that its distinctive petal-shaped arms were simply an adaptation for life on an unsteady sandy seafloor. His opinion swiftly gained support from other researchers (notably Dr. D.B. Blake) investigating ossicle structure in starfishes; so by the early 1970s *Platasterias* was not only reclassified as a euasteroid, but was also renamed, becoming *Luidia latiradiata*—thereby jettisoning the subclass Somasteroidea back into Ordovician obscurity.[67]

EXPOSING THE INVISIBLE IRUKANDJI

Generations of bathers and swimmers in the shallow seawater off Australia's northern coasts have been very severely stung during this country's summer months (December to February) by a mystery creature that in many ways seemed more akin to a phantom than to any form of corporeal animal. Known as the irukandji, after the Irukandji aboriginal tribe that formerly lived on the coast north of Cairns, Queensland (in whose shallow waters irukandji incidents most commonly occur), it was never seen, and eluded capture even in fishing nets of the finest mesh. Yet it left its victims in no doubt whatsoever of its physical reality.[53,68]

The first indication of an irukandji sting is a sharp prickling sensation in the precise area of the victim's flesh that has come into direct contact with the creature. This is followed a few minutes later by the appearance of 'goose-flesh' and then by a much more extensive region of reddening. Deceptively, this vanishes fairly swiftly afterwards, leaving the skin looking normal once more—but after 20 minutes or so, excruciating backache begins, followed by overall weakness, sudden (often intense) abdominal pains and contractions, headache, sweating, and sometimes laboured breathing, culminating in violent spasms of coughing and vomiting that can last for some hours if the victim is left untreated. Prior to 2002 (see below) there had been no documented fatalities, but as can be appreciated from the above description the effects of an irukandji sting are nonetheless quite devastating, and require intravenous injection of pethidine for effective control. The characteristic set of symptoms engendered by such a sting has been well known in medical circles for at least 70 years, and since 1952 it has been officially referred to as the irukandji syndrome.[53,68]

Yet for many years the irukandji itself eluded every attempt made to discover its identity or to snare a specimen—though this is not too surprising really. After all, how could anyone capture a creature that could not be seen, and that could not be trapped in a net?[53,68]

The answer to this riddle came in 1961. For much of the 1950s, Cairns marine biologist Dr. John H. Barnes had been investigating box jellies (cubomedusans), and had been particularly keen to ascertain the identity of the seemingly intangible irukandji. In an attempt to capture one, he had been employing a specially-constructed piece of apparatus that functioned by snapping shut on any fast-moving underwater creature that the operator wished to catch. Even so, this was still only a partial solution to the problem—needless to say, it could only be used on creatures that the operator could see, and the irukandji was famous for staying *unseen*![53,68]

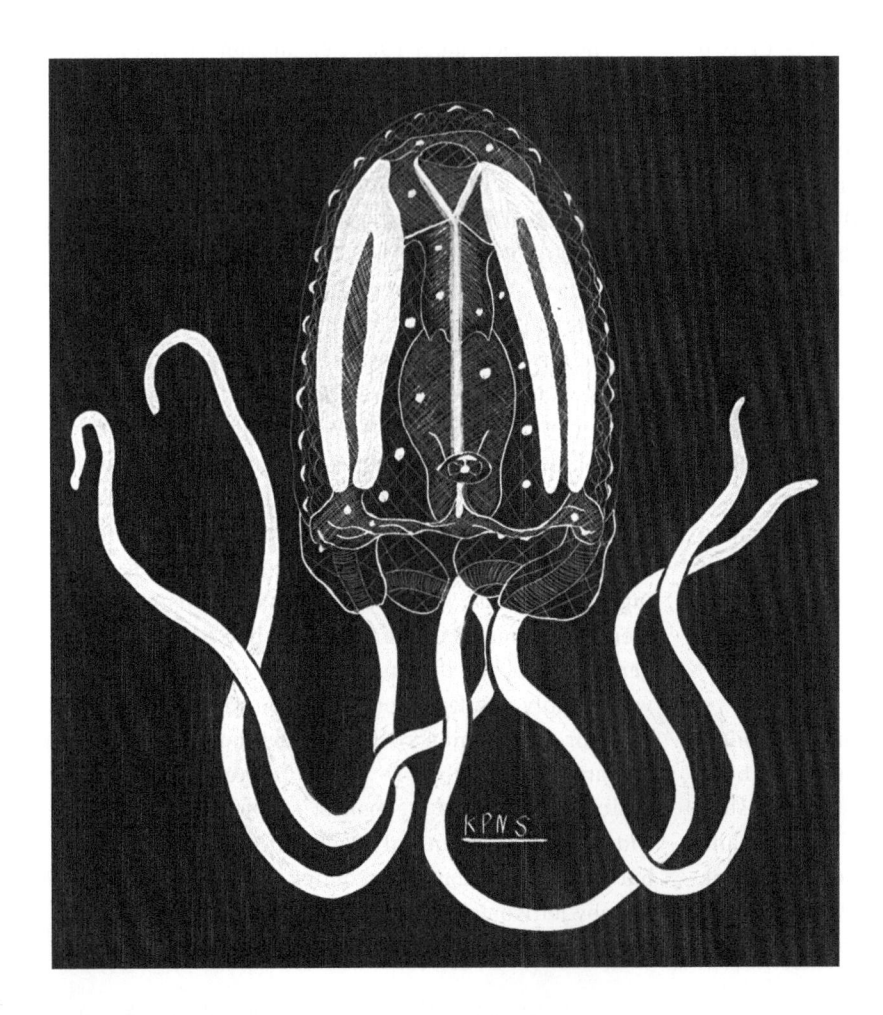

Irukandji—the unpleasant effects of its stinging cells were well known to medical science several decades before the identity of this 'invisible' species was finally exposed by zoology. (Dr. Karl P.N. Shuker)

On 10 December 1961, however, the irukandji made a fatal mistake, which cost it its scientific anonymity. On that day, Barnes was exploring the shallow sea off Queensland's Palm Beach, when suddenly something small swam right in front of his face. Even at such close range as that, Barnes was unable to spot the creature's body, but he could discern some thin, white tentacles. So he immediately brought his 'catcher' into operation, and deftly snapped it shut on the unidentified sea animal. It proved to be an exceedingly small, transparent box jelly, of a type hitherto unknown to him.[53,68]

Little more than 0.5 in across its bell, it was rather unusual in appearance because its bell resembled a rounded rectangle in shape, quite unlike the characteristic cuboidal version possessed by most other box jellies. Moreover, whereas other box jellies sported an entire bunch of tentacles at each of the four lowermost corners of their bell's cube, each of the four lowermost corners of this creature's rectangular bell bore only a single tentacle, thus adding up to just four in total, each measuring no more than 16-20 in long.[53,68]

Clearly this was an important new species, but could it also be the legendary irukandji? Certainly its transparent body would ensure that it remained unseen by swimmers, and its minute size would enable it to pass through fine-meshed nets—but there was only one way to determine *conclusively* whether or not it was truly the irukandji. So Barnes stoically applied its tentacles directly onto his own skin, to record the effect! And sure enough, he experienced the well-documented, diagnostic irukandji syndrome. The experiment was repeated on various brave volunteers, and they too experienced those very same symptoms. The mystery was solved—science finally had a specimen of the elusive irukandji.[53,68]

Later that same day, a small fish that seemed to be in some way afflicted was scooped out of the water near the original irukandji's site of capture, and when examined it was found to be attached to an irukandji, which Barnes also preserved. Others were captured later, and in 1965 the type specimen and one additional example were passed by Barnes to Australian jellyfish expert Dr. R.V. Southcott, who described their species in 1967. He named it *Carukia barnesi*—commemorating Dr. Barnes's valiant determination to expose the irukandji's identity, and recognising its morphological singularities by housing it in a new genus.[69]

The secret of why the irukandji is so notoriously effective as a stinging organism has now been revealed too. Its formidable stinging cells (nematocysts) are not just confined to its tentacles as with most jellyfishes—instead, there is also a generous sprinkling of wart-like nematocyst clusters all over its bell. In view of this devastating arsenal, it is fortunate that the irukandji does not reside permanently in Queensland's coastal waters; it is primarily an oceanic species, only invading the coasts if swarms are diverted by an appropriate combination of topographical features, prey, and water temperatures offshore. Moreover, in recent years the irukandji's zoogeographical range has been shown to be far greater than previously suspected, with specimens having been recorded off Japan, the coast of Florida, and even the British Isles (amazingly, some had been discovered in the water supply of the English city of Derby!).[53,68-9]

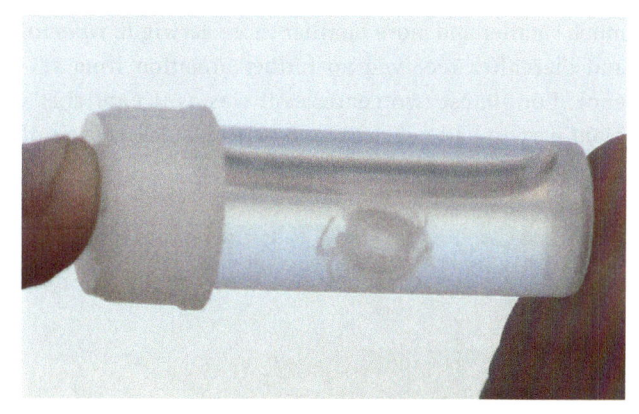

Malo kingi is a second species (and genus) of irukandji.

Tragically, during late January 2002, while holidaying in Australia, British tourist Richard Jordan became the first officially confirmed human fatality resulting from the sting of an irukandji. He inadvertently brushed against one of these tiny, transparent box jellies while swimming near Hamilton Island, northern Queensland, on 30 January. Within an hour, suffering excruciating pain, he lapsed into a coma, and later died from a brain haemorrhage. There is currently no antidote to this species' sting.[69a]

During the first decade of the 21st century, two additional species of irukandji have been recognised. These have been housed in a new genus, *Malo*, of which the smaller but much more dangerous of the two, *Malo kingi*, was formally described in 2007. The larger *M. maxima* was described two years earlier, along with a second *Carukia* species, *C. shinju*, but this latter irukandji is nowadays not widely recognised as a separate species. *M. kingi* derives its specific name from American tourist Robert King, who was killed in 2002 by a specimen of this then-unnamed species while swimming in waters off Port Douglas, Queensland. *M. kingi*, a tiny species the size of a human thumbnail, is readily distinguished from all other jellyfishes by a unique series of halo-like rings encircling its tentacles. And it is believed that there may be further irukandji species still awaiting detection.[69b]

AN ENORMOUS EARWIG FROM ST. HELENA

People who mistakenly believe in the unfounded old wives' tale that earwigs take a particular delight in entering the ears of the unwary have nothing to fear from *Labidura herculeana*—unless they have exceptionally large ears! Attaining a formidable total length (for an earwig) of 2.5-3 in, this robust resident of the South Atlantic island of St. Helena is the world's largest species of earwig, and has a highly unusual history.

Shiny black with red legs and no hind wings, it was officially described and named as long ago as 1798, by the famous Danish entomologist Johan Christian Fabricius, but somehow it later became confounded with the much smaller and more familiar shore earwig *L. riparia*, and thereafter received no further attention from science. For almost two centuries it was as if Fabricius's giant earwig had never existed, with his description of it long since forgotten.[6,70]

Extra-large model of St Helena giant earwig. (Zoo Operations Ltd)

In 1962, while seeking bird bones in the sands of St. Helena's Prosperous Bay, ornithologists N.P. Ashmole and D.F. Dorward discovered some dried, tail-end pincers from what must have been some form of enormous earwig—as confirmed by zoologist Arthur Loveridge, to whom they were given. These were subsequently examined by a number of other zoologists, all of whom assumed that they must be from a wholly new species, never before documented by science. And so it was christened *L. loveridgei*, honouring the man who had first studied its remains.[71] From their condition, they seemed to be of recent origin (i.e. they were not fossilised), inciting speculation that this dramatic species might still be alive. In 1965, a team of entomologists arrived at St. Helena to look for it—and discovered quite a few, inhabiting deep burrows hidden beneath heavy boulders at Horse Point Plain (near Prosperous Bay), but that was not all that they found.[6,70]

Back home, researching these specimens, they uncovered details concerning the long-forgotten giant earwig *L. herculeana*, and realised that this was the very same species as the newly-described *L. loveridgei*. And so, in accordance with the rules of nomenclatural precedence in relation to scientific names, the latter version was suppressed, and *L. herculeana* was reinstated as the species' official scientific name.[70]

Tragically, however, the issue of what it should be called may now be somewhat academic. Searches since the 1960s have failed to locate any living specimens, either at Horse Point Plain or anywhere else on the island. A well-publicised search was made in spring 1988 by Project Hercules—a two-man expedition launched by London Zoo, comprising Dave Clarke and Richard Veal. Another search took place in September 1993, and a third in 2003. All were unsuccessful. Nevertheless, bearing in mind its predominantly subterranean lifestyle, it is not impossible that the giant earwig of St. Helena still survives, undetected by humans, and it is currently classified as critically endangered by the IUCN.[72]

MIDGARDIA—WRAPPING ITS ARMS AROUND A WORLD RECORD

On 18 August 1969, while dredging in the Gulf of Mexico's southern region at a depth of 1500 ft, Texas A & M University's research vessel *Alaminos* collected a female starfish, bright red in colour, that was exceedingly large and extremely fragile. Both of these conditions were due to its twelve extraordinarily long and slender arms, which yielded a maximum armtip-to-armtip span of just under 54.5 in—effortlessly surpassing that of any other starfish specimen in record. In stark contrast, it weighed less than 2.5 oz dry, and its central body disc was only just over an inch in total diameter—thereby creating the illusion of a bodiless sea monster bristling with endless serpentine arms![73]

This, at least, may well have been the image that it conjured up in the mind of Maureen E. Downey of the Smithsonian Institution—because on 29 February 1972, when she officially described this striking new species (sufficiently removed from all others to require its own genus), she christened it *Midgardia xandaros*. *Midgardia* is an allusion to the famous Midgard Serpent of Norse mythology, which encircled the entire world in its limitless coils, and *xandaros* is Greek for 'sea monster'.[73]

Despite a superficial similarity to some giant form of brittle star (ophiuroid), *M. xandaros* belongs to the brisingid starfish family, whose members only occur in deep water, thus making its type specimen's capture a notable event. A second specimen of Midgard starfish had been caught during the same dredge, but it broke into fragments before being brought to the surface.[73]

TRICHOPLAX—THE RETURN OF AN ENIGMA

One of the smallest yet most mysterious of all multicellular animals is a tiny creature known scientifically as

Trichoplax adhaerens, but lacking a common name as even its very existence remains unknown to most people (including the majority of zoologists!). Yet it is so different from all other animals that in 1971 its principal researcher, Prof. Karl G. Grell of Tübingen University, housed it within a phylum of its own—variously termed Phagocytellozoa or (most commonly) Placozoa. For convenience, this phylum's sole occupant is referred to in general terms as a placozoan—'flat animal'. (A second species, discovered in the Gulf of Naples, was christened *Treptoplax reptans* by F.S. Monticelli in 1896, but has never been recorded again; most authorities today doubt that it is a genuinely distinct species.)[74]

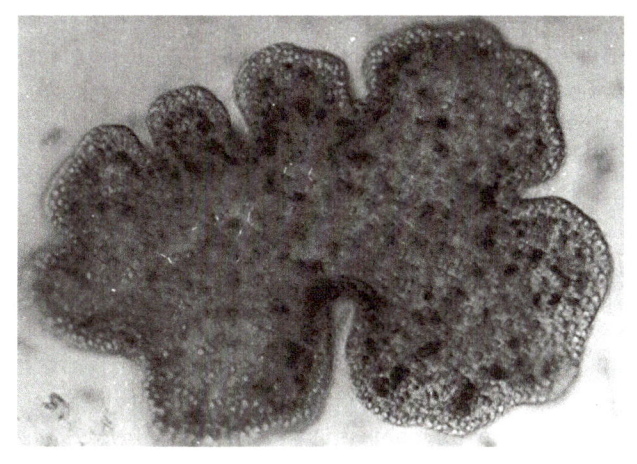

Placozoan, Trichoplax adhaerens.
(Prof. Karl Grell)

Trichoplax was first discovered in 1883, when observed by German zoologist Franz Eilhard Schulze in a marine aquarium at the Zoological Institute in Graz, Austria, and was spasmodically recorded from other laboratory aquaria around the world in the years to come. But it had not been reported for a long time until it was formally rediscovered in 1971 by Dr. Richard Miller, who observed it adhering to the walls of some seawater tanks at Philadelphia's Temple University.[74]

Structurally, *Trichoplax* is among the simplest of all multicelled animal forms—a flat, amorphous mass superficially resembling an oversized, grey-coloured amoeba that continually changes its shape, and measuring up to about 0.4 mm in diameter. Unlike the unicellular amoebae, however, its body is composed of several thousand cells, arranged in two clearly delineated layers (separated from each other by a cavity filled with gelatinous fluid and held open by star-like fibres), which provide it with a distinct dorsal and ventral surface, but it lacks organs and tissues. Moreover, as it has neither a fixed front and rear end nor a left-hand and right-hand side, it can move in any direction—by crawling (juveniles can also swim).[55,74]

Reproduction can occur by simple division of one animal into two (fission), or by budding, or by the production of eggs subsequently fertilised by released sperm. Feeding consists of enclosing any encountered algae or organic debris within a rapidly-formed pocket on its ventral surface; it then discharges enzymes into this pocket to initiate digestion, eventually absorbing the resulting 'soup' directly through its body surface, as it has no mouth. All in all, a truly exceptional species, whose rediscovery in 1971 was aptly referred to by Miller as 'the return of an enigma'.[55,74]

THE NO-EYED BIG-EYED WOLF SPIDER—AN EIGHT-LEGGED CONTRADICTION FROM KAUAI
In 1973, a most peculiar species of Hawaiian spider was described—currently known only from the deeper regions of Kauai's Koloa Cave and a few others on this island's southeastern coast, yielding six populations in total. These caves consist of lava tubes resulting from an eruption of the volcano Koloa.[75]

The curiously-named no-eyed big-eyed wolf spider of Kauai, one of the Hawaiian islands. (USFWS)

Referred to scientifically as *Adelocosa anops*, this spelaean spider (sole member of its genus) delights in a very contradictory common name—the no-eyed big-eyed wolf spider! The reason for this etymological enigma stems from *Adelocosa*'s membership of a taxonomic family of wolf spiders whose species are generally typified by very large, well-developed eyes, and are thus called big-eyed wolf spiders. In the case of *Adelocosa*, however, its ancestors apparently abandoned a traditional above-ground lifestyle in favour of a highly-specialised subterranean one instead—in which eyes were superfluous. Consequently, during the resulting evolution of this much-modified cave-dwelling species, they were eventually lost, thus explaining the apparent paradox of a no-eyed big-eyed spider.[6,75]

Although made known to science only fairly recently, this distinctive spider is familiar to Kauai's indigenous people, who call it *pe'e pe'e maka'ole*. It is easily identified by its long and semi-transparent, orange-coloured legs, its orange-brown cephalothorax (combined head-and-body section), and its white abdomen, as well as, of course, its lack of eyes.[6,75]

In 1991, many young specimens of *Troglodiplura lowryi*, a South Australian eyeless spider known before only from dried carcases, were found alive in a Nullarbor Plain cave.[76]

THE BAY OF BISCAY'S LIVING FOSSILS
Despite their name and resemblance to underwater palm trees, sea lilies are definitely animals—known scientifically as crinoids and related to starfishes, sea urchins, brittle stars, and sea cucumbers. Once thought to have died out millions of years ago, the world's first living sea lily was discovered in 1755, in deep water off the Caribbean island of Martinique. Since then, many others have been obtained, belonging to more than 80 different modern-day species.

In 1973, a collection of marine specimens was made in southwestern France's Bay of Biscay (Gascogne Gulf) by the French research vessel *Thalassa*. Three years later, these were scrutinised by French biologist Dr. Michel Roux from the University of South Paris, who found to his delight that amongst the collection's sea lilies were some examples that obviously belonged to the genus *Conocrinus*, previously known only from fossil forms. In short, not only had a new species been discovered, but also an entire fossil genus had been restored to life.[77]

Furthermore, the examples were shown to belong to two separate *Conocrinus* species, later named *C. cherbonnieri* and *C. cabiochi*, and proved remarkably similar to fossil *Conocrinus* from southwestern France that dated back to the Eocene epoch, 54-36 million years ago. Additional studies disclosed that the modern-day *Conocrinus* species seemed to have been ousted from the bay's muddy bottom by a more advanced sea lily, *Democrinus parfaiti*, but still survived at higher levels, at depths of 1000-1830 ft.[77]

NEOGLYPHEA—RESCUING ITS LINE FROM 50 MILLION YEARS OF EXTINCTION
Among the vast assortment of marine specimens collected during summer 1908 in the vicinity of the Philippines by the oceanographic ship *Albatross* was a small, pink, crab-like animal with large eyes. Caught on 17 July, it had been scooped up from the sandy bottom of the South China Sea, at a depth of 623 ft. It attracted little attention from scientists, but was dutifully preserved before taking its place alongside other *Albatross* specimens in the collections of zoological material housed in Washington's Smithsonian Institution.[78]

There it remained, unnamed and unstudied, for the next 67 years, until in spring 1975 it captured the interest of two French scientists, Drs. Jacques Forest and Michèle de Saint Laurent. And with good reason, because they recognised it to be a glyphid—a type of crustacean apparently ancestral to modern-day crabs, lobsters, and crayfishes, and believed until then to have died

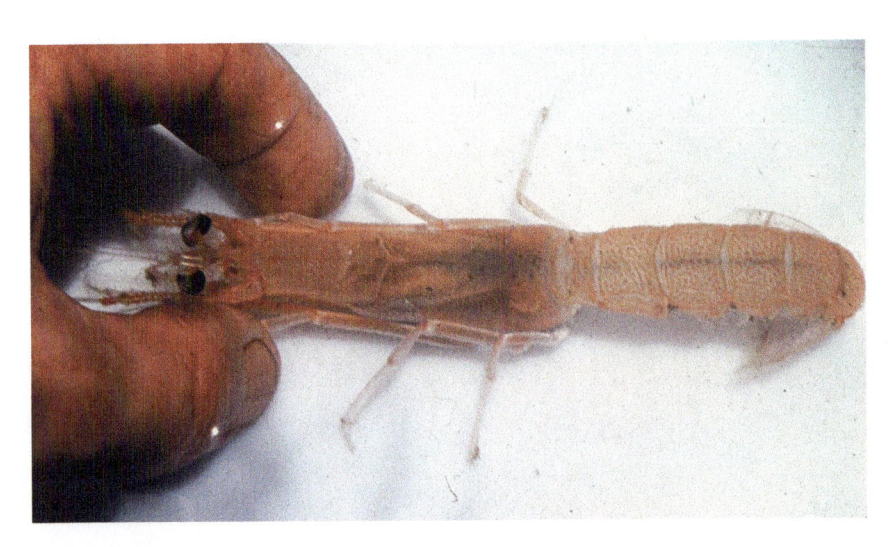

Neoglyphea inopinata—overlooked for 67 years, in 1975 its type specimen revived a group of crustaceans previously believed extinct for 50 million years. (Dr. Jacques Forest)

Penultimate instar of a female Cooloola monster—this grasshopper species is so extraordinary that its discovery in 1976 required the creation of a totally new taxonomic family. (Dr. David Rentz)

out around 50 million years ago! They named its species *Neoglyphea inopinata* ('unexpected near-glyphid'), and wondered if other specimens existed.[78]

The answer came a year later, when in March 1976 they succeeded in catching nine specimens at a depth of approximately 660 ft in the same area as the provenance of the *Albatross* individual. Another 'living fossil' had entered the zoological annals.[79]

As an unexpected bonus, *Neoglyphea* was found to harbour a new species of crustacean parasite, the copepod *Nicothoe tumulosa*, described in 1976 by biologist Dr. R.F. Cressey.[80]

In 2006, moreover, it was announced that during a recent exploration of Coral Sea waters off the Chesterfield Islands, northwest of New Caledonia, French workers had caught at 1312 ft a female specimen of a second, previously-unknown species of living glyphid, which was duly christened *Neoglyphea neocaledonica*. Almost 5 inches long, it looked like a strange hybrid of mud lobster and shrimp, and its huge eyes, thickset body, and reddish spots distinguish it from the more lissom, smaller-eyed, undecorated *N. inopinata*. More recently, it has been transferred to a new genus, and is now known as *Laurentaeglyphea neocaledoniae*, allying it more closely with various fossil species.[80a]

THE COOLOOLA MONSTER

The year 1976 also saw the discovery of a creature so extraordinary that it was initially assumed to be a manmade hoax (see this section's opening quote, p. 221).

It was spotted by Queensland Museum researcher Ted Dahms amongst the insects collected in February by a colleague, V. Davies, in a pitfall trap set on the rainforest floor near Poona Lake, in southern Queensland's Cooloola National Park—from which the new animal derived its striking vernacular name, 'the Cooloola monster'. An adult male, on first sight it resembled a very robust form of cricket, but detailed studies by orthopteran (cricket/grasshopper) specialist Dr. David C.F. Rentz (of Canberra's Australian National Insect Collection) revealed it to be so overwhelmingly different from all others known that it could not even be satisfactorily housed within any existing orthopteran family, let alone any genus or species. Hence a completely new family—Cooloolidae—had to be erected; and in 1980, the unique species that it housed was officially described, and named *Cooloola propator*.[81]

Other specimens have been collected, and used in morphological research that showed this important new insect to be most closely related to the ensiform orthopterans (the crickets and long-horned grasshoppers). Attaining a length of just over an inch in adult males, with its shovel-like head and limbs, very short antennae and near-sightless eyes, vestigial wings (absent altogether in females), and broad, powerful body (with prominently swollen abdomen in females) the Cooloola monster was quite evidently a burrower. Studies with living specimens verified this, disclosing that it inhabits the sandy, moist soil of Queensland's coastal rainforests, spending most of its time underground. As for food, its mouthparts' structure indicated a predatory existence, probably devouring other subterranean fauna.[81]

Three additional species of 'monster' were later discovered, including one from South Percy Island, a continental island 53 miles southeast of MacKay, Queensland. They are: the dingo monster *C. dingo* (the smallest 'monster' species), described in 1986; the sugarcane monster *C. ziljan* (the largest species), 1986; and Pearson's monster *C. pearsoni* (from South Percy Island), 1999.[81a]

A VISTA OF VESTIMENTIFERANS—MANKIND'S FIRST VIEW OF THE SPECTACULAR VENT WORLD

The year 1977 secured a momentous place in zoological history for witnessing not just the disclosure of some very significant new species, but also the totally unexpected discovery of an entirely new world—a world never before seen by mankind.

The ocean bottom had traditionally been thought of as a barren, pitch-black zone never penetrated by the sun's life-giving light and warmth (and therefore virtually bereft of living organisms), a world perpetually maintained at icy temperatures throughout its vast expanse. Then in 1972, a huge rift valley on the stretch of seafloor about 210 miles northeast of the Galapagos Islands became the subject of a detailed oceanographic survey, which surprised geologists by revealing unexpected temperature variations across the valley. To explain this anomaly, it was postulated that in seafloor regions experiencing upwelling of magma (molten rock) during the creation of new oceanic crust, hot water may be rising through the seafloor via surface cracks or vents. But how could this hypothesis of hydrothermal activity be tested? There was only one way to settle the matter conclusively.[82]

Giant vent clams. (Prof. John M. Edmond)

During February and March 1977, a scientific team led by Dr. John Corliss from Oregon State University and Dr. Robert Ballard from Massachusetts's Woods Hole Oceanographic Institution travelled over 8000 ft down through the sea to the Galapagos Islands' seafloor rift valley, transported in a 23-ft-long, deep-diving U.S. Navy submarine named *Alvin*.[82]

Sure enough, they found their predicted hydrothermal vents—but that was not all. To their amazement, they also discovered that the vents supported a thriving animal community, teaming with an immense variety of forms—which in almost every case comprised species totally unknown to science! To quote team member Kathleen Crane: "It was like a page out of the Jules Verne novel *20,000 Leagues Under the Sea*".[82]

One vent was surrounded with blind yellow-white crabs (a new species ultimately named *Bythograea thermydron* and allocated its own taxonomic family),[83] plus white-shelled clams up to 1 ft long (a new species again, later dubbed *Calyptogena magnifica*, with bright-red, haemoglobin-laden tissues).[84] Another vent was populated with peculiar creatures resembling dandelions gone to seed, and attached to the sea bottom via a network of fine filaments. Nicknamed 'dandelions' (and constituting yet another new species, christened *Thermopalia taraxaca*), these were siphonophores—relatives of the Portuguese man-o'-war *Physalia*.[85]

One vent particularly rich in fauna (and aptly named 'the Garden of Eden' by the *Alvin* team) bore crabs, clams, 'dandelions', squat lobsters (galatheids), limpets, amber-coloured mussels, thread-like enteropneusts nicknamed 'spaghetti worms' (a new species, dubbed *Saxipendium coronatum*[86]), plus the hydrothermal vents' *pièce de resistance*—great clusters of 8.5-ft-long, tube-dwelling worms.[82]

Even by the vent world's own exceptional standards for remarkable fauna, these worms were truly extraordinary. Thousands of their tall, vertical tubes clustered around the vent, and from the top of each tube a spectacular, vivid red plume of petal-like tentacles emerged—yielding a breathtaking display so reminiscent of magnificent flowerbeds that in 1979 a new expedition referred to another vent also bearing their tubes as 'the Rose Garden'.[82,87]

Specimens were collected for comprehensive laboratory examination. Nothing like these tubicolous vent worms had ever been seen before, and it soon became clear that their unique external appearance was more than matched by their internal anatomy. Painstaking dissection and microscopical studies of every aspect of their morphology were carried out at the Smithsonian Institution, headed by its Curator of Worms, Dr. Meredith Jones.[82,87]

When taken out of its tube, each such worm was found to possess four distinct body regions. Uppermost was the striking plume or obturaculum (whose red colouration was due to the presence of haemoglobin), in turn composed of numerous tentacles and attached to a central support. Beneath it was the vestimentum, a solid body region largely composed of muscle and connective tissue, and bearing wing-like projections externally. Internally, it housed the worm's brain, and possessed a dorsal chamber into which the worm's genital aperture opened; the sexes were separate in this species. At its lowermost end the vestimentum connected to the trunk, the third body region, which contained the worm's

Colony of the giant tubicolous vent worm Riftia pachyptila on Galapagos sea floor. (Prof. John M. Edmond)

single ventral nerve cord, dorsal and ventral blood vessels, the gonad (egg- or sperm-producing organ) and gonoducts, plus a very curious structure termed the trochosome (well-supplied with blood vessels, and proving to be a great mass of closely-packed bacteria). The final portion of the worm's body was the opisthosome, a segmented region bearing bristle-like structures called setae, and terminating in a rounded knob.[87]

And as if all of that were not strange enough, this peculiar worm—surely as alien as any life form from Outer Space—provided an extra surprise. It totally lacked a digestive system—having neither mouth nor gut. It also lacked any vestige of eyes or comparable structures (though in view of its lightless habitat on the ocean floor, this is less startling).[87]

Jones's studies revealed that the tubicolous vent worm's closest relatives were none other than *Lamellibrachia barhami* and *L. luymesi*, those remarkable,

Riftia pachyptila colony. (Prof. John M. Edmond)

afrenulate species of pogonophoran (see p. 234) constituting the taxonomic order Vestimentifera. Hence it was within this same order that Jones placed their giant cousin from the vent world, when he formally described it in 1980, naming it *Riftia pachyptila* ('thick-plumed rift dweller').[87]

Moreover, by 1985 he had elevated Vestimentifera from a mere order within the phylum Pogonophora to an entirely separate phylum in its own right, emphatically underlining its three known species' fundamental distinction from all other animals.[88] (Most other workers, however, preferred to retain Vestimentifera within Pogonophora, albeit as a separate class; and nowadays Pogonophora itself has been demoted from a discrete phylum to a mere family, Siboglinidae, within the polychaete class of annelid worms.)

Returning to its anatomy: as *Riftia* has neither gut nor mouth, how does it feed? Its plume's large surface area and numerous tentacles, each richly supplied with blood vessels, enable it to function as an efficient organ for the absorption of organic molecules from the surrounding seawater. Direct absorption across the worm's body surface within its tube is also likely. It may well obtain nutrients from its internal, trochosomal bacteria too. But what of the other members of the vent fauna? What were they feeding on?[87]

Until the discovery of the vent world, all known life forms on Earth ultimately obtained their energy from the sun. Plants and certain bacteria harnessed sunlight to manufacture their own food, in the form of carbohydrates—the familiar process of photosynthesis. Herbivorous organisms obtained their energy by eating these photosynthesising species, and carnivorous organisms obtained theirs by devouring the herbivores. The vent world's discovery, however, unveiled the operation of a radically new, autonomous energy chain, one that was completely independent of the sun. Instead of sunlight, the energy source being used was chemical energy—released during the combination of sulphates with hydrogen to yield hydrogen sulphide, a process occurring when sulphate-laden warm water rising up through the seafloor's crust makes contact with the cold, oxygenated water present at the vents' openings. The released energy was being utilised by sulphur-oxidising bacteria to manufacture their own food. In turn, they were being eaten by larger life forms, which were themselves being preyed upon by others, and so on.[82]

In more recent times, hydrothermal vent systems have also been discovered elsewhere. One, further north than the Galapagos version, is sited on a zone called the East Pacific Rise, and was discovered in 1979, by another scientific team exploring in *Alvin*. It was distinguished by its impressive chimney-shaped vents—nicknamed the chimneys of hell. Some of these belched out smoky water rich in metals, such as iron, copper, zinc, and nickel, and became known as black smokers, but few organisms have been spied near them. Others, dubbed white smokers, released warm creamy-white water, and bore a rich community of creatures on their outer walls and the surrounding seabed. These included a strange 4-in-long pink polychaete dubbed the Pompeii worm *Alvinella pompejana*, a very primitive short-stalked barnacle *Neoverruca brachylepadoformis*, and a living Palaeozoic-type scallop *Bathypecten vulcani*.[82,89]

The Pompeii worm is a truly extraordinary species, inhabiting tubes attached to the chimneys' walls, where its posterior body region is able to withstand direct contact with heat from the walls measured at 80°C—the equivalent of sitting upon a hot plate! No less remarkable is its profuse dorsal covering of white filamentous bacteria, living symbiotically with this species. Moreover, it is just one of a whole taxonomic family of new polychaetes, Alvinellidae, named in 1986—whose members all live exclusively around vents, currently number ten species in two genera (*Alvinella* with two species, *Paralvinella* with eight), and are formally referred to as alvinellids.[82,89]

A few years later, vents borne upon exotic pagoda-like edifices were discovered on the seafloor 120 miles south of Mexico's Baja California. These were surrounded by tube worms, and visited by milky-white octopuses, as well as by three species of endemic vent fishes, all new to science. Two of these fishes, *Thermarces cerberus* (also reported from the Galapagos rift system) and *T. andersoni*, were eelpouts—slender, eel-like species measuring just under 1 ft long, with pinkish-white skin and small prickly teeth. The third, *Bythites hollisi*, was allied to the cusk-eels (ophidiids).[90]

Clearly, vent worlds were not only more abundant, but also more diverse, than originally assumed. Continued researches led to the intriguing proposal that decomposing whale carcasses on the seafloor may be serving as 'stepping

Giant tubicolous vent worms.

stones' for vent world animals, enabling them to migrate across regions of the floor that would otherwise be too inhospitable for their passage. As whale carcases release sulphur compounds during their disintegration, these compounds would feed the sulphur-oxidising bacteria characterising hydrothermal vents, and in turn provide the larger vent world creatures with their sustenance.[91]

Certainly, vent animals are very adaptable, and in recent years they have turned up in some highly unexpected localities. During the 1980s, a diver discovered that miniature hydrothermal vents emanating from sewers in Los Angeles, California, were sustaining mats of white bacteria off the Palos Verde Peninsula, and these in turn were being dined upon by sizeable quantities of black abalones. Similarly, in June 1992 scientists announced the finding of 6-ft-long tubicolous worms that were utilising hydrogen sulphide released from rotting beans in the hold of a ship that had sunk 13 years earlier off the Spanish coast.[92]

Perhaps the most dramatic recent discovery was that of an enormous vent system—possibly the largest anywhere on Earth—on the Atlantic Ocean's floor at a site called Broken Spur, about 800 miles southwest of the

Black smokers, like this one, harbour few life forms.

Azores. Its existence was disclosed in spring 1993, after it had been visited by a team of British scientists working from the research ship *RRS Charles Darwin*. Here, colossal chimneys standing up to 30 ft high were disgorging great quantities of metals, including iron and even gold—and vast numbers of vent creatures were also present, such as *Riftia*, several different forms of deepsea crustacean, and a distinctive species of mesogastropod known as the hairy snail *Alviniconcha hessleri*. This last-mentioned mollusc had previously been recorded from the Philippines' Mariana Trench vent system, where it takes the place of *Riftia* tube worms.[93] And speaking of extraordinary vent molluscs: another species of note is the scaly-footed snail *Crysomallon squamiferum*—discovered on the bases of black smokers in 2001 during an expedition to the Indian Ocean's Kairei hydrothermal vent field—whose foot is reinforced by scales made of iron sulphides (pyrite and greigite) instead of the usual calcium carbonate.[93a]

No less interesting is the curious case of the glowing vents. These were first reported in 1988, when Washington University oceanographer Prof. John Delaney and his team aboard the submarine *Alvin* visited two vents more than 7000 ft below the surface of the Pacific Ocean about 180 miles west of Washington's Olympic Peninsula. The mysterious glow was not visible to the human eye, but was detected with an ultra-sensitive electronic camera, and came as a total surprise—because as sunlight cannot reach such depths, there should not be any source of light present here.[94]

Its existence has still not been explained, but Delaney has suggested that its source seems to be the water of the vent systems, which may be hot enough to emit light. Alternatively, as noted by ecologist Dr. Cindy Lee Van Dover from Massachusetts' Woods Hole Oceanographic Institution, it could be due to rapid crystallisation of minerals ejected from the vents.[94]

Whatever its source, however, its presence has led to speculation that photosynthesis might actually occur within vent systems—which, if true, would be the first known example of this complex biochemical process that does not utilise sunlight.[94]

Pompeii worm.

Also of relevance here is the existence near thermal vents along the Mid-Atlantic Ridge of a strange species of eyeless shrimp called *Rimicaris exoculata*. Although it has no eyes, it does possess light-sensitive receptors in the middle of its back, which contain a substance called rhodopsin—normally associated with visual perception in dim light. Naturally, therefore, the fact that this species possesses such adaptations as these implies that vent-associated illumination is not restricted to the Olympic Peninsula system. Moreover, however faint its glow may be, it must be sufficient to serve some function in the shrimp's lifestyle—after all, why else would the species have evolved photoreceptors?[94]

Another notable member of the vent fauna came to public attention in April 1994, when scientists announced the discovery of a remarkable new species of brittle star—a slender-limbed relative of starfishes. It had been found by Surrey geologist Bramley Murton while diving in *Alvin* to the Broken Spur vent system of giant chimneys—and, unlike most other vent species, it was common in the vicinity of the black smokers. What rendered this new species even more distinctive, however, were its mouthparts—containing several elongated teeth that resembled chopsticks, and which the brittle star may use to scoop up bacteria and transfer them to its mouth. A number of other new species were also recorded here, including three shrimps, three sea anemones, and two tube-dwelling worms.[95]

Last, but certainly not least: traditionally, all life forms have been divided into two fundamental categories—prokaryotes and eukaryotes. Bacteria and blue-green algae are prokaryotes; the DNA inside their cells is not organised into chromosomes or contained within a nucleus. All higher forms of life, whose cells contain chromosomes within a discrete nucleus, are eukaryotes. Sensationally, however, a third fundamental category has lately been established, in order to accommodate a unique group of primitive organisms known as archaea—initially assumed to be a form of bacterium when first discovered in 1982. And where can the amazing archaea be found? Where else but the equally amazing vent world![96]

Evidently, there is a great deal of fundamentally significant biological knowledge to be gleaned from this revolutionary ecosystem residing in darkened splendour on the ocean bottom. How ironic, then, that while man has been fascinated for so long by the possibility of new life forms existing on extraterrestrial worlds in Outer Space, this new world in every sense of the phrase has been awaiting discovery here on his own planet!

VACELETIA—ALL SPHINCTOZOANS ARE NOT THE SAME!

First appearing in the Carboniferous Period (359-299 million years ago), the sphinctozoans constituted an odd group of segmented, sponge-like animals that seemingly died out over 200 million years later, at the end of the Cretaceous Period, without giving rise to any modern-day descendants. This time-honoured palaeontological presumption was decisively disproven in April 1977, however, when several dozen specimens of a *living* sphinctozoan species were recovered by the French oceanographic vessel *Suroît* just off the Îles Glorieuses, a collection of tiny islets north of the Mozambique Canal. This provided scientists with a unique chance to uncover the precise relationship of sphinctozoans to other animal forms.[97]

With at least some of the fossil species, palaeontologists had obtained what appeared to be insurmountable proof from anatomical and other comparisons that their correct taxonomic position was alongside the calcareous sponges (class Calcarea). Yet when Dr. J. Vacelet—a researcher at the Endoume marine station (Marseilles)—studied this newly-disclosed 'living fossil', which she named *Neocoelia crypta*, she came to a very different conclusion. Her findings strongly implied that, albeit a genuine sphinctozoan, this anachronistic animal was actually more closely related to the sclerosponges (p. 225) than to the calcareous ones.[97]

If neither party were wrong, how could this paradox be resolved? Vacelet's studies suggested to her an acceptable solution—namely, that the sphinctozoans constituted a diphyletic group. That is to say, instead of all sphinctozoans sharing a common ancestry there were really *two*, totally distinct lines, traditionally classed together because of superficial similarities in morphology but actually having *separate* origins from one another. One of these lines had shared its ancestry with the calcareous sponges, whereas the other had shared its ancestry with the sclerosponges. This demonstrates well that however painstaking the form taken by palaeontological work on fossilised animal specimens, only a living representative can offer conclusive evidence for their correct taxonomic categorisation.[97]

In 1982, *Neocoelia* was renamed *Vaceletia* ('*Neocoelia*' had already been used for another animal); and in 1984, it was recorded for the first time from the western Pacific. Moreover, nowadays the sphinctozoans are deemed not merely diphyletic but polyphyletic (as, indeed, are the sclerosponges too), with *Vaceletia* housed within the Demospongiae class of sponges.[98]

Pausing with strange sponges: in 1979 Vacelet described *Cryptosyringa membranophila*, a peculiar tetractinomorph sponge from underwater Jamaican caves, and associated with some very unusual membranous structures. A brand new family was originally created to accommodate it, but it is nowadays housed within a previously-existing family, Ancorinidae, which contains many different genera of sponge.[99]

REMIPEDES—OAR-FOOTED CRUSTACEANS ROWING INTO ZOOLOGICAL HISTORY

The discovery of *Hutchinsoniella* in the 1950s (see p. 238) brought the total of modern-day subclasses of crustacean to eight. In late 1979, American biologist Dr. Jill Yager, from Virginia's Old Dominion University at that time, was responsible for the creation of a ninth.[100]

A keen novice diver, with fellow biologist and diving instructor Dennis Williams she had been exploring Lucayan Cavern—a pitch-black, water-filled cave beneath the island of Grand Bahama—when the beam of light from her underwater torch disclosed a myriad of tiny specks flitting in every direction through the surrounding water. At first she assumed that they were simply dust particles, but closer observation revealed that they were minute life forms. One tiny worm-like animal seemed particularly unusual—a very graceful but totally blind swimmer measuring 0.4 in long, with a series of paired, oar-like limbs, a slender body, and very long antennae—so she carefully captured it in a plastic collecting bottle.[100]

Upon her return, she took the strange little creature to cave biologist Dr. John Holsinger for identification. To say that he was delighted with it would be putting it mildly. The tiny animal in her bottle was a crustacean, but so unlike any other species known that it became sole member of a wholly new genus, family, order, and subclass! Named *Speleonectes lucayensis* by Dr. Yager in 1981, this extraordinary species became known as a remipede ('oar-footed') after its most noticeable characteristic; its subclass was named Remipedia.[100]

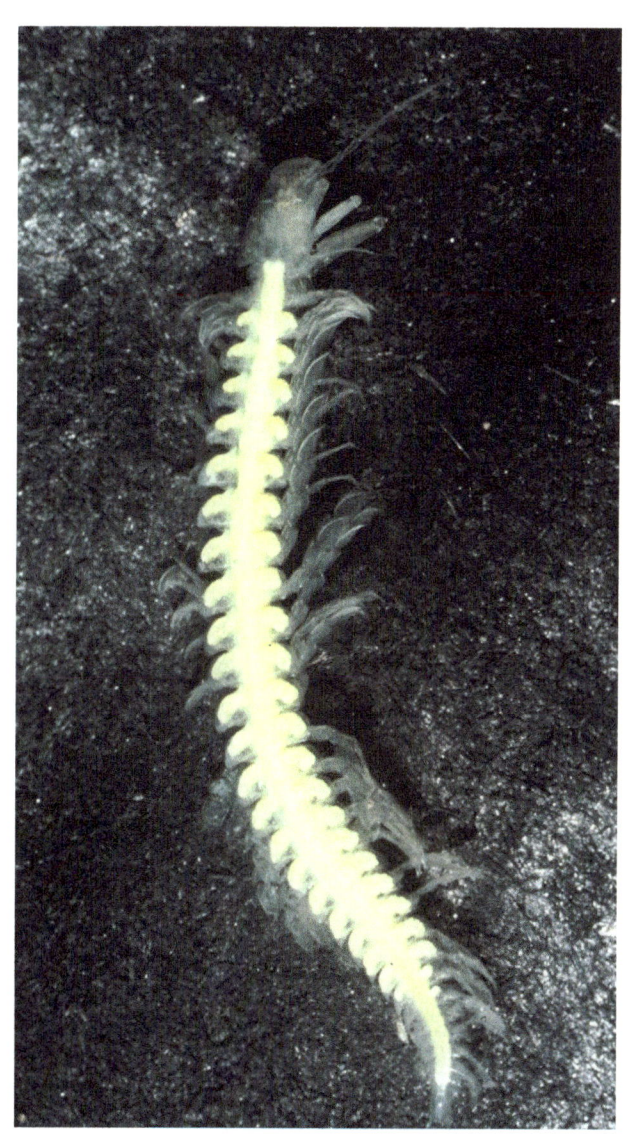

Lasionectes entrichoma.
(Dennis Williams, courtesy of Dr. Jill Yager)

Quite a number of additional species of remipede were discovered within the next few years (in February 1992, when I was preparing this present book's first incarnation, *The Lost Ark*, 1993, Dr. Yager informed me that the total at that time stood at nine); the current tally, as of August 2011, is 20 (housed in eight genera within three families), plus a single-known fossil species. They have been recorded from the Bahamas, Turks and Caicos, western Australia, Cuba, the Dominican Republic, Mexico, and the Canary Islands.[25,101]

The first of the Canary Island species to be discovered, during the early 1980s, was particularly interesting, because it occurred on the opposite side of the Atlantic from all of the others known at that time—thereby revealing that the remipedes had a much wider distribution range than first suspected (they are now known to occur globally in the neotropical zone).[25,101]

Discovered in Lanzarote's underwater lava tunnel of Jameos del Agua, and described in 1984, this particular remipede received the intriguing scientific name *Morlockia ondinae*—morlocks were subterranean entities in H.G. Wells's novelette *The Time Machine*, and Ondine was a water nymph. Even more memorable, however, was another genus of Canary Island remipede, whose sole, 2-in-long member is *Godzillus robustus*! A related genus, *Godzilliognomus*, contains two species, one of which, *G. frondosus*, is notable among remipedes for possessing a highly-organised, well-differentiated brain, suggesting that these crustaceans are not as primitive as originally thought.[25,101]

In September 1997, scientists announced the finding of a hitherto-unexplored remipede-inhabited cavern beneath a hotel in Majorca, accidentally discovered by workmen digging there, who had originally planned to pump it full of waster water and sewage. Happily, this plan was abandoned once the zoological treasures that it contained were unfurled.[102]

Incidentally, since the late 1970s many authorities elevate Crustacea from a class to a phylum, with all of its former subclasses (including Remipedia) thus becoming classes.

Morlockia ondinae. (Dennis Williams, courtesy of Dr. Jill Yager)

TAMBUSISI TREE-NYMPH—
A GHOSTLY GIANT FROM SULAWESI

Idea constitutes a genus of Asian, forest-dwelling butterflies known as tree-nymphs, with very large, black-and-white, semi-transparent wings capable only of sustaining a weak, fluttering flight—giving these delicate insects a ghostly, disembodied appearance that has inspired fanciful myths and superstitions throughout their distribution range.[6,103]

Some tree-nymphs, such as the Moluccan *I. idea*, can attain very sizeable wingspans, up to 5.5 in. Yet even these were dwarfed by the monstrous examples that were sometimes seen flitting high overhead by the team of young explorers trekking across Mount Tambusisi's southwestern slopes, during an Operation Drake expedition to the southeast Asian island of Sulawesi (Celebes) in March 1980. As they seemed much too large to be specimens of *I. blanchardii*, the only species of tree-nymph *known* to inhabit Sulawesi, every effort was made to secure some for identification and study—but each attempt met with failure. Consequently, when two of these immense *Idea* fluttered into view on 10 March, just as the team was about to pack up camp and move to another section of the island, they instantly found themselves being pursued *en masse* by a flurry of flailing butterfly nets brandished by every member of the expedition in the vicinity, until one of the two, a female, was successfully caught by the team's entomological expert, Major Anthony Bedford Russell.[103]

Not only did it belong to a completely new species, which Bedford Russell named *Idea tambusisiana*, the giant Tambusisi tree-nymph, but with a massive wingspan of 6.5 in, it was also one of the largest butterflies ever to be recorded by science, as well as the first new species of tree-nymph to be described for 97 years.[6,104]

KING BEE—BUZZING BACK WITH A CAPITAL 'B'!

Although it hardly compares with the gigantic versions beloved by science-fiction film producers, the world's largest bee attains a very respectable total length of 1.75 in (more than twice the size of the familiar *Bombus* bumblebees of field and garden). Commonly called the king bee, it was discovered in 1859, when naturalist Alfred Russell Wallace collected a female specimen on the Indonesian island of Bacan, in the north Moluccas.[6,105]

Except for its pale cheeks (genae), and a white band encircling the front edge of its abdomen, it was predominantly black in colour, with long, smoky wings, and was readily distinguished from all other bees not only by its huge size, but also by its enormous mandibles—formidable plier-shaped mouthparts projecting forwards and outwards like a pair of sturdy, ebony antlers, and accompanied by an equally enlarged labrum or 'upper lip'. This extraordinary insect was sent to the British Museum (Natural History), and its species was described in 1861, when it was named *Chalicodoma* [now *Megachile*] *pluto*. A short time later, a second female was obtained—after which, despite its notable size, the king bee proved to be expertly adept at remaining concealed, because it was not reported again for over a century.[6,105]

In February 1981, entomologist Dr. Adam Messer was studying the insect life on the Moluccan island of Halmahera. During a visit one day to some primary lowland forest about 5 miles southeast of Kampung Pasir Putih, he was startled to see two enormous bees, busily engaged in gathering resin from tree trunks—loosening it using their immensely large mandibles and then scraping it up, bulldozer-style, with the long labrum, until a solid ball of resin was produced. Closer inspection revealed beyond any doubt that these were female king bees! Further observations by Messer led to this species' discovery on the island of Soasiu (Tidore) as well, and to its rediscovery on Bacan. Messer also uncovered the secret of its surprising success at concealment—unlike so many other species of bee, it does not build external nests, but instead inside those of various species of tree-inhabiting termite.[6,105]

King bee (with honey bee, left, for comparison). (Dr. Adam Messer)

Equally important was his capture of some male king bees, the first ever recorded. At just under 1 in long, these proved to be noticeably smaller than the females. Similarly, their mandibles and labrum were shorter, and their cheeks were rufous, not white.[6,105]

Lastly, anyone finding the concept of a 1.75 in bee somewhat traumatic will be comforted to learn that the extra-large size of the king bee's body is not paralleled by the potency of its sting. In fact, because its sting is not barbed, it is actually *less* painful than that of many much smaller and more familiar bee species.[6,105]

THE TANTALISING TANTULOCARIDS
In 1975, biologist Dr. K-H. Becker described a bizarre little creature found as an external parasite on several deepsea copepod crustaceans. Their parasite was itself a crustacean, but an extremely unusual one—little more than a sac of organs equipped at one end with piercing mouthparts. Some miniscule limbs are present behind these in the larval form (tantulus), but are shed at its exoskeleton's final moult before attaining the adult, reproductive stage. There are no limbs comparable to other crustaceans' abdominal limbs.[25,106]

Becker named this highly modified species *Basipodella harpacticola*, and classed it as an aberrant copepod. However, when a similar species, *Deoterthron dentatum*, was described in 1980, comparisons of the two in relation to other crustaceans suggested that they were more closely allied to the barnacles. This controversy was resolved in 1983, when further studies concluded that these perplexing parasites, together with some obscure forms documented as long ago as 1903, warranted their own crustacean subclass—dubbed Tantulocarida, which is nowadays housed within the crustacean class Maxillopoda. Just over 30 species are currently recognised, one of which, *Tantulacus dieteri*, described in 2010, is the world's smallest known species of arthropod, measuring a minuscule 85 μm (0.0033 in) in total body length.[25,106-106a]

LORICIFERA—A PREDICTION COME TRUE
Giving one's name to an uncommonly ugly form of animal larva may not be everyone's idea of obtaining scientific immortality, but it is nonetheless an effective way to achieve this—especially when that larva's species is so utterly different from all others that a completely new *phylum* has to be created to accommodate it.

Its story began in 1961 when, as a student at Washington's National Museum of Natural History, Robert Higgins predicted the existence of a remarkable little creature unlike any known to science at that time. By a sadistically ironic twist of fate, in May 1974 he actually found a real-life specimen of his hitherto-hypothetical creature—but failed to recognise it for what it was! Instead, he deemed it to be nothing more than a larval priapulid worm.[107]

The following year, however, another specimen was found, this time by Danish zoologist Dr. Reinhardt

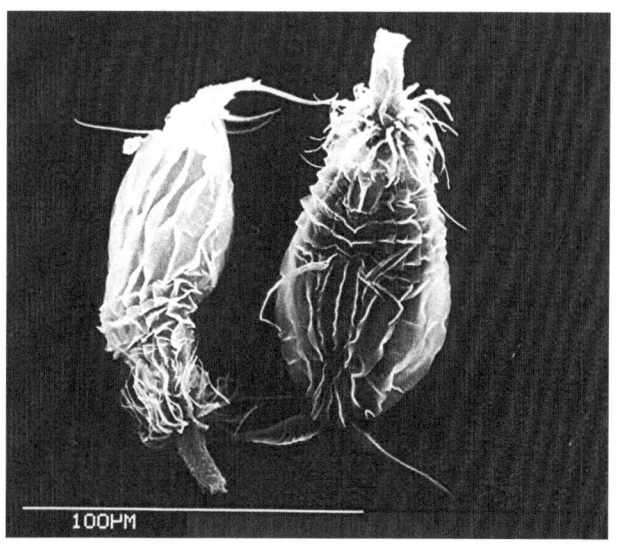

Larvae of Nanaloricus mysticus. (Dr. Reinhardt Kristensen)

Møbjerg Kristensen, from the University of Copenhagen. Yet as bad luck would have it, the tiny animal was destroyed during its preparation for transmission electron microscopy. Happily, between 1976 and 1979 Dr. Kristensen discovered some larvae, in shell gravel obtained from depths of 330-365 ft outside western Greenland's Godhavn Harbour. And finally, in April 1982, an adult turned up—completely by accident.[107]

Kristensen had obtained a huge sample of shell gravel from a depth of 83-100 ft during field work at the Marine Biological Station in Roscoff, France, and was in a hurry to examine the minute creatures living between the gravel particles, as this was his last day there before leaving for Denmark again. Consequently, instead of employing the usual sophisticated but somewhat protracted techniques for dislodging animals from the particles, lack of sufficient time spurred him to use a cruder but much quicker method—simply washing the gravel in freshwater. The change in salt concentration experienced by the tiny marine organisms in the gravel shocked them into loosening their grip on its particles, and they could then be collected in the surrounding water. Among the creatures obtained in this way was an adult of Higgins's postulated animal form, plus others from every stage in its life history. Shortly afterwards, specimens belonging to a slightly different species were obtained from Greenland gravel samples, using this same technique.[107]

By now, Kristensen and Higgins had learnt about each other's interest in these mysterious minute creatures, and had teamed up to work on them. They discovered that the individual (a larva) collected by Higgins in 1974 was indeed of the same group, but sufficiently different from Kristensen's species to warrant separation within a new genus and family. As for the creatures *in toto*, true to Higgins's expectations they required a brand new phylum. In 1983, Kristensen named it Loricifera, and formally described its first species, the Roscoff one, which he christened *Nanaloricus mysticus*.[107]

A tiny creature, no more than 0.01 in long, with a fairly squat body and a head section bearing a collar of radiating spines, it leads a sedentary existence—quite unlike its free-swimming larva, whose striated, pear-shaped body has a rear pair of flipper-like appendages. Anatomically, the species combines features from several different phyla, but is characterised by a unique mouth, consisting of a long tube that can be retracted completely within the creature's body in a manner not previously recorded from any other type of animal.[107]

As for Higgins, although he did not have the honour of describing the first real-life species of his conjectured creature he was given an unusual consolation prize—ever afterwards, the basic larval type produced by loriciferans would be officially referred to in zoological parlance as the Higgins larva. Higgins's reaction to this accolade was to comment: "I'm very pleased of course, even though it is such an ugly creature." Twenty-two years after his student prognostication, his hypothetical animal was hypothetical no longer.[107]

Incidentally, in 1986 Higgins was able to describe the species to which his lost specimen belonged; its scientific name is *Pliciloricus enigmaticus*, and it was just one of eight new species that Higgins and Kristensen described within a single paper. The other seven species were: *P. dubius*, *P. gracilis*, *P. orphanus*, *P. profundus*, *Rugiloricus carolinensis*, *R. cauliculus*, and *R. ornatus*.[108] In 1988, Kristensen described a notably important species, *P. hadalis*—the first loriciferan recorded from fine sediment (red clay), from a depth (27,082 ft) great enough to be included within the hadal bathymetric zone, and from the western Pacific.[109]

In February 1992, I was delighted and honoured to learn from Dr. Kristensen that in due course he would be naming a new species of loriciferan after me (at that time, he had *nearly 70* undescribed species in his

Pliciloricus enigmaticus.
(Dr. Reinhardt Kristensen)

collection!), in recognition of the very significant contribution made by this present book's first edition, *The Lost Ark* (1993), to the zoological literature. He later informed me that 'my' loriciferan was currently the most interesting species known, because it is neotenous, i.e. its Higgins larval stage becomes sexually mature precociously, developing an ovary. Shuker's loriciferan was collected in the Faroe Bank, and was formally described by Kristensen and fellow loriciferan researcher Dr. Iben Heiner in 2005, when it was duly christened *Pliciloricus shukeri*. Quoting from their paper the etymological derivation of this species' name[110a]:

> The name of this species epithet is in honor of Dr. Karl Shuker, a prominent expert in cryptozoology. The new species is dedicated to Dr. Shuker for his outstanding book "The Lost Ark, New and Rediscovered Animals of the 20th Century". In this book, the discovery of Loricifera received much credit as one of the major events of the 20th Century.

As of August 2011, 22 species of loriciferan in eight genera have been described, but at least a hundred more have been discovered and are currently awaiting description.[110-110a]

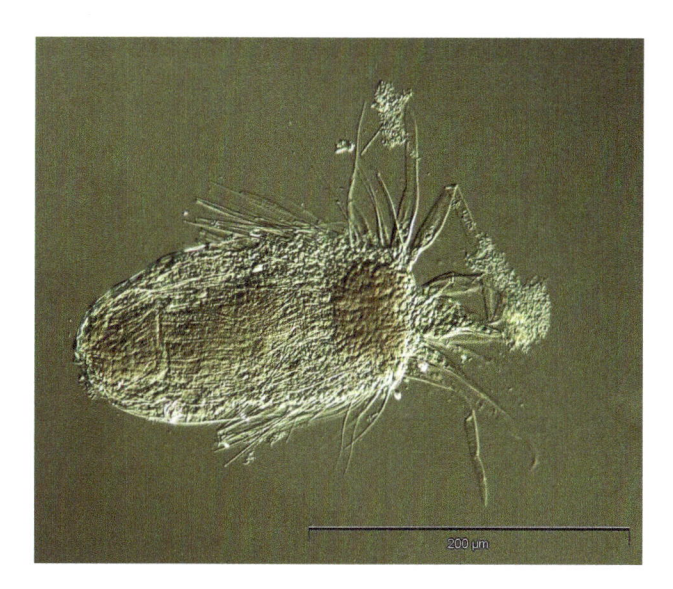

Shuker's loriciferan Pliciloricus shukeri.
(Dr. Reinhardt Kristensen)

SEA DAISY—A DRIFTWOOD-DERIVED DISCOVERY

In the early summer of 1985, biologist Dr. Bruce Marshall at New Zealand's National Museum was studying some tiny marine snails, obtained from crevices in samples of driftwood collected at the sea bottom around the country's coasts. The first samples had been gathered on 9 July 1983, just west of Hokitika, South Island, at a depth of about 3800 ft, with further samples having been obtained between April and May 1985, off Castlepoint, North Island, at depths of 3523-4027 ft. Among the snail specimens contained within these were some small starfishes too, so in August 1985 three echinoderm specialists called by to take them for study. When they examined the vessels containing the starfishes, however, they noticed that some much smaller, unfamiliar-looking animals were also present.[111]

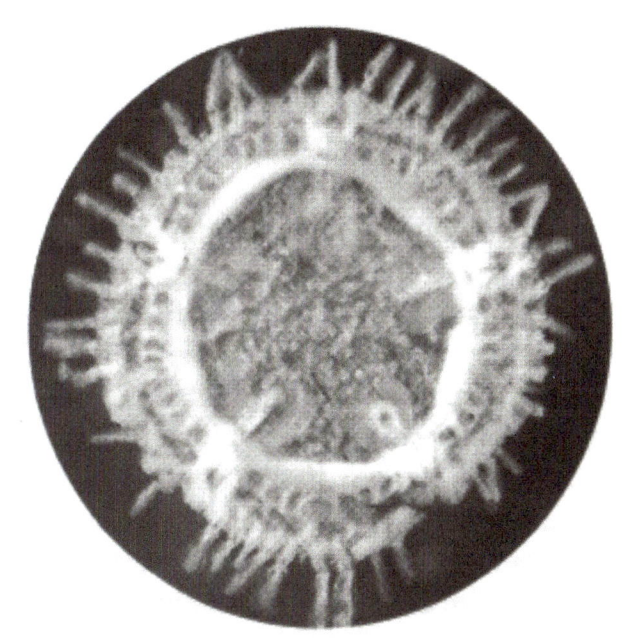

Sea daisy Xyloplax medusiformis.
(Dr. Alan Baker)

To most observers, these minute mysteries, none of which had a total diameter exceeding 0.35 in, would seem to be nothing more exciting than some form of miniature jellyfish medusa. Conversely, the echinoderm experts—Drs. Alan Baker and Helen E.S. Clark (from New Zealand's National Museum) and Francis W.E. Rowe (from Sydney's Australian Museum)—recognised them to be the zoological find of a lifetime.[111]

The overtly novel species that they represented soon became known as the sea daisy, and in June 1986 it was officially named *Xyloplax medusiformis*—'medusa-like plated animal found on wood'. Belying its superficial outward appearance, however, the tiny creature was not a jellyfish at all, but a highly significant new member of the echinoderm phylum. Previously, there had been five classes of modern-day echinoderm; now, thanks to the salvaged samples of driftwood, there was a sixth—a newly-created class christened Concentricycloidea, with the sea daisy *Xyloplax* as its sole occupant.[111]

Although the sea daisy betrays its echinoderm identity via such characteristics as a calcite-based skeleton, a body structure founded upon a basically five-rayed (pentamerous) configuration, and peculiar little appendages called tube-feet, it also has several features that readily delineate it from all previously-described members of the phylum.[111]

For example, the perimeter of its disc-shaped, plated body bears a fringe of petal-like spines (the reason why it became known as the sea daisy). For hydrostatic purposes, its body contains two, concentric water-containing rings (hence its class's name, Concentricycloidea) instead of just a single ring like other echinoderms. The sea daisy's underside is covered by a thin membrane, through which it probably absorbs all its nutrients and excretes all its waste products, as it has neither a mouth nor an anus (it does not have a stomach either). And its tube-feet are aligned in a single ring around the edge of its body's underside, instead of in five or more pairs of rows radiating outwards from the centre of its underside (like the spokes of a wheel) in the style of all other echinoderms.[111]

Its overall structural anatomy indicates that the sea daisy is most closely related to the starfishes and brittle stars, although it has no arms of its own—just another contradictory feature of this enigmatic little echinoderm.[111]

In 1988, the three echinoderm specialists who described *X. medusiformis* brought to scientific attention a second species of sea daisy, which they christened *X. turnerae*. This was based upon specimens obtained from wooden plates that had been submerged for 12-40 months in the Tongue of the Ocean, east of Andros Island, Bahamas, at a depth of just over 6885 ft. Its most striking differences from *X. medusiformis* are its possession of a stomach, and its apparent oviparity (*X. medusiformis* is viviparous).[112]

A third species, *X. janetae*, was described in 2006 from 103 specimens collected on 31 August 2004 by Chicago Field Museum researcher Janet Voight (after whom it was named), using the submarine *Alvin*. They had been recovered from samples of wood that had been experimentally deployed at a depth of 8776 ft in the northeastern Pacific Ocean (near 42° 45.2′ N 126° 42.5′ W) two years earlier by the Monterey Bay Aquarium Research Institute's remotely-operated vehicle *Tiburon*. Following this latest species' description, in which phylogenetic affinities between the sea daisies and the starfishes were highlighted, Concentricycloidea was downgraded from a taxonomic class in its own right to an infraclass within the starfish class, Asteroidea.[112a]

Staying with echinoderms but returning to 1985, this was also the year in which Dr. Michel Roux described a quite exceptional 'living fossil', five specimens of which had been obtained at depths of 5000-6660 ft off the Mascarene island of Reunion in the Indian Ocean during 1982. A species of stalked sea lily named *Guillecrinus reunionensis*, what makes this particular species so significant is that it constitutes the very first modern-day representative of the sea lily subclass Inadunata—believed until then to be wholly extinct since the Palaeozoic Era, which ended 251 million years ago. A second living species, *G. neocaledonicus*, was described in 1991.[113]

EXSUL SINGULARIS—REAPPEARANCE OF THE WORLD'S RAREST FLY

The year 1987 ended on an entomological high note, with the return of an uncommonly handsome species of fly—one that spurns the traditional buzzing bluebottle image in favour of a mighty 2-in-long body, and in the male (but not the female) a pair of enormous, butterfly-like wings of shimmering, white-edged, ebony hue, which heat up in sunlight, enabling it to take flight. Endemic to New Zealand but a member of the house fly family, Muscidae, it was originally described in 1901 by Captain F.W. Hutton from a specimen captured at Milford Sound by a Professor Wall, and was named *Exsul singularis*, but this attractive insect, nowadays known as the bat-winged fly, had not been reported since 1942.[114]

However, it reappeared alive and well 45 years later in the Homer Tunnel, Dunedin, in the southernmost region of South Island (in 1996, New Zealand conservationist B.H. Patrick revealed that a single post-1942 specimen had also been collected here back in 1975, as well as one at Tutuko Beach in 1980, but these had not attracted widespread attention). Its discoverer was a five-year-old boy—Jamie Morris, who lived close to the tunnel. The specimen that he captured was photographed and conclusively identified by entomologists—no doubt with great excitement, as *E. singularis* was generally considered at that time to be the world's rarest fly. Happily, a number of additional specimens from various localities have since been recorded.[114]

ANTS IN THE PLANTS, AND AUSTRALIAN AMAZONIANS

Sometimes, notable invertebrate finds can be made in the most surprising places.

One day in 1989, Kathryn Fuller, president of the U.S.A. division of the World Wide Fund For Nature (WWF), happened to notice some ants running across her desk in her Washington office. By good fortune, Harvard University zoologist Prof. Edward O. Wilson, a renowned authority on ants and author of a newly-completed definitive tome on the ants of the world, was visiting Fuller at that time, so she asked him to identify her desk's tiny trespassers. To his great surprise, they did not seem to correspond to any species on record! Eager to learn more about them, Wilson and Fuller investigated their origin, and discovered that they had come from a colony in a potted plant behind Fuller's desk. Some worker specimens were collected, and later, when back at Harvard, Wilson received from Fuller some soldier specimens too. Their secretive, hitherto-undocumented species belonged to the Neotropical *Pheidole* genus, and in 2003 Wilson formally named it *P. fullerae* in honour of its discoverer.[115]

In spring 1991, bulldozing work on a building site in Sydney, Australia, was abandoned after workers discovered that it was inhabited by some highly unexpected occupants. Most people are familiar with the sight of millipedes scuttling along the ground on their multitude of legs, for there are numerous species, found throughout the world. In contrast, only a very few species of fully aquatic, freshwater millipede had been recorded—the most famous of these, described in 1985, being *Myrmecodesmus adisi*, from the blackwater swamps of the Amazon River near Manaus in Brazil—or at least this was true until the bulldozers' discovery in Sydney. For during their preliminary work, they encountered a colony of pin-sized aquatic millipedes.[116]

Moreover, in 1991 a second species was discovered at the very same site! Whereas the first had been a siphonotid (and has yet to be formally described?), this second one was a pyrgodesmid, related to the aforementioned Amazon species, and was subsequently identified as *Aporodesminus wallacei*, originally described in 1904 and previously reported from St. Helena, Tahiti, Polynesia, and Hawaii, but not Australia. Unexpectedly, the building work recommenced shortly afterwards, and for a time it seemed that both species had been destroyed, as several dead specimens but no living ones were discovered. Happily, living examples of both were subsequently found.[116]

THE BIGGEST SPANISH DANCER IN THE WORLD?

In stark contrast to their morphologically unappealing land-dwelling relatives, the sea slugs or nudibranchs offer their fortunate eyewitnesses a visual extravaganza—with each species presenting such a dazzling pageant of sumptuous hues and so flamboyant a bouquet of plume-like gills that it could readily be mistaken for a surrealistic, sea-dwelling bird of paradise. There are numerous species currently known to science, of worldwide distribution, and until fairly recently the largest on record was the northwestern Pacific's *Tochuina tetraqueta*, up to 1 ft long.

In May 1990, however, artist Tamara Double was diving in the Red Sea, near to Djibouti's Les Sept Frères islands, when she saw some huge sea slugs almost twice as long as *T. tetraqueta*, each bearing six branches of deep red, profusely-plumed gills, and undulating its pinkish-red mantle edges like a flamenco dancer. They seemed to be a giant form of the aptly-named Spanish dancer *Hexabranchus sanguineus*, not believed to exceed 9 in long in the Red Sea. Double captured one, which was subsequently studied by Dr. Nathalie Yonow at Swansea University to discover if it were indeed an outsized Spanish dancer, or an exciting new species in its own right. Others have been seen near Fiji and in the South China Sea, but all of these giant forms are currently deemed conspecific with their smaller, more familiar counterparts.[117]

VULTURE BEES AND DRACULA ANTS

Vultures come in all shapes and sizes—but few are any stranger than the version brought to zoological and public attention during the early 1980s by Dr. David W. Roubik from the Smithsonian Tropical Research Institute, based in Panama. Named *Trigona hypogea* and scientifically described back in 1902 from Brazil by Dr. Filippo Silvestri, it lacked feathers, but possessed two pairs of wings, three pairs of legs, and was no bigger than a bee—which is hardly surprising, because that is precisely what it was. A bee—the world's first-known species of obligate necrophagous bee, i.e. feeding exclusively upon dead flesh.[118]

Roubik first suspected the dark but dramatic dietary secret of *T. hypogea*, a social stingless species, when he opened one of its hives and, instead of finding pollen in its storage cells, discovered a strange substance which, when analysed, was found to have an unexpectedly high protein content. A further clue came to light when he noticed that this species' workers lacked pollen baskets on their legs, proving that they did not collect pollen—the normal source of protein in the diet of bees. The final piece of this entomological jigsaw was provided by a rather unusual source—the Roubik family's Thanksgiving turkey. After completing their meal, they placed the remains of their turkey outside—and watched.[118]

Like a veritable flock of vultures, a swarm of slender black *T. hypogea* workers swiftly descended and promptly began stripping every last vestige of flesh from its bones, excavating deep inside the carcase too in order to reach the internal tissues, and reducing it to a skeleton over the next few hours. Later studies revealed that these bees strip off flesh by coating it with an enzyme to break it down, after which they masticate and partially digest it. Following their return to the hive, they regurgitate this meaty mush, enabling other workers to metabolise it and then store it in the hive's brood cells for feeding to the larval bees.[118]

Roubik's remarkable discovery had an equally unexpected sequel. He documented his findings concerning necrophagy in bees during 1982, but later researches revealed that although *T. hypogea* was indeed necrophagous, the species that Roubik had been studying was not *T. hypogea* at all! Instead, it was a species totally new to science, which was officially described in 1991 by Roubik and Brazilian entomologist Dr. João Camargo, who aptly dubbed it *T. necrophaga* ('carrion eater'). Moreover, their researches showed that a third species, *T. crassipes* (first described in 1793 by Fabricius), was also necrophagous. Investigating their favoured meat sources revealed a marked preference for dead reptiles and amphibians, but even deceased earthworms would suffice if large enough.[119]

On 15 February 1993, while conducting field work in western Madagascar's Zombitse Forest, California University entomologist Dr. Philip S. Ward collected 21 worker specimens of an extremely unusual but hitherto-unrecorded species of small, pale, subterranean ant. Formally describing it in 1994, he designated it the sole occupant of a new genus, and dubbed it *Adetomyrma venatrix*. Among its many remarkable, primitive features was a wasp-like single-joint waist or petiole joining its thorax to its abdomen (rather than the more familiar two- or three-joint waist of modern ants), its lack of eyes, and its enormous barbed sting (bigger relative to body size than that of any other species of ant known to science).[119a]

Yet the most extraordinary feature of all exhibited by this strange ant, a relict species like many other native Madagascan animals, remained undisclosed for several more years. During January 2001, however, Dr. Brian Fisher from the California Academy of Sciences reported the recent discovery—inside a rotten rainforest log—of an entire colony belonging to this species, containing the first recorded queens (wingless), larvae, and males (winged). When the colony was studied, an unexpected and decidedly gruesome facet of *A. venatrix* behaviour was exposed. If in need of nutrition, the queen and workers will visit the colony's nursery containing the ant larvae, bite into them using their mandibles, and chew until haemolymph (insect blood) begins oozing from the larvae, which the adults then suck, but without killing the larvae.[119a]

This bizarre, vampirish activity is known as non-destructive cannibalism, and Fisher believes that it may have given rise during ant evolution to the practice, exhibited today by various advanced ant species, of larvae regurgitating semi-digested food for the workers to feed upon. Previously without a common name, *A. venatrix* is now known as the Dracula ant, in keeping with its grisly blood-sucking behaviour.[119a]

AFTER 300 MILLION YEARS—HAVE THE GRAPTOLITES RETURNED?

Constituting a once-diverse group of colonial, polyp-budding hemichordates related to the modern-day colonial pterobranchs, those fossil marine invertebrates known as graptolites were traditionally believed to have died out entirely by the end of the Carboniferous Period, approximately 300 million years ago—but did they?

On 27 February 1989, specimens of a hitherto unknown pterobranch were collected at a depth of 830 ft near the New Caledonian island of Lifou, by the submersible *Cyana* during the French Calsub Expedition. When these were examined by pterobranch expert Prof. P. Noel Dilly, then based at St George's Hospital Medical School, London, he recognised that this new species, which in 1993 he named *Cephalodiscus graptolitoides*, was very special.[120]

Each individual within a colony of graptolites or pterobranchs is known as a zooid, and zooid colonies

Fossil graptolites—Pendeograptus fruticosus—dating back more than 470 million years, from Australia's Lower Ordovician; could Cephalodiscus graptolitoides truly be a living relative?

are created via the secretion of tubes by these individuals. Interestingly, pterobranch zooids can actually come out of their tubes if need be, in order to repair damaged tissue or feed. But what of graptolite zooids? Fossils reveal that graptolites bore spine-like structures called nemata, yet how these were produced has long remained controversial. Some researchers have suggested that the creation of a nema must have been carried out by special external tissue whose remains have not been preserved by fossilisation. Others have proposed that a nema was produced directly by one or more graptolite zooids after leaving their tubes, by secreting collagen from special lobes as they crawled over the outer surfaces of their tubes. Now, with the discovery of *C. graptolitoides*, these and other proffered hypotheses could finally be tested.[120]

This was because, uniquely among pterobranchs, this species actually produced spine-like structures that closely resembled the nemata of graptolites—and not only externally but also internally, when examined on an ultrastructural level via transmission electron microscopy (in the mid-1970s, Polish palaeobiologist Dr. Adam Urbanek had revolutionised graptolite research by utilising TEM techniques on graptolite fossils for the first time). And it turned out that the spines of *C. graptolitoides* are indeed produced when zooids leave their tubes and crawl over their surface, secreting collagen.[120]

In view of the apparent homology of its spines and the graptolite nemata and also of the mechanism for producing these structures, Dilly therefore concluded: "There is little if any reason for not considering *C. graptolitoides* as a living fossil and a member of the graptolites previously thought to be extinct for over 300 million years.... With *C. graptolitoides* we now have a pterobranch that is almost indistinguishable from the graptolites".[120]

Not everyone shared his view, however, with some researchers claiming that the fossilised tubes and nemata of graptolites were too complex to have been created by the zooids themselves and hence were produced by some still-undiscovered layer of enveloping external tissue. Moreover, Dr. Adam Urbanek opined that the graptolite nema's spiny shape was merely a secondary development, and that the nema may have originally been derived from the basal disc of the primary zooid in the graptolite colony. Responding to these claims, Dilly, however, reaffirmed that *C. graptolitoides* and graptolites can and did produce spines in an identical manner, concluding: "It is important to remember that graptolites were once living animals; to study them exclusively from the remains of their homes is like studying humans from a derelict housing estate. We have in *C. graptolitoides* a zooid that secretes tissue indistinguishable even at the electron-microscope level from that of the graptolites".[121]

Evidently, Dilly had no doubt that, after 300 million years, one of the most famous 'prehistoric' groups of creatures had finally been resurrected, with *C. graptolitoides* deemed by him to be a bona fide living graptolite. Other researchers, conversely, remained unconvinced, and this enigmatic species is currently still housed by the vast majority of zoologists within the pterobranch class of hemichordates rather than the graptolite class.

MIMIC OCTOPUS—AN EIGHT-LEGGED MASTER OF DISGUISE!

One of the most extraordinary species of animal to be discovered by science during the 20th century first came to zoological attention as recently the mid-1990s. While swimming one day across a shallow portion of Maumare Bay off the Indonesian island of Flores, Australian underwater photographer Rudie Kuiter spied a swimming flounder—or what seemed to be a swimming flounder. Moments later, however, and to Kuiter's amazement, the 'flounder' transformed itself into an octopus! What Kuiter had seen was an astonishingly accurate impersonation of a flounder, which had been skilfully accomplished by the octopus wrapping its arms behind its body, raising its eyes up in faithful imitation of a flounder's eyes, utilising its funnel to skim above the sand in just the way a flounder would swim, and even changing its colour to the brown, sandy shade sported by flounders.[122-3]

But that was only the beginning. As Kuiter and a colleague, fellow underwater photographer Roger Steene, continued to watch this remarkable octopus, it executed a stunning repertoire of impersonations—transforming with protean ease into a sea snake, cuttlefish, stingray, brittle star, and several other diverse forms of marine creature. More than fifteen different disguises have been recorded from what has very aptly become known as the mimic octopus. In addition to those already listed, they include a lionfish, snake eel, jawfish, blenny, mantis shrimp, ghost crab, hermit crab, jellyfish, sand anemone, sea cucumber, and feather star. Of these, the most extreme disguise adopted must surely be the sea snake, for which the mimic octopus buries itself in the sand with just two tentacles undulating sinuously above the surface in an 'S'-shaped curve.[122-3]

When glimpsed in its true form, the mimic octopus is revealed to be an exceedingly long-limbed species (its slender tentacles are considerably longer than its body), boldly patterned in black-and-white stripes, though it can change colour at will to suit whichever disguise it is adopting. Kuiter and Steene also found it off northern Sulawesi and Bali, and it is now also known from New Guinea, the Philippines, Malaysia, New Caledonia, and even as far afield as the Red Sea.[122-3a]

Mimic octopus (upper left) and wunderpus (lower right)—two extraordinary new species of octopus impressionists! (William M. Rebsamen)

During autumn 1998, Australian octopus expert Dr. Mark Norman of James Cook University in Queensland and his colleague Julian Finn succeeded in finding and filming eight specimens of this dramatic new species over a total of 3 weeks for a BBC-Discovery TV documentary, entitled *Animal People: Octopus Hunter*. They discovered that whereas other octopuses normally emerge at night, to avoid predators, the mimic octopus is active during the day, frequenting soft sandy areas where it can be readily spied by all manner of potential predators. Clearly, therefore, its ability as a master impersonator of often very dangerous, venomous animals, notably sea snakes and lionfishes, is very successful—otherwise it would have become extinct long ago. Moreover, it may actually be venomous itself.[122-3]

In September 2001, the mimic octopus finally made its debut within the formal scientific literature. This was when Dr. Mark Norman and two colleagues co-authored a short report documenting the diversity of this species' amazing powers of impersonation.[123a]

On 31 August 2005, the mimic octopus's long-awaited naming and description was finally published, in a *Molluscan Research* paper authored by Norman and Santa Barbara Museum of Natural History colleague Dr. F.G. Hochberg. In their paper, they formally christened this invertebrate impressionist *Thaumoctopus mimicus*, the first member of an entirely new genus of octopus.[123a]

Another very interesting, related species, discovered by Norman during the early 1990s, has been dubbed the wunderpus—for good reason. Its tentacles, wider than those of the mimic octopus, are emblazoned with such dazzling black-and-white stripes that it is almost psychedelic in appearance! Furthermore, just like the mimic, the wunderpus can perform a very convincing sea snake impression, but this seems to be the limit of its impersonation ability. And whereas the mimic is diurnal, the wunderpus is crepuscular. It first came to Norman's attention when he was shown photographs taken by Alex Kierstich of specimens from the Philippine aquarium trade. Norman and Hochberg subsequently tracked down specimens from the Philippines, Indonesia, New Guinea, and Vanuatu, and have observed living wunderpuses off northern Sulawesi.[122]

In late 2006, Norman and Hochberg teamed up with fellow octopus researcher Dr. J. Finn to describe and formally name the wunderpus. In their paper, published in *Molluscan Research*, Hochberg, Norman, and Finn dubbed this spectacular species *Wunderpus photogenicus*, which, like the mimic octopus, makes it the first species in a brand-new genus.[123b]

Norman, Finn, and Hochberg are unquestionably the most successful discoverers of new octopuses in modern times. During the 1990s alone, they discovered no less than 150 new species in western Pacific and Indian Ocean waters, including 68 off Australia, which range from pygmies the size of a fingernail to giants with 12-ft tentacle spans.[122]

FLESH-EATING SPONGES IN THE MEDITERRANEAN

Known technically as poriferans, sponges have traditionally been looked upon by zoologists as innocuous aquatic animals of harmless, filter-feeding predilection—but no more! January 1995 will forever be celebrated by poriferan pundits as the month that witnessed the unmasking of the world's first-known species of carnivorous sponge.[124]

Belonging to the demospongid family Cladorhizidae, and formally christened *Asbestopluma hypogea* in 1996, it was discovered in a shallow-water Mediterranean cave by researchers from the Centre d'Océanologie de Marseille, and flagrantly eschews the passive filter-feeding lifestyle characterising all other sponges, in which a single layer of flagellated cells called choanocytes (absent in this new species) is utilised in pumping a unidirectional water current though the sponge's body. Instead, it employs long tendril-like structures to seize tiny crustaceans swimming by—which are then hauled inside its body and digested by what has been vividly described as "a crawling chaos of cells". Carnivory is now known to be both common and typical for cladorhizid species.[124]

ENTERING THE LOST WORLD OF MOVILE CAVE

Not everyone can emulate Sir Arthur Conan Doyle's fictitious Professor Challenger and discover a Lost World teeming with new species never before seen by mankind—but in 1986 this spectacular achievement was accomplished by Romanian biologist Dr. Serban Sarbu and colleagues from Bucharest's Speleological Institute when they became the first scientists ever to penetrate the stygian gloom of a subterranean system of bell-shaped grottoes called Movile Cave, sited 80 ft below Dobrogea, near southeastern Romania's Casimcea Valley.[125]

Unfortunately, it was not possible for Sarbu to investigate fully its remarkable contents during the regime of Nicolae Ceaucescu, and he later moved to the U.S.A., taking up a post at Cincinnati University. During the mid-1990s, however, his work at Movile Cave resumed, and its amazing secrets finally became known to the west.[125]

Within its limestone confines, Sarbu and his team discovered a wholly autonomous realm sealed off from the rest of the world for 5.5 million years, and populated by an array of dark-adapted denizens whose evolutionary pathway had diverged dramatically from that of their brethren elsewhere on the planet. In the eternal night of Movile Cave's interior, where sunlight has been banished for countless millennia, eyes and sun-shielding body pigments are surplus to functional requirements. Consequently, its inhabitants were blind and pallid; many had evolved extra-long antennae or analogous organs instead, supplanting sight with tactile sensory structures—a speleozoological phenomenon called troglomorphy.[125]

As expected from so specialised a kingdom, most of its animals were hitherto unknown to science—over 30 new species have currently been documented. Its terrestrial complement of newcomers included four beetles, five spiders, three pseudoscorpions, two springtails, four woodlice, and one acarine mite. The cave's internal lake contained several new species too, including three nematode worms, a blind species of 'water scorpion' bug (*Nepa anophthalma*), one leech, one water snail (*Semisalsa dobrogica*), and one copepod crustacean.[125]

Incredibly, however, these findings were only the tip of the biological iceberg. What was so remarkable about Movile Cave is that it contained any life at all—because its air and its lake's waters were suffused with hydrogen sulphide, which is normally poisonous to living things. Moreover, its atmosphere contained 100 times as much carbon dioxide and less than ten percent as much oxygen as the air breathed by living things in the outside world.[125]

Far from proving lethal to life, however, within the confines of Movile Cave hydrogen sulphide had become the very source of its unique ecosystem's success, Sarbu's researches revealed. Elsewhere on Earth, sunlight is the primary energy source in food chains—but not here. Within the independent kingdom of Movile Cave, sunlight energy is replaced by chemical energy, creating what is termed a chemo-autotrophic (as opposed to a photo-autotrophic) system. Its chemical energy is released during the breakdown (oxidation) of hydrogen sulphide within the lake's waters by chemosynthetic bacteria known as beggiatoaceans. This energy is in turn harnessed by other types of bacteria, which utilise it in the manufacture of glucose. As these latter bacteria multiply, they become imprisoned by a network of fungal filaments that gain nourishment from them; the final result is a gelatinous floating 'sail', up to 1 inch thick, covering the lake's surface.[125]

This extraordinary microbial sail is greedily grazed by the lake's smaller species of animal, such as protozoan ciliates, rotifers (wheel animals), and copepods, which in turn are preyed upon by larger ciliates and leeches, as well as the blind water scorpion. This sightless aquatic insect, unique to Movile Cave, is related to the familiar water boatman and earns its name from its long 'tail', which in reality is a breathing snorkel, not a sting. Dense microbial mats also carpet the lake's rocky shores, where they are browsed upon by snails and woodlice, and even on the cave walls, where blind spiders and pseudoscorpions lurk, preying upon grazing springtails and mites. Interestingly, one of these spiders, a radically new species dubbed *Lascona cristiani*, is born with eight eyes like most normal spiders, but as it matures they gradually atrophy until they eventually vanish completely.[125]

Science only knows of one other autonomous chemo-autotrophic ecosystem—and that was not discovered until 1977, when the U.S. Navy submarine *Alvin* took scientists down to the sea floor near the Galapagos Islands. Here they were astonished to find a thriving world populated by such exotic, previously unrecorded species as giant tubicolous worms with long scarlet tentacles, filamentous enteropneusts resembling living strands of spaghetti, ethereal relatives of the Portuguese man o'war that recalled underwater dandelion clocks, and luxuriant microbial mats harnessing energy released during the combination of hydrogen with sulphates present in warm water rising up through hydrothermal vents in the sea floor[82] (see p. 250).

With the disclosures from Movile Cave, however, we now know that a comparable regime can exist in freshwater too. Indeed, there may be caves elsewhere in the world that sustain similar ecosystems—caves whose secrets have yet to be brought out of the darkness of scientific anonymity and subjected to the spotlight of formal examination.

A NEW PHYLUM, FOUND ON THE LIPS OF LOBSTERS!

Taxonomists, who spend their entire working lives classifying species, are often lucky enough to be able to describe new species, especially if they are studying insects or microscopic animals. Yet even among these fortunate few, none can compare with the unique record set by Dr. Reinhardt Møbjerg Kristensen from Copenhagen University's Zoological Museum. In little more than a decade, he has described two new species so drastically different from all other animals on Earth that each has required the creation of a brand new phylum—the highest taxonomic category within the animal kingdom.

As documented on p. 257, the first of his two new phyla was Loricifera, erected in 1983 to house a microscopic creature discovered in shell gravel at Roscoff, France, called *Nanaloricus mysticus*. Many other loriciferan species have since been discovered.[107-10]

In December 1995, however, Kristensen formally announced in the journal *Nature* the creation of a second new phylum, dubbed Cycliophora, in order to accommodate a diminutive but extraordinary multicellular parasite exclusive to the lips of Norwegian lobsters! First discovered in August 1991 colonising the mouthparts of *Nephrops norvegicus* specimens taken off Frederikshavn, North Kattegat, Denmark (and first documented in a short Danish report published two years later[126]), its species was dubbed *Symbion pandora* by Kristensen and co-discoverer Dr. Peter Funch Andersen. Like the loriciferans, it is microscopic, less than 0.014 in (350 mm) long and under 0.005 in (120 mm) wide.[127]

The diagnostic characteristics of *S. pandora* include its body's bilateral symmetry and acoelomate form (lacking an internal body cavity), a well-differentiated cuticle, a U-shaped gut with mouth and anus, compound cilia engaged in filter feeding operating as a downstream-collecting system, and several different, distinctive stages in its life cycle. The largest and most conspicuous of these stages is a sessile feeding female. Others include smaller, mobile, non-feeding forms— Pandora larvae, dwarf males, and dispersive females that become chordoid larvae. Indeed, the lifecycle of *S. pandora* is very complex.[127]

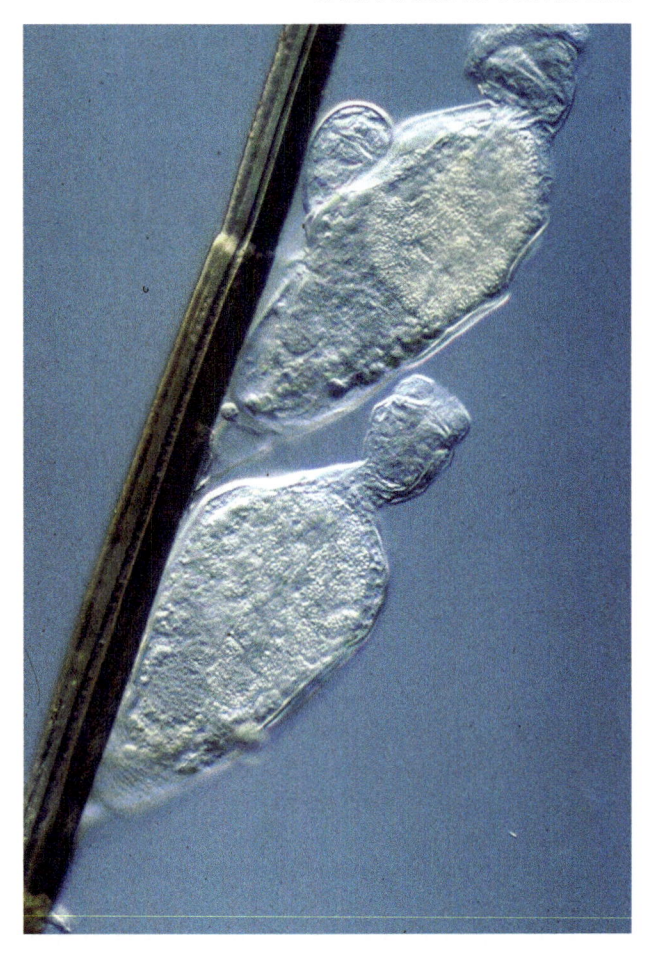

Symbion pandora. (Dr. Reinhardt Kristensen)

It begins with the chordoid larva, a small free-swimming stage that constitutes a modified trochophore larva and contains internal buds. This larva eventually settles on the mouthparts of a lobster specimen, and its body then degenerates, but one of its internal buds develops a bell-shaped buccal (mouth) funnel with a circular wheel-like mouth ring (hence Cycliophora—'carrying a small wheel') and begins feeding. It has now become the sessile feeding female stage, whose body consists of a buccal funnel, an ovoid trunk, and a stalk ending in a disc that attaches the female to its lobster host.[127]

Not only does the female feed, however, but it also continually produces internal buds, which eventually replace the buccal funnel. When mature, the sessile female asexually develops a short-lived, non-feeding, dispersive larva known as a Pandora inside a brood chamber. Moreover, budding taking place inside the Pandora while still inside the brood chamber creates a small new feeding stage. At the same time, an internal bud inside the sessile female transforms into a new buccal funnel. As for the Pandora, this larval form finally escapes out of the sessile female, settles, degenerates, its internal bud transforms into a sessile female, and the above-described asexual cycle commences again.[127]

In addition, there is also a sexual cycle, which is initiated when the host lobster is near to the end of its moult cycle. Instead of producing a Pandora larva, some sessile females either produce a dwarf male or a mobile female (both are short-lived, dispersive, non-feeding forms), though the mode of production in both cases is similar to the process that results in a Pandora. If a male is created, it ultimately escapes from the sessile female and, filled with sperm, settles upon another sessile female if the latter contains a developing female.[127]

After this has taken place, the developing female escapes from its own sessile female, but settles on the mouthparts of the same host. Within this female is a zygote, which develops into a chordoid larva encased within a shell. The female finally dies, and the chordoid larva hatches from its shell, swims away, develops internal buds, and settles on a new host. The larva's own body then degenerates, but one of its internal buds transforms into a sessile feeding female, ready to begin the complicated life cycle of *S. pandora* all over again.[127]

Kristensen initially deemed *S. pandora* to be allied to the ectoprocts (moss animals or bryozoans) and the entoprocts, but in 1998 he and two colleagues published the results of a phylogenetic analysis of 18S ribosomal RNA sequence data. These suggested a sister-group relationship between the phylum Cycliophora and a Rotifera-Acanthocephala (i.e. wheel animal/thorny-headed worm) clade, but still fully warranting separate phylum status. Two additional cycliophoran species have lately been discovered—*S. americanus*, whose host is the American lobster *Homarus americanus*; and an as-yet-unnamed species, whose host is the European lobster *H. gammarus*.[128-128a]

SYNALPHEUS REGALIS— FIRST KNOWN SOCIAL SHRIMP

In 1996, crustacean expert Dr. J. Emmett Duffy from the School of Marine Science and Virginia Institute of Marine Science, Virginia, described a new sponge-dwelling shrimp from the Belize Barrier Reef, in the Caribbean Sea off Central America. Named *Synalpheus regalis*, what made this particular half-inch-long species so interesting was that it proved to be the first known eusocial marine animal, with a communal lifestyle paralleling that of eusocial insects, such as certain bees, wasps, ants, and termites. Each colony of *S. regalis* contains a breeding queen shrimp, and up to 300 non-breeding shrimps from overlapping generations. These act as workers, which maintain the colony and raise the queen's offspring; or as soldiers (usually males), which are larger than the workers and utilise their huge claws to defend the colony from potential attackers. In recent years, eusocial behaviour has also been revealed in a few other species of *Synalpheus* shrimp.[129]

Synalpheus regalis. (Dr. J. Emmett Duffy)

METHANE ICE WORMS OF THE MEXICAN GULF

The methane ice worm is nothing if not aptly named. The ocean floor of the Mexican Gulf, about 150 miles south of Louisiana's coast, bears some sizeable mounds of yellow and white methane ice (gas hydrates), also known as methane clathrate. Traditionally, these mushroom-shaped mounds, measuring 6-8 ft across and forming naturally at high pressure and low temperature, have been deemed too noxious to sustain any form of animal life. In July 1997, however, a team of scientists from Pennsylvania State University, led by Prof. Charles Fisher, journeyed 1800 ft down to the Gulf's sea bottom in a mini-submarine and were amazed to discover considerable numbers of a seemingly unknown species of vermiform invertebrate thriving upon and burrowing within the methane mounds. Pink, flat, 1-2 in long, and superficially resembling centipedes, these unexpected excavators were found to constitute a new species of hesionid polychaete—segmented, multi-legged worms related to the familiar ragworms and lugworms—which in 1998 was formally dubbed *Hesiocaeca*

Methane ice worm.

methanicola. It apparently survives by grazing upon chemosynthetic bacteria growing upon the methane ice.[130]

A JELLYFISH FROM THE MIST

In 1926, an article by W. Crowder concerning the milky-hued, semi-transparent moon jellyfish *Aurelia aurita*, which was published in the *National Geographic*, included a photograph of an unusual, unidentified specimen captioned merely as a 'black jellyfish'. Nothing more was heard of this strange-looking scyphozoan for almost 40 years, until some more photos of what was clearly the same mysterious species appeared in Vol. 1 of Prof. B.W. Halstead's monumental work *Poisonous and Venomous Marine Animals of the World* (1965). Here, however, it was categorised as *Cyanea capillata*, the lion's mane jellyfish, even though it exhibited obvious differences from this latter, familiar species.[131]

Another 24 years were to pass before it surfaced again—but this time its appearance was very dramatic. Hundreds of specimens were spied in the sea and were subsequently washed ashore off southern California and Mexico's Baja California during late 1989. Ironically, however, only four of these were actually salvaged and preserved, though colour photos and footage of living specimens were also obtained.[132]

That was fortunate, because when the morphology of those preserved and filmed examples was compared with that other jellyfishes, scyphozoan experts realised that although this species evidently belonged to the sea-nettle genus *Chrysaora*, it did not correspond with any known *Chrysaora* species. After more than 60 years, Crowder's enigmatic 'black jellyfish' had finally been recognised for what it was—a wholly new species undescribed by science.[133]

With a bell diameter estimated to exceed 3 ft in the fully mature medusa, plus 24 long delicate tentacles of pale pinkish-white shade and four fairly thick oral arms with frilly edges and purple corkscrew-shaped form extending possibly as much as 20 ft beneath the bell when alive, this is a relatively enormous species, made even more spectacular by virtue of the dark purple to black colouration sported by its huge bell. In August 1997, it was finally described and named, by a team of American researchers who aptly dubbed it *Chrysaora achlyos*—'jellyfish of the mist' (or 'dark'). This name not only refers to its dark colouration but also alludes to its elusive, shadowy history of rare, unrecognised appearances in the past, emerging only briefly from the mist of zoological obscurity before vanishing again. In common parlance, it is known as the black sea-nettle or simply as the black jellyfish, and is one of the largest species of invertebrate to have been discovered during the 20th century.[133]

LAND OF THE GIANTS—UNDERWATER!

Prior to the all-annihilating progression of ice across Antarctica during the late Pliocene and Pleistocene a few million years ago, this southern continent had supported a thriving terrestrial ecosystem. Today, glacially encased, it has become for many people the forgotten continent, the last place where one might expect to find major new forms of life. In reality, however, although the face of Antarctica is indeed a bleak vista, the seas encircling it are among the world's richest, zoologically, and have lately unfurled some amazing surprises.

On 8 September 1997, at a meeting of the British Association held in Leeds, biologist Dr. Lloyd Peck of the British Antarctic Survey revealed that Antarctica's encompassing ocean was a veritable kingdom of hitherto-undiscovered monsters . . . big monsters! Recent iceberg studies conducted by the survey had also incorporated investigations of the diverse life-forms inhabiting this continent's chilling maritime world. These had uncovered many previously unsuspected and remarkable cases of gigantism—the growth of creatures to sizes far in excess of their more familiar, modest-proportioned counterparts found elsewhere.[134]

Take, for instance, the small, humble woodlice. These terrestrial isopod crustaceans have a number of marine relatives known as sea slaters, and those living off the temperate coasts of Britain and continental Europe measure little more than an inch in length. The sea slaters frequenting the waters off Antarctica, conversely, are almost 7 in long.[134]

Equally dramatic are the size differences recorded between the sea spiders of Antarctica and Europe. Known technically as pycnogonids, sea spiders are only very distantly related to true spiders, and on first sight seem to constitute an animated assemblage of slender limbs, with little if any head or body. Europe's sea spiders tend to be decidedly small, and even when measured from leg tip to leg tip they often span less than an inch—whereas those of Antarctica are up to a thousand times bigger, and boast spans sometimes exceeding 1 ft.[134]

Perhaps the most awesome giants of this southern continent, however, are its ribbon worms or nemerteans. These very elongate invertebrates are also known as proboscis worms, on account of their characteristic, and extremely lengthy, eversible mouthpart—the proboscis—lying coiled up in a sheath on the upper side of the gut, and sometimes measuring twice as long as the worm's entire body. Nemerteans are carnivorous creatures, and catch their prey by shooting out their proboscis at it. Depending upon the nemertean species in question, this lethal apparatus either pierces its victim via armed stylets on its surface and secretes poisonous mucus inside, or coils around it rather like a constricting snake, and is then retracted, still grasping its helpless prey, back inside the worm.[134]

Many nemerteans are no more than 8 in long, but those of Antarctica have been found to grow up to 9 ft, with bodies as thick as an average human thumb. The hunting technique of these serpentine monsters is evidently a sight to behold, as revealed by Dr. Peck: "They have been seen grabbing fish by the head and swallowing them whole, like a snake".[134]

Much less formidable but no less imposing are the giant sponges of Antarctica. Eschewing the nondescript proportions attained by relatives elsewhere, these icy-water versions grow up to 10 ft tall, enabling divers to clamber inside their hollow architecture.[134]

The secret of these and other Brobdingnagian beasts dwelling here is the icy temperature of the Antarctic sea, for as Peck points out: "At low temperatures, it costs less to keep a given amount of tissue alive, because the metabolism runs more slowly. You can get bigger on the same amount of resource. You can also live longer".[134]

EVOLUTION IN ACTION IN THE LONDON UNDERGROUND

One of the most amazing animal discoveries in recent times features *Culex pipiens* var. *molestus*—the name given to a strange variety of blood-sucking mosquito inhabiting the tunnels of the London Underground (aka the Tube) train system in England's capital city. First entering the London Underground when it was being excavated during the 19th century, what were then normal gnats have transformed dramatically over successive generations, acquiring a taste for human blood. In August 1998, a study conducted by London University biologists Kate Byrne and Dr. Richard Nichols revealed that the London Underground mosquito possesses sufficient genetic differences from its aboveground ancestors of only a century ago to warrant possible reclassification in the foreseeable future as a separate taxonomic form in its own right.[135]

Indeed, it is nowadays classified as a distinct subspecies—*C. p. molestus* (some researchers even name it as a distinct species, *C. molestus*), which is characterised by breeding all-year-round, cold-intolerance, and by biting rats, mice, and humans, whereas the aboveground mosquito hibernates in the winter, is cold-tolerant, and bites only birds. Moreover, when these two mosquitoes were cross-bred in the laboratory, the resulting eggs were infertile, indicating reproductive isolation.[135]

A BRAWNY 'NEW' PRAWN DOWN UNDER

In late 1998, a decidedly brawny new species of mantis shrimp from Sydney Harbour, Australia, was formally described by Australian Museum research fellow Dr. Shane T. Ahyong. Up to 8 in long (*not* 16 in, as erroneously claimed in various media reports), purple in colour with banded antennae, and christened *Erugosquilla*

Erugosquilla grahami. (Dr. Shane Ahyong)

grahami, this formidable species is armed with razor-sharp pincers, with which it seizes small fishes at speeds of 5-8 milliseconds—one of the fastest movements known in the animal world.[136]

Equally interesting is that although it is new to science, this sizeable prawn-like stomatopod crustacean has been turning up in local fishermen's nets and fish markets for many years. It may even have been dined upon by Captain James Cook and his crew in the 1700s, when they trawled for food here. A few specimens have also been recorded from the South China Sea, near Taiwan. Yet until the late 1990s, no-one had realised that it had never been zoologically documented—a truly remarkable oversight, bearing in mind that Ahyong considers this species to be "probably one of the most distinctive invertebrates ever found". Within his paper, co-authored by fellow researcher Dr. S.B. Manning, Ahyong also described a second new *Erugosquilla* species, this time from the portion of the South China Sea off Vietnam, which was dubbed *E. serenei*.[136]

JAPANESE JELLIES

Distantly related to true jellyfishes are the ctenophores or comb jellies, named after their comb-like plates providing locomotion. Numerous new species and genera of marine creature are discovered each year. However, the two comb jellies caught back in 1992 at Toba port in western Japan's Mie Prefecture, but only formally described in 2000, by Toba Aquarium biologist Dr. Takushi Horita, were found to constitute a hitherto-unknown species so different from all other ctenophores that an entirely new taxonomic family had to be created to accommodate it. This remarkable new species (and genus) has been named *Lobatolampea tetragona*, on account of its four gonads—fewer than in other ctenophores.[137]

DISCOVERING THE DISKO BEAST— KRISTENSEN'S THIRD NEW PHYLUM

Already immortalised in zoological history as the only contemporary scientist to create two new animal phyla

(Loricifera and Cycliophora) based upon dramatically new species described by him, in the opinion of some researchers Danish zoologist Dr. Reinhardt Kristensen can lay claim to an astonishing hat-trick.

Limnognathia maerski.
(Dr. Reinhardt Kristensen)

In 1995, he and co-worker Dr. Peter Funch documented a strange new species of minute ciliated invertebrate that had been found in a cold spring at Disko Island, western Greenland. Only around 0.004 in (.100 mm) long, making it one of the world's smallest animals known, but possessing a large ganglion as a brain in its head, it was notable for its complicated jaws, whose ultrastructure was very similar to that of the jaws of certain gnathostomulids (p. 240), but also seemed to share certain characteristics exhibited by the mouthparts of rotifers. A photo of this creature's jawparts appeared in a short paper by Kristensen, again in 1995, within a symposium entitled *Body Cavities: Function and Phylogeny*.[138]

In 2000, Kristensen and Funch formally described this remarkable new creature, naming it *Limnognathia maerski*, and housing it within a specially-created taxonomic class of its own, dubbed Micrognathozoa. However, gnathostomulids and rotifers constitute two separate phyla. Consequently, because, as suggested by Kristensen and Funch in their description of it, the minibeast from Disko Island (and now also recorded from the subantarctic Crozet Islands) seemingly constitutes a 'missing link' between Gnathostomulida and Rotifera, some researchers prefer to house it in an entirely new phylum all to itself.[139]

In other words, thanks once again to the revolutionary, groundbreaking discoveries and researches of Dr. Kristensen—truly the Linnaeus of modern times!—yet another new phylum of animals has entered the zoological textbooks.

PART 2:
21ST-CENTURY DISCOVERIES AND REDISCOVERIES

Bornean clouded leopard. (Paolo Philippidus)

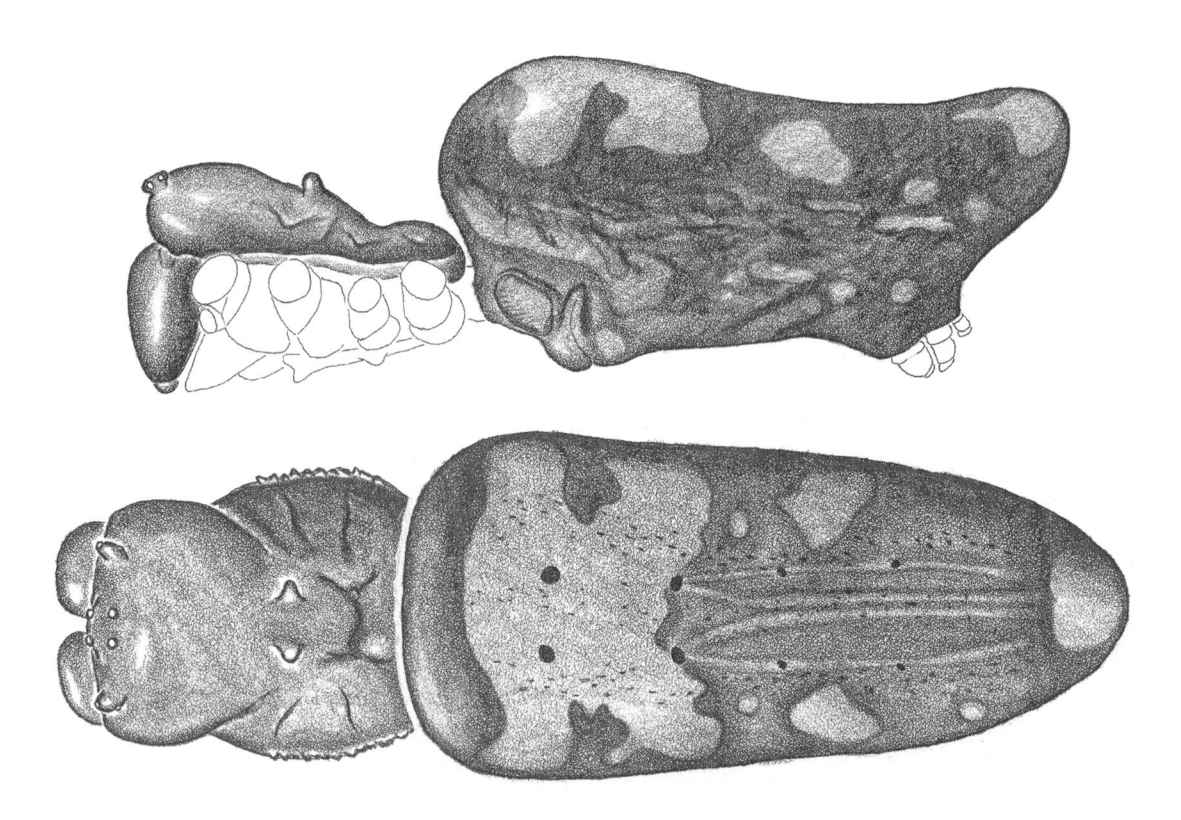

Nephila komaci, female. (Kuntner and Coddington, PLoS 2009)

Burrunan dolphin, *Tursiops australis*. (Charleton-Robb et al, PLoS 2011)

GIANT PECCARIES AND ZOMBIE WORMS

The borderland of zoology is very extensive; the number of animals still to be discovered on this small planet is much greater than is popularly realized or science is willing to advertise. Nor are all of these microscopic worms or tiny, obscure tropical beetles. . . . A notion has somehow gained popular credence that the surface of the earth is now fully explored and for the most part well known and even mapped. There was never a greater misconception.

IVAN T. SANDERSON—*MORE "THINGS"*

Those lines are as true today—in the second decade of the 21st century—as they were back in 1969, when Sanderson's book, *More "Things"*, was published. Naturally, it would be impossible to predict the inventory of major new and rediscovered animals still destined to be unveiled during this century, but those already unveiled include such remarkable finds and refinds as the world's largest species of peccary, the world's smallest sloth, a chameleonesque snake, an entire new order of highly distinctive insects aptly dubbed the gladiators, a brand-new genus of African monkey (the first for over 80 years), bizarre zombie worms that live upon (and inside) the bones of dead whales, and the world's largest species of spitting cobra (which is also the world's second largest species of venomous snake). And those are just a few from the first decade alone, as now revealed here.

21st-CENTURY MAMMALS

A FLURRY OF PHASCOGALES
Several years ago, a mammalogist colleague said to me that researchers in the field would eventually identify mammalian species not with a field guide but rather with a DNA analysis kit. Lately, that time seems increasingly to have arrived. In June 2000, Dr. Peter Spencer at Australia's Marsupial Co-operative Research Centre announced that this technique had shown that the brush-tailed phascogale or tuan *Phascogale tapoatafa*, an arboreal marsupial carnivore, was not a single widely-distributed species as traditionally believed. Instead, it was three taxonomically distinct species—one inhabiting eastern Australia, one in western Australia, and one in northern Australia. Moreover, despite their outward similarity, these species had been genetically separate for over 3 million years. To date, however, this reclassification of the tuan into three separate, formally named and described species has not appeared in the scientific literature.[1]

RIGHT AT LAST
The right whales living in the North Pacific have lately been shown to constitute a separate species from the few hundred right whales in the North Atlantic. Genetic research substantiating this categorisation was conducted using a specially-developed technique for extracting DNA from various long-preserved whale remains, some over a century old, carried out by Dr. Howard Rosenbaum of New York's Wildlife Conservation Society and a team from the American Museum of Natural History. Thus there are now three species in total—the southern right whale *Eubalaena australis*, the North Atlantic right whale *E. glacialis*, and the newly-differentiated North Pacific right whale *E. japonica*.[2]

LEMURS ABOUNDING
In 2000, three new species of mouse lemur were described from western Madagascar by an international team of scientists. Only one species was originally thought to live here, but by the beginning of the 21st

Some mammalogists now believe that the brush-tailed phascogale is not one species but three, and should therefore be split accordingly.

Century seven had been found, including the new trio. These were named: *Microcebus berthae*, *M. sambiranensis*, and *M. tavaratra*, and their taxonomic status as valid species was confirmed by independent genetic analysis. Moreover, news emerged from the 18th biennial Congress of the International Primatological Society of even more freshly-revealed lemurs. These comprised five new species of dwarf lemur (*Cheirogaleus adipicaudatus*, *C. crossleyi*, *C. sibreei*, *C. minusculus*, and *C. ravus*), plus a new woolly lemur (*Avahi unicolor*), which was a much larger lemur. With the total number of lemur species at that point having reached 61 (and still more to come in subsequent years), only Brazil, with 79 by then, could boast more primate species than Madagascar.[3]

THE THIRD CAMEL?

Traditionally, the few wild Bactrian camels inhabiting Mongolia's Gobi desert and arid regions of northwestern China plus the countless domesticated Bactrians have all been classed together as a single species—*Camelus bactrianus*. Genetic tests conducted by scientists with the United Nations Environment Programme (UNEP), however, have revealed significant genetic differences between the wild and domestic Bactrians. Indeed, in 2001 UNEP biodiversity official Rob Hepworth announced that all of the evidence so far obtained from this study seemed to points towards the wild Bactrian camel constituting a valid, third camel species, *C. ferus*, in its own right (alongside the domestic Bactrian and the dromedary). Sadly, it is also an extremely rare one—only about 1000 are believed to exist.[4]

LITTLE SLOW-FOOT—
THE PYGMY THREE-TOED SLOTH

In 2001, a new species of three-toed sloth was revealed—and in a very unexpected locality, at least for sloths. Formally described by Kansas University zoologist Dr. Robert P. Anderson and Smithsonian Institution colleague the late Dr. Charles O. Handley, Jr., it was found amid the mangrove swamps of a tiny island called Isla Escudo de Veraguas—the most remote member of the Bocas del Toro archipelago situated just off the northwestern Caribbean coast of Panama. This island has been cut off from mainland Panama for around 9000 years, during which time its sloths have changed noticeably from their mainland counterparts. They weigh 40 per cent less than the latter, are about 20 per cent smaller in overall size, and sport a distinctive fringe of long hair, giving this island species a hooded appearance. It was christened the pygmy three-toed sloth *Bradypus pygmaeus*—translating as 'little slow-foot".[5]

THE COLOUR PURPLE

A famously controversial Aussie animal hit the headlines in October 2001 when a paper published by a team of Macquarie University researchers provided persuasive mitochondrial DNA-based genetic evidence that the purple-necked rock wallaby was a valid species in its own right, as originally proposed when first described by Australian zoologist and Taronga Zoo director Albert S. Le Souef in 1924—and not, as widely maintained since then, merely a subspecies of the black-footed rock wallaby *Petrogale lateralis*. Consequently, it was duly reinstated as *P. purpureicollis*, commemorating its most celebrated claim to fame—the very distinctive purple colouration that appears on its neck and upper back at certain times of the year, but whose precise production mechanism and the reason for it remain unresolved. Indeed, because it is readily washed off in rain and also fades rapidly once the animal is dead, many scientists had long (but erroneously) supposed that this pigment's origin was external, caused possibly by the wallaby rubbing its neck on mossy rocks, rather than genuinely being secreted by the wallaby itself.[6]

HUNTING HIGH—AND LOWE!
After 70 years, one of the world's most elusive mammals was finally tracked down, thanks to a camera trap set on the eastern side of Tanzania's Udzungwa Mountains National Park in 2002. When a scientist from the Wildlife Conservation Society (WCS) examined this camera's film, he discovered to his great surprise that it had photographed a live specimen of Lowe's servaline genet *Genetta lowei*. Until then, this reclusive species (sometimes classed as a subspecies of the common servaline genet *G. servalina*, but distinguished by the orange colour in its white facial spots and also by its pale feet and legs) had been known to science only from its type specimen—a spotted pelt obtained in the Udzungwas by British explorer-naturalist Willoughby Lowe in 1932.[7]

A WHALE OF A FIND—TWICE!
The latest in a long line of new beaked whale species discovered during the past 100 years was formally named in the July 2002 issue of the journal *Marine Mammal Science*. Dubbed Perrin's beaked whale *Mesoplodon perrini* after cetologist William F. Perrin, it measures 13 ft long, is dark grey dorsally and white ventrally, has a shorter beak than most other beaked whale species, and was identified via DNA analyses conducted upon a series of specimens found washed ashore in California during the 1970s. Originally assumed to be specimens of Hector's beaked whale *M. hectori*, their separate taxonomic status was revealed by a team of researchers featuring Dr. Merel L. Dalebout.[8]

DIGGING UP A GOLDEN REDISCOVERY
Golden moles constitute a taxonomic family (possibly even an entire discrete order) of insectivorous mammals exclusive to southern Africa, and are among the world's most mysterious, cryptic mammals. Only distantly related to true moles, they nonetheless greatly resemble them superficially on account of convergent lifestyles. Just like true moles, golden moles are subterranean species with short limbs but powerful digging feet and claws, and most species construct permanent burrows beneath forests, grassy plains, swamps, and montane terrain. Conversely, those species that are desert dwellers spend their lives 'swimming' directly through loose sand. The eyes of golden moles are non-functional and covered by a thin layer of skin, their ears are merely tiny holes, and their nostrils are protected by an enlarged leathery pad. As their name indicates, their fur is often golden in colour, sometimes imbued with a greenish or mauve sheen.

Golden mole. (Killer-18/Wikipedia)

Although 21 species are currently recognised, on account of their small size (ranging between 3 in and 8 in long) and subterranean lifestyle, golden moles are rarely seen, and in some cases have remained virtually unknown since their formal scientific discovery. A case in point is Van Zyl's golden mole *Cryptochloris zyli*. Officially described and named in 1938, a year after its discovery (which was assisted by South African landowner Gideon Van Zyl), and measuring little more than 3 in long, this aptly-dubbed cryptic species remained resolutely hidden from further detection for many years, known only from its type specimen—which had been obtained at Compagnies Drift near Lambert's Bay in South Africa's Western Cape Province. In November 2003, however, it made a welcome if unexpected return when a second specimen was collected at Groenriviermond, about 93 miles further north along the Namaqualand coast. No other specimens have been recorded since then, however, so this mysterious little mammal remains scarcely known even today.[8a]

So too does Visagie's golden mole *Chrysochloris visagiei*, which is represented only by its type specimen, obtained on the estate of a landowner called I.H.J. Visagie at Gouna in South Africa's northern Cape Province, and was formally described and named in 1950. It has been sought by several field trips here in subsequent years, but always unsuccessfully; and because its type locality has now been greatly changed by extensive agricultural development, this elusive animal may in reality be extinct. As for the Somali golden mole *Calcochloris* [formerly *Amblysomus*] *tytonis*, this barely tangible species, described in 1968, is known only from a single partial skeleton found at Giohar, Somalia, four years earlier—inside a regurgitated barn owl pellet![8a]

AND THEN THERE WERE EIGHT
Traditionally deemed to constitute six species, in 2003 those large baleen whales known as rorquals, belonging to the genus *Balaenoptera*, saw their numbers boosted

to eight. In 1970, eight adult rorqual specimens (five females, three males) were killed by Japanese whalers for research purposes in the eastern Indian Ocean and the Solomon Sea, and were assumed to be undersized fin whales *B. physalus*. Eighteen years later, a female rorqual was accidentally killed in the Sea of Japan. DNA samples from these nine whales were subsequently studied by Japanese scientists, as were their anatomical features, which included fewer baleen plates in the mouth than the fin whale, and a relatively broad, flat skull.[9]

In a *Nature* article of November 2003, the team revealed that the nine specimens represented a hitherto-undescribed, distinct species, which they dubbed *B. omurai*—Omura's whale, in honour of Japanese whale researcher Dr. Hideo Omura. They also proclaimed that Bryde's whale actually constituted two separate species, thereafter to be referred to respectively as *B. brydei* and *B. edeni* (Eden's whale), and thus restoring the original two-species classification of such whales that ended in 1950 when a Norwegian scientist lumped the two into one.[9]

WELCOMING THE WOODPECKER MOUSE

Diminutive in size but very dramatic in taxonomic stature was a bizarre new mouse lately revealed in the cloudforests and lowlands of Peru's Manu National Park. So distinct was this rodent from others that it became only the second species known from the genus *Rhagomys*. Externally it is quite striking, courtesy of its orange back, pink belly, and flattened face, but by far its most remarkable feature is its extraordinarily long tongue, which, after tearing a hole through the bark of a hollow branch with its unusually robust teeth, the mouse thrusts inside in search of insects that it then draws back out and engulfs—thereby adopting a feeding mode remarkably reminiscent of a woodpecker's! Accordingly, in 2003 this extraordinary little mammal was officially dubbed *Rhagomys longilingua*—the woodpecker mouse or long-tongued arboreal mouse.[10]

PIGGING OUT ON A NEW SPECIES

Until the mid-1970s, only two living species of tropical America's pig-like peccaries were recognised by science—the collared and the white-lipped. Then the Chacoan peccary *Catagonus wagneri*, a third, bigger species, previously known only in fossil form, was discovered alive and well in southern South America's Chaco area during the early 1970s (see p. 73). Less than four decades later, moreover, a even larger, fourth peccary species was revealed—and duly devoured![11]

Dubbed the giant peccary due to its notable size—4-4.5 ft long and weighing up to 110 lb—what might otherwise have been its type specimen was spied by German wildlife cinematographer Lothar Frenz and Dutch zoologist Dr. Marc van Roosmalen (famous for having discovered several new species of South American monkey during the past decade, and who had sighted this undescribed peccary form on previous expeditions).[11]

They were conducting an expedition to the Amazon region of Rio Aripuanã in 2000 when they came across a giant peccary struggling in the clutches of some Indian villagers, who promptly killed it, flayed it, roasted its corpse upon a spit, and ate it. Fortunately for science, the two Westerners were able to rescue some of the peccary's remains, which were genetically analysed.[11]

Collared peccary skin (top) compared with giant peccary skin. (Dr. Marc van Roosmalen)

In 2003, moreover, Marc was able not only to see but also to film three specimens. One complete mitochondrial D-loop and two nuclear SINE PRE-1 DNA sequences from giant peccary material compared with that of the collared peccary *Pecari tajacu* supported full-species status, with an estimated divergence time of 1-1.2 million years BP.[11]

Accordingly, in their *Bonner Zoologischen Beiträge* paper, published in June 2007, Marc and his four co-authors formally dubbed the giant peccary *P. maximus*, which now takes over from the previous record-holder, the Chacoan peccary, as the world's largest living species of peccary. Having said that, however, not all zoologists are convinced that the giant peccary is taxonomic distinct from the collared, some claiming that the latter species' notable intraspecific morphological variation might account for the outward differences between it and the giant peccary.[11]

A VOLE REVEALED

The Bavarian pine vole *Microtus bavaricus* once had the decidedly unfortunate distinction of being one of the few European mammal species to have become extinct in recent times. First described in 1962, from an alpine meadow in southern Germany, it later vanished and remained unrecorded thereafter—until in 2004 a population was revealed in the Austrian Alps' Northern Tyrol by Dr. Friederike Spitzenberger from Austria's Naturhistorisches Museum.[12]

A MOMENTOUS MACAQUE

On 16 December 2004, media services worldwide carried details concerning the formal description of a notable primate—a new species of large macaque monkey. Formally dubbed the Arunachal macaque *Macaca munzala*, it had been encountered by an expedition from the New York-based Nature Conservation Society and various allied organisations while exploring the mountainous Indian state of Arunachal Pradesh, bordering Tibet and Myanmar—thus making it one of the world's highest-dwelling species of primate, living at 5,250-11,500 ft above sea level.[13]

It is a true cryptozoological discovery, for although previously undescribed by science (though first noted by Westerners in 1997), this creature is very familiar to the local Dirang Monpa people, who refer to it as *mun zala* ('deep-forest monkey'), duly reiterated in its scientific name. No less than 14 troops of this new species were encountered, although most had ten or fewer members in them. The Arunachal macaque, quite large and brown-furred, is distinguished from other such species by the dark hair on its head, its distinctive facial markings, and its relatively short tail.[13]

UNVEILING A COUPLE OF OVERLOOKED UNGULATES

In 2004, Australian National University zoologists Drs. Erik Meijaard and Colin Groves announced that their morphological researches involving four unusual museum specimens of chevrotain or mouse-deer collected in the vicinity of Nhatrang, Vietnam, during the early 20th century had revealed that they represented a well-delineated but hitherto-undescribed species, which they formally named *Tragulus versicolor*, the silver-backed chevrotain. It is distinguished from all other chevrotains by its skin colouration pattern, the roughness of its skin hair, and various craniometrical characteristics. Meijord and Groves called for field surveys to determine whether this long-overlooked species still existed. Later in 2004, a fifth museum specimen, an adult male, was uncovered, in the Zoological Museum of Moscow University. It had been obtained from local hunters in January 1990 near Tra River in Vietnam's Gia Lai Province, thus confirming that the silver-backed chevrotain was still alive only 14 years previously, and thereby giving hope that it still survives today.[13a]

A second notable 'new' ungulate was belatedly unveiled a year later, when in 2005, based upon a study of 35 museum specimens obtained in 1926 and during 1947-8 in the Upemba wetlands within what is now the Democratic Republic of Congo, a new species of lechwe was formally described. Named the Upemba lechwe *Kobus anselli*, only a thousand or so individuals survive today in Katanga Province, its numbers having been severely depleted by poachers during its previous taxonomic anonymity.[13b]

STRANGER OFF THE SHORE

Often mistaken initially for a dugong on account of its very rounded bulbous brow, a shy three-shaded variety of dolphin with a very small stubby dorsal fin has been regularly seen for many years in the shallow waters off Australia's Great Barrier Reef. Until recently, scientists assumed that these were merely Irrawaddy dolphins *Orcaella brevirostris* visiting the Queensland waters.[14]

When researchers from James Cook University, the Museum of Tropical Queensland, and California's National Oceanic and Atmospheric Administration's Southwest Fisheries Science Center at La Jolla conducted DNA and other studies, however, they were startled to discover that the Australian animals actually belonged to a hitherto-undescribed, unrecognised *Orcaella* species, taxonomically discrete from the Irrawaddy dolphin (which, in addition, is uniformly slate-grey, rather than tricolored). Now known to be a permanent resident in the waters off northern Australia and possibly also Papua New Guinea, in 2005 this freshly-revealed species was officially dubbed *O. heinsohni*, the Australian snubfin dolphin.[14]

MONTY PYTHON'S WOOLLY LEMUR

There is, I suppose, a kind of weird logic in naming a creature given to making silly jumps after a man forever associated with the Ministry of Silly Walks—which is why in 2005 a recently-revealed species of Madagascan woolly lemur was officially dubbed *Avahi cleesei*, in honour of veteran *Monty Python's Flying Circus* star John Cleese. Weighing about 2 lb, and distinguished from related species by its lighter fur colour and facial markings, Cleese's woolly lemur (also known as the Bemaraha woolly lemur after its species' provenance) has long limbs that it puts to good effect when leaping, and was named after John Cleese by Zurich University anthropologists Drs. Urs Thalmann and Thomas Geissman.[15]

This was in recognition of Cleese's promotion of lemur conservation in a major documentary and also in the film *Fierce Creatures*, famously featuring a ring-tailed lemur. Thalmann and Geissman had actually discovered Cleese's woolly lemur 15 years earlier, in central western Madagascar's remote Tsingy de Bemaraha Strict Nature Reserve, but only in 2005 was it formally verified as a distinct species in its own right.[15]

SPORTING A STAMPEDE OF SPORTIVES

In 2006, a veritable stampede of new species of sportive lemur was reported from Madagascar. After over six years of fieldwork, coupled with DNA analyses conducted at Omaha's Henry Doorly Zoo, genetics researcher Edward Louis from the zoo's Grewcock Center for Conservation and Research and a team of colleagues published a major revision of these lemurs, in which, on the basis of molecular characteristics, they differentiated no less than eleven new species! However, this taxonomic revision has not won universal acceptance, with lemur specialist Dr. Ian Tattersall in particular advising extreme caution when assessing these new species until supporting data is forthcoming.[16]

UNMASKING THE ANOMALOUS APES OF BILI

Until quite recently, even amid the many remote regions of darkest Africa, the possibility of an unknown form of great ape existing there yet still eluding scientific recognition seemed ludicrous—but then came the Bili ape.[17]

The saga of this remarkable, highly controversial primate began more than a century ago, when in 1898 a Belgian army officer returned home from what is now the Democratic Republic of Congo with some gorilla skulls obtained by him in a forested region near the village of Bili, on the Uele River in northern Congo's Bondo area—even though no other gorillas had been found within hundreds of miles of Bili before (or since). He donated them to Belgium's Congo Museum in Tervueren, where in due course they were examined by its curator, Henri Schouteden. He was sufficiently struck by their anatomical differences from other gorilla skulls as well as by their unique provenance (roughly halfway between the extreme edges of the western and eastern distribution of any gorilla populations) to classify them as a new subspecies of gorilla, which he dubbed *Gorilla gorilla uellensis*.[17]

Less convinced of their separate taxonomic status, conversely, was mammalogist Prof. Colin Groves, whose examination of these skulls in 1970 led him to announce that they were indistinguishable from western lowland gorillas. Thereafter, the Bili ape sank back into obscurity—until 1996, when Kenyan-based conservationist and wildlife photographer Karl Ammann, intrigued by its strange history and apparent disappearance, set out on the first of several Congolese quests to rediscover this mysterious primate.[17]

And rediscover it he did, bringing back such compelling evidence for its presence that several other notable investigators launched their own searches, and returned with equally fascinating clues concerning the Bili ape's nature. Such researchers included primatologist Dr. Shelly Williams from Maryland's Jane Goodall Institute, Dr. Richard Wrangham from the Leakey Foundation, Dr. Christophe Boesch from Leipzig's Max Planck Institute for Evolutionary Anthropology, Dr. Esteban Sarmiento from New York's American Museum of Natural History, and Dr. George Schaller from New York's Wildlife Conservation Society.[17]

What made their various finds so especially interesting was the ambivalent identity that they collectively yielded for the Bili ape—because, uniquely, it deftly yet bemusingly combines characteristics of gorillas with those of chimpanzees, creating a shadowy anthropoid that is at once both yet neither. For instance: if the Bili ape is a chimpanzee, it is a veritable giant, because videos of living specimens and photographs of dead ones suggest a height of 5-6 ft—a mighty stature supported by the discovery of enormous footprints, some measuring almost 14 in long, and therefore nearly 2 in longer even than those of the mountain gorilla![17]

Also, very large ground nests constructed by Bili apes have been found that compare with those created by gorillas; normal chimps build smaller, tree-borne nests. Further evidence of the Bili ape's great size comes from local Bondo hunters, who distinguish two distinct apes—'tree-beaters' (normal chimps) and 'lion-killers' (the Bili apes). The latter earn their name from their

This photograph of a shot Bili ape appeared in Vol. 1 of Adolf Friedrich Herzog von Mecklenburg's tome "Vom Kongo zum Niger und Nil: Berichte der Deutschen Zentralafrika-Expedition 1910/1911", which was published in 1912—more than a century before this highly distinctive ape form's reality was formally recognised by science.

combined size and ferocity, a mix potent enough to ensure their terrestrial safety even in a jungle profusely populated by lions and leopards.[17]

Indeed, so unafraid of these great cats are the Bili apes that according to media claims they hoot loudly when the moon rises and sets—an activity unknown among normal, smaller chimps, who avoid doing so in case they attract predators. However, these latter claims have been denied by Amsterdam University field researcher Cleve Hicks, who spent a year with colleagues tracking Bili apes during from mid-2005 to June 2006, followed by a second study spanning July 2006-February 2007.[17]

Particularly noticeable is the presence of a pronounced sagittal crest running along the top of one of the original skulls collected by the Belgian army officer, and also on a Bili ape skull found by Ammann in 1996—because this crest, normally an indication of powerful jaws as the jaw muscles are attached to it, is characteristic of gorillas, not of chimps. Conversely, the facial anatomy of the Bili skulls is decidedly chimp-like, not gorilla-like. In addition, hair samples taken from Bili ape ground nests have been shown to contain mitochondrial DNA similar to that of chimps, and the fruit-rich content of examined faecal droppings is again consistent with a chimp identity—although, perplexingly, the droppings themselves outwardly resemble those of gorillas.[17]

So what *is* the Bili ape—a gorilla-sized chimp (freak population?/new subspecies?/new species?), an aberrant form of gorilla (freak population?/new subspecies?/new species?) that has evolved certain chimp-like anatomical and behavioural characteristics, or even possibly a genuine chimpanzee-gorilla hybrid? No confirmed crossbreeding between chimp and gorilla has ever been recorded, but the two species are sufficiently similar genetically to engender viable offspring. Mitochondrial DNA is passed down exclusively from the maternal parent, so if such interspecific matings are indeed occurring they must involve female chimps and male gorillas, to explain why the mitochondrial DNA from the Bili ape samples is chimp-like.[17]

Happily, however, the Bili ape's identity was eventually unmasked. Comprehensive DNA analyses, including nuclear DNA (thus shedding light on both the maternal and the paternal lineages of the Bili ape), had been underway since autumn 2003 at Omaha's Henry Doorly Zoo, under the auspices of conservation geneticist Dr. Ed Louis, and involving DNA comparisons with gorillas, chimps, and also bonobos (pygmy chimps).[17]

So too had analyses of mitochondrial DNA taken from faecal samples conducted by Dr. Cleve Hicks and other Amsterdam University colleagues, who had also examined these primates' behaviour in the field. And in 2006, this latter team announced that their findings all confirmed that the Bili ape belongs to a known subspecies of chimp—the eastern chimpanzee *Pan troglodytes schweinfurthii*. Presumably, therefore, the Bili ape's very distinctive morphological features have evolved through its population's isolation from others of this subspecies but involve relatively little change at the genetic level. After years of mystery and intrigue, the riddle of the Bili ape had at last been solved.[17]

ROCK RAT IS SHOCK RAT!

In 2005, the scientific world was, understandably, shocked by an extremely unexpected revelation of the rodent kind in Laos. For the newly-revealed species in question proved to be so dramatically different from all other known rodents that an entirely new taxonomic family needed to be created in order to accommodate it! Bearing in mind that the last new family of rodents was erected in 1941 (for Selevin's dormouse *Selevinia betpakdalaensis*) and that the last family of any type of mammal was established in 1974 (for Kitti's hog-nosed bumblebee bat *Craseonycteris thonglongyai*), this was a truly historic discovery—but an even more startling twist to the tale was to occur just a year later, in 2006.[18]

*Kha-nyou or Laotian rock rat—
a living diatomyid. (Markus Bühler)*

It all began one morning in 1996. Wildlife Conservation Society scientist Dr. Robert Timmins was strolling through a market in the Khammouan region of Laos when he spotted the body of an unusual-looking rodent for sale on a food stall, next to some vegetables. Unable to identify it, he enquired what it was, and learnt from the locals that it was a *kha-nyou* (pronounced 'ga-nyou') or rock rat, and that such creatures were trapped in the limestone karst close by. Timmins obtained some specimens, and so too, albeit totally independently and two years later, did fellow zoologist Dr. Mark Robinson. A series of specimens was then sent for identification to the Natural History Museum in London, and tissue samples were sent for DNA analysis to Vermont University.[18]

After extensive studies, the scientists working upon this material agreed that the kha-nyou was wholly distinct from anything previously documented by science, and in 2005 it was formally described by Timmins, Robinson, and two co-workers, who named it *Laonastes aenigmamus*. Recalling an unlikely hybrid of squirrel and large dark rat, the kha-nyou has long whiskers, large paws, sturdy limbs, and a thick furry tail, and in total length measures approximately 16 in. Moreover, unlike most rodents, instead of giving birth to a litter of young it apparently gives birth to a single young at a time. It seems to be nocturnal, and to be at least predominantly herbivorous.[18]

Most notable of all, however, is that based upon its cranial and skeletal structure, plus its DNA profile, the kha-nyou appears to have branched off from all other modern-day rodents many millions of years ago, and may be ancestral to the hystricognaths (porcupines, chinchillas, guinea pigs, capybaras, etc).[18]

However, this was still far from the last surprise offered up by the curious kha-nyou, because its classification as the sole member of a brand-new family of rodents subsequently proved to be premature, inasmuch as the kha-nyou was subsequently found to belong to a family already known to science—but only from fossil species several million years old! After studying the kha-nyou following its formal description, Carnegie Museum of Natural History palaeontologist Dr. Mary Dawson realised that this enigmatic species is a living diatomyid, and dubbed it "the coelacanth of rodents".[18]

Prior to this, the most recent diatomyid known was a species very like the kha-nyou, named *Diatomys shantungensis*, but which had died out 11 million years ago. As stated by Dawson, this is an excellent instance of what is known as the Lazarus effect, a concept formulated in the 1980s by Chicago University palaeontologist Dr. David Jablonski—whereby a species of animal is discovered that belongs to a lineage previously assumed to have become extinct millions of years ago because no fossils of more recent age had been found, but now, thanks to the new discovery, has been literally resurrected.[18]

The finding of a living species of coelacanth, *Latimeria chalumnae*, in the 1930s was a classic example of the Lazarus effect, as noted by Jablonski: "Coelacanths have a whopping Lazarus effect in terms of time—they're found in deep water and don't have an easily accessible fossil record on land". So who knows—perhaps, for this same reason, continuing cryptozoological investigations of various mystery beasts will expose additional examples of the Lazarus effect in the future.[18]

BARKING UP THE RIGHT TREE

Just as the kha-nyou was discovered independently by two different researchers, a remarkable new monkey was revealed independently by two different scientific teams, working hundreds of miles apart from one another in Tanzania. One team was based on Mount Rungwe and in Kitulo National Park, the other in Ndundulu Forest Reserve in the Udzungwa Mountains.[19]

Between December 2003 and July 2004, both teams encountered an unfamiliar type of tree-dwelling mangabey, which voiced a very distinctive, low-pitched, but loud honking bark, wholly unlike the whooping gobble sounds produced by other mangabey species. The scientists were first alerted to the new mangabey's existence by locals, who spoke of a shy monkey known to them as the *kipunji*, and when the researchers were able to observe some specimens they realised that their species was totally new to science.[19]

Dubbed the highland mangabey or kipunji, during 2005 it was duly described in a paper jointly written by the two teams, who named it *Lophocebus kipunji*. Roughly 3 ft long, it has long brownish fur, a noticeable crest and profuse whiskers that give its head a strange triangular-looking appearance, an off-white belly and (long) tail, plus a black face, hands, and feet. This striking animal is the first new species of African monkey to have been discovered for over 20 years, but only approximately 11000 individuals currently exist, and it is categorised by the IUCN as critically endangered.[19]

Moreover, the results of a detailed genetic and morphological study subsequently conducted on a dead specimen found in a farmer's trap revealed the highland mangabey to be even more remarkable than initially supposed. So discrete taxonomically from other species of mangabey is it now known to be that in 2006 it was formally reclassified—by being removed from the genus *Lophocebus* housing other mangabeys and rehoused in a brand-new genus, *Rungwecebus*, as *Rungwecebus kipunji* (after its Rungwe-Livingstone forest domain).[19]

This elevation of its taxonomic status makes the highland mangabey the first new genus of Old World monkey to be named since 1923—which was when Allen's swamp guenon, a species first discovered in 1907 and originally housed in the genus *Cercopithecus* with most other guenons, was reclassified in its own separate genus as *Allenopithecus nigroviridis*.[19]

WHAT AM I BID FOR THIS MONKEY'S NAME?

The kipunji was not the only new primate from the mid-2000s to attract media interest. So too did the Golden Palace titi *Callicebus aureipalatii*, a Bolivian species with orange-brown fur, golden crown, a white tip to its tail, and dark red hands and feet—on account of the unusual way in which it received its common and formal scientific names. In 2005, a charity auction was held, with the prize being that whoever won the auction would have this monkey species (discovered in 2004 by a Wildlife Conservation Society research expedition) named after them. The auction was won with a bid of

Kipunji—a new species of mangabey so distinct from all others that it required the creation of a brand-new genus.

$650,000 by an online casino, Goldenpalace.com—so in 2006 the titi was christened accordingly, with the money being used to help maintain its home in western Bolivia's Madidi National Park.[19a]

NEW MOUSE IS A LITTLE BIG-HEAD

One little grey mouse may seem very like another, until you look more closely, that is. And when visiting Durham University archaeologist Dr. Thomas Cucchi did just that with the mice on Cyprus in 2004, he had a great surprise. Some of them seemed to have bigger heads, larger eyes and ears, and more primitive teeth than those of the others, and appeared more similar to various Stone Age fossil Cypriot mice. Moreover, genetic analyses confirmed that these little big-heads were indeed so distinct from the 'normal' common house mice *Mus musculus* on the island that they clearly constituted a new species, which in 2006 was duly christened *M. cypriacus* . . . except that, in reality, this 'new' species wasn't new at all.[20]

For whereas the common house mouse did not reach Cyprus until a mere 10,000 years ago, when it was accidentally introduced by the first human settlers, the 'new' mouse had actually been resident here for far longer, roughly 100,000 years, because it was conspecific with the fossil mice, and had been waiting ever since for scientific recognition of its taxonomic individuality. Incidentally, although many media reports (and even some mammalogists) loudly trumpeted about how the Cypriot mouse was the first new species of European mammal to have been discovered for a century, it was actually just the latest in a fairly long list (containing no less than 32 species, in fact, at that time), which included quite a range of new European rodents, insectivores, and bats.[20]

GIANT TONGUES AND STICKY FEET

Two very distinctive new species of bat were revealed during the mid-2000s. Bat #1, known as the tube-lipped nectar bat *Anoura fistulata* on account of its unusual tubular lower lip, was discovered amid the tropical Andean cloudforests of Ecuador, and described in 2005. Moreover, it was soon on every bat researcher's tongue—on account of its own, amazingly long tongue, which, measuring 3.2 in, is one and a half times as long as its entire body. Indeed, so lengthy is it that the bat has to store it in its thoracic cavity (ribcage) when not extending it out to reach nectar deep within the extra-long-necked blooms of the lobelia-related flower *Centropogon nigricans*.[21]

In fact, relative to body size, this bat's inordinate tongue is longer than that of any other vertebrate except for certain chameleons. Putting its size into perspective: if a cat's tongue were of the same proportion, it could lap milk out of a bowl placed 2 ft away![21]

Bat #2, the western sucker-footed bat, which was christened *Myzopoda schliemanni* in 2007, is doubly distinctive. Not only does this Madagascan species possess flat adhesive suckers attached to its thumbs and hind feet, but in addition it proved to be a second, hitherto-unknown member of a bat family previously believed to house just a single, rare species, *M. aurita*, also Madagascan.[22]

This latter species is found only in the island's humid eastern forests, whereas the new species is found only in its dry western forests, further delineating it from *M. aurita*, along with differences in pelage colouration, cranial characteristics, and external measurements. Both species use their suckers to climb and stick to the flat, slippery surfaces of the broad leaves of certain plants, particularly the traveller's palm *Ravenala madagascariensis*.[22]

SPOTTING THE DIFFERENCE

It's not always easy to see the wood for the trees—or, in this particular case, the new leopard for the old leopard. So it was that although the Bornean or Sunda clouded leopard (also found on nearby Sumatra) has long been known to science, it had traditionally been classed merely as *Neofelis nebulosa diardi*, i.e. as a subspecies of the single species of clouded leopard, *Neofelis nebulosa*.[23]

Thanks to a comparison of its DNA with that of other clouded leopards by an American research team led by felid genetics expert Prof. Stephen O'Brien and published in December 2006, however, it was sensationally upgraded to the status of a full species in its own right, as *Neofelis diardi*, standing apart as a newly-unfurled, second *Neofelis* species. The reason for this shock recategorisation was the revelation that the Bornean clouded leopard's DNA exhibited differences from the DNA of mainland Asian clouded leopards that were almost as numerous as those genetic differences separating lions and tigers.[23]

Bornean clouded leopard (1834 engraving).

Moreover, comparative re-examination of clouded leopard pelage morphology exposed some noticeable differences too. The characteristic *Neofelis* cloud-shaped coat markings are smaller (and contain more spots) in Bornean specimens than in mainland ones, and Bornean specimens exhibit a pronounced double dorsal stripe whereas mainland ones only have a partially-formed dorsal stripe. In addition, the Bornean clouded leopard's coat is darker and greyer than that of the tawnier mainland version.[23]

BRIGHT-EYED AND VERY BUSHY-TAILED

Boasting an exceptionally bushy tail that would put any self-respecting squirrel to shame, and sporting a veritable Mohawk that would turn any self-respecting punk emerald with envy, one of the most distinctive new rodents to have been discovered in recent times is *Isothrix barbarabrownae*, Barbara Brown's brush-tailed rat. This arboreal, nocturnal rodent was revealed high in the cloudforests of southern Peru's Manu National Park and Biosphere Reserve Mountains in 1999, but remained scientifically undescribed until 2006.[24]

Characterised by its thickly-furred tail, and the unusual crest of black fur on its broad head's crown as well as down its nape and shoulders, despite its squirrel-like appearance this novel mammal actually belongs to the spiny rat family, Echimyidae. However, it is so different from its closest relatives, which are all lowland species, that in the words of Chicago Field Museum mammalogist Dr. Bruce Patterson: "[It] casts a striking new light on the evolution of an entire group of arboreal rodents".[24]

BLOND CAPUCHIN—A PAINTING COME TO LIFE

Way back in 1774, German naturalist Johann C.D. von Schreber had included in his book *Die Säugethiere in Abbildungen nach der Natur, mit Beschreibungen* a description and accompanying colour painting of a golden-haired capuchin monkey supposedly native to northeastern Brazil. Schreber formally dubbed its hitherto-unnamed species *Simia flavia*. Moreover, it had previously been described even further back, in 1648, by Georg Marcgrave in the sixth volume of his monumental work on Brazilian natural history, *Historiae Rerum Naturalium Brasiliae*, referring to it as the caitaia.[24a]

The blond capuchin monkey was first brought to public attention in 1774 via this painting, but nothing further was heard until 2006.

THE ENCYCLOPÆDIA OF NEW AND REDISCOVERED ANIMALS

Nothing further was heard about this pale-furred mystery monkey, however, for over 300 years—not until 2006, in fact, when an extremely similar form was described from Brazil's Pernambuco state as a newly-discovered species, *Cebus queirozi*—the blond capuchin. But at much the same time, a different research team resurrected *S. flavia*, recognising it to belong to the modern genus, *Cebus*, and thus renaming it *Cebus flavius*. Moreover, as the only notable difference between the *C. flavius* monkey in the painting from 1774 and the modern-day blond capuchin *C. queirozi* is that the latter possesses a white band running from ear to ear that is absent in the *C. flavius* painting (but an absence that may have simply been due to the age of the painted individual anyway), many primatologists consider that *C. flavius* and *C. queirozi* are almost certainly one and the same species. So as *C. queirozi* is a much later name, it is nowadays usually deemed to be a junior synonym of *C. flavius*. As for this lately-rediscovered species itself, only around 180 individuals are believed to exist, and it is categorised as critically endangered by the IUCN.[24a]

MARC VAN ROOSMALEN'S MARVELLOUS MENAGERIE OF NEW MAMMALS

During the 1990s, Vietnam was the place to be for cryptozoological discoveries—hosting a startling number of new mammalian finds. Now, during the 'noughties', a new locality has gained ascendancy in the cryptozoological world—Amazonia, the vast green heartland of Brazil.

One man has achieved a unique distinction in unveiling an unparalleled number of new mammals in this verdant wilderness. Winner of one of *Time Magazine*'s illustrious 'Heroes for the Planet' environmental awards in 2000, he is Dutch zoologist Dr. Marc van Roosmalen, who has been conducting field research in Brazil for several years. For much of that time, he and I have been in communication with one another, and when he knew that I was preparing this present book's second edition, *The New Zoo* (2002), Marc very kindly provided me with a welter of information and illustrations concerning the several new species and subspecies of Amazonian monkey that he had already discovered and formally named, as well as a new species of tree porcupine.

Amazingly, however, even these have now proven to be little more than the tip of an incredible crypto-iceberg. During the past few years, many rumours have circulated online and in the cryptozoological community concerning much bigger mammals, all allegedly new to science, that Marc had uncovered, but no-one seemed very sure whether any of them were real. In early 2007, however, Marc contacted me to update me on what he had been finding, and I was stunned to learn that the truth was far more extraordinary than any rumour.

Dr. Marc van Roosmalen holding up a giant peccary skin, alongside a skin and skull of the mysterious onça-canguçú jaguar (foreground left) and a mystery orange coati skin (foreground right) (Dr. Marc van Roosmalen).

Marc's website—marcvanroosmalen.org[25]—not only documents his recently-described new monkeys and details of his other fieldwork, but also includes descriptions and photos of over two dozen totally new, large (sometimes very large) mammals that he has lately discovered and which now await formal description and naming by him. Of these, one, the giant peccary, has already been documented by me here (see p. 276)[11].

For the others, this present account (adapted from my earlier exclusive *Fortean Times* report[25]) marks their debut in book form, and is written with full support from Marc—who has also most generously provided me with exclusive access to a series of papers co-authored by him that officially describe and name various of the new mammalian taxa documented here.

Faced with such an exceptional array of new mammals, space permits only the most concise series of descriptions, and I strongly recommend readers to check out Marc's website for further information. So without more ado, permit me to introduce Marc's truly marvellous menagerie of newly-disclosed wonders from the Amazonian rainforests:

ARBOREAL GIANT ANTEATER

Unlike the tamandua and silky anteater, the giant anteater *Myrmecophaga tridactyla* is terrestrial. However, Marc has discovered a tree-climbing form of giant anteater, previously undocumented, which is almost as big as the familiar ground-dwelling version but climbs by grasping with its hindfeet, and has distinctive neck markings. Marc is not releasing any details concerning its distribution until he has obtained holotype material for formal description.[25]

BLACK DWARF LOWLAND TAPIR

Much smaller and darker than Brazil's familiar reddish-brown lowland tapir *Tapirus terrestris* (and indeed, seemingly the smallest living species of tapir presently known to science), the black dwarf lowland tapir is referred to locally as the *anta pretinho* ('little black tapir'), and appears to be restricted to the lower and middle part of the Rio Aripuanã basin. Data contained in Marc's official description of this new species, in which he formally named it *T. pygmaeus*, disclosed that it is not only considerably smaller than other tapirs but also has a unique dentition, and has taxonomically significant differences in its mitochondrial and nuclear DNA. Outwardly, it is recognisable by virtue of its size, colouration, and also the lack of white tips to its ears. This new species' holotype is the skull of an adult female killed for food by a local hunter on 2 May 2006, close to the settlement of Tucunaré, along the Rio Aripuanã's left bank.[25-6]

BLACK GIANT OTTER

Slightly smaller than the known giant otter or saro *Pteronura brasiliensis*, this new form is also readily distinguished by its near-black pelage, differing markedly from the saro's brown fur. Further details concerning it are classified until enough evidence for its existence has been amassed for publication.[25]

DWARF MANATEE

This is a very controversial animal. In his formal description of the dwarf manatee as a discrete new species, Marc dubbed it *Trichechus pygmaeus*, but according to the ICZN Code this is an invalid name. Less than 4.5 ft long and only around 132 lb in weight, this freshwater mini-manatee is the smallest of all living sirenians, and is adapted for an existence in clear, fast-flowing, shallow streams where it browses horizontally on bottom-dwelling, non-floating plants. In grave danger of extinction, the only known viable population is limited to the Rio Arauazinho, a 75-mile-long left bank tributary of the Rio Aripuanã. In 2004, Marc was able to film, photograph, and examine a living adult male dwarf manatee in its natural habitat. Physically, it differs from the larger Amazonian manatee *T. inunguis* not only in terms of body size but also by virtue of its very dark, almost black colouration and white throat patch, as well as its body proportions. Conversely, the two species are indistinguishable from one another with regard to their mtDNA, which has led various mammalogists, including manatee evolution expert Dr. Daryl Domning of the Smithsonian Institution, to favour the likelihood that dwarf manatees are merely immature Amazonian manatees.[25, 27]

Dwarf manatee. (Dr. Marc van Roosmalen)

GIANT PACA

The two currently-recognised species of paca are almost-tailless rodents up to 2 ft long, 11 lb in weight, and adorned with usually four longitudinal rows of white spots on each side of their blackish-brown-furred body. However, Marc has encountered—and collected—a much larger form of paca, known locally as the *paca concha*. It appears to have a very wide distribution range, and is distinguished from the two recognised species by its greater size, weighing up to 27 lb, its lighter fur colour, and the merging of most of its spots into longitudinal lines. In a paper awaiting publication, Marc has named this form as a new species. Several suspected specimens of giant paca are held at Brazil's Museu Paraense Emilio Goeldi, where Marc's holotype of this new species, killed for food by a local hunter on 28 May 2006 near Tucunaré, has now been deposited.[25]

Giant paca—an undescribed extra-large paca species? (Dr. Marc van Roosmalen)

GREY AGOUTI, AND AGOUTI-FURRED ACOUCHY

Pacas are closely related to agoutis—speaking of which, Marc has also discovered an apparently new agouti, distinguished from all previously-described species by its pale grey fur (the other agoutis' fur ranges from pale orange through several shades of brown to near-black), and found only along the Rio Aripuanã's left bank. Also related to pacas and agoutis are the two known species of acouchy, one red-furred, the other olive-green, but in the Rios Tapajós/Amazonas/Madeira interfluve region Marc has on many occasions observed an agouti-shaded acouchy readily distinguishable from both of the two currently-recognised species.[25]

MONKEYS OF MANY KINDS

Continuing his astounding success at discovering new species of South American monkey, Marc currently lists in his website no fewer than 14 more presently awaiting formal description. These include two new woolly monkeys (most dramatic of which is a bright ginger-orange form originally made known to science as far back as 1935 but afterwards forgotten), two new spider monkeys, a uakari, a marmoset, a squirrel monkey, two tamarins, three titis, and two sakis. He names them as: the orange woolly monkey, black woolly monkey, long-limbed black spider monkey, silvery-bellied spider monkey, Rio Pauiní white bald-headed uakari, Upper Xingú Amazonian marmoset, Rio Aripuanã green-backed squirrel monkey, Cruz Lima's saddleback tamarin, eastern saddleback tamarin, Rio Mamurú titi, Rio Purús collared titi, Upper Rio Xingú titi, grey saki, and southbank Rio Negro saki.[25]

ORANGE COATI

As documented in Marc's formal description, awaiting publication, this striking creature is distinguished from the familiar ring-tailed coati *Nasua nasua* by its bright orange fur, larger size, tendency to move about only in pairs rather than large troops, and fiercer demeanour. Several hitherto-unrecognised specimens of this form are present in the zoological collections of Brazil's Museu Paraense Emilio Goeldi.[25]

ORANGE PECCARY

As its name suggests, this peccary is recognisable by way of its orange pelage, as well as by its modest size. It lives in small groups in Amazonia, and appears distinct not only from the two previously-known peccary species living here (collared peccary *Pecari tajacu* and white-lipped peccary *Tayassu pecari*) but also from the two other, new ones lately revealed by Marc (giant peccary *P. maximus* and white-hoofed peccary—see later entry here).[25]

ORANGE TAYRA

This is yet another orange-furred novelty, thereby differing noticeably from the normal black-furred tayra *Eira barbara* (a large member of the mustelid family), and also lacking the latter's familiar yellow throat patch. Marc is not releasing any distribution details concerning this creature until he has obtained holotype material.[25]

RIO ARIPUANÃ RIVER DOLPHIN

Known to occur only in the clear-water Rio Aripuanã as far upstream as the Periquito Falls, and as far downstream as its river mouth, this freshwater dolphin is swiftly distinguished from the pink-coloured Amazonian boto *Inia geoffrensis* by its grey skin, and by lacking the boto's long beak and swollen brow. This new dolphin also has a different breathing rhythm from the boto, and is said by locals to be more aggressive.[25]

VAN TIENHOVEN'S FAIR BROCKET

Also known as the white brocket, this small new deer was named by Marc after Dutch naturalist-lawyer Pieter Gerbrand van Tienhoven, who founded the international nature conservation movement in the Netherlands. It differs from the other two brockets—the red brocket *Mazama americana*, and the Amazonian brown brocket *M. nemorivaga*—that co-habit lowland Amazonia not only by virtue of its whitish-brown pelage but also by being intermediate in size between them, and by way of its shorter but thicker, spike-like antlers. Ironically, although not previously known to science, van Tienhoven's fair brocket is the most commonly hunted of the three brockets inhabiting lowland Amazonia. Marc has prepared a formal description of it, naming it as a new species, *M. tienhoveni*, based upon material collected.[25, 28]

WHITE-HOOFED PECCARY
This new form is similar to the white-lipped peccary *Tayassu pecari*—except that it lacks a white lip! It is also distinguished by its white hooves. Marc has seen entire herds of white-hoofed peccaries, and actually maintained a female specimen, named Piggy, in captivity in his compound until she was killed by some intruding local hunters.[25]

WHITE-THROATED BLACK JAGUAR
Last, and most mysterious of all, this unclassified big cat, which is known locally as the *onça-canguçú* ('bigger jaguar that goes in pairs'), resembles a very large black (melanistic) jaguar *Panthera onca*, but, uniquely, has a white throat and a tufted tail. Moreover, unlike normal melanistic jaguars, which, when viewed at certain angles, can be seen to be rosetted, the *onça-canguçú* is pitch-black with no coat patterning whatsoever. Marc has yet to see this creature personally, and also narrowly missed the opportunity to inspect one pelt—a hunter who had killed one of these cats threw its pelt away shortly before Marc arrived asking about this feline cryptid. Happily, he later obtained both a pelt and a skull, which should greatly assist in determining the *onça-canguçú*'s zoological status.[25]

White-hoofed peccary. (Dr. Marc van Roosmalen)

The mysterious onça-canguçú or white-throated jaguar. (William M. Rebsamen)

How many of these scientifically-undescribed mammalian novelties will prove to be new species or new subspecies, or merely local colour morphs or varieties, has still to be seen. Yet whatever they are, they are all sufficiently different from their closest previously-known counterparts to deserve formal investigation and documentation.

Equally, Marc deserves international recognition and acclaim for providing dramatic, unequivocal evidence of the extraordinary diversity of wildlife still awaiting discovery within the Brazilian rainforests, which in turn underlines how great the loss will be to our world if deforestation in this uniquely rich ecosystem continues. Marc urgently needs support and sponsorship to continue his invaluable studies in Amazonia, so if the amazing discoveries documented here have excited and alerted you to the immense potential for future finds here, please visit Marc's website, and offer him whatever assistance you can.

After all, it's not every day that offers you the opportunity to be instrumental in securing the continuing existence of such fascinating creatures as these (and no doubt others too), already in danger of extinction even as they stand at last on the very brink of scientific acceptance.

THE GIANT RAT OF FOJA
Irian Jaya (aka Papua), or Indonesian New Guinea, occupies the western half of the island of New Guinea, and has vast tracts of little-explored, remote terrain—but few such regions are more inaccessible than the mist-shrouded slopes of the 7,000-ft-high Foja Mountains, uncolonised by local tribes, and never previously visited

by a major modern-day scientific expedition. During November and December 2005, however, its exile from human scrutiny came to an end when a global team, co-led by Dr. Bruce Beehler of Conservation International, spent a month here—exploring, examining, and identifying its rich diversity of wildlife. In December 2007, a return visit yielded a sizeable surprise of the rodent variety.[29]

The creature in question was a giant rat measuring approximately 22 in long and weighing in at a hefty 3 lb. Wandering fearlessly into the scientists' camp in search of food, it caused quite a stir when it was first spotted—because due to its extraordinary size (roughly five times as big as a typical city-inhabiting brown rat *Rattus norvegicus*), it was initially mistaken for a cat! Laying claim to the title of the world's biggest known form of rat, this very rangy rodent is further distinguished by its long grey hair and very wide underside. Belonging to the *Mallomys* genus of woolly rats, it may well constitute a totally new species.[29]

During a six-week expedition in 2009, a very similar but even larger specimen of *Mallomys* woolly rat, measuring an extraordinary 32.2 in and weighing approximately 3.3 lb, was discovered by a team of scientists and accompanying film makers from the BBC's Natural History Unit inside the 2-mile-diameter crater of Mount Bosavi—an extinct volcano within the Southern Highlands province of Papua New Guinea. With long dense silver-brown fur, a captured specimen of this giant rat species showed no fear of humans.[29]

SING A SONG OF SENGIS

Also known as elephant shrews on account of their superficially shrew-like form and long trunk-like snout, sengis constitute an exclusively African taxonomic order of mammals, Macroscelidea. Relatively small, forest-floor inhabitants, only 15 species were known until recently, when a sixteenth, and, in sengi terms, positively huge species was dramatically unveiled to science. The size of a small dog, roughly 25 per cent bigger than all other sengis, and brightly coloured with orange and grey fur, in 2008 this veritable goliath was formally dubbed the grey-faced giant sengi *Rhynchocyon udzungwensis*.[30]

It is the first new species of giant sengi to have been discovered for 126 years, and inhabits the Ndunlulu forest in Tanzania's remote Udzungwa Mountains. Its existence was first made known to science in 2005, when one was caught on film by a camera trap set up in the forest by Italian zoologist Dr. Francesco Rovero. Then in March 2006, Rovero returned there with fellow zoologist Dr. Galen Rathbun from the California Academy of Sciences, and captured four specimens alive. Two separate populations are now known to exist in this still little-explored locality.[30]

In September 2010, news emerged that what may be a second new, giant species of sengi had been discovered in northeastern Kenya's Boni-Dodori rainforest. Measuring 2 ft long, this unexpected find has grizzled yellow-brown shoulders and flanks, black thighs and rump, and what appears to be a dark mane. Its existence was revealed by motion sensor cameras set up by the Zoological Society of London's 'Edge of Existence' research program, and various dead specimens were later procured for study purposes.[30]

A PLETHORA OF PRIMATES

And still they came. Continuing a seemingly endless series of Neotropical simian discoveries that began during the early 1990s, yet another new species of monkey lately unveiled in South America was Ayres's black uakari *Cacajao ayresi*, named after Brazilian biologist Dr. Marcio Ayres, who pioneered field studies of uakaris. It was discovered by Auckland University primatologist Dr. Jean-Phillipe Boubli, while conducting field studies around Brazil's Araçá River in Amazonas, and is only the second species of black uakari known to science. Moreover, also named by Boubli in the same paper was another new black species—the Neblina uakari *C. hosomi*, native to far northwest Brazilian Amazon and southern Venezuela.[31]

In 2010, two years after its discovery, a new and very distinctive but critically-endangered titi monkey was described from Colombia's Department of Caquetá region. Christened the Caquetá or red-bearded titi *Callicebus caquetensis*, it had first been reported back in 1976, by animal behaviour expert Martin Moynihan. And in August 2011, the discovery by zoologist Julio Dalponte of a new, currently unnamed titi species during an expedition the previous December to Brazil's Guariba-Roosevelt Extractive Reserve (in Mato Grosso State) was publicly announced by the WWF. It is distinguished from other titi species in this region by its bright orange cheeks and tail.[31]

Two additional species of large lemur were recently unmasked in Madagascar, thanks to DNA studies. Although outwardly very similar to closely-related species, their respective DNA profiles were sufficiently distinct to warrant their classification as separate species in their own right. The scientists responsible hail from Henry Doorly Zoo's Center for Conservation and Research, based in Omaha, whose Madagascar Biodiversity and Biogeography Project has identified no less than 18 new species of lemur in the previous three years alone. These latest two were Moore's woolly lemur *Avahi mooreorum*, named in 2008 in honour of the Gordon and Betty Moore Foundation in San Francisco, native to Madagascar's Masaola National Park, and characterised by its thick brown-grey fur and reddish tail; and Scott's sportive lemur *Lepilemur scottorum*, an agile species with greyish-brown fur and a black-tipped tail, which was named in 2008 in honour of the Suzanne and Walter Scott, Jr., Foundation, and is also native to the Masaola National Park.[32]

UNMUDDLING A MUNTJAC

In 2002, a muntjac was snared in a hunter's trap set high on a remote mountainside in Sumatra's densely-forested Kerinci-Seblat National Park. When the deer was discovered soon afterwards by a Flora and Fauna International scientist and park officials, it was photographed and released, but nothing more was thought about it at the time, because it was assumed merely to be a specimen of the common red muntjac *Muntiacus muntjak*. In 2008, however, the photographic evidence was reassessed, revealing that the deer was actually a far rarer species, the Sumatran muntjac *M. montanus*. This elusive deer was last reported in 1930, only 16 years after its original discovery on 29 April 1914 high on the slopes of Mount Kerinci during the Robinson and Kloss expedition to western Sumatra. Two other Sumatran muntjac were subsequently photographed elsewhere in the park.[33]

WELCOMING FLIPPER THE THIRD

TV's most famous dolphin, Flipper, was a bottlenose dolphin, of which until recently only two species were known—the common bottlenose *Tursiops truncatus*, and the coastal Indo-Pacific bottlenose *T. aduncus*. During DNA testing of coastal dolphins off Victoria, Tasmania, and South Australia, however, Macquarie and Monash University scientists were startled to discover that although they were all alike outwardly, some were very different genetically from the known coastal species—different enough, in fact, to warrant their reclassification as a third, previously-unrecognised species of bottlenose, dubbed the burrunan, and formally described and named *Tursiops australis*.[34]

DURRELL'S VONTSIRA, A NEW MADAGASCAN CARNIVORE

As long ago as 2004, a mysterious form of mongoose-like carnivore had been photographed swimming in Madagascar's marsh-surrounded Lake Alaotra by workers from the Durrell Conservation Wildlife Trust, but it had been dismissed as a brown-tailed vontsira *Salanoia concolor*—a species already known from this island continent's eastern rainforests. In 2010, however, following morphological and molecular analyses of a male specimen that had died in a trap, a team of scientists from the Durrell Wildlife Conservation Trust, London's Natural History Museum, Jersey's Nature Heritage, and Conservation International revealed that this speckled-brown cat-sized creature (likened to a "scruffy little ferret") was a hitherto-undocumented, distinct species. Accordingly, it has been fittingly christened Durrell's vontsira *S. durrelli*,

Three separate species of bottlenose dolphin are now recognised. (William M. Rebsamen)

in honour of the celebrated conservationist and author Gerald Durrell. Cranial, dental, and paw modifications that distinguish it from its brown-tailed relative demonstrate adaptations to life in an aquatic environment.[35]

CRYING OUT FOR ATTENTION— A NEW GIBBON FROM INDOCHINA

In 2010, news emerged that a new species of gibbon—distinguished by the unique frequency and tempo of its tree-top vocalisations as well as via genetic comparisons—had been discovered in the tropical rainforests of the central Annamite mountain range between Vietnam, Laos, and Cambodia. Formally dubbed the northern buff-cheeked gibbon *Nomascus annamensis*, it was distinguished from the outwardly similar yellow-cheeked gibbon *N. gabriellae* (which inhabits the same region too) by a scientific team from the German Primate Center in Göttingen working in conjunction with American and Vietnamese researchers, and it is the seventh species of crested gibbon to be recognised. Having said that, only the males, which are mostly black in colour but shine silver in sunlight, bear a crest; the females, with orange-beige fur, are crestless.[36]

Now that Egypt's so-called 'jackal' has been exposed as a hitherto-unrecognised form of wolf, the ancient Egyptian deity Anubis should be referred to hereafter as wolf-headed, not jackal-headed. (Dr. Karl P.N. Shuker)

AFRICA'S 'NEW' WOLF—
AN IDENTITY CRISIS FOR ANUBIS?

One of the most familiar members of ancient Egypt's pantheon of gods is Anubis, the deity of mummification. For untold ages, Anubis has always been referred to as jackal-headed, on account of being portrayed with a canine head traditionally identified as that of Egypt's jackal *Canis aureus*. However, this tradition has now come to a very abrupt end, thanks to the remarkable discovery that Egypt's jackal is nothing of the sort. Recent genetic research by a team of workers led by Dr. Eli Rueness has revealed that it is not a jackal at all but is in fact a hitherto-unrecognised form of the grey wolf *Canis lupus*. As such, it is, therefore, the only form of wolf native to the continent of Africa, has been duly dubbed the African wolf, and is most closely related to the Himalayan wolf. This in turn means that from now on, Anubis should be referred to not as jackal-headed but as wolf-headed.[37]

21st-CENTURY BIRDS

GROUSING ABOUT SAGE-SPLITTING

The sage grouse *Centrocercus urophasianus* is famous in North America for the male's elaborate communal displays to attract females. At the turn of the century, however, it also attracted ornithological attention, for good reason. Studies into the size, genetics, and behavioural attributes of the southwestern Colorado and extreme southeastern Utah contingent of sage grouse revealed that these birds, numbering around 5000 in total, were sufficiently distinct from other sage grouse to warrant reclassification as a separate species in their own right. In 2000, this newly-delineated species was formally named *C. minimus*, the Gunnison sage grouse, after Colorado's Gunnison Basin, which contains the largest known population. A third smaller in overall body size than *C. urophasianus*, the Gunnison sage grouse is also readily distinguished by the much thicker plumes behind its head, and by the male's less elaborate courtship display.[1]

TUCKING INTO A
REDISCOVERED BRUSH TURKEY

Brush turkeys or megapodes are not true turkeys (although they are distantly related), and are famous for a number of species erecting sizeable mounds of vegetation inside which their eggs are incubated by the heat generated when the vegetation decomposes. Several species are scarcely known, and prior to the 21st Century, one of these, Bruijn's megapode *Aepypodius bruijnii*, native to the New Guinea island of Waigeu, had not been sighted alive by western scientists since 1938, causing many to fear that it had become extinct.[2]

On 23 February 2001, however, an expedition from Rotterdam's Natural History Museum obtained conclusive evidence that it had survived until much more recently. A Waigeu hunter showed the team the remains of his recent dinner—which just so happened to be the head and some bones of a Bruijn's megapode! Further searches in 2002 confirmed this elusive species' continuing existence via sightings of birds and discoveries of 28 incubation mounds, but it is currently categorised as endangered by the IUCN.[2]

REDISCOVERIES DOWN SOUTH

Two distinctive species of South American bird written off as extinct for several decades also made a welcome return in 2002. In May of that year, while surveying a portion of rainforest in the Para region of Brazil, a team of BirdLife International scientists rediscovered the golden-crowned manakin *Lepidothrix vilasboasi*, a small but highly attractive species of green and yellow passerine (perching bird) not seen since its original discovery back in 1957.[3]

And on 28 July 2002, while ascending a volcano in the forests of central Colombia, a team of ornithologists led by Dr. Jorge Velasquez spied a flock of 14 bright-green parrots with indigo wings, pale blue crowns, and red shoulders descending from the mists to alight on some nearby trees—and realised to their amazement that these brilliantly-plumaged wonders were none other than Fuertes's parrot *Hapalopsittaca fuertesi*, a scarcely-known species not reliably seen since 1911.[4]

BALD BUT BEAUTIFUL—A
REMARKABLE NEW PARROT

Staying with South American parrots: a quite extraordinary new species was formally described and named in 2002, but its very existence had not been made known publicly until earlier in that same year. That was when a single specimen that had been captured in September 1999 within Brazil's Mato Grosso region, near the Madeira River and a tributary of the upper Tapajos, by two zoology graduate students—Renato Gaban-Lima and Marcos Raposo—from São Paulo University was photographed for a Brazilian bird magazine and also filmed by a local television station.[5]

Predominantly green in body plumage except for its red underwing coverts, and measuring approximately 10 in from head to tail, what made this new parrot so outstanding was the total absence of feathers on its head. At first, some scientists wondered whether it was just a freak individual, but Mato Grosso locals confirmed that there were others, constituting a completely separate, valid form in its own right, and they even had their own Portuguese name for it—'papagaio careco' ('bald parrot').[5]

Another initial line of scientific scepticism was that perhaps such birds merely represented an immature version of the vulturine parrot *Pyrilia* [formerly *Pionopsitta*] *vulturina*—another bare-headed species (and named after the vulture-like appearance of its bald head), which the new bird greatly resembled, except for the orange colour of its head's skin (the skin of the vulturine parrot's head is black). However, once some specimens of this orange-headed bald parrot were obtained and examined, it was discovered that they were all sexually mature, so the local people had been correct all along—it really was a genuine species. Consequently, Gaban-Lima, Raposo, and co-researcher E. Hofling duly christened it *Pionopsitta* [now *Pyrilia*] *aurantiocephala*, basing their formal description on four recently-collected specimens from the Rio Cururu-açu region as well as upon a re-examination of seven preserved specimens discovered in various museum collections.[5]

Moreover, it was subsequently discovered that in both of these species, the heads of immature birds are green and fully-feathered, thereby rendering them very different in appearance from their respective adult forms. Ironically, celebrated Brazilian ornithologist Dr. Helmut Sick, who rediscovered Lear's macaw (see p. 142,) actually captured a specimen of this 'new' parrot back in the 1950s, but he too mistakenly assumed that it was merely a juvenile vulturine parrot.[5]

A similar situation was revealed in 2005 with the formal description of a new species of Brazilian parakeet—*Aratinga pintoi*, the sulphur-breasted parakeet. Although this bird had been known about for over a century, and was well-represented in museum collections, it had hitherto been mistakenly looked upon as a juvenile form of the sun parakeet *A. solstitialis*, even though the two species have very different plumages. Unfortunately, however, in 2009 it was discovered that this 'new' parakeet had actually been described earlier—a lot earlier, in fact—back in 1776, when it had been named *A. maculata*, but until the 2005 description appeared, this classification had long been dismissed as invalid. As it is now recognised again, however, nomenclatural precedence means that the sulphur-breasted parakeet has been renamed *A. maculata*, with *A. pintoi* becoming a junior synonym.[5a]

A LITTLE SOMETHING FROM LITTLE SUMBA

Continuing a steady stream of discoveries on the owl front in recent years, in 2002 a team of ornithological researchers from the University of Canberra announced the finding of a new species of Indonesian hawk owl, belonging to the genus *Ninox*. Formally reported in the journal *Emu*, the owl in question was discovered on the island of Little Sumba, and possibly just in time, as its researchers fear that it may be an endangered species, and therefore in need of official protection. Commemorating its island provenance, it was christened *Ninox sumbaensis*, the Little Sumba hawk owl.[6]

TWO NEW MINI MEAT-EATERS

In August 2003, a new species of small forest-falcon was announced from the rainforests of southeastern Amazonian Brazil. Aptly dubbed the cryptic forest-falcon *Micrastur mintoni* by its discoverer Andrew Whittaker, it was first spotted by Whittaker after hearing an unfamiliar forest-falcon call while birdwatching at Caxiuanã, and he later uncovered several preserved specimens at the Museu Paraenese Emilio Goeldi in Belém.[7]

Two months earlier, in June 2003, Brazil had seen the formal naming of a new species of owl—the Pernambuco pygmy owl *Glaucidium mooreorum*. This tiny bird, measuring a mere 6 in from beak-tip to tail-tip, and weighing just 2 oz, was first known from a specimen collected back in 1990, but it had not been recognised at that time as a new species.[8]

REFOUND IN FIJI

In 2003, after a century of supposed extinction, the Fijian long-legged warbler or thicketbird *Trichocichla rufa*—sole member of its genus—was rediscovered, and also photographed alive for the first time as well as having its dulcet song recorded. Twelve pairs were observed near Mount Tomaniivi on Viti Levu by a team of BirdLife International researchers—the first specimens spied there since 1894.[9]

ANEW IN NEW ZEALAND, AND CHEERY NEWS FROM CHILE

Another notable rediscovery in 2003 was the New Zealand storm petrel *Oceanites maorianus*. This small, reclusive seabird, allied to shearwaters and albatrosses, had last been reported over 150 years earlier, but on 17 November 2003 birdwatchers Bob Flood and Bryan Thomas not only sighted but also succeeded in photographing and videoing 10-20 individuals off North Island's Great Barrier and Little Barrier Islands. And during late 2005/early 2006, four specimens were captured in this same locality and subsequently released, three with radio transmitters attached in the hope that this elusive species' undiscovered breeding location would finally be traced, but as yet this has not happened.[10]

The existence off Chile's Puerto Montt coastal community of an apparently new, currently-undescribed species of *Oceanites* storm petrel—characterised by the very extensive expanse of white upon its underparts, and represented by many hundreds of individuals—was made public in March 2011 by British seabird specialist Peter Harrison. He had observed this scientifically-overlooked bird on many occasions, and had photographed it too.

Finally, however, in mid-February 2011, he and the other members of his five-man expedition in search of it also succeeded in netting 12 specimens for formal examination by Chilean scientists.[10]

BOWING TO THE INEVITABLE

One of the most enigmatic Australian birds on record is Rawnsley's bowerbird—traditionally known from just a single specimen shot by surveyor Henry Charles Rawnsley near Indooroopilly, Brisbane, in Queensland on 14 July 1867, and named later that same year as a new species, *Ptilonorhynchus rawnsleyi*. Sadly, this priceless specimen was later lost, and as no others were ever reported its ultra-mysterious species was dismissed as a unique intergeneric hybrid of the satin bowerbird *P. violaceus* and the regent bowerbird *Sericulus chrysocephalus*, as it shares the former's predominantly blue plumage but also displays the latter's distinctive patch of gold on each wing.[11]

In November 2003, however, environmental consultant Daniel Blunt photographed a very unusual bird perched upon his chicken coop at Beechmont, in southeastern Queensland, adjacent to the Lamington National Park. Unable to identify its species, he posted the photos on the Birding-Aus website, where they came to the attention of bowerbird expert Dr. Clifford Frith—who was

Rawnsley's bowerbird.

delighted to confirm that the mystery bird was a Rawnsley's bowerbird, the first recorded since 1867 and the first-ever living specimen documented by science. After studying Blunt's photographs, Frith co-authored a paper with him, published online in 2004, in which they reaffirmed the traditional belief that this enigmatic bowerbird was not a valid species but instead was a natural hybrid of the satin and regent bowerbirds.[11]

ON THE TRAIL OF A RAIL—OR TWO
During the past 250 years, a sizeable number of species of island-inhabiting flightless rail have become extinct, their lack of flight making them highly vulnerable to hunting by humans and various attendant introduced species (cat, dog, and rat in particular), as well as to habitat modification or destruction. Consequently, it was with great surprise that a British-Filipino ornithological expedition to the small, remote, but nonetheless inhabited northern Philippine island of Calayan (last visited by scientists in 1903) discovered on 11 May 2004 a small group of unfamiliar birds that proved to be a totally new species of rail—and one, moreover, that appeared to be, if not wholly flightless, at least nearly so.[12]

Related to the well-known moorhens, and sharing these latter birds' familiar, contrasting appearance of bright red beak and legs against much darker body plumage, the crow-sized Calayan rail was subsequently described in the journal *Forktail*, and was officially christened *Gallirallus calayensis*. One specimen was sacrificed as the new species' type specimen, and when dissected its wing muscles were found to be too weak to carry it far.[12]

This corresponds with statements derived from the island's local population—who know the rail well, terming it the *piding* in their language—that it has never been seen to fly. Around 200 pairs at most of these birds are estimated to live on Calayan, so any new development of the island or introduction of predators would soon put the entire species' continuing survival at great risk.[12]

In 2011, a new species of fairly large, predominantly brown rail was described from Madagascar. Named *Mentocrex* [now *Canirallus*] *beankaensis*, the Tsingy wood rail, it was first reported scientifically during a biological survey within the Beanka Forest (after which it was named) in late 2009, and it inhabits areas containing limestone karst and dry deciduous forests in the lowlands of this island's central-western region.[12]

A CACKLING OF CANADA GEESE
Taxonomically speaking, one of North America's most familiar birds has lately developed a split personality. In July 2004, the American Ornithologists' Union's (AOU) Committee on Classification and Nomenclature formally divided the Canada goose *Branta canadensis*

Cackling goose.

(traditionally classed as a single species made up of eleven subspecies) into two distinct species, distinguished by size and vocal characteristics. The larger of these species continues to be known as the Canada goose, retains seven of the eleven original Canada goose subspecies, and breeds in the interior of North America. The smaller new species comprises the other four subspecies, nests mostly in the Arctic, and due to its voice has been dubbed the cackling goose, formally christened *Branta hutchinsii*.[13]

FEATHERED REVELATIONS
IN THE FOJA MOUNTAINS
The November-December 2005 expedition to the remote, uncolonised Foja Mountains of Irian Jaya or Indonesian New Guinea (see p. 144) made some noteworthy avian discoveries—and rediscoveries.[14]

The first of these also turned out, taxonomically-speaking, to be the best, because the first bird to attract the scientists' attention after being dropped in the area by helicopter proved to be a totally new species. Readily distinguished from all related forms by its wattled, orange-hued face, it was a previously unknown species of honeyeater, which in 2007 was formally christened the wattled smoky honeyeater *Melipotes carolae*, and is the first new species of bird discovered in New Guinea for several decades.[14]

Moreover, two very important avian rediscoveries were made here during this same expedition. One was a famously elusive species of bowerbird called the yellow-fronted gardener *Amblyornis flavifrons* (see p. 144). The

Indonesian commemorative postage stamps from the author's collection, celebrating the wattled smoky honeyeater's discovery (bottom left) and the yellow-fronted gardener's rediscovery (bottom right) amid the Foja Mountains of Indonesian New Guinea. (Dr. Karl P.N. Shuker)

other one—which until now had been equally evanescent—was Berlepsch's six-wired bird of paradise *Parotia berlepschi*, sometimes classed as a subspecies of *P. carolae* but nowadays usually given distinct specific status. Formally described in 1897 and named after German ornithologist Hans von Berlepsch who had chanced upon some taxiderm museum specimens, all male, this beautiful bronze-plumaged bird was another 'homeless' New Guinea enigma—its stuffed specimens had definitely originated somewhere in this huge island, but where?[14]

Once again, California University zoologist Dr. Jared Diamond, co-rediscoverer of the yellow-fronted gardener, had helped to fill in a gap, when in 1979 he had encountered what he considered to be some female specimens of this rarity, alive and well and living in the Foja Mountains. In 2005, however, Beehler's team achieved a notable first—the first scientists to observe a living male Berlepsch's parotia, its throat adorned in gleaming metallic plumage, and its head sporting a striking crest of six disc-tipped, wire-like plumes.[14]

REDISCOVERING A REED WARBLER AND OUTING THE ODEDI

In 1867, a reed warbler was collected in India's Sutlej Valley, near Rampoor in Himachal Pradesh, which became the type specimen of a new species, the large-billed reed warbler *Acrocephalus orinus*. It also became an enigma, because no further specimen was ever reported—until 27 March 2006, that is, when one was captured alive in a wastewater treatment centre outside Bangkok, Thailand, by Mahidol University ornithologist Philip Round—thereby resurrecting this long-lost species from almost 140 years of extinction.[15]

Moreover, with stunning irony, just 6 months after Round's rediscovery, a second, hitherto-unknown 19th-century specimen came to light—discovered misclassified in a drawer of preserved Blyth's reed warblers *A. dumetorum* at the British Museum (Natural History)'s ornithological department in Tring, where it had lain ever since being collected in 1869 in India's Uttar Pradesh. And in 2009, a breeding site of this twice-lost, long-beaked, short-winged species was discovered in the Wakhan Corridor of the Pamir of northeastern Afghanistan by Dr. Robert Timmins of the Wildlife Conservation Society.[15]

A new species of warbler formally described in 2006 was the odedi, a species of *Cettia* bush warbler native to Bougainville Island in the Solomons group within the southwestern Pacific. As *Cettia* warblers are famous not only for their extremely nondescript appearance (epitomising the 'little brown job' description commonly applied by birdwatchers to small undistinguished-looking passerines) but also for skulking out of sight in dense riparine foliage, with only their loud cries betraying their existence, normally I wouldn't have deemed this bird significant enough to include here. However, its long-standing existence as a bona fide ethnoknown cryptid, i.e. a creature familiar to local inhabitants but unknown to science, is such that at least a concise documentation of it is merited.[16]

The odedi's cryptozoological history dates back at least as far as 1972, when American zoologist Dr. Jared

Diamond discovered that the Rotokas-speaking natives of Bougainville spoke of a scientifically-unidentified mountain-dwelling passerine known to them as the *kopipi*, and also known to this island's Nasioi speakers, who called it the *odedi*. Between 1977 and 1980, fellow ornithologist Don Hadden heard this mystery bird calling on several occasions, mostly during misty or at least damp weather; and in 1983, Dr. Bruce Beehler (destined to make some remarkable avian discoveries in New Guinea's Foja region two decades later) documented it in the journal *Emu*, in which he deemed it most likely to be a species of bush warbler.[16]

In January 2000, however, speculation was finally replaced by facts, when Hadden succeeded in obtaining a specimen of this elusive bird, followed by others later that same year and also in 2001. These confirmed that it was indeed a bush warbler, of the genus *Cettia*, and dark chestnut in colour but typically plain in overall appearance, though relatively large in comparison with other species of southwestern Pacific *Cettia* warbler, with a wider beak and longer toes. In recognition of Hadden's diligent investigation of this previously cryptic species, in 2006 ornithological researchers Drs. Mary LeCroy and F. Keith Barker formally christened the odedi *Cettia haddeni*.[16]

FANFARE FOR A FROGMOUTH
It had been more than a century since a frogmouth—a large-mouthed nightjar-related bird—had been caught anywhere in the Solomon Islands. That lone specimen, now apparently lost, had been classified in 1901 as a subspecies of the Australian marbled frogmouth *Podargus ocellatus*, after which nothing more whatsoever on the frogmouth front had been recorded from this island group—until 1998, that is.[17]

During a collecting trip that year to the island of Santa Isabel in the Solomons, Florida Museum of Natural History ornithologist Dr. Andrew Kratter was alerted by local hunters to the presence there of a frogmouth, and a single specimen was collected. Once this unique individual was examined and studied, Kratter and fellow Florida ornithologist Dr. David Steadman were startled to discover that it was dramatically different from all known species—so much so, in fact, that when finally described in the April 2007 issue of *Ibis*, it required the creation of an entirely new genus to accommodate it.[17]

Formally named *Rigidipenna inexpectata* ('unexpected stiff-feathered'), the Solomon Island frogmouth differs markedly from others in several notable ways. Most significant of all is that it is far less adept at flying, due to only possessing 8 tail feathers (all other frogmouth species sport 10 or 12), thereby reducing its lift, and to its much coarser feathers, which diminish its powers of manoeuvrability. It also exhibits distinctive barring on its primary wing feathers and tail feathers, and its plumage boasts larger speckles and more pronounced ventral white spots. Moreover, genetic comparisons fully support this morphological evidence for its taxonomic delineation—thereby making the Solomon Island frogmouth one of the most important ornithological discoveries of the 21st Century to date.[17]

SOMETHING TO CROW ABOUT
Also of note in 2007 was the rediscovery of the Banggai crow *Corvus unicolor*, when two specimens were obtained and additional ones sighted on the remote Indonesian island of Peleng in the Banggai Archipelago by local ornithologist Mochamad Indrawan and other members of the Celebes Bird Club. Prior to their find, this small, all-black, forest-dwelling species, presently categorised as critically endangered by the IUCN, had been known only from the two specimens upon which its formal description back in 1900 had been based.[18]

BECK'S IS BACK
Formally described in 1928, and last reported in the late 1920s too, when the only two specimens then known to science were collected and held thereafter at the American Museum of Natural History in New York, Beck's petrel *Pseudobulweria becki* had long been written off as extinct by ornithologists. During an expedition in summer 2007, however, to the Bismarck Archipelago, in the southwest Pacific Ocean, northeast of Papua New Guinea, led by Israeli ornithologist and writer Hadoram Shirihai, this small but highly significant seabird was dramatically rediscovered.[19]

Shirihai and his team not only came upon and photographed over 30 individuals but also observed young specimens in flight (suggesting a nearby breeding site), and obtained a dead individual in the sea, which became only the third recorded museum representative of its species. Shirihai thought that he had spied a Beck's petrel when visiting this area back in 2003, but he was not certain, so he planned the 2007 expedition specifically to search for evidence of the species' survival. In March 2008, he published a description of its rediscovery in the British Ornithologists' Union's *Bulletin* journal.[19]

A BALD-HEADED BULBUL FROM LAOS
First recorded scientifically back in May 1995 by Dr. Robert Timmins (discoverer of several new Indochinese mammals as well as the first known breeding site of the large-billed reed warbler *Acrocephalus orinus*), in July 2009 a most unusual and highly distinctive species of songbird was finally officially described and named—*Pycnonotus hualon*, the bare-faced bulbul, so-called because of its almost featherless pink face ('hualon' is a Lao word for 'bald-headed'). Native to Pha Lom, a

rugged limestone karst outcrop in Laos, this thrush-sized, predominantly olive-grey arboreal bird is the first new species of Asian bulbul to have been described for many years.[20]

SMALL BUT SUFFICIENT— THE WORLD'S TINIEST SHEARWATER

Back in 1963, a small, unusual specimen of shearwater was discovered and collected in a burrow within a colony of petrels in the Hawaiian Islands' Midway Atoll during the Pacific Ocean Biological Survey Program, and was initially assumed to be a little shearwater *Puffinus assimilis*. Subsequent DNA analysis, however, revealed that it actually belonged to a hitherto-undocumented species, which in 2011 was formally described and named—*P. bryani*, Bryan's shearwater. Moreover, its very distinct DNA profile is mirrored by its equally discrete morphology. Although predominantly black and white in colour, it has a bluish-grey beak and blue legs, and it constitutes the world's smallest shearwater species. Sadly, however, it also appears to be very rare, if not already extinct, as this lately-confirmed species is still known only from the single specimen collected in 1963, and its breeding grounds are currently undiscovered.[21]

21ST-CENTURY REPTILES AND AMPHIBIANS

PYTHONS APLENTY

In 2000, it was revealed that assessment of various lately-collected island-dwelling specimens of the Indonesian scrub python *Morelia amethistina* and reassessment of some museum specimens had increased the world's known species tally of these particular pythons. Two former subspecies of *M. amethistina* were elevated to full species status, and three brand-new species were described. These latter are: the Moluccan python *Morelia clastolepis*, the Tanimbar python *M. nauta*, and the Halmahera python *M. tracyae*.[1]

WHAT A LOAD OF COBRAS!

In 2003, zoo staff at London Zoo were very startled to discover that a haul of spitting cobras donated there by Customs, following their seizure of specimens originating in Egypt, did not belong to that country's familiar species, the red spitting cobra *Naja pallida*, but instead constituted a wholly new species, hitherto undescribed by science! Distinguished from *N. pallida* by differences in its throat and neck scale pattern, its overall body colour, and its DNA, this previously overlooked serpent has been formally dubbed *Naja nubiae*, the Nubian spitting cobra. Moreover, its London representatives lost no time in swelling the ranks of captive individuals of this newly-exposed species by laying a clutch of 10 eggs, all of which hatched successfully on 5 June 2003.[2]

I'VE NEVER SEEN A PURPLE FROG...

The discovery of a remarkable new species of frog in the Western Ghats Mountains of southern India was also announced in 2003. Its notable nature was twofold. Firstly, its skin colour was a spectacular, vivid purple, which, combined with its small head, stalk-like snout, and tiny eyes but bloated body, made it look more like a plum than a frog. Secondly, it was a bona fide 'living fossil', a very primitive relict species whose closest living relatives taxonomically were the sooglossid frogs found only in the Seychelles, some 2000 miles from India. Indeed, its discoverers, Drs. S.D. Biju and Franky Bossuyt, initially deemed it sufficiently distinct to merit creating an entirely new taxonomic family to accommodate it—which would have been the first since 1926— but it was subsequently housed within Sooglossidae, and was formally named *Nasikabatrachus sahyadrensis*, the purple frog.[3]

Purple frog. (Karthickbala/Wikipedia)

THE MISSING SKINKS

During September 2003, four specimens (including two adults) of a superficially worm-like limbless lizard known as the Barkuda (aka Badakuda) Island skink *Barkudia insularis* were observed and also videoed on Barkuda/Badakuda Island (its only known home) in Chilka Lake, which is a brackish water lagoon within the eastern Indian state of Orissa. What makes these sightings so interesting is that only two previous records of *B. insularis* are known, and both occurred more than 80 years earlier. The first was when its holotype was discovered in some loose earth near the roots of a banyan tree here during 1917. Its species (housed in a genus to itself) was formally described that same year by

Nelson Annandale. During the rainy season of 1919, a second specimen was spied in the same locality by Indian Museum zoologist Dr. Frederic H. Gravely, but it swiftly burrowed into the ground. Even today, little is known of this reclusive species' lifestyle, though it may be nocturnal.[4]

In December 2003, herpetologist Dr. Ivan Ineich from the National Museum of Natural History in Paris, France, discovered a living specimen of Bocourt's giant skink *Phoboscincus bocourti*, measuring 20 in long, on a tiny isle just off the Île des Pins, close to New Caledonia. Significantly, this is only the second specimen ever recorded of Bocourt's sizeable lizard (also known as the terror skink on account of its notably large size for a skink). Prior to it, the single known specimen was a preserved individual in the museum's collections that had been obtained on the same isle back in 1870; it became the holotype of this species, which was formally described and named in 1876.[5]

THE CHAMELEON SNAKE OF BORNEO

Captured in 2005 but only formally described as a species new to science in 2006, the 1.6-ft-long mildly-venomous species of rear-fanged water snake procured by WWF researchers in a swamp near Borneo's Kapuas River in the Betung Kerihun National Park seemed nothing overly distinctive—until one of the team members placed it for safe keeping in a bucket of water. Within just a few minutes, the reddish-brown snake turned almost entirely white![6]

Although this quick-change ability had been previously reported from various lizards, most famously the chameleons, in the few species of snake known to be capable of changing colour the change tends to be much slower, occurring over days or even weeks, as during the mating season, for instance. Swifter transformations, involving either rapid changes in skin pigment cells or by elevating body scales, are much rarer in snakes.[6]

The reason for this chameleonic ability in the new snake, which has been officially dubbed the Kapuas mud snake *Enhydris gyii*, is presently undetermined. It does not seem to be linked to camouflage, unlike the chameleon's own talent, so perhaps it serves as a warning or defence mechanism instead. Nor is it yet known whether this new snake can turn other colours too, or what biological processes power its colour change.[6]

IN SPITTING DISTANCE OF AN EARLIER REVELATION

One of the world's most spectacular 'new' species of snake was actually discovered four decades ago. Back in the 1960s, James Ashe, founder of the Bio-Ken Snake Farm in Watamu on Kenya's tropical coast, captured an unexpectedly large spitting cobra, and suggested that it may belong to a new, undescribed species. However, his suggestion was not pursued, and the subject remained closed until some additional, extra-large specimens were more recently obtained from the same coastal region.[7]

These were duly investigated by Bangor University researcher Dr. Wolfgang Wüster and colleague Dr. Donald Broadley from the Biodiversity Foundation for Africa, and their DNA confirmed that they belonged to a wholly distinct species. Moreover, attaining a length of almost 9 ft, this new species, in 2007 formally christened Ashe's spitting cobra *Naja ashei* in honour of its original discoverer, now deceased, is the world's largest spitting cobra, and is not only very dramatic but also very dangerous, possessing sufficient venom to kill at least 15 people.[7]

GREEN BLOOD AND TURQUOISE BONES— IT'S THE SAMKOS BUSH FROG

Also in 2007, one of the most distinctive new frogs discovered in a long while was formally described and named—*Chiromantis samkosensis*, the Samkos bush frog. Native to the jungles of southwestern Cambodia's Cardamom Mountains, it is very small (less than an inch long from snout to vent), and so well camouflaged that only its trilling call betrays its presence. Once examined more closely, however, this remarkable little amphibian is found to display two very noteworthy characteristics—turquoise-coloured bones, and green blood. These are due to the green bile pigment biliverdin, which in other species in normally processed by the liver as a waste product, but in the Samkos bush frog is released back into the bloodstream, and by being visible through its translucent skin apparently assists in camouflaging it.[8]

IN THE PINK IN THE GALAPAGOS— A REMARKABLE NEW LAND IGUANA

One of the most distinctive yet unexpected lizards to have been discovered for many years was officially, and very belatedly, described and named in January 2009. Its story should have begun back in 1835. That was when Charles Darwin visited several of the Galapagos Islands, including the largest one, Isabela. Unfortunately, however, he never explored Volcan Wolf in Isabela's northernmost region, and so he never realised that whereas all of the land iguanas that existed elsewhere in the Galapagos archipelago were a typical yellow-brown colouration, those living exclusively on this volcano were pink, striped with black. Moreover, no-one else realised it either—not until as recently as 1986, in fact, when some Galapagos park rangers visiting Volcan Wolf saw pink iguanas there. Even then, these were initially dismissed as freak individuals of the familiar Galapagos land iguana *Conolophus subcristatus*.[9]

In 2009, however, following close scrutiny of living specimens and genetic samples taken from a live individual designated as the pink (aka rosada) iguana's holotype, this highly distinctive yet long-overlooked lizard was formally classified as a separate species and was christened *C. marthae*. Moreover, as if its unique colouration and exceedingly limited distribution range were not significant enough (not to mention certain head-bobbing behavioural idiosyncrasies), genetic research revealed that far from being a recently-evolved species (as originally presumed), the pink iguana had actually diverged from the ancestral stock for all Galapagos land iguanas some 5.7 million years ago—which is one of the oldest evolutionary divergences ever recorded from this archipelago.[9]

Galapagos pink land iguana. (Tim Morris)

A PAIR OF NOVEL MONITOR LIZARDS FROM SOUTHEAST ASIA

At the beginning of April 2010, a quite spectacular yet hitherto-unrecognised species of monitor lizard from the Philippine island of Luzon was formally described and named—*Varanus bitatawa*, the northern Sierra Madre forest monitor. It can reach a total length in excess of 6 ft, but rarely weighs more than 22 lb as, in spite of its size, it is a tree-dweller, and is confined entirely to the Sierre Madre forest on the island's northeastern coast. Like its closest relative, Gray's monitor *V. olivaceus* (see p. 177) of southern Luzon, yet unlike all but one other monitor species, it is frugivorous, its preferred diet being the fruit of *Pandanus* trees. Making this novel lizard's belated scientific discovery all the more surprising is its eyecatching colouration—combining a dark blue-black background body colouration with bright gold flecks arranged in transverse stripes, and a lengthy tail banded in black and green.[10]

Although this species is regularly killed and eaten by the Aeta and Ilongot tribes, who refer to it as the *butikaw*, it only first attracted scientific attention in 2001, when a team of biologists surveying northern Luzon took photos of various monitor lizard carcasses belonging to this species that were being carried by some hunters. Then in 2009, a near-dead adult male specimen was brought by hunters into the camp of an expedition team of graduate students. Genetic tests confirmed that it belonged to a species new to science.[10]

By the end of April 2010, moreover, a second remarkable new species of southeast Asian monitor, this time from Indonesia, had also been described and named—*V. obor*, the torch monitor, referring to its bright orange head and glossy black body ('obor' also translates as 'torch' in Indonesian). Confined entirely to the small island of Sanana in the western Moluccas group, measuring up to 5 ft long, and unusual among monitors in possessing red pigmentation, it was originally discovered in spring 2009, by Finnish graduate research student Valter Weijola. He returned to Sanana towards the end of that same year with fellow biologist Prof. Sam Sweet, from the University of California Santa Barbara, spending five weeks there photographing and studying it in its native habitat of coastal sago palm swamps and rainforests.[11]

VAMPIRE TADPOLES FROM VIETNAM

Traditionally, the most famous attribute of many *Rhacophorus* tree frogs from Asia has been their ability to glide through the air, by leaping from trees and then spreading out the strong, wide membranes linking the long toes of their feet. In December 2010, however, a new species was formally described and named that has an equally startling surprise. For whereas the mouthparts of tadpoles are normally beak-like, this new species' tadpole uniquely possesses a pair of hard black curved fang-like hooks projecting from the underside of its mouth. Not surprisingly, therefore, it has been dubbed the vampire flying (or gliding) frog *R. vampyrus*.[12]

Recorded only from the cloudforests of southern Vietnam, where it was first observed scientifically in 2008, this species' 2-inch-long adult, which is brown dorsally and white ventrally, spends its entire life in the trees, even laying its eggs in pools of water trapped within nooks or hollows in tree trunks. When these eggs hatch, the fanged tadpoles feed upon additional, unfertilised eggs laid specially by their mother to serve as food for them. So it is possible that the tadpoles' fangs have evolved to slice open these eggs, or may even assist them in clinging to the sides of tree hollows in which they swim after hatching—otherwise there is no clue as yet to their function.[12]

21st-CENTURY FISHES

AN EEL OF A DISCOVERY—TWICE!

New species, and even new genera, of fishes are still described quite frequently by zoologists, but the creation of an entirely new taxonomic family to house a new fish is a far, far rarer occurrence. Yet it may happen, thanks to a peculiar eel-like fish discovered in 1999 by researcher Ilse Walker. Working with Brazil's National

Amazon Research Institute (INPA) in flood plains near Manaus, she came upon an unfamiliar long-bodied 6-in fish that was later found to combine characteristics from a range of quite separate fish groups.[1]

Among its most notable features are its full set of fins (unlike true eels), a tail resembling that of the huge freshwater arapaima (pirarucu) *Arapaima gigas*, and up to 10 air chambers (instead of the usual two-to-three in other fishes, used for maintaining their position underwater). This latter feature incited researchers to speculate in 2003, when this species' discovery was first made public, that their still-undescribed 'mystery fish' (as they call it) may breathe at the water surface.[1]

And proving that, contrary to traditional dictum, history does indeed repeat itself sometimes, the year 2011 saw the formal description of another small but extraordinary new species of eel that was so far removed, taxonomically speaking, from all others that it too required the creation not only of a new genus but also of a new family to accommodate it. Dubbed *Protoanguilla palau*, it first came to attention in March 2010, when some specimens were discovered swimming in a deep underwater cave in a fringing reef off the coast of Palau—a tiny island nation in the Pacific Ocean. And because it is believed to have diverged from other eels as far back as 200 million years ago, during the Mesozoic Era, *P. palau* is considered to be the world's most primitive species of living true eel (i.e. belonging to the order Anguilliformes), and is thus being referred to as a bona fide 'living fossil'. Among its unique characteristics are its distinctive collar-like gill openings, disproportionately large head, slightly produced caudal fin rays, and short compressed body.[1]

UNICORNS AND DRAGONS

For four weeks in 2003, a deepsea research ship called *Tangaroa* probed the Tasman Sea separating Australia from New Zealand, and snared 500 species of fish and 1300 species of invertebrate. These were subsequently studied, and were found to include some remarkable new species. One of the most distinctive, however, must surely be the unicorn batfish. This deepwater fish quite literally walks along the ocean bed, using specially-modified fins that resemble and function like true legs, and has an extraordinary head that tapers to a point (the rostrum) so that it resembles a unicorn.[2]

Staying in deepwater but moving to those of Bear Seamount, off New England, U.S.A., in 2004 a new species of dragonfish was formally christened *Eustomias jimcraddocki* (after a famous deepsea fish biologist). It was obtained during an expedition to these waters sponsored by the National Oceanic and Atmospheric Administration's Office of Ocean Exploration, and was retrieved from a sample net by fish ecologist Dr. Tracey Sutton. Further specimens, hitherto unidentified or misidentified, were subsequently uncovered by Sutton in the collections of the Smithsonian Institution and Harvard University. Although somewhat larger than most previous species, just like those Jim Craddock's dragonfish sports a series of bioluminescent organs along its belly, as well as a trailing chin-attached barbel that also bears a bioluminescent organ at its tip.[3]

ITSY BITSY INFANTFISH!

Aficionados of zoological superlatives will confirm that the traditional, longstanding holder of the 'world's smallest fish' (and vertebrate) title has been the dwarf goby *Trimmatom nanus* of the western Indian Ocean, measuring a minuscule 0.4 in (10 mm) long—but no more, thanks to a lately-unveiled usurper known as the stout infantfish. Formally christened *Schindleria brevipinguis* in 2004, this minute goby-related free-living (i.e. non-parasitic) fish has males that measure no more than 0.28 in (7 mm) long, with the slightly burlier females reaching 0.33 in (8.4 mm).[4]

Ironically, despite its tiny size the stout infantfish first came to light as long ago as 1979, when a specimen was collected by Jeff Leis of the Australian Museum during fieldwork in the Great Barrier Reef's Lizard Island region. However, it remained unrecognised as a new, undescribed species until re-examined more recently by Dr. H.J. Walker from California's Scripps Institution of Oceanography and William Watson of the Southwest Fisheries Science Center in La Jolla. By the time of its formal description, six specimens had been recorded, revealing this diminutive species to be paedomorphic—retaining many juvenile characteristics even when adult, which explains its common name. Even its lifespan is suitably brief—estimated to be a mere two months.[4]

In 2006, the stout infantfish faced a serious challenge to its title as the world's smallest free-living fish from a comparably minute, newly-described species—*Paedocypris progenetica*, native to the dark peat swamps of Sumatra. Its smallest mature female specimen on record measured just 0.32 in (7.9 mm), which is smaller than the female of any other vertebrate species. A member of the carp family, Cyprinidae, *P. progenetica* was originally discovered and identified in 1996, by Swiss ichthyologist Dr. Maurice Kottelat and fellow ichthyologist Dr. Tan Heok Hui from the Raffles Museum of Biodiversity Research and the National University of Singapore, but a further decade would pass by before it was formally described and named. Even when fully mature, this species' transparent miniature-sized body and head lack a number of features typical of adult fishes while retaining various larval characteristics; moreover, its skull is so rudimentary that the brain is exposed.[5]

Incidentally, although *P. progenetica* when mature is a claimant for the title of the smallest free-living adult fish, it is not the world's smallest adult fish of any kind. That title goes to the ectoparasitic dwarf male of *Photocorynus spiniceps*—a deepsea anglerfish living in the Gulf of Panama, whose tiny males spend their entire adult lives permanently fused by their jaws to specimens of the much bigger female form. Sexually mature males of just 0.25 in (6.2 mm) and 0.26 in (6.5 mm) have been recorded by Washington University anglerfish expert Dr. Ted Pietsch. This remarkable species was originally described and named back in 1925, by eminent ichthyologist Dr. Charles Tate Regan.[6]

CUDDLES—A CARTILAGINOUS CURIOSITY

One of the most surprising animal discoveries—in many ways—of recent times features a shark with the unlikely nickname of Cuddles. Her story began at the beginning of this present century, when she was sold privately to an Austrian pet shop, which in turn sold her to a fitness studio in Upper Austria. Later, the studio was destroyed by flooding, but fortunately for Cuddles she was rescued and taken to an animal rescue centre, where, not long afterwards, she was moved on yet again, this time to Vienna's Schoenbrunn Zoo. Here she stayed for two years, housed in an aquarium where she could only be readily perceived from the top, from which angle she seemed to be nothing more than an ordinary 2-ft-long nurse shark *Ginglymostoma cirratum*.[7]

In 2004, however, the much-travelled Cuddles was given by the zoo to the Sea Star Aquarium in Coburg, Germany, where she was observed closely—and found to be unlike any shark, nurse or otherwise, previously recorded by science! When viewed from the side, Cuddles is seen to be covered in hairy bristle-like structures, to which algae are attached, turning the bristles red in colour. She also has a skin patterning of small black spots. Moreover, unlike those of other sharks, the iris in each of Cuddles's eyes is fixed open (suggesting that in the wild state her species is adapted for a shadowy rather than a brightly-lit environment), her nasal apertures are unusually large, she lacks the array of sensory organs at the front of her head so characteristic of most other sharks, and her fins are smaller but more muscular than those of other similar-sized sharks.[7]

In addition, by clapping these fins together she readily hops along the bottom of her tank like a frog, instead of gliding along gracefully in typical shark mode. According to media reports, she also has one pair of gills designed for filtering plankton. Yet unlike other plankton-feeding sharks, Cuddles has a full set of sizeable teeth too, of which the principal ones are extraordinarily long, and so far she has eaten everything offered to her at the aquarium, in so doing revealing an exceptionally powerful bite for her size.[7]

In the opinion of the aquarium's curator, Peter Faltermeier (not Faltermeer, as incorrectly given in media reports), Cuddles's singular assemblage of characteristics points to a dark underwater cave-dwelling rather than deepwater lifestyle, quite probably in waters around southern Africa. In particular, her bristles, if sensory, may well assist her in locating approaching prey by detecting water vibrations caused by their movements, and this ability might be augmented by the algae growing on the bristles, as these would extend their length and thus their sensitivity.[7]

At the time of the original media coverage regarding Cuddles, staff at the aquarium planned to prepare a formal description and name for Cuddles, and were also, in conjunction with Schoenbrunn Zoo, hoping to discover where Cuddles originated. Meanwhile, their hairy hopping shark has become a popular attraction, living blissfully unaware of her unique ichthyological status—the type specimen of the only shark species discovered by science in a fish tank.[7]

On 2 July 2011, I learnt from German cryptozoological researcher Markus Hemmler that German researchers plan to x-ray Cuddles later in July to ascertain her identity, and also that Faltermeier had already christened her hitherto-unnamed species the dotted nurse shark *Pseudoginglymostoma nigropunctatum*, thereby making it only the second member of this genus. However, neither Markus nor I have been able to trace any publication containing the above binomial name as part of an official description for this species, so whether it is a scientifically-recognised name, or merely a nomen nudum, is presently unclear.[7]

CHOMPING THE CHIAPAS CATFISH

The official discovery of the Chiapas catfish in 1996 by Mexican ichthyologist Dr. Rócio Rodiles-Hernández—who collected the first scientifically-recorded specimens from the Lacantún River on the border of Guatemala with the Mexican state of Chiapas—was highly unexpected, for two very different reasons. Firstly, it is so dissimilar from all other catfishes known to science that it is not merely a new species, but a new genus, *Lacantunia*, and one that in addition required the creation of a brand-new taxonomic family to accommodate it. Secondly, despite its singular novelty to the scientific world, this fish is such a commonplace creature locally that for generations it has been eaten fried or in soups when caught by fishermen in the Lacantún River, flowing through the Chiapas highlands.[8]

Nevertheless, jaw muscle attachments, the number of rays in its fins, the number and position of its facial barbels, not to mention its DNA, were all sufficiently

The ostensibly familiar manta ray has recently been shown to constitute at least two (possibly three) separate species.

distinct to warrant in 2005 the Chiapas catfish's dramatic classification, following examination of a range of preserved specimens by ichthyologist Dr. John Lundberg from Philadelphia's Academy of Natural Sciences and other co-workers. In its official scientific description, it was formally, and fittingly, christened *Lacantunia enigmatica*.[8]

IS THERE A NEW MANTA... OR TWO?

Traditionally, only one species of manta ray, *Manta birostris*, has been recognised by science. In 2008, however, manta researcher Dr. Andrea Marshall announced at the American Elasmobranch Society's annual meeting in Montreal, Canada, that some smaller, reef-dwelling mantas with distinctive appearance spied off southeast Asia, Mozambique, and elsewhere may well warrant formal reclassification as a second, separate species.[9]

Unlike the known, larger manta, which is migratory and relatively shy, swimming away from divers, the new, smaller manta is a residential species, staying in much the same coastal regions, and is not intimidated by the presence of human divers nearby. Moreover, the known, migratory manta has a non-functioning barb on its tail, whereas the smaller, residential manta does not. Marshall (sponsored by the Save Our Seas Foundation) has also noted differences in reproductive behaviour, skin texture, and colour, and believes that there may even be a third species of manta present, represented by specimens currently under study in the Caribbean portion of the Atlantic Ocean.[9]

In 2009, the new manta was formally christened *M. alfredi*, the reef manta (with the larger, familiar species, *M. birostris*, now being known as the giant manta). This naming thereby resurrects a 19th-century binomial that until now had been synonymised with *M. birostris*.[9]

DRACULA FISH—A JAW-FANGED ANOMALY

The year 2009 also saw the formal description and naming of a much smaller but no less surprising species of fish—*Danionella dracula*, the Dracula fish. Two years earlier, a consignment of aquarium fishes shipped into the U.K. from Myanmar contained some tiny but highly unusual cyprinids that were clearly related to the familiar zebrafish *Brachydanio rerio* but could not be assigned to any known species. With a maximum length of little more than half an inch, a scaleless translucent body, remarkably few bones, and a superficially larval appearance even in the adult form, this hitherto-unknown species was already very distinctive, but was rendered unique among modern-day cyprinids by the notably large fang-like structures protruding from its jaws, because all other cyprinids since at least the Upper Eocene epoch (roughly 35 million years ago) have been toothless.[10]

Upon closer observation, however, the new fish's vampiresque fangs (explaining its subsequent scientific and common names) were found not to be genuine teeth at all. Instead, they were sharp, pointed bones projecting from this fish's jaw bones, and were most prominent in male specimens—which have been observed using them to spar with one another. To date, the Dracula fish has only been recorded in the wild state from a single small stream, at Sha Du Zup in northern Myanmar.[10]

BREAKING ALL THE RULES—
THE PSYCHEDELIC FROGFISH

Yet another iconoclastic new fish described in 2009 was the psychedelic frogfish *Histiophryne psychedelica*, a species of anglerfish belonging to the taxonomic family Antennariidae. It was first photographed in the wild during January 2008 (two specimens collected back in June 1992 from Bali and dubbed 'paisley anglerfishes' had been misidentified as a known, related species, and duly forgotten), and is currently recorded only from the waters off the Indonesian islands of Ambon and Bali.[11]

This eyecatching species, garbed resplendently in a flamboyant array of swirling white and yellow-brown stripes, is, uniquely, an anglerfish that lacks these species' most characteristic feature—the illicium or luring appendage (see p. 202). Present in a variety of different forms (depending upon the species in question), the illicium is normally suspended above the anglerfish's open jaws (or occasionally even inside them), and entices its unsuspecting victims within reach.[11]

Moreover, whereas some related species remain virtually immobile on the seafloor, or, if they move around, do so by jettisoning themselves off the seafloor before swimming normally, the psychedelic frogfish often uses its jet-propulsion skills (shooting water through its gills) to hop from place to place while its body expands into a spherical form, so that it almost bounces, like a child's rubber ball. And although its relatives often change colour to match their surroundings, this garishly-patterned species does not, even when its stripes render it strikingly visible—and therefore vulnerable to attack by potential predators. No less novel is its thick, loose skin, which hangs in deep fleshy folds and even envelops its unpaired fins.[11]

Most remarkable of all, however, are its eyes. In stark contrast to those of almost all other fishes, which are located on the sides of the head, in this species they are positioned at the front of its unusually small, flat face, and are directed forward, thereby affording the fish binocular vision and enabling depth perception, as in primates and certain other mammals.[11]

A PAIR OF STINGLESS STINGRAYS FROM SOUTH AMERICA

In 2011, a new genus of freshwater stingray, containing two previously-undescribed species, was formally named and documented—*Heliotrygon*. The two species, Gomes's round ray *H. gomesi* and Rosa's round ray *H. rosai*, both inhabit the Nanay River, near Iquitos in Peru, grow up to 1.6 ft long, and are pancake-shaped, but, intriguingly, lack the most infamous characteristic of many stingrays—their sting. Instead, their tails only bear a tiny, degenerate, entirely harmless spine. Equally interesting is that despite having remained undescribed by science until as recently as 2011, they have been familiar fishes in the international pet trade for many years, having been sold under a variety of different trade names.[12]

21st-CENTURY INVERTEBRATES

LOOK OUT FOR THE MILLENNIUM BUG!

Just when everyone thought it was safe to go back to the keyboard in 2001, scientists announced that the millennium bug was alive and well, and living in Australia! Fortunately, however, it was not the computer-crashing variety. Instead, it was a hemipteran—the taxonomic order of insects known not only colloquially but also scientifically as bugs. Discovered during the late 1990s, the new species in question was a small (less than 0.1 inch long) but predatory water strider belonging to the family Veliidae, and was closely related to the larger, more familiar pond skaters that race across the water surface of ponds and lakes.[1]

Native to freshwater streams in at least eight known mountain locations of northeastern New South Wales and southeastern Queensland, entomology's millennium bug constituted not only a new species but also a new genus. When formally described in 2001 by Dr. Nils Møller Andersen from the Zoological Museum of Copenhagen University and Australian entomologist Dr. Tom Weir from CSIRO, it was christened *Drepanovelia millennium*.[1]

WELCOMING BACK THE WALKING SAUSAGE AFTER 83 YEARS

Lord Howe Island, a small island located 370 miles east of New South Wales, Australia, once possessed its very own species of stick insect. Nocturnal and wingless, *Dryococelus australis* was also characterised by its long legs, hooked claws, and half-inch-thick, 6-in-long, reddish-brown sausage-like body—which earned it the memorable nickname of 'walking sausage' in media reports of its highly unexpected 21st-century comeback.[2]

During the first week of February 2001, a team of Australian scientists from the New South Wales National Parks and Wildlife Service discovered three female specimens of the Lord Howe Island stick insect (plus a number of its eggs) feasting upon a single tea-tree bush on a volcanic rock called Ball's Pyramid, situated 14 miles southeast of Lord Howe Island itself. Lord Howe Island was home to *D. australis* (known locally as the land lobster) until 1918, when it was wiped out by invading rats following the running aground there of the supply ship S.S. *Makambo*. Since then, this spectacular stick insect was assumed to be extinct—until the 1960s, that is, when some rock climbers visiting Ball's Pyramid claimed to have seen some strange insects there

Lord Howe Island stick insect. (Tim Morris)

that fitted the description of the supposedly demised *D. australis*.[2]

Their claims were now vindicated, and the scientific team hoped to re-establish a viable colony on Lord Howe Island. With this in mind, two breeding pairs were collected—one of which was given to a private breeder, the other to Melbourne Zoo. By 2008, the total captive population had risen to approximately 450 individuals, and 20 had been repatriated to a special habitat on Lord Howe Island, but their species is still critically endangered, as categorised by the IUCN.[2]

SPONGING SECRETS FROM THE SEA

Up until the late Cretaceous Period, 65 million years ago, some sponges existed in massive underwater reefs, analogous to coral reefs. Today, however, sponge reefs are long gone, except for their ancient, Mesozoic fossil remains—or so it was thought.

In March 2001, however, Dr. Kim Conway from the Geological Survey of Canada formally announced the discovery of four colonies of hexactinellids or glass sponges constituting giant reefs covering more than 420 square miles in the sea off British Columbia's Queen Charlotte Islands, forming part of the western Canadian continental shelf. Occurring at a depth of approximately 330 ft, with each individual sponge standing 4-6 ft high, and spanning at least 9 ft across, their presence was first suspected following sonar surveys back in 1984. However, they were not actually viewed until 1999, during an 18-day submarine foray.[3]

In 2005, three more hexactinellid reefs were discovered, again on the western Canadian continental shelf, but this time in the Georgia Basin. Incorporating three predominant, siliceous species, they take a variety of forms, resembling massive domes 6 ft high and 12-16 ft across, or consisting of a mass of delicate radiating tubes, each only 0.04-0.08 in thick. Due to their extreme fragility, however, they are in danger of being severely damaged by bottom trawling. In 1999, what has since become a major ongoing scientific study of the British Columbia sponge reefs was jointly launched by Canada and Germany, and is known as the Sponge Reef Project.[3]

IT'S BIGFIN!

In December 2001, a scientific team that included Dr. Michael Vecchione of Washington DC's National Museum of Natural History, Hawaii University researcher Dr. Richard Edward Young, and oceanographer Prof. William Sager of Texas A&M University, published a short but fascinating report in *Science*. It concerned a truly extraordinary species of bathypelagic squid, which, although apparently undescribed by science and unrepresented by even a single preserved specimen, was nevertheless unquestionably real—thanks to a series of

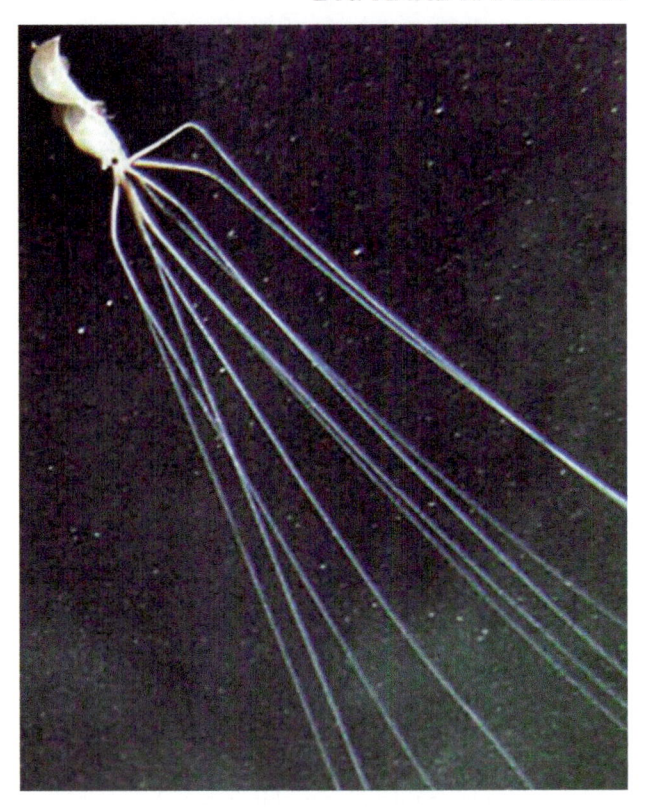

Long-arm squid.

excellent video frames accompanying their report, taken from video film segments (some up to 10 minutes long) obtained by several different submersibles near the sea bottom in a wide range of different locations between 1988 and 2001.[4]

These locations included the western Atlantic Ocean off Brazil, the eastern Atlantic Ocean off Africa, the Indian Ocean, the Gulf of Mexico, and the Central Pacific. Unnamed it might be, and seen only eight times so far, but this species was clearly of worldwide distribution—and unlike any known squid.[4]

Although it had ten limbs like other squids, they were all identical—whereas in other squids, two are much longer than the remaining eight. Moreover, this squid held its limbs in a unique position—spread outward from the body axis, then abruptly bent anteriorly, making them look as if each had an elbow. These limbs were also extremely slender and capable of stretching considerably. More eyecatching still, however, was this squid's extremely large, terminally-sited pair of fins, which resembled huge wings. When swimming normally or hovering, these fins undulated slowly, but in rapid swimming or when the squid made physical contact with an object they flapped vigorously.[4]

In their report, Vecchione and his colleagues suggested that this bizarre creature, which has been dubbed the long-arm squid, may be the hitherto-unknown adult form of a decidedly odd form of squid originally represented only by a handful of immature specimens—all obtained in epipelagic waters in the central and eastern Pacific. Due to their noticeably large terminal fins, this squid form was christened 'bigfin', when first documented by Young back in 1991. Moreover, in 1998 Young and Vecchione joined forces to create not only a new genus, *Magnapinna* ('big fin') but also a new taxonomic family, Magnapinnidae, in order to house bigfin and various subsequently-discovered, related species.[4]

Now, at least according to Vecchione and colleagues, with the videoed observations of the long-arm squid science may finally have witnessed bigfin's hitherto-elusive adult form. However, only direct study of long-arm specimens (of which none has so far been procured) can determine conclusively whether it and the bigfin larval form belong to the same species or, indeed, even closely-related ones.[4]

ENTRY OF THE GLADIATORS—
A NEW ORDER OF INSECTS!

Thousands of new insect species are discovered every year, but it had been almost a century since any proved so dramatically different from all others known to science that a wholly new taxonomic order needed to be created to house them. However, such an extraordinary event did indeed take place in 2002.[5]

In 1909, two insects were collected by a priest in what is now Namibia and deposited in a Berlin collection, only to be forgotten afterwards—until 2001, when found there by Danish zoologist Dr. Niels Peder Kristensen. By an amazing coincidence, in November 2001 an identical but freshly collected insect specimen had been shown to Dr. Eugène Marais, head of entomology at Namibia's National Museum at Windhoek. So he readily recognised their significance when, just a few days later, Kristensen alerted him to the Berlin duo's existence. Measuring 4-8 in long, this enigmatic species resembled a cross between a praying mantis and a stick insect

Postage stamp commemorating the discovery of the gladiators. (Dr. Karl P.N. Shuker)

(which belong to separate taxonomic orders), but lacked the mantis's triangular head and seizing forearms, and was wingless (as indeed all subsequently-discovered species have been found to be).[5]

Three specimens had been found in southern Namibia's Huns Mountains and Palmwag district during 1994, but their species was initially believed to be extremely rare, yet was exceptionally significant, because as boldly proposed by Marais, it did indeed require a new taxonomic order—the first new order erected since Zoraptera and Grylloblattodea, both more than 80 years earlier. In March 2002, an expedition to western Namibia's remote Brandberg region, sponsored by Oxford University Museum and London's Natural History Museum, announced that it had collected some additional specimens of this insect.[5]

In a paper published by the journal *Science* later in 2002, a joint Danish-German team formally named the new order Mantophasmatodea, emphasising the insects' unique combination of mantid and phasmatodean (stick insect) features. Two modern-day species were named—*Mantophasma zephyrum*, based upon a female specimen from Namibia, and *M.* [now *Tanzaniophasma*]

Mantophasma zephyrum.
(Dr. Philip E. Bragg/Wikipedia)

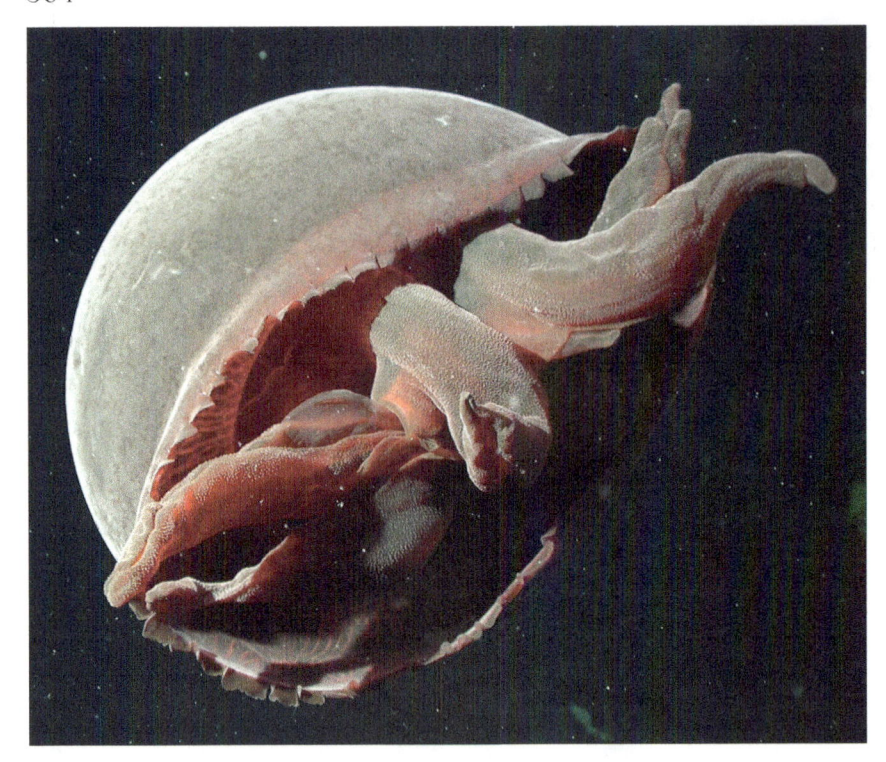

Big Red—so distinctive a new species of jellyfish that it required the creation of a brand-new subfamily to accommodate it. (NOAA)

subsolanum, based upon a male specimen from Tanzania. Moreover, a specimen embedded in a piece of 45-million-year-old Baltic amber that had previously been classified as an orthopteran (i.e. belonging to the grasshopper and cricket order) was subsequently reclassified as a mantophasmatodean, and was formally dubbed *Raptophasma zompro*, after German entomologist Dr. Oliver Zompro from the Max Planck Institute of Limnology who has officially described several of these newly-revealed insects. He was also the first scientist (and while still a postgraduate student) to recognise the affinity of the amber-embedded specimen with these modern-day species.[5]

Scientifically, these new insects are known as mantophasmatodeans, but they have also received a rather more readily pronounceable common name—gladiators, on account of their formidable armoured forelimbs, reminiscent of the fighting armour borne upon the arms of Roman gladiators, and also due to the armour-like covering exhibited by juvenile specimens (nymphs) of these insects.[5]

In 2004, the gladiators' claim to separate ordinal status was re-evaluated, as a result of which a previously-abandoned insect order, Notoptera, was reinstated, with both Mantophasmatodea and Grylloblattodea (housing the ice-crawlers and also hitherto deemed a separate order) being placed within it, and from 2006 onwards relegated to the level of suborders. Nevertheless, both of them remain among the most taxonomically-significant insect groups to have been uncovered within the past 100 years, with so recent a discovery and recognition of an entire living suborder of insects as the past decade (in the case of the gladiators) being truly extraordinary. Moreover, as of August 2011, numerous species, belonging to at least eleven genera, have been described.[5]

BIG RED, DOWN DEEP
A fairly recent marine surprise was Big Red, or, more formally, *Tiburonia granrojo*, a new species of cannonball jellyfish so distinct from all others that it required the creation in 2003 by the Monterey Bay Aquarium Research Institute's jellyfish expert Dr. George Matsumoto of a brand-new taxonomic subfamily to house it. Named after its size (measuring up to almost 10 ft across) and blood-red colouration, Big Red's most striking feature is the absence of the most famous characteristic of jellyfishes—their long stinging tentacles. Instead, it boasts only a series of short thick oral arms (gut extensions), and even these vary in number from specimen to specimen, with anything between four and seven having been recorded from those individuals reported so far.[6]

Big Red was first discovered at depths of 2115 ft in the Monterey Submarine Canyon, California, but has also been sighted off the coasts of Oregon, Washington, Hawaii, and Japan. Its lifestyle, however, is virtually unknown—in particular, how it kills its prey (bearing in mind that it has no stinging tentacles), and what that prey consists of, remain open to debate.[6]

WEIRD WINGS!
Small but exceedingly strange was a certain grotesque species of fly described from Arizona in 2004. With dozens if not hundreds of new flies revealed every year, it takes something very odd to attract media interest in

the finding of yet another one, but this particular fly, formally christened *Erebomyia exalloptera*, is certainly nothing if not odd. Its left and right wings are consistently different not only in shape but also in size, yielding a discrepancy of at least six percent, which is a degree of difference far greater than anything previously recorded from any known species of winged creature. Indeed, so great is this fly's wing asymmetry that entomologists were amazed that it could even fly at all, and that, if it could, it didn't merely go round and round in circles![7]

Remarkably, however, it can indeed fly—and normally too—as confirmed by field observations in spring 2003, which also established that the first specimens observed, back in 2001, were not malformed freak individuals of some previously-known species but instead represented a bona fide species in its own right. In addition, as only the males exhibit these mismatched wings, it is now thought that they may well play some active role in attracting females.[7]

A NEW STAR FROM THE SEA

Not all new species are described shortly after being discovered. Back in 1990, Dr. Kevin Raskoff and colleagues from the Monterey Bay Aquarium Research Institute (MBARI) in California were using a remote-operated vehicle dubbed *Ventana* to explore the depths of Monterey Bay when they spied a highly unusual jellyfish. Translucent blue-white in colour, it lacked the tentacles characterising most jellyfishes, but had four long oral arms trailing from its mouth, and these arms as well as its bell were covered in clumps of nematocysts (stinging cells), thus earning this mysterious sea creature the nickname 'Bumpy'.[8]

By 2004, a further six sightings of 'Bumpy' jellyfishes had occurred—four in the bay, and two in the Gulf of California during a MBARI expedition there in spring 2003. A year later, in February 2004, Drs. Kevin Raskoff and George Matsumoto from MBARI (and who had already co-described another notable new jellyfish, Big Red *Tiburonia granrojo*, back in 2003) gave Bumpy a more dignified, official scientific name—*Stellamedusa ventana* ('Ventana star jellyfish'), because its appearance reminded the researchers of a shooting star. Moreover, it is sufficiently different from all other jellyfishes to warrant the creation of not only a new genus but also an entirely new taxonomic subfamily to house it.[8]

The reason for the delay in describing such a remarkable species was to enable the researchers to observe not just its morphology but its behaviour and distribution too. It is now known to inhabit depths of 490-1800 ft, just below the level of sunlight penetration, and may feed mostly upon other jellyfishes.[8]

HAIRY IN OREGON

Hailing from Oregon, in 2004 a very small but taxonomically significant species of cladoceran crustacean finally received its formal description and name— *Dumontia oregonensis*. It is a water flea allied to the tiny freshwater *Daphnia* familiar to aquarists everywhere, but has been dubbed the hairy water flea on account of its bristly limbs, and is so different from all other water fleas that it has required the creation of a totally new taxonomic family to accommodate it.[9]

Discovered in 1999 lurking in seasonal ponds north of Medford, Oregon, by Wisconsin University freshwater ecologist Prof. Stanley Dodson, it was not until his graduate student Carlos Santos-Flores examined Dodson's specimens, and saw that their hitherto-overlooked species uniquely combined features separating all other water fleas into two wholly distinct taxonomic groups, that its dramatic significance was finally realised.[9]

AN IMR INTANGIBLE

Decidedly mysterious is the 1-ft-long worm-like creature with vivid red anterior and long, curved, sand-buried posterior photographed while crawling around the sea bottom at a depth of 6500 ft, during the first comprehensive deepsea probes of the Mid-Atlantic Ridge, conducted in 2004 by a 2-month-long Norwegian-led international expedition. The Mid-Atlantic Ridge is a chain of undersea mountains running between Iceland and the Azores, and over 80,000 specimens were collected, but sadly the mystery worm was not one of them, as it defied attempts to bring it to the surface.[10]

Scientists from Bergen's Institute of Marine Research (IMR) have likened the creature to the acorn worms or enteropneusts, but concede that it may even represent an entirely new phylum of animals. Based upon its photos, it recalls those equally enigmatic vermiform beasts the pogonophorans or beardworms, which include the giant tube-dwelling, red-tentacled hydrothermal vent worms that were themselves made known to science as recently as the mid-1970s.[10]

Another mystery unveiled by this expedition that may or may not be related to this strange worm was the presence of never-before-seen, perfectly straight, evenly-spaced lines of 2-inch-diameter holes that look as if they have been stitched into the seabed at a depth of 6000 ft. Presumably the burrows of some marine animal, what makes them so perplexing is how any creature could create such straight lines.[10]

ZOMBIE WORMS ON DEAD WHALES' BONES

No less strange were some decidedly weird worms first discovered in February 2002, by researchers from California's Monterey Bay Aquarium Research Institute

(MBARI), at a depth of 9,400 ft living upon and inside the bones of dead whales in Monterey Bay. Assigned in 2004 to a brand-new genus, *Osedax* ('bone-eating'), respectively dubbed *O. rubiplumus* and *O. frankpressi*, and nowadays known colloquially as zombie worms, whale worms, or boneworms, they range in length from 1 in to 2.5 in, sport bright-red feathery gills that function as respiratory organs, and also possess green root-like structures that penetrate the whales' bones. Bacteria living inside the worms then proceed to digest fats and oils present within the whalebone.[11]

Also inhabiting these worms, which were all found to be female, are their respective species' males—up to 100 microscopic males inside each female. The males contained large quantities of sperm and appear able to achieve sexual maturity while still larvae, and the females were all full of eggs. Strangest of all, however, is that the worms have no mouth or digestive system, no visual sensory organs, and no body segmentation. Instead, they most closely resemble, albeit in miniature form, the hydrothermal vent worms (see p. 250)—and, indeed, just like the vent worms, the zombie worms belong to the pogonophoran taxonomic family.[11]

Separated from the Pacific Ocean floor by two ocean basins and over 8200 ft in the water column, a new zombie worm was revealed in late 2005 within the North Sea region of the north Atlantic—this time dwelling on the bones of a dead minke whale at a depth of just 400 ft near Tjarno Marine Laboratory on the Swedish coast. Only 0.4-0.8 in long, this Atlantic species closely resembles its Pacific counterparts, anchoring itself to the whale bones via strong root-like processes that also probe into the bones, from which they absorb oils with the assistance of symbiotic bacteria, while extracting oxygen from the surrounding seawater by means of long petal-like plumes. The exposed portion of the worm is covered by a ball of mucus (probably for defence purposes), which has earned it an apt if somewhat nauseating scientific name—*Osedax mucofloris*, loosely translated by its describers as 'bone-eating snot-flower'![11]

In late November 2009, researchers revealed that as many as 15 different species of zombie worm had been reported living in Monterey Bay, thereby increasing very considerably the total number on record. Moreover, in 2007, the first known zombie sea anemone was formally described—*Anthosactis pearseae*, a small white tooth-shaped species that lives upon the bones of dead whales in California's Monterey Canyon, almost 2 miles below sea level.[11]

GIANT LEECH IS MEGA-FIND

Mowing the lawn one sunny morning in July 2003 at his home in Salem County, New Jersey, Bill Ott spotted what looked like a small black snake, roughly 10 inches long, in a dirt channel. Picking it up for a closer look, he thought that it was an unusual worm—which it was, but far more unusual than he could ever have imagined. It was a leech, but quite unlike any species of leech known to science in North America.[12]

To begin with, whereas most leeches are aquatic, this one, dubbed Piwi, was terrestrial. This is a much rarer leech trait usually exhibited only by tropical species (other than Piwi, there is just one species of terrestrial leech in North America, *Haemopis terrestris*, whose grey body with black and yellow markings is visibly distinct from Piwi's uniform inky-black colouration). And whereas leeches are famous for being blood-drinkers, Piwi preferred to eat earthworms. Most notable of all, however, was Piwi's size.[12]

After being transferred to a tank under the watchful scrutiny of leech expert Dr. Dan Shain at Rutgers-Camden University (following an email to him by Ott's wife, Carol, alerting him to her husband's discovery), Piwi grew to a length of just under 17 in when fully extended. This is of great significance, because the world's biggest species of leech, the Amazonian *Haementeria ghilianii*, is only known to reach a maximum fully-extended length of 18 in. Moreover, James Parks, of Franklin Township, who in October 2004 succeeded in finding four more Piwi-type leeches, in a cedar bog near his home, and gave them to Dr. Shain, claims to have once found one specimen measuring roughly 20 in long, curled up in a coil the size of a tennis ball. If true, this would dethrone *H. ghilianii* and bestow the superlative size status upon the Piwi leech.[12]

At the time, however, no-one knew that these leeches were anything special, so the 20-in monster was not collected. Now, thanks to Piwi, plus those newly-found specimens that survived (one was swiftly eaten by Piwi after being placed in his tank!), science became well aware of their importance. Swain's plan was to establish a breeding colony in order to uncover this enigmatic form's identity—once he had found out how to stop them from eating each other![12]

In January 2010 in hard-copy format (and September 2009 online), this remarkable, and very sizeable, new species was formally described and named by Shain and then-graduate student Beth Wirchansky—becoming *Haemopis ottorum*, in honour of its discoverers, the Otts.[12]

LURED BY THE BRIGHT LIGHTS

Siphonophores are astonishing beasts at the best of times. Most famously represented by the Portuguese man o' war, they superficially resemble jellyfishes, but in reality each single siphonophore is actually a colony of several individuals, each of which has become an organ or some other structure (float, tentacle, etc) of what is in effect a super-individual. But in 2005, marine biologist

Dr. Steven Haddock and a team of fellow researchers from the Monterey Bay Aquarium Research Institute (MBARI) announced the discovery of an even more amazing siphonophore.[13]

Inhabiting the waters of California's Monterey Bay at depths of a mile to a mile and a half, this new species belongs to the genus *Erenna*, whose members are characterised by utilising bioluminescent (light-emitting) tentacles to defend themselves against predators. This latest (and seemingly still-unnamed) *Erenna*, however, has effectively turned the tables upon this convention, by utilising the flashing red lights at its writhing tentacles' tips not to repel but rather to lure—enticing small fishes that normally feed upon tiny phosphorescent red crustaceans called copepods.[13]

And when these hapless fishes brush against the siphonophore's tentacles, they are swiftly killed by the tentacles' lethal stinging cells (nematocysts). Moreover, this is the first-known example of a marine invertebrate emitting light at the red end of the spectrum rather than at the blue/green end.[13]

YETI LOBSTER—A BLONDE BOMBSHELLFISH

The discovery in March 2005 of an extraordinary crustacean in 7200-ft-deep waters at a site 900 miles south of Easter Island in the South Pacific was nothing short of a bombshell, taxonomically speaking. So bizarre was this freshly-revealed shellfish—a new species, and genus, of squat lobster—that it required the creation of an entirely new zoological family in order to accommodate it.[14]

White in colour and 5.9 inches long, what makes it instantly distinguishable from any previously-described lobster is the mass of silky, blonde, hair-like strands covering its thoracic legs, in particular its large pincers or chelae. This very curious crustacean was officially described in March 2006, and was christened *Kiwa hirsuta*—commemorating not only its hairy claws but also Kiwaida, a Polynesian goddess of marine life—but it is referred to informally as the yeti lobster.[14]

It had been discovered by a scientific team organised by Dr. Robert Vrijenhoek of the Monterey Bay Aquarium Research Institute (MBARI) in Monterey, California, and Dr. Michel Segonzac of the French Research Institute for Exploitation of the Sea (Ilfremer), using the submarine DSV *Alvin*, who found it living on hydrothermal vents along the Pacific-Antarctic Ridge. These vents are surrounded by mineral-rich water, and scientists believe that the 'hairs' or setae on the pincers of *Kiwa*, which contain filamentous bacteria, serve to detoxify this water, and as an auxiliary nutritional aid. In addition, like so many other species living in this inky zone, *Kiwa* has very reduced, non-pigmented eyes, as it has no need for vision here.[14]

THE LOST WORLD OF AYALON

In early June 2006, a team of scientists from Israel's Hebrew University announced the existence of a veritable lost world—a 300-ft-deep underground cave system containing a lake and a unique ecosystem and array of species, which until revealed in April of that same year by a drilling operation at a cement quarry near Ramle, central Israel, had remained sealed off from sunlight and the outside world for millions of years. Deriving its energy not from the sun (hence devoid of photosynthetic organisms) but most probably from the oxidation of subterranean hydrogen-containing minerals and hydrogen sulphide by chemosynthetic bacteria that are then eaten by larger organisms, this spelaean system, now named Ayalon Cave, was found to harbour at least eight previously unknown species of invertebrate.[15]

These include a large omnivorous blind crustacean at the apex of the system's food chain, as well as smaller aquatic and terrestrial forms, also blind, some of which are seawater species, others freshwater or brackish. This indicates that their ancestors may have had a marine link in prehistoric times.[15]

Intriguingly, newspaper reports of the cave's discovery all showed a photograph of a distinctive arthropod invertebrate invariably referred to in the reports as a blind scorpion. Yet the creature depicted was unquestionably a shrimp-like crustacean—as confirmed by its possessing two pairs of antennae (a uniquely crustacean characteristic; arachnids, conversely, have no antennae at all), five pairs of limbs (identifying it as a decapod crustacean; arachnids have only four pairs), and a typically shrimp-like fan-shaped tail (wholly distinct from the long curved sting-tipped tail of scorpions).[15]

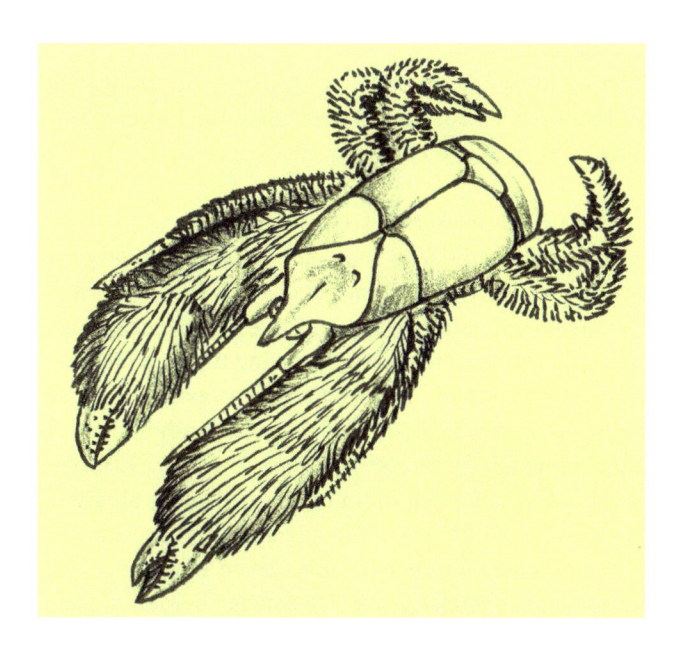

Yeti lobster. (Tim Morris)

WHAT AM I BID FOR THIS NEW SPECIES?

The internet auction site eBay is well known for its diversity of items available for bidding upon, but 2006 may well have been the first time that a hitherto unknown species made its scientific debut there. That was the year when Dr. Simon Coppard of the International Commission on Zoological Nomenclature based at London's Natural History Museum was alerted by several collectors to some odd-looking sea urchins listed for bidding on the site's pages. Originating from the Pacific waters around New Caledonia, their species had been named by the sellers as *Coelopleurus interruptus*, but Dr. Coppard realised that in reality these sea urchins' appearance did not match that of any known species, having unusually bright, attractive colours (including red, purple, lavender, and pale green) on their spines and outer casing. Accordingly, he named this newly-revealed multicoloured species *Coelopleurus exquisitus*, specimens of which could still be found on eBay in August 2006.[16]

A NEW PHYLUM FOR XENOTURBELLA

The highest taxonomic category within the animal kingdom is the phylum—of which the most recent (until now), Cycliophora, was created in 1995 to house *Symbion pandora*, a bizarre parasite of lobster mouthparts. In 2006, however, an additional, entirely new phylum was established, in order to accommodate a lone genus of equally extraordinary mini-beasts, *Xenoturbella*, whose singularity had been overlooked for over half a century.[17]

This half-inch-long worm-like creature was first discovered in 1915, trawled up from the depths of the Baltic Sea, but was not formally described until 1949, when it was dubbed *Xenoturbella bocki* after its discoverer, scientist Sixten Bock. A second species, *X. westbladi*, was described in 1999, and was named after E. Westblad, who had described *X. bocki* 50 years earlier.[17]

Xenoturbella translates as 'strange flatworm', and it was indeed originally classed as an aberrant flatworm, but was later dismissed as a weird bivalve mollusc, based upon genetic research and the finding of bivalve eggs inside it. However, as lately revealed by a 14-strong scientific team that included neuroscience zoologist Prof. Leonid Moroz from Florida University's Whitney Laboratory for Marine Bioscience, this research was flawed, because the genetic material that was utilised, and also the eggs, were shown to have originated not from *Xenoturbella* itself but from its close association to clams and its likely habit of consuming bivalve eggs.[17]

Instead, new DNA research confirmed that *Xenoturbella* is fundamentally distinct genetically from all other animal species. Moreover, it is no less bizarre anatomically, lacking a gut, organised gonads, excretory system, and central brain. Of especial interest is that it appears to share a common ancestor with echinoderms (starfishes, sea urchins, etc), hemichordates (acorn worms and pterobranchs), and chordates (vertebrates, lancelets, and tunicates), thereby shedding light on the likely anatomy and morphology of the chordates' antecedents. The new phylum for *Xenoturbella* has been christened Xenoturbellida.[17]

STICKING OUT A MILE

An insect measuring a substantial 22.3 inches long may seem difficult to overlook by science, but that is exactly what happened with what is now recognised to be the world's longest species of insect. Amid the rainforest at Ulu Moyog, in Sabah's Penampang district within northern Borneo, in February 1989 a local farmer came upon an extraordinarily lengthy female stick insect, wingless and dark green, which he procured and added to his own private collection. There it remained, wholly unknown to the wider entomological community, until the farmer gave it to Malaysian entomologist Datuk Chan Chew Lun. He in turn realised that it constituted a dramatically new species, and eventually sent it in late 2007 to British entomologist Dr. Philip Bragg for a thorough examination. A much smaller (5.13 in long) male specimen, winged and mostly brown in colour, but definitely belonging to this same species, had also been collected, back in 1983.[18]

Formally named Chan's megastick *Phobaeticus chani* by Bragg in 2008, females of this remarkable species are not only longer than any previously documented stick insect but also longer than any other insect of any kind. Its once-unheralded but now scientifically-priceless type specimen (whose body alone measures a stupendous 14 in long) has been donated by Chan to the Natural History Museum in London, as has the male from 1983, and four further specimens (one female and three males) now exist in Sabah collections.[18]

GHOST SLUG—ALL WHITE IN WALES

Trigonochlamydidae is a taxonomic family of very distinctive terrestrial slugs, inasmuch as its species are subterranean and carnivorous, killing earthworms at night with blade-like teeth and then sucking them up into their jaws like spaghetti. Traditionally, however, they were known only from Turkey and Georgia, with no records from anywhere in Western Europe. Then on 29 December 2004, a specimen of an unfamiliar white slug with eyeless eyestalks was collected in the churchyard of Brecon Cathedral in Wales, followed on 29 October 2006 by a second specimen found in a lane in Caerphilly, South Wales, which was photographed and then released.[19]

A year later, a third specimen was obtained, this time in a garden near Cardiff, which when examined proved not only to be a trigonochlamydid, but also one that was hitherto unknown to science. Other specimens have since been discovered in Gorseinon near Swansea, Hay-on-Wye on the Welsh-English border, and Knowle near Bristol in England. Believed to have been introduced into the U.K. rather than being native here, this unexpected new species was swiftly dubbed the ghost slug, and when formally described in 2008 by then-postgraduate student Ben Rowson from the National Museum Wales and Cardiff University ecologist Dr. Bill Symondson it was officially christened *Selenochlamys ysbryda*—'ysbryd' is Welsh for 'ghost'.[19]

A WEB OF DISCOVERY
In 2000, while perusing a collection of biological specimens belonging to the Plant Protection Research Institute in Pretoria, South Africa, Slovenian biologist Dr. Matjaz Kuntner noticed among them a truly huge female orb-web spider—but which, as his subsequent researches revealed, did not correspond with any known form. In his official description of this spider's hitherto-unrecognised species in 2009, he christened it *Nephila komaci*—Komac's golden orb-web spider, after the late Andrej Komac.[20]

Significantly, it constitutes the world's largest species of web-spinning spider, which makes the fact that it had been overlooked for such a long time by science so remarkable. Moreover, until three additional specimens were obtained in 2007 during a field survey conducted in South Africa's Tembe Elephant Park, the Pretoria individual was this elusive mega-species' only known example. Indeed, Kuntner had feared that it was extinct. However, it is now on record from Maputaland in South Africa and also from Madagascar. Whereas the male is tiny, the female has a leg span of up to 4.75 in, and her web can measure up to 3.3 ft across.[20]

GREEN BOMBER WORM—
LETTING LOOSE ITS GLOWING GILLS
In 2009, a small but highly unusual new species—and genus—of marine polychaete worm was described from the deep trenches of Monterey Bay, off the coast of California. It was named the green bomber worm *Swima bombiviridis* in recognition of its very remarkable mode of defence—when disturbed, this 0.75-4-in-long worm sheds from its body a series of spherical modified gills that are bioluminescent, and which continue to emit their brilliant green glow for several seconds after being cast off, like a scattering of ghostly bombs. Various additional *Swima* species have since been discovered too, and are currently awaiting description and naming.[21]

TENTACLES APLENTY FOR THE SQUID WORM
In 2010 online (and in 2011 in hard-copy format), an equally strange species that proved to be closely related to the green bomber worm was formally described and named—*Teuthidodrilus samae*, the squid worm (aka squidworm). It derives its name from the unique assortment of vaguely squid-like tentacular structures borne upon its head. Ten of them are as long as or even longer than its body (which measures up to 3.7 in), and are known as notochaetae; eight of these are used for breathing, and the other two for feeding. There are also six pairs of free-standing and oppositely-paired sensory structures called nuchal organs, which enable the worm to taste and smell underwater.[22]

The first scientifically-recorded specimen of *T. samae* was observed and videoed in the Celebes Sea off the Philippines during October 2007 via an undersea robot, and it has since been found at a depth of 1.7 miles here, feeding upon plankton as well as cast-off mucus, faecal material, and other nutritious detritus. This intriguing worm, a cirratuliform polychaete, which swims upright using two body-length rows of iridescent paddle-shaped limbs, may represent a transitional organism between benthic and pelagic creatures.[22]

DIGGING DEEP FOR THE DEVIL WORM
In 2011, a zoological team that included geoscientists Drs. Gaetan Borgonie and Tullis Onstott announced the discovery of a new species of microscopic nematode worm—an event that would not normally be considered momentous, because more than 28,000 nematode species have already been described, with researchers believing that as many as a million species exist in total. What makes this one so very special, however, is that it appears to be the deepest-living multicellular animal known to science, because it was found underground living in fracture water within a series of gold mines in South Africa, at depths of 0.56-2.24 miles into the Earth's crust.[23]

Until now, such deep, inhospitable zones were not deemed capable of sustaining multicellular life forms due to the very high temperatures, immense pressure, confined space, and lack of oxygen. However, this newly-revealed, relatively long-tailed species has confounded the experts on all counts, so in recognition of its ability to survive in this infernal subterranean realm, where it feeds upon bacteria and reproduces sexually, it has been formally named *Halicephalobus mephisto*, and informally dubbed the devil worm of hell![23]

Is there an American black panther? (William M. Rebsamen)

William Beebe's white-banded manta, still undescribed. (William M. Rebsamen)

THE FUTURE

With another nine decades-worth of zoological discoveries and rediscoveries still waiting to be unfurled, and in view of such remarkable recent finds as the Vu Quang ox, dingiso, giant muntjac, *Xenoperdix*, hololissa, Thai giant stingray, mainland Javan rhinos, Indonesian coelacanth, giant peccary, gladiators, kipunji, zombie worms, pink iguana, and giant spitting cobra, it may well include the names of at least a few of the many exciting forms documented in the numerous books and articles dealing with cryptozoological animals (officially referred to as cryptids).

Of course, some cryptids may simply be based upon misidentifications of known native animals, or of exotic species that have escaped from captivity—the latter prospect undoubtedly providing the explanation for the plethora of reports on file variously describing puma-like or black panther-like cats sighted in Britain, continental Europe, North America, Australia, and even New Zealand.

Conversely, there are many cryptids that genuinely seem to be undiscovered or officially extinct species. Currently comprising the world's most controversial creatures—those whose very existence has yet to be formally accepted by science (in spite of substantiation by eyewitness accounts, photographs, and sometimes even a preserved specimen or two)—they are exceedingly varied in type.

Equally, they range from such famous, extensively-investigated examples as North America's bigfoot (*Gigantopithecus*?), the Himalayan yeti (terrestrial orang-utan?), sea serpents (a motley assemblage of strange seals, primitive whales, giant eels, and aquatic reptiles?), lake monsters (ditto?), and the Congolese mokele-mbembe (living dinosaur?), to such little-investigated mystery beasts as the New Guinea devil pig (palorchestid diprotodont?), Chad's 'mountain tiger' (sabre-tooth?), the crowing crested cobra (African mystery snake), tatzelworm (large, unidentified Alpine lizard?), Timor Sea ground shark (giant wobbegong?), and Bahamian lusca (gigantic cirrate octopus?).

Representation of the true yeti, based upon eyewitness descriptions. (Richard Svensson)

Today, these are animals of ambiguity; tomorrow, they could be taking their places alongside the okapi, Komodo dragon, Congo peacock, coelacanths, vestimentiferans, megamouth, Jerdon's courser, Vu Quang ox, forest spotted owlet, giant peccary, and all of the many other zoological success stories of the 20th and early 21st centuries. Similarly, although the vast majority of new and rediscovered animals will nonetheless be relatively small, seemingly trivial ones, there is no reason why some of these should not prove in reality to be as significant in scientific stature as the loriciferans, Selevin's dormouse, *Neoglyphea*, irukandji, *Burramys*,

Based upon eyewitness accounts, the mysterious mountain tiger of Chad has been likened to a living sabre-tooth. (Tim Morris)

po'o-uli, sea daisy, *Neopilina*, pogonophorans, *Symbion pandora*, Cooloola monster, *Xenoturbella*, the gladiators, Disko beast, and *Indostomus*.

If this book of mine can not only sweep away the dusty, discredited fallacy that the 20th and early 21st centuries have been conspicuously devoid of zoological surprises, but also serve as a tangible rebuff to anyone denying that the future may also hold many such surprises, then it will more than successfully have achieved all of my ambitions for it when preparing it. Provided that the expanses of tropical rainforest still existing can be comprehensively explored before mankind succeeds in its terrifying attempts to destroy them, provided that the vast volumes of ocean, sea, lake, and river can be investigated before mankind succeeds in polluting them thoroughly and irrevocably, or—in a more optimistic vein—provided that all types of habitat can not only be searched but also be conserved for the foreseeable future, there really can be no doubt whatsoever that many additional new and rediscovered animals of note will be added to these pages in the decades to come.

Just to give an idea of the sheer quantity of new animal species discovered lately: As noted in Matthew Bille's *Shadows of Existence* (2001), the annual rate of species discovery for mammals in recent years has sometimes been as high as 45 (including reclassifications—which, incidentally, may soon include splitting the giraffe into several separate species and the African lion into at least two). At much the same time Field Museum mammalogist Dr Lawrence Heaney was quoted as predicting that the total of 4600 mammalian species currently known to science may ultimately expand to 8000 (moreover, in less than a decade since his statement, this number has been reassessed at just under 5500). In the year 2000, the annual rate of species discovery globally was estimated at 60 for reptiles, 80 for amphibians, and 200 for fishes, with at least two new species of bird having been described annually since 1980.

Indeed, in 2009 it was announced that no fewer than 25 new species of bird had been described since 2000 in the neotropics alone (similarly, in October 2010 the WWF published a report in which it revealed that during the period 1999-2009, 39 new species of mammal, 16 birds, 55 reptiles, 216 amphibians, and 257 fishes had all been discovered in the Amazon ecosystem). Nor should we forget the several thousand new species of invertebrates (especially insects) that are formally described and named globally each year, as well as the considerable number of rediscovered animals recorded.

The latest significant update on this subject came as recently as 24 August 2011, when a team of Canadian researchers from the Census of Marine Life publicly announced that using an advanced mathematical model to calculate the total number of species currently alive on Earth, the figure that they had obtained was 8.7 million. Bearing in mind that so far only 1.2 million species have been formally described, this means that a staggering 86.2 per cent of all living species on Earth today remain undiscovered and undescribed (with less than 10 per cent of all species in the world's vast oceans presently documented)—all of which demonstrates only too vividly that the capacity for uncovering remarkable new and supposedly extinct animals is very far from exhausted.

A LAST WORD

It was in 1886, within his book *Mythical Monsters*, that naturalist Charles Gould asked: "Can we suppose that we have at all exhausted the great museum of nature? Have we, in fact, penetrated yet beyond its antechambers?" The wealth of major new and rediscovered animals unveiled during the 20th and early 21st centuries, as documented in this book, emphatically verifies that by 1886 we had indeed far from exhausted that peerless museum. And if only a small proportion of the extraordinary animals currently documented within the bulging files of cryptozoology are found during the 21st century's remaining nine decades, these will more than adequately demonstrate that, more than two hundred years on from Gould's enquiry, we will *still* be positioned far from that museum's inner sanctum.

How much closer our steps will take us towards this depends not only upon how eager science is to seek those creatures still hidden from its view, but also upon how eager its disciples are to recognise that even an extensive, widely-respected knowledge of science is still infinitely less than omniscience.

A wise scientist does not take pride in how much he knows, but rather takes heed of how little he knows.

Golden poison dart frog, Phyllobates terribilis. (Brian Gratwicke)

Okapi. (Marcel Schauer)

Golden hamster. (Andrey Davidenko)

APPENDIX
THE SCIENTIFIC CLASSIFICATION OF ANIMALS

Taxonomy, the scientific process of classifying animals (and all other living organisms), employs a hierarchial system of categories. At the base of this hierarchy is the most important taxonomic category of all—the species. Although it has been defined in a number of different ways, a species is considered by most authorities to comprise one or more populations of organisms whose members can readily interbreed with one another to yield viable, fertile offspring. In some cases, such populations may actually be separated from one another by physical geographical boundaries, such as a mountain range, river, sea, or some other topographical barrier; but if members from these populations are brought together and found to be capable of mating with one another to yield fit progeny, they are still classed as members as the same species. For example, the red foxes of Britain, continental Europe, and North America are separated from each other geographically, but if brought together will all freely interbreed to produce fertile, fit offspring—hence they are all considered to belong to one and the same species.

Members of populations belonging to *different* species can sometimes mate too, and even occasionally produce offspring, but these are usually sterile, and are known as hybrids.

It was the famous 18th-century Swedish botanist Carolus Linnaeus (=Carl Linné) who first devised the binomial system of nomenclature still used today in scientific classification, a system that provides every species with its own unique two-part name—its scientific name. This is generally of Latin or Greek origin (or both), as these were the universal languages of scholars worldwide in the time of Linnaeus, and is always printed in a different typeface (normally italics) from the main text surrounding it.

For example, the wolf's scientific name is *Canis lupus*; *lupus* is the wolf's specific or trivial name, and *Canis* is its generic name or genus (plural: genera). By devising the binomial system, Linnaeus created a means by which a species mentioned in any given book, scientific paper, article, etc, could be instantly identified by the reader, regardless of the language in which the publication had been written. The problem with common names is that they change—often radically—from one language to another.

For instance, the species of bird referred to in Britain as the common gull is known in America as the mew gull, in France as the ash-grey gull (goéland cendré) and in Germany as the storm gull (Sturmmöwe)—four fundamentally different names. As scientific names, conversely, stay exactly the same in all languages, this species' binomial name in Britain, America, France, Germany, and everywhere else is *Larus canus*, thereby eliminating any inter-linguistic ambiguities.

Incidentally, whereas specific names always begin with a small (lower case) letter, genera always begin with a capital (upper case) letter. Moreover, if a given genus is referred to more than once within a single passage or section of a publication, it is often abbreviated after the first instance of usage to its initial letter. For example, once the wolf's name has been given as *Canis lupus*, for the remainder of the passage or section in question it can be referred to simply as *C. lupus*.

The binomial system of nomenclature also greatly facilitates taxonomy—because although each will possess a different specific (trivial) name, closely related species are housed together within the same genus. Consequently, as the wolf, domestic dog, coyote, and golden jackal, for instance, are all closely related to one another, their scientific names are: *Canis lupus*, *C. familiaris*, *C. latrans*, and *C. aureus* respectively. Rather more distantly related to these species—but very closely related to each other—are the fox species, such as the red fox, swift fox, kit fox, and sand fox. These therefore share a different genus, *Vulpes*, thus becoming *Vulpes vulpes*, *V. velox*, *V. macrotis*, and *V. pallipes* respectively.

In the same way that closely related species are housed together within the same genus, closely related genera are housed together within the same family. In the case of the wolf, domestic dog, red fox, swift fox, dhole, bushdog, Cape hunting dog, fennec, and all other dogs, this is the dog family, known scientifically as Canidae. Similarly, closely related families are housed together within the same order—the next category on the ascending hierarchy of scientific classification—and further progression upwards in the hierarchy follows the same, standard approach. That is, closely related orders are housed together within the same class, closely related classes within the same phylum (plural: phyla), and, reaching the apex of the hierachy, closely related phyla within the same kingdom. In the case of animals, the animal phyla are housed together within the kingdom Animalia; the plant phyla are housed within the kingdom Plantae; the fungi phyla within the kingdom Fungi; and so on.

To illustrate the hierarchial approach to the scientific classification of animals, Table 1 constitutes a complete classification of the wolf, from species to kingdom. And to demonstrate the diversity of animal life, a complete classification of the animal kingdom down to the level of individual classes (for the major phyla) is provided in Table 2.

It will be seen from these that the higher the level within the hierarchy, the less closely related and more diverse are the animals present. For example, two animal species housed within separate classes of the same phylum are less closely related to one another and are (usually!) more different from one another in appearance than are two species housed within separate orders of the same class. Put another way, a pheasant and a wolf are visibly less closely related to one another than are a pheasant and a sparrow.

Sometimes, to facilitate classification (especially when dealing with taxonomically complex animal groups), additional categories, prefixed by 'sub' or 'super', are inserted between existing ones. Thus, the category of subfamily is sometimes used, inserted between genus and family; so too is the category of superfamily, inserted between family and order. This in turn leads us to one final but important category, the subspecies—because even a species can be subdivided, if it consists of several geographically separate populations that can be readily distinguished from one another in some way (e.g. visually, aurally, behaviourally).

Each such population is termed a subspecies of that species. In nature, the subspecies of any given species are, by definition, separated from one another geographically, so that interbreeding cannot usually occur—but if individuals of separate subspecies are deliberately brought together, they will breed and produce viable, fertile offspring (thereby revealing that despite their distinguishing features they do indeed belong to the same species, rather than comprising separate species in their own right).

In scientific classification, subspecies have trinomial (three-part) names. All subspecies within a given species share the first two names, but each has its own unique third name, to distinguish it from the other subspecies. For instance, the wolf is split into many different subspecies, which include the timber wolf *Canis lupus lycaon*, the Texas grey wolf *C. l. monstrabilis*, the Himalayan wolf *C. l. chanca*, and the European wolf *C. l. lupus*.

TABLE 1:
SCIENTIFIC CLASSIFICATION OF THE WOLF *Canis lupus*

KINGDOM: Animalia
Other kingdoms include: Plantae (plants), Fungi (fungi).

PHYLUM: Chordata
Other phyla within the kingdom Animalia include: Mollusca (molluscs), Platyhelminthes (flatworms), Uniramia (insects and relatives), Echinodermata (spiny-skinned animals), Porifera (sponges).

CLASS: Mammalia
Other classes within the phylum Chordata include: Aves (birds), Osteichthyes (bony fishes), Reptilia (reptiles), Ascidiacea (sea squirts), Amphibia (amphibians).

ORDER: Carnivora
Other orders within the class Mammalia include: Cetacea (toothed and toothless whales), Chiroptera (bats), Primates (primates, including man), Proboscidea (elephants), Rodentia (rodents), Artiodactyla (even-toed ungulates).

FAMILY: Canidae
Other families within the order Carnivora include: Felidae (cats), Mustelidae (weasels and relatives), Hyaenidae (hyaenas), Ursidae (bears and giant panda).

GENUS: *Canis*
Other genera within the family Canidae include: *Vulpes* (foxes), *Cuon* (dhole), *Lycaon* (hunting dog), *Chrysocyon* (maned wolf), *Speothos* (bushdog), *Fennecus* (fennec).

SPECIES: *lupus*
Other species within the genus *Canis* include: *C. latrans* (coyote), *C. aureus* (golden jackal), *C. mesomelas* (black-backed jackal), *C. familiaris* (domestic dog).

TABLE 2:
THE 38 PHYLA OF THE ANIMAL KINGDOM
(including the classes of the major phyla)

N.B.—There is still much disagreement regarding animal classification (especially in relation to the protozoans, which is split into several separate phyla by some taxonomists): the following table follows the version most widely accepted. Also: until the mid-1970s, the classes of the phyla Chelicerata, Uniramia, and Crustacea were grouped together within a single phylum, Arthropoda ('jointed-legged animals'); in that system, the present-day classes of Crustacea were treated merely as subclasses.

Abbreviation: a.k.a. = also known as.

1 PHYLUM: CHORDATA—Chordates (with a notochord at some developmental stage)

  Class: Mammalia—mammals
  Class: Aves—birds
  Class: Reptilia—reptiles
  Class: Amphibia—amphibians
  Class: Osteichthyes—bony fishes; a.k.a. Pisces
  Class: Chondrichthyes—cartilaginous fishes; a.k.a. Selachii
  Class: Agnatha—jawless fishes
  Class: Leptocardii—lancelets
  Class: Thaliacea—salps
  Class: Larvacea—larvaceans
  Class: Ascidiacea—sea squirts
  Class: Sorberacea—sorberaceans (sometimes included in Ascidiacea)

2 PHYLUM: HEMICHORDATA—Hemichordates

  Class: Pterobranchia—pterobranchs
  Class: Enteropneusta—acorn worms

3 PHYLUM: ECHINODERMATA—Spiny-skinned animals

  Class: Holothuroidea—sea cucumbers
  Class: Echinoidea—sea urchins
  Class: Ophiuroidea—brittle stars
  Class: Asteroidea—starfishes and sea daisies
  Class: Crinoidea—sea lilies and feather stars

4 PHYLUM: XENOTURBELLIDA—Xenoturbellidans

5 PHYLUM: CHAETOGNATHA—Arrow worms

6 PHYLUM: BRYOZOA—Moss animals; a.k.a. ECTOPROCTA and POLYZOA

7 PHYLUM: PHORONIDA—Phoronid worms

8 PHYLUM: BRACHIOPODA—Lamp shells

9 PHYLUM: TARDIGRADA—Water bears

10 PHYLUM: PENTASTOMIDA—Tongue worms; a.k.a. LINGUATULIDA (sometimes included in Crustacea)

11 PHYLUM: CRUSTACEA—Crustaceans (biramous arthropods)

  Class: Malacostraca—crabs, lobsters, shrimps, woodlice, amphipods
  Class: Cirripedia—barnacles
  Class: Branchiura—fish 'lice'
  Class: Copepoda—copepods
  Class: Tantulocarida—tantulocarids
  Class: Remipedia—remipedes
  Class: Mystacocarida—mystacocarids
  Class: Ostracoda—seed shrimps
  Class: Branchiopoda—fairy shrimps, brine shrimps, water 'fleas'
  Class: Cephalocarida—cephalocaridans

12 PHYLUM: CHELICERATA—Chelicera-jawed arthropods

  Class: Pycnogonida—sea spiders; a.k.a. Pantopoda
  Class: Arachnida—spiders, scorpions, ricinuleids, ticks, mites
  Class: Merostomata—king (horseshoe) crabs

13 PHYLUM: UNIRAMIA—Uniramous arthropods

  Class: Insecta—insects; a.k.a. Hexapoda
  Class: Diplura—diplurans
  Class: Protura—proturans
  Class: Collembola—springtails
  Class: Symphyla—symphylans
  Class: Pauropoda—pauropods
  Class: Chilopoda—centipedes
  Class: Diplopoda—millipedes

14 PHYLUM: ONYCHOPHORA—Velvet worms

15 PHYLUM: MOLLUSCA—Molluscs

- Class: Cephalopoda—squids, octopuses, vampire squid, nautilus
- Class: Scaphopoda—elephant tusk shells
- Class: Bivalvia—two-shelled molluscs; a.k.a. Lamellibranchia
- Class: Gastropoda—single-shelled 'seashells', slugs, snails
- Class: Polyplacophora—chitons
- Class: Aplacophora—solenogasters and caudofoveatans
- Class: Monoplacophora—monoplacophorans (including *Neopilina*)

16 PHYLUM: PRIAPULIDA—Priapulid worms

17 PHYLUM: ECHIURA—Spoon worms (sometimes included in Annelida)

18 PHYLUM: SIPUNCULA—Peanut worms (sometimes included in Annelida)

19 PHYLUM: ANNELIDA—Segmented (Ringed) worms

- Class: Hirudinea—leeches
- Class: Oligochaeta—earthworms, pond worms, and tubeworms
- Class: Polychaeta—marine bristleworms and beard worms

20 PHYLUM: GNATHOSTOMULIDA—Gnathostomulid worms

21 PHYLUM: ACANTHOCEPHALA—Thorny-headed worms

22 PHYLUM: MICROGNATHOZOA—*Limnognathia maerski*, the only known species

23 PHYLUM: ROTIFERA—Wheel animalcules

24 PHYLUM: CYCLIOPHORA—Cycliophorans

25 PHYLUM: ENTOPROCTA—Entoprocts; a.k.a. KAMPTOZOA

26 PHYLUM: LORICIFERA—Loriciferans

27 PHYLUM: KINORHYNCHA—Kinorhynch worms; a.k.a. ECHINODERA

28 PHYLUM: NEMATOMORPHA—Horsehair worms

29 PHYLUM: NEMATODA—Round worms

30 PHYLUM: GASTROTRICHA—Gastrotrich worms

31 PHYLUM: NEMERTEA—Ribbon worms; a.k.a. RHYNCHOCOELA

32 PHYLUM: PLATYHELMINTHES—Flatworms

- Class: Cestoda—tapeworms
- Class: Trematoda—internal parasitic flukes
- Class: Monogenea—external parasitic flukes
- Class: Turbellaria—turbellarians

33 PHYLUM: CTENOPHORA—Comb jellies

34 PHYLUM: CNIDARIA—Stinging-celled animals

- Class: Anthozoa—sea anemones, soft corals, hard corals
- Class: Cubozoa—box jellies
- Class: Scyphozoa—true jellyfishes
- Class: Hydrozoa—hydras, sea firs, siphonophores

35 PHYLUM: PLACOZOA—Placozoans (only one species, *Trichoplax adhaerens*)

36 PHYLUM: MESOZOA—Mesozoans

37 PHYLUM: PORIFERA—Sponges

- Class: Demospongiae—demosponges
- Class: Hexactinellida—six-rayed sponges
- Class: Calcarea—calcareous sponges

38 PHYLUM: PROTOZOA—Single-celled animals (amoebae, ciliates, flagellates, etc)

BIBLIOGRAPHY

In all of the forthcoming reference lists, each asterisked (*) reference is a paper comprising a given species' formal scientific description.

Abbreviations: *A.M.N.H.* = *Annals and Magazine of Natural History*; *B.B.M.N.H.Z.* = *Bulletin of the British Museum (Natural History)–Zoology*; *B.B.O.C.* = *Bulletin–British Ornithologists Club*; *C.R.A.S.* = *Comptes Rendus de l'Académie des Sciences*; O.c. = Op. cit.; *P.B.S.W.* = *Proceedings of the Biological Society of Washington*; *P.Z.S.L.* = *Proceedings of the Zoological Society of London*; *R.Z.B.A.* = *Revue de Zoologie et la Botanique Africaines*; *S.G.N.F.B.* = *Sitzungsberichte Gesellschaft Naturforschender Freunde zu Berlin*.

PART 1:
20TH-CENTURY NEW AND REDISCOVERED ANIMALS

Section 1: The Mammals

1 THOMAS, O. (1900). The white rhinoceros on the Upper Nile. *Nature*, 62 (18 December): 599; *LYDEKKER, R. (1908). The white rhinoceros. *The Field*, 111 (22 February): 31.
2 HEUVELMANS, B. (1958). *On the Track of Unknown Animals*. Rupert Hart-Davis (London).
3 WENDT, H. (1959). *Out of Noah's Ark*. Weidenfeld & Nicolson (London); SHUKER, K.P.N. (Consultant) (1993). *Secrets of the Natural World* (Vol. 11 of Quest For the Unknown). Reader's Digest (Pleasantville).
3a GROVES, C.P. et al. (2010). The sixth rhino: A taxonomic re-assessment of the critically endangered northern white rhinoceros. *PLoS ONE*, 5(4): e9703. doi:10.1371/journal.pone.0009703
4 STANLEY, H.M. (1890). *In Darkest Africa*. Sampson Low (London).
5 JOHNSTON, H. (1901). The okapi. Newly-discovered beast living in Central Africa. *McClure's Magazine*, 17 (September): 497-501.
6 JOHNSTON, H. (1900). Letter from, containing an account of a supposed new species of zebra inhabiting the Congo forest. *P.Z.S.L.*, (20 November): 774-5.
7 SCLATER, P.L. (1900). Exhibition of, and remarks upon, two bandoliers made from the skin of a supposed new species of zebra. Ibid., (18 December): 950; *SCLATER, P.L. (1901). On an apparently new species of zebra from the Semliki Forest. Ibid., (5 February): 50-2.
8 BURTON, M. & BURTON, R. (Eds.) (1968-70). *Purnell's Encyclopedia of Animal Life* (6 vols). BPC (London).
9 SCLATER, P.L. (1901). Exhibition of water-colour painting of okapi, and extracts of letter written by Sir Harry Johnston. *P.Z.S.L.*, (7 May): 3-6; LANKESTER, E.R. (1901). Exhibition of two skulls and the skin of the new mammal, the okapi. Ibid., (18 June): 279-81; LANKESTER, E.R. (1902). On *Okapia* [sic], a new genus of Giraffidae [sic] from Central Africa. *Transactions of the Zoological Society of London*, 16: 279-307.
10 GRZIMEK, B. (Ed.) (1972-5). *Grzimek's Animal Life Encyclopedia* (13 vols). Van Nostrand Reinhold (London).
11 BRIDGES, W.M. (1937). Okapi comes to the zoological park. *Bulletin of the New York Zoological Society*, 40 (September-October): 135-47.
12 JOHNSTON, H. (1901). The five-horned giraffe. *The Times* (London), 20 June; THOMAS, O. (1901). On the five-horned giraffe obtained by Sir Harry Johnston near Mount Elgon. *P.Z.S.L.*, (19 November): 474-83.
13 ROTHSCHILD, M. (1983). *Dear Lord Rothschild*. Hutchinson (London).
14 *THOMAS, O. (1901). On the more notable mammals obtained by Sir Harry Johnston in the Uganda Protectorate. *P.Z.S.L.*, (21 May): 85-90; KINGDON, J. (1977). *East African Mammals. An Atlas of Evolution in Africa*. Vol. IIIA. Academic Press (London).
15 *MATSCHIE, P. (1903). Ueber einen Gorilla aus Deutsch-Ostafrika. *S.G.N.F.B.*, No. 6 (9 June): 253-9; http://www.berggorilla.de/english/gjournal/texte/24beringe.html accessed 22 November 2009.
16 GROVES, C. (1970). Population systematics of the gorilla. *Journal of Zoology*, 161: 287-300.
16a ANON. (1996). Bwindi gorillas—a new subspecies? *Oryx*, 30 (April): 94; SUTER, J. & OATES, J. (2000). Sanctuary in Nigeria for possible fourth subspecies of gorilla. Ibid., 34 (January): 71; CONSTABLE, T. (2000). Primate shake-up. *BBC Wildlife*, 18 (July): 34.
17 *ELLIOT, D.G. (1913). *A Review of the Primates*. American Museum of Natural History (New York).
18 HEUVELMANS, B. (1980). *Les Bêtes Humaines d'Afrique*. Plon (Paris).
19 ALLEN, G.M. (1942). *Extinct and Vanishing Mammals of the Western Hemisphere*. American Committee For International Wild Life Protection (Washington D.C.).

20 DAY, D. (1989). *The Encyclopedia of Vanished Species*. Universal Books (London).
21 *PRENTISS, D.W. (1903). Description of an extinct mink from the shell-heaps of the Maine coast. *Proceedings of the U.S. National Museum*, 26: 887-8; MANVILLE, R.H. (1966). The extinct sea mink, with taxonomic notes. Ibid., 122: 1-12.
22 *MILLER, G.S. (1903). Seventy new Malayan mammals. *Smithsonian Miscellaneous Collections*, 45: 1-73.
23 SHUKER, K.P.N. (1991). *Extraordinary Animals Worldwide*. Robert Hale (London).
24 *PETERS, W. (1873). Ueber *Dinomys*, eine merkwueridge neue Gattung der stachelschweinartigen Nagethiere aus den Hochgebirgen von Peru. *Auszug Monatsberichte der Konigl.-preuss. Akademie der Wissenschaften zu Berlin*, (10 July): 551-2.
25 GOELDI, E.A. (1904). On the rare rodent *Dinomys branickii* Peters. *P.Z.S.L.*, (7 June): 158-65.
26 RIBEIRO, A. DE M. (1919). *Dinomys pacarana*. *Arch. Escola Superiore Agricultura e Medicina Veterinário*, 2: 13-15; ANTHONY, H.E. (1921). New mammals from British Guiana and Colombia. *American Museum Novitates*, No. 19: 6-7; SANBORN, C.C. (1931). Notes on *Dinomys*. *Publications of the Field Museum of Natural History (Zoology Series)*, 18: 149-55.
27 *THOMAS, O. (1904). On *Hylochoerus*, the forest-pig of Central Africa. *P.Z.S.L.*, (15 November): 193-9.
28 ANON. (1909). Forest hog from the Upper Congo. *The Field*, 114: 193.
29 *THOMAS, O. (1904). New *Callithrix*, *Midas*, *Felis*, *Rhipidomys*, and *Proechimys* from Brazil and Ecuador. *A.M.N.H.*, (Series 7), 14: 188-96; RIBEIRO, A. DE M. (1912). Zwei neue Affen unserer Fauna. *Rundschau*: 21-3.
30 HILL, W.C.O. (1957). *Primates. Comparative Anatomy and Taxonomy. Volume III: Pithecoidea–Platyrrhini*. Edinburgh University Press (Edinburgh).
31 *MATSCHIE, P. (1905). Eine Robbe von Laysan. *S.G.N.F.B.*: 254-62.
32 STONEHOUSE, B. (1985). *Sea Mammals of the World*. Penguin (Middlesex).
32a BOYD, I.L. & STANFIELD, M.P. (1998). Circumstantial evidence for the presence of monk seals in the West Indies. *Oryx*, 32 (October): 310-316.
33 DENIS, A. (1963). A new seal. *Animals*, 3 (24 December): 57.
34 *GOELDI, E.A. (1907). On some new and insufficiently known species of marmoset monkeys from the Amazonian region. *P.Z.S.L.*: 88-99.
35 *BONHOTE, J.L. (1907). On a collection of mammals made by Dr. Vassal in Annam. Ibid.: 3-11; *TIEN, D.V. (1960). Sur une nouvelle espèce de *Nycticebus* au Vietnam. *Zoologischer Anzeiger*, 164: 240-3; ANNANDALE, N. (1908). An unknown lemur from the Lushai Hills, Assam. *P.Z.S.L.*, (17 November): 888-9.
36 *MILLER, G.S. & HOLLISTER, N. (1921). Twenty new mammals collected by H.C. Raven in Celebes. *P.B.S.W.*, 34, 93-104; MUSSER, G.C. & DAGOSTO, M. (1987). The identity of *Tarsius pumilus*, a pygmy species endemic to the montane forests of central Sulawesi. *American Museum Novitates*, No. 2867 (12 March): 1-53.
36a *NIEMITZ, C. et al. (1991). *Tarsius dianae*: A new primate species from central Sulawesi (Indonesia). *Folia Primatologica*, 56: 105-116.
37 *ANDREWS, R.C. (1908). Description of a new species of *Mesoplodon* from Canterbury Province, New Zealand. *Bulletin of the American Museum of Natural History*, 24: 205-15.
38 WATSON, L. (1985). *Whales of the World* (Rev. Edit.). Hutchinson (London).
38a PITMAN, R.L. et al. (2006). Shepherd's beaked whale (*Tasmacetus shepherdi*): information on appearance and biology based on strandings and at-sea observations. *Marine Mammal Science*, 22 (July): 744-55.
39 LYDEKKER, R. (1910). A new African antelope. *The Times* (London), 23 September; *LYDEKKER, R. (1910). The spotted kudu. *Nature*, 84 (29 September): 397; LYDEKKER, R. (1910). The spotted kudu and the Arusi bushbucks. *The Field*, 116 (22 October): 798; LYDEKKER, R. (1911). On the mountain nyala, *Tragelaphus buxtoni*. *P.Z.S.L.*, (21 February): 348-53.
40 BROWN, L. (1969). Ethiopia's elusive nyala. *Animals*, 12 (December): 340-4.
41 *OUWENS, P.A. (1910). Contribution à la connaissance des mammifères de Celebes. *Bulletin de la Department d'Agriculture Indes Neerlandes*, 38: 1-7; GROVES, C.P. (1969). Systematics of the anoa (Mammalia, Bovidae). *Beaufortia*, 17 (7 November): 1-12.
42 FISHER, J., SIMON, N., & VINCENT, J. (1969). *The Red Book: Wildlife in Danger*. Collins (London).
43 *THOMAS, O. (1910). A new genus of fruit-bats and two new shrews from Africa. *A.M.N.H.* (Series 8): 6: 111-14; *THOMAS, O. (1915). New African rodents and insectivores, mostly collected by Dr C. Christy for the Congo Museum. Ibid., 16: 146-52.
44 ALLEN, J.A. (1917). The skeletal characters of *Scutisorex* Thomas. *Bulletin of the American Museum of Natural History*, 37 (26 November): 769-84; BURTON, M. (1951). The hero shrew. *Illustrated London News*, 219 (17 November): 812.
45 *MORTON, S.G. (1849). Additional observations of a new living species of hippopotamus of Western Africa. *Journal of the Academy of Natural Sciences of Philadelphia* (Series 2), 1 (August): 3-11; LEIDY, J. (1852). On the osteology of the head of hippopotamus and a description of the osteological characters of a new genus of Hippopotamidae. Ibid., (new series), 2 (7 May): 207-34.
46 MACALLISTER, A. (1873). On the visceral anatomy and myology of a young female hippopotamus which died in the Dublin Zoological Society. *Proceedings of the Royal Irish Academy*, 2: 494-500.
47 SCHOMBURGK, H. (1912). On the trail of the pygmy hippo. *Bulletin of the New York Zoological Society*, 16: 808-84; SCHOMBURGK, H. (1913). Distribution and habits of the pygmy hippopotamus. *Report of the New York Zoological Society*, 17: 113-20.
48 NOWAK, R. (Ed.) (1999). *Walker's Mammals of the World* (6th Edit.). Johns Hopkins University Press (Baltimore).
48a ELTRINGHAM, S.K. (1999). *The Hippos*. Academic Press (London).
49 *LAHILLE, F. (1912). Nota preliminar sobre una nueva especie de Marsopa del rio de la Plata (*Phocaena dioptrica*). *Anales del Museo Nacional de Buenos Aires*, 23: 269-78.
50 *TRUE, F.W. (1913). Description of *Mesoplodon mirum*, a beaked whale recently discovered on the coast of North Carolina. *Proceedings of the U.S. National Museum*, 45: 651-7.
50a STOKES, J. (2001). True happiness in Biscay. *Birdwatch*, No. 112: 48.

51 *THOMAS, O. (1912). Two new genera and a new species of viverrine Carnivora. *P.Z.S.L.*, (19 March): 498-503; ANON. (1991). Photo first. *Wild About Animals*, (October): 14; PRINGLE, A. (1999). Personal communication, 1 December.

52 *ALLEN, J.A. (1919). Preliminary notes on African Carnivora. *Journal of Mammalogy*, 1: 25-6.

53 HART, J.A. & TIMM, R.M. (1978). Observations on the aquatic genet in Zaire. *Carnivore*, 1: 130-2.

53a CARROLL, Helen (1996). Found, the living legends of Africa. *Daily Mail* (London), 15 July.

54 PILLERI, G. (1979). The Chinese river dolphin (*Lipotes vexillifer*) in poetry, literature and legend. *Investigations Cetacea*, 10: 335-49; *MILLER, G.S. (1918). A new river-dolphin from China. *Smithsonian Miscellaneous Collections*, 68: 1-12.

55 KAIYA ZHOU (1988). The baiji. In: HARRISON, R. & BRYDEN, M.M. (Eds.), *Whales, Dolphins and Porpoises*. Merehurst (London). pp. 82-3.

55a ANON. (2006). Rare Yangtze dolphin may be extinct. *The Hankyoreh* (Seoul), 5 December; ANON. (2007). Rare dolphin seen in China, experts say. *New York Times* (New York), 30 August.

56 *ALEXANDER, W.B. (1918). A new species of marsupial of the sub-family Phalangerinae. *Journal of the Royal Society of West Australia*, 4: 31-6.

57 TROUGHTON, E. (1965). *Furred Animals of Australia* (8th Edit.). Angus & Robertson (Sydney).

58 SALVADORI, F.B. & FLORIO, P.L. (1978). *Wildlife in Peril*. Westbridge Books (Newton Abbot).

59 *POCOCK, R.I. (1927). Description of a new species of cheetah (*Acinonyx*). *P.Z.S.L.*, (22 February): p. 18, and (6 April): 245-52.

60 SHUKER, K.P.N. (1989). *Mystery Cats of the World: From Blue Tigers to Exmoor Beasts*. Robert Hale (London).

61 BOTTRIELL, L.G. (1987). *King Cheetah: The Story of the Quest*. Brill (Leiden).

62 GUGGISBERG, C.A.W. (1975). *Wild Cats of the World*. David & Charles (Newton Abbot).

63 *LONGMAN, H.A. (1926). New records of cetacea. *Memoirs of the Queensland Museum*, 8: 266-78.

64 AZZAROLI, M.L. (1968). Second specimen of the rarest living beaked whale. *Monitore Zoologico Italiano*, 2: 67-79.

65 MOORE, J.C. (1968). Relationships among the living genera of beaked whales. *Fieldiana Zoology*, 53: 209-98.

66 ANON. (2002). Rare beaked whale found on S. Africa beach. Reuters, 4 August 2002; DALEBOUT, M.L. et al. (2003). Appearance, distribution, and genetic distinctiveness of Longman's beaked whale, *Indopacetus pacificus*. *Marine Mammal Science*, 19 (July): 421-461.

66a PITMAN, R.L. et al. (1987). Observations of an unidentified beaked whale (*Mesoplodon* sp.) in the eastern tropical Pacific. *Marine Mammal Science*, 3 (October): 345-52; PITMAN, R.L. et al. (1999). Sightings and possible identity of a bottlenose whale in the tropical Indo-Pacific: *Indopacetus pacificus*? *Marine Mammal Science*, 15: 531-549.

67 *GRANDIDIER, G. (1929). Un nouveau type de mammifère insectivore de Madagascar, le *Dasogale fontoynonti* G. Grand. *Bulletin de l'Académie Malgache (new series)*, 11: 85-90.

68 MACPHEE, R.D.E. (1987). Systematic status of *Dasogale fontoynonti* (Tenrecidae, Insectivora). *Journal of Mammalogy*, 68: 133-5.

69 *SCHWARZ, E. (1929). Das Vorkommen des Schimpansen auf den linken Kongo-Ufer. *R.Z.B.A.*, 16: 425-6; SCHOUTEDEN, H. (1931). Quelques notes sur le chimpanze de la rive gauche du Congo, *Pan satyrus paniscus*. Ibid., 20: 310-14.

70 SUSMAN, R.L. (Ed.) (1984). *The Pygmy Chimpanzee: Evolutionary Biology and Behavior*. Plenum Press (London).

71 MORRIS, R. & MORRIS, D. (1981). *The Giant Panda*. Kogan Page Ltd (London).

72 MILNE-EDWARDS, A. (1869). Extrait d'une lettre de même (M. l'Abbé David) datée de la principalité Thibetaine (independante) de Moupin, le 21 mars 1869. *Nouvelles Archives du Museum d'Histoire Naturelle (Bulletin)*, 5: 13.

73 MILNE-EDWARDS, A. (1870). Sur quelques mammifères du Thibet oriental. *Annales des Sciences Naturelles (Series 5)*, 70: 341-2.

74 ROOSEVELT, T. & ROOSEVELT, K. (1929). *Trailing the Giant Panda*. Scribner's (New York).

75 HARKNESS, R. (1938). *The Baby Giant Panda*. Carrick & Evans (New York).

76 BELSON, J. & GILHEANY, J. (1981). *The Giant Panda Book*. Collins (London).

77 APPLETON, L. (1987). Who's that girl? It's Blue Sky! *Daily Mail* (London), 16 November.

78 LEONE, C.A. & WIENS, A.L. (1956). Comparative serology of carnivores. *Journal of Mammalogy*, 37: 11-23.

79 O'BRIEN, S. (1987). The ancestry of the giant panda. *Scientific American*, 257 (November): 82-7.

80 *WAN, Q-H. et al. (2005). A new subspecies of giant panda (*Ailuropoda melanoleuca*) from Shaanxi, China. *Journal of Mammalogy*, 86: 397-402; ANON. (2006). [Qinling giant panda as separate subspecies.] *People's Daily Online* (China), 13 January.

81 SHUKER, K.P.N. (1989). Mystery bears. *Fate*, 43 (April): 30-8.

82 HEUVELMANS, B. (1986). Annotated checklist of apparently unknown animals with which cryptozoology is concerned. *Cryptozoology*, 5: 1-26.

83 AHARONI, I. (1942). *Memoirs of a Hebrew Zoologist*. Am Oved Limited (Tel Aviv); SIEGEL, H.I. (Ed.) (1985). *The Hamster: Reproduction and Behavior*. Plenum Press (London).

83a GATTERMANN, R. et al. (2001). Notes on the current distribution and the ecology of wild golden hamsters (*Mesocricetus auratus*). *Journal of Zoology*, 254 (July): 359-365.

84 FINLAYSON, H.H. (1932). Rediscovery of *Caloprymnus campestris* (Marsupialia). *Nature*, 129 (11 June): 871.

85 ANON. (1933). [Shepherd's beaked whale reports.] *Hawera Star* (Hawera–New Zealand), 9 November and 12 December.

86 *OLIVER, W.R.B. (1937). *Tasmacetus shepherdi*: a new genus and species of beaked whale from New Zealand. *P.Z.S.L.*, (4 May): 371-81.

87 HEUVELMANS, B. (1968). *In the Wake of the Sea-Serpents*. Rupert Hart-Davis (London).

88 PODUSCHKA, W. (1975). Solenodon story. *Wildlife*, 17 (March): 108-11.

88a SNOW, A. (2003). Furry find in Cuba has a nose for worms—"Alejandrito" proves rare species isn't extinct. *Seattle Times* (Seattle), 25 September.
88b TURVEY, S. & HELGEN, K. (2008). *Solenodon marcanoi*. In: IUCN. (2008). *IUCN Red List of Threatened Species 2009*. IUCN (Gland).
89 MILLER, G.S. (1930). Three small collections of mammals from Hispaniola. *Smithsonian Miscellaneous Collections*, 82: 1-10.
89a MORGAN, G.S. & WOODS, C.A. (1986). Extinction and the zoogeography of West Indian land mammals. *Biological Journal of the Linnean Society*, 28 (May): 167-203.
90 KEMF, E. (1988). Fighting for the forest ox. *New Scientist*, 118 (30 June): 51-3.
91 *URBAIN, A. (1937). Le kou prey ou boeuf gris Cambodgien. *Bulletin de la Société de France*, 62 (8 June): 305-7.
92 GRIGSON, C. (1988). Complex cattle. *New Scientist*, 119 (4 August): 93-4.
93 COOLIDGE, H.J. (1940). The Indo-Chinese forest ox or kouprey. *Memoirs of the Museum of Comparative Zoology, Harvard College*, 54 (August): 417-531.
94 SITWELL, N. (1970). Is this the rarest animal in the world? *Animals*, 13 (November): 306-7.
95 FITTER, R. (1976). The kouprey survives. *Oryx*, 13 (July): 249; ANON. (1980). Kouprey alive, well, and protected. Ibid., 15 (July): 238.
96 ANON. (1988). The cow quest. *BBC Wildlife*, 6 (November): 635; SEITRE, R. (2000). Return to the killing fields. Ibid., 18 (August): 60-4; MCDERMID C. & SOKHA, C. (2006). Search for the kouprey: trail runs cold for Cambodia's national animal. *Phnom Pen Post* (Phnom Penh), 21 April-4 May.
96a GALBREATH, G.J. et al. (2006). Genetically solving a zoological mystery: was the kouprey (*Bos sauveli*) a feral hybrid? *Journal of Zoology*, 270 (April): 561-4; GALBREATH, G.J. et al. (2007). An evolutionary conundrum involving kouprey and banteng: A response from Galbreath, Mordacq and Weiler. *Journal of Zoology*, 271 (March): 253-4; HASSANIN, A. & ROPIQUET, A. (2007). Resolving a zoological mystery: the kouprey is a real species, *Proceedings of the Royal Society, Series B*, 274 (22 November): 2849-55.
96b COLEMAN, L. (2006). Kouprey debate. http://www.cryptomundo.com/cryptozoo-news/kouprey-debate/ 17 September.
97 *BELOSLUDOV, B.A. & BASHANOV, V.S. (1939). A new genus and species of rodent from the Central Kazakhstan. *Uchenye Zaipiski Kazakkskii Universitet Alma-Ata Biol.*, 1: 81-6; BASHANOV, V.S. & BELOSLUDOV, B.A. (1941). A remarkable family of rodent from Kazakhstan, U.S.S.R. *Journal of Mammalogy*, 22 (14 August): 311-15.
98 JOHNSON, D.H. (1948). A rediscovered Haitian rodent, *Plagiodontia aedium*, with a synopsis of related species. *P.B.S.W.*, 61 (16 June): 69-76.
99 *MILLER, G. (1927). The rodents of the genus *Plagiodontia*. *Proceedings of the U.S. National Museum*, 72: 1-8.
99a TURVEY, S. & HELGEN, K. (2008). *Plagiodontia ipnaeum*. In: *IUCN (2008). IUCN Red List of Threatened Species 2009*. IUCN (Gland).
100 *ALLEN, G.M. (1917). An extinct Cuban *Capromys*. *Proceedings of the New England Zoological Club*, 6 (28 March): 53-6.
101 *BARBOUR, T. (1926). A remarkable new bird from Cuba. Ibid., 9: 73-5; *BARBOUR, T. & PETERS, J.L. (1927). Two more remarkable new birds from Cuba. Ibid., 9: 95-7.
102 *VARONA, L.S. (1970). Descripcion de una nueva especia de *Capromys* del sur de Cuba. *Poeyana Instituto de Biologia, Cuba (Series A)*, No. 74: 1-16.
103 BURTON, J. & PEARSON, B. (1988). *Collins Guide to the Rare Mammals of the World*. Collins (London).
104 ANON. (1991). [Hutias rediscovered.] *Oryx*, 25 (July): 133.
105 *VARONA, L.S. & GARRIDO, O.H. (1970). Vertebrados de los Cayos de San Felipe, Cuba, incluyendo una nuevo especie de jutia. *Poeyana Instituto de Biologica, Cuba (Series A)*, No. 75.
106 *VARONA, L.S. (1970). Subgenero y especie nuevas de *Capromys* (Rodentia: Caviomorpha) para Cuba. *Poeyana Instituto Zool. Acad. Cienc. de Cuba*, 194: 1-33.
107 MCWHIRTER, R. & MCWHIRTER, N. (Eds.) (1974). *The Guinness Book of Records (21st Edit.)*. Guinness Superlatives (London); MCWHIRTER, N. (Ed.) (1978). *The Guinness Book of Records (25th Edit.)*. Guinness Superlatives (London).
108 DAVIES, G. & BIRKENHÄGER, B. (1990). Jentink's duiker in Sierra Leone: evidence from the Freetown Peninsula. *Oryx*, 24 (July): 143-6.
109 *KRUMBIEGEL, I. (1949). Der Andenwolf—ein neuentdecktes Grosstier. *Umschau*, 49: 590-1; KRUMBIEGEL, I. (1953). Der "Andenwolf", *Dasycyon hagenbecki* (Krumbiegel, 1949). *Säugetierkundliche Mitteilungen*, 1: 97-104.
110 DIETERLEN, I. (1954). Über den Haarbau des Andenwolfes, *Dasycyon hagenbecki* (Krumbiegel, 1949). Ibid., 2: 26-31; CABRERA, A. (1957). Catalago de los mamiferos de America del Sur. 1. (Metatheria–Unguiculata–Carnivora). *Revista–Museo Argentino de Ciencas Naturales Rivadavia (Zoologia)*, 4: 1-307; SHUKER, K.P.N. (1996). A wolf in sheepdog's clothing? *All About Dogs*, 1 (July-August): 72; EBERHART, G.M. (2002). *Mysterious Creatures: A Guide to Cryptozoology*, 2 vols. ABC-CLIO (Santa Barbara, California).
111 *SANBORN, C.C. (1951). Two new mammals from southern Peru. *Fieldiana Zoology*, 31: 473-7.
112 *IZOR, R.J. & TORRE, L. DE LA (1978). A new species of weasel (*Mustela*) from the highlands of Colombia, with comments on the evolution and distribution of South American weasels. *Journal of Mammalogy*, 59: 92-102.
113 SCHREIBER, A. et al. (1989). *Weasels, Civets, Mongooses, and Their Relatives*. IUCN (Gland–Switzerland).
114 YAMADA, M. (1954). An account of a rare porpoise *Feresa* from Japan. *Scientific Reports of the Whales Research Institute*, 9: 59-88.
115 *MIKHALEV, YU. A. et al. (1981). The distribution and biology of killer whales in the southern hemisphere. *Report of the International Whale Commission*, 31: 551-65.
116 *BERZIN, A.A. & VLADIMIROV, V.L. (1983). A new species of killer whale (Cetacea, Delphinidae) from the Antarctic waters. *Zoologicheskii Zhurnal*, 62: 287-95; BIGG, M.A., et al. (1987). *Killer Whales, a Study of Their Identification, Genealogy and Natural History in British Colombia and Washington State*. Phantom Press and Publications (Nanaimo).

116a PITMAN, R.L. et al. (2007). A dwarf form of killer whale in Antarctica. *Journal of Mammalogy*, 88: 43-8.

117 *HEIM DE BALSAC, H. (1954). Un genre inedit et inattendu de mammifère (Insectivore Tenrecidae) d'Afrique Occidentale. *C.R.A.S.*, 239: 102-4; *DE WITTE, G-F & FRECHKOP, S. (1955). Sur une espèce encore inconnue de mammifère africain, *Potamogale ruwenzorii*, sp. n. *Bulletin–Institut Royal des Sciences Naturelles de Belgique*, 31: 1-11.

118 GEE, E.P. (1964). *The Wild Life of India*. Collins (London).

119 *KHARJURIA, H. (1956). A new langur (Primates: Colobidae) from Goalpara District, Assam. *A.M.N.H.* (Series 12), 9: 86-9.

120 *T'AN PANG-CHIEH (1957). Rare catches by Chinese animal collectors. *Zoo Life*, 12 (Winter): 61-3; NAPIER, J.R. & NAPIER, P.H. (1967). *A Handbook of Living Primates*. Academic Press (New York).

121 *FRASER, F.C. (1956). A new Sarawak dolphin. *Sarawak Museum Journal*, 7: 478-503.

122 *NISHIWAKI, M. & KAMIYA, T. (1958). A beaked whale *Mesoplodon* stranded at Oiso Beach, Japan. *Scientific Reports of the Whales Research Institute*, 13: 53-83.

123 DERANIYAGALA, P. (1963). Mass mortality ... and a new beaked whale from Ceylon. *Spolia Zeylandica*, 29: 79-84; MOORE, J.C. & GILMORE, R.M. (1965). A beaked whale new to the western hemisphere. *Nature*, 205 (20 March): 1239-40.

123a KIRIONA, R. (2003). Whale-sized interest in giant mammal on beach. *Daily News* (New Zealand), 23 April.

124 *NORRIS, K.S. & MCFARLAND, W.N. (1958). A new harbor porpoise of the genus *Phocoena* from the Gulf of California. *Journal of Mammalogy*, 39: 22-39.

125 *HAYMAN, R.W. (1958). A new genus and species of West African mongoose. *A.M.N.H.* (Series 13), 1: 448-52.

126 SCHLITTER, D.A. (1974). Notes on the Liberian mongoose, *Liberiictis kuhni* Hayman, 1958. *Journal of Mammalogy*, 55 (May): 438-42; TAYLOR, M.E. (1992). The Liberian mongoose. *Oryx*, 26 (April): 103-6.

127 ANON. (1959). Protection for "friendly" monster. *Guardian* (Manchester), 4 March; ANON. (1961). Camera captures unknown ape. *Popular Science Monthly*, 179 (July): 83.

128 HILL, W.C.O. (1963). The ufiti: the present position. *Symposia of the Zoological Society of London*, 10: 57-9.

128a ANSELL, W.F.H. & DOWSETT, R.J. (1988). *Mammals of Malawi*. Trendrine Press (Zennor); ANSELL, W.F.H. & DOWSETT, R.J. (1991). Addenda and corrigenda to Mammals of Malawi (Ansell & Dowsett, 1988). *Nyala*, 15(1): 43-46; ANSELL, W.F.H. (1994). Personal communication. 1 March.

128b ANON. (1997). DNA tests identify possible new chimp subspecies. *ENN Daily News*, 24 July, at: http://www.enn.com/enn-news-archive/1997/07/072497/07249713.asp/

129 BLYTH, E. (1863). *Catalogue of the Mammalia in the Museum of the Asiatic Society*. (Calcutta); BAILEY, F.M. (1912). Journey through a portion of south-eastern Tibet and the Mishmi Hills. *Geographical Journal*, 39: 334-47; DOLLMAN, J.G. (1932). Mammals collected by Lord Cranbrook and Captain F. Kingdom Ward in Upper Burma. *Proceedings of the Linnaean Society of London*, 145: 9-11.

130 *HAYMAN, R.W. (1961). The red goral of the North-East Frontier region. *P.Z.S.L.*, 136: 317-24.

131 HLA AUNG, S. (1967). Observations on the red goral ... at Rangoon Zoo. *International Zoo Yearbook*, 7: 225-6.

132 *POCOCK, R.I. (1914). Description of a new species of goral (*Nemorhaedus*) shot by Captain F.M. Bailey. *Journal of the Bombay Natural History Society*, 23: 32-3.

133 GROVES, C.P. & GRUBB, P. (1985). Reclassification of the serows and gorals (*Nemorhaedus*: Bovidae). In: LOVARI, S. (Ed.), *The Biology and Management of Mountain Ungulates*. Croom Helm (London). pp. 45-50.

134 *SCHALDACH, W.J. & MCLAUGHLIN, C.A. (1960). A new genus and species of glossophagine bat from Colima, Mexico. *Contributions to Science, Los Angeles*, No. 37: 1-8.

135 PIZZEY, G. (1963). Lost and found. *Animals*, 2 (26 November): 626-8; WILKINSON, H.E. (1961). The rediscovery of Leadbeater's possum, *Gymnobelideus leadbeateri* McCoy. *Victorian Naturalist*, 78: 97-102; ANON. (2009). A million native animals may have died in Victorian bushfires. *The Australian* (Sydney), 11 February.

136 *MOORE, J.C. (1963). Recognizing certain species of beaked whales of the Pacific Ocean. *American Midland Naturalist*, 70: 396-428.

137 SCHAEFER, E. (1937). Über das Zwergblauschaf (*Pseudois* spec. nov.) und das Grossblauschaf (*Pseudois nahoor* Hdgs.) in Tibet. *Zoologische Garten*, 9: 263-78; *HALTENORTH, T. (1963). Klassifikation der Säugetiere: Artiodactyla. *Handbüche der Zoologie*, 8: 1-167.

138 GROVES, C.P. (1978). The taxonomic status of the dwarf blue sheep (Artiodactyla; Bovidae). *Säugetierkundliche Mitteilungen*, 26: 177-83.

139 ANON. (1968). New mammal discovered. *Animals*, 10 (March): 501-3.

140 *IMAIZUMI, Y. (1967). A new genus and species of cat from Iriomote, Ryukyu Islands. *Journal of the Mammalogical Society of Japan*, 3: 75-106.

141 WURSTER-HILL, D.H. et al. (1987). Banded chromosome study of the Iriomote cat. *Journal of Heredity*, 78: 105-7.

142 *KURODA, N. (1924). *On new mammals from the Riu Kiu Islands and vicinity*. Self-published (Tokyo); IMAIZUMI, Y. (1973). Taxonomic study of the wild boar from the Ryukyu Islands, Japan. *Memoirs of the National Science Museum*, 6: 113-29; ANON. (1975). Iriomote cat survey finds a new pig. *Wildlife*, 17 (February): 87-8.

143 JACKSON, P. (1989). New cat discovered. *Cat News*, No. 10 (January); LOXTON, H. (1990). *The Noble Cat*. Merehurst (London).

144 FITTER, R. (1968). *Vanishing Wild Animals of the World*. Midland Bank/Kaye & Ward (London).

145 DECHAU, C-P. (1990). Rainforest yields 'extinct' lemur. *New Scientist*, 125 (27 January): 33.

146 CARWADINE, M. (1995). *The Guinness Book of Animal Records*. Guinness Publishing (Enfield); ANON. (1996). Found again. *Fortean Times*, No. 86 (May): 6.

147 WODZICKI, K. & FLUX, J.E.C. (1967). The rediscovery of the white-throated wallaby on Kawau island. *Australian Journal of Science*, 29: 429-30.

148 SEEBECK, J. (1967). Burramys—only known from fossils. *Animals*, 10 (October): 271-2; TROUGHTON, E. (1967). Broom's pygmy possum. *Proceedings of the*

Royal Zoological Society of New South Wales (for 1966-67): 20-4.
149 MORCOMBE, M.K. (1967). The dibbler—unseen for 83 years. *Animals*, 10 (October): 273-4; MORCOMBE, M.K. (1967). The rediscovery after 83 years of the dibbler *Antechinus apicalis* (Marsupialia, Dasyuridae). *Western Australian Naturalist*, 10: 103-11; DICKMAN, C. (1986). Return of the phantom dibbler. *Australian Natural History*, 22 (winter): 33.
150 TESSIER-YANDELL, J. (1971). Rediscovery of the pygmy hog. *Animals*, 13 (December): 956-8.
151 MALLINSON, J. (1989). *In Search of Endangered Species*. David & Charles (Newton Abbot).
152 *MISHRA, A.C. & SINGH, K.N. (1978). Description of *Haematopinus oliveri* sp. nov. (Anoplura: Haematopinidae) parasitizing *Sus salvanius* in India. *Bulletin of the Zoological Survey of India*, 1: 167-9.
153 *HILL, J.E. (1974). A new family, genus and species of bat (Mammalia: Chiroptera) from Thailand. *B.B.M.N.H.Z.*, 27: 301-36.
154 ANON. (1985). The world's smallest mammal. *World Wildlife News*, (spring): 5.
155 GOULD, A.B. (1986). Smallest mammal still at large. *BBC Wildlife*, 4 (November): 566-7.
155a PEREIRA, M.J.R. et al. (2006). Status of the world's smallest mammal, the bumble-bee bat *Craseonycteris thonglongyai*, in Myanmar. *Oryx*, 40 (October): 456-63.
156 *THONGLONGYA, K. (1972). A new genus and species of fruit bat from South India (Chiroptera: Pteropodidae). *Journal of the Bombay Natural History Society*, 69: 151-8; WATKINS, M. (1993). Second sight. *BBC Wildlife*, 11 (August): 59; GOULD, E. (1978). Rediscovery of *Hipposideros ridleyi*. *Biotropica*, 10: 30-2.
157 WETZEL, R.M. et al. (1975). *Catagonus*, an "extinct" peccary, alive in Paraguay. *Science*, 189 (1 August): 379-81; WETZEL, R.M. (1977). The Chacoan peccary *Catagonus wagneri* (Rusconi). *Bulletin of Carnegie Museum of Natural History*, No. 3: 1-36; WETZEL, R.M. (1981). The hidden Chacoan peccary. *Carnegie Magazine*, 55: 24-32.
158 ANON. (1988). Chacoan peccary project. *Oryx*, 22 (April): 120.
159 MITTERMEIER, R.A., MACEDO-RUIZ, H. DE, & LUSCOMBE, A. (1975). A woolly monkey rediscovered in Peru. *Oryx*, 13 (January): 41-6.
160 ANON. (1982). Monkey discovered. *Wildlife*, 24 (May): 164; OurAmazingPlanet Staff (2010). Surprise! Hidden yellow-tailed monkey colony discovered. http://www.livescience.com/animals/woolley-monkey-colony-found-101123.html 23 November.
161 *ARCHER, M. (1975). *Ningaui*, a new genus of tiny dasyurids (Marsupialia) and two new species, *N. timealeyi* and *N. ridei*, from arid Western Australia. *Memoirs of the Queensland Museum*, 17: 237-49; ANON. (1975). Animal like a womble is new find. *The Times* (London), 13 January.
162 *KITCHENER, D.J. et al. (1983). A taxonomic appraisal of the genus *Ningaui* Archer (Marsupialia: Dasyuridae), including description of a new species. *Australian Journal of Zoology*, 31: 361-70.
163 *SPENCER, W.B. (1908). Description of a new species of *Sminthopsis*. *Proceedings of the Royal Society of Victoria* (Series 2), 21: 449-51.
164 DAVEY, K. (1983). *Our Arid Environment*. Reed (New South Wales).
165 ANON. (1981). 'Extinct' rodent found in desert. *Globe & Mail* (Toronto), 21 September; ANON. (1982). Rare marsupial found. *Oryx*, 16 (October): 313.
166 ANON. (1977). Survey team finds new rock wallaby. *Wildlife*, 19 (November): 491; *MAYNES, G.M. (1982). A new species of rock wallaby, *Petrogale Persephone* ... from Proserpine, Central Queensland. *Australian Mammalogy*, 5: 47-58.
167 *MENZIES, J.I. (1977). Fossil and subfossil fruit bats from the mountains of New Guinea. *Australian Journal of Zoology*, 25: 329-36.
168 HYNDMAN, D. & MENZIES, J.I. (1980). *Aproteles bulmerae* (Chiroptera: Pteropodidae) of New Guinea is not extinct. *Journal of Mammalogy*, 61: 159-60; FLANNERY, T.F. & SERI, L. (1993). Rediscovery of *Aproteles bulmerae* (Chiroptera: Pteropodidae). Morphology, ecology and conservation. *Mammalia*, 57: 19-25; BONACCORSO, F.J. (1998). *Bats of Papua New Guinea*. Conservation International (Washington D.C.).
169 WOOD, G.L. (1982). *The Guinness Book of Animal Facts and Feats* (3rd Edit.). Guinness Superlatives (London).
170 MATTHEWS, P. & MCWHIRTER, N. (Eds) (1992). *The Guinness Book of Records 1993*. Guinness Publishing (London).
171 ANON. (1989). New species of deer in China. *Oryx*, 23 (April): 109; *MA SHILAI, WANG YINGXIANG, & SHI LIMING. (1990). A new species of the genus *Muntiacus* from Yunnan, China. *Acta Zoologica Sinica*, 11: 52-53 [English translation abstract].
172 *GROVES, C.P. & GRUBB, P. (1982). The species of muntjac (genus *Muntiacus*) in Borneo: unrecognized sympatry in tropical deer. *Zoologische Mededelingen*, 56: 203-16; PUTMAN, R. (1988). *The Natural History of Deer*. Christopher Helm (London).
173 ANON. (1978). Giant civet rediscovered. *Oryx*, 14 (October): 309; ANON. (1979). The 'arboreal dog' re-appears. *World Wildlife News*, (summer): 5.
174 *SEEBECK, J.H. & JOHNSTON, P.G. (1980). *Potorous longipes* ... a new species from eastern Victoria. *Australian Journal of Zoology*, 28: 119-34.
175 PERRIN, W.F. et al. (1981). *Stenella clymene*, a rediscovered tropical dolphin of the Atlantic. *Journal of Mammalogy*, 62 (August): 583-98.
176 KINGDON, J. (1986). An embarrassment of monkeys. *BBC Wildlife*, 4 (February): 52-7.
177 REDMOND, I. (1986). New monkey puzzle. Ibid., 2 (August): 384; ANON. (1988). Monkey makes its debut. *New Scientist*, 118 (23 June): 31.
178 *HARRISON, M.J.S. (1988). A new species of guenon (genus *Cercopithecus*) from Gabon. *Journal of Zoology*, 215 (July): 561-75.
179 *THYS VAN DEN AUDENAERDE, F.E.T. (1977). Description of a monkey-skin from east-central Zaire as a probably new monkey species (Mammalia, Cercopithecidae). *Revue de Zoologie Africaine*, 91: 1000-10.
180 *SCHWARZ, E. (1932). Der Vertreter der Diana-Meerkatze in Zentral-Afrika. *R.Z.B.A.*, 21: 251-4; COLYN, M. et al. (1991). *Cercopithecus dryas* Schwarz 1932 and *C. salongo* Vanden Audenaerde, Thys 1977

are the same species with an age-related coat pattern. *Folia Primatologica*, 56: 167-70.
181 *POCOCK, R.I. (1907). A monographic revision of the monkeys of the genus *Cercopithecus*. *P.Z.S.L.*: 677-746.
182 *GROVES, C.P. & LAY, D. (1985). A new species of the genus *Gazella* (Mammalia: Artiodactyla: Bovidae) from the Arabian Peninsula. *Mammalia*, 49: 27-36; SCOTT, M. (1986). Gazelle comes out of the cupboard. *BBC Wildlife*, 4 (August): 378.
183 BURTON, J.A. (1988). No yeti yet. Ibid., 3 (October): 461; ZHOU GUOXING. (1987). The big wildman and the little wildman. Ibid., 5 (September): 442-4.
184 ZHOU GUOXING. (1984). Morphological analysis of the Jiulong Mountain "manbear" (wildman) hand and foot specimens. *Cryptozoology*, 3: 58-70.
185 GREENWELL, J.R. (1985). Groves joins editorial board. *ISC Newsletter*, 4 (winter): 10; TISDALE, L. (1986). Hair today, gone tomorrow. *BBC Wildlife*, 4 (May): 206-7.
186 *AYRES, J.M. (1985). On a new species of squirrel monkey, genus *Saimiri*, from Brazilian Amazonia (Primates, Cebidae). *Papeis Avulsos de Zoologia*, 36: 147-64; ANON. (1986). New monkey from old. *BBC Wildlife*, 4 (January): 10.
187 GREENWELL, J.R. (1986). Onza specimen obtained—identity being studied. *ISC Newsletter*, 5 (spring): 1-6; GREENWELL, J.R. (1987). Is this the beast the Spaniard saw in Montezuma's zoo? *BBC Wildlife*, 6 (July): 354-9.
188 TINSLEY, J.R. (1987). *The Puma: Legendary Lion of the Americas*. Texas Weston Press (El Paso); CARMONY, N.B. (1995). *Onza! The Hunt For a Legendary Cat*. High-Lonesome Books (Silver City, New Mexico); MARSHALL, R. (1961). *The Onza*. Exposition Press (New York); DOBIE, J.F. (1935). *Tongues of the Monte*. Doubleday, Doran, & Co (New York); SHUKER, K.P.N. (1986). Cryptoletter [onza]. *ISC Newsletter*, 5 (winter): 11.
189 GREENWELL, J.R. (1985). Two new onza skulls found. Ibid., 4 (winter): 5-6.
190 GREENWELL, J.R. (1988). Onza identity still unresolved. Ibid., 7 (winter): 5-6.
191 SHUKER, K.P.N. (1998). Unmasking the onza—from Aztecs to Arizona. *All About Cats*, 5 (January-February): 44-45; DRATCH, P.A., et al. (1993-96 [published 1998]). Molecular genetic identification of a Mexican onza specimen as a puma (*Puma concolor*). *Cryptozoology*, 12: 42-9; PALMEROS, R.A.L. (1995). Was an onza shot in early 1995? In: DOWNES, J. (Ed.), *CFZ Yearbook 1996*. CFZ (Exwick). pp. 167-8.
192 *MEIER, B. et al. (1987). A new species of *Hapalemur* from South East Madagascar. *Folia Primatologica*, 48: 211-15; CHERFAS, J. & DECHAU, C. (1988). Zoologist discovers new species of primate. *New Scientist*, 117 (28 January): 33.
193 *WOZENCRAFT, W.C. (1986). A new species of striped mongoose from Madagascar. *Journal of Mammalogy*, 67: 561-71
194 KIRBY, T. (1987). Rudimentary porcupine returns to earth. *BBC Wildlife*, 5 (March): 140-1.
195 ANON. (1992). New species of porcupine. *Oryx*, 26 (October): 198.
196 ANON. (1987). Sumatran rhino rediscovery. Ibid., 21 (April): 120; RUSSELL, C. (1986). The rhino's return. *Daily Mail* (London), 22 October.
197 SCHALLER, G. et al. (1990). Javan rhinoceros in Vietnam. *Oryx*, 24 (April): 77-80; ANON. (1999). Pictured at last, the shyest beast on earth. *Daily Mail* (London), 16 July.
198 *VAN DYCK, S. (1988). The bronze quoll, *Dasyurus spartacus* (Marsupialia: Dasyuridae), a new species from the savannahs of Papua New Guinea. *Australian Mammalogy*, 11: 145-56.
199 FLANNERY, T. (1998). *Throwim Way Leg: Adventures in the Jungles of New Guinea*. Weidenfeld & Nicolson (London).
200 *FLANNERY, T.F. & SERI, L. (1990). *Dendrolagus scottae* n. sp. (Marsupialia: Macropodidae): a new tree-kangaroo from Papua New Guinea. *Records of the Australian Museum*, 42: 237-245.
201 ANON. (1988). New whale type found off Peru. *Cincinnati Enquirer* (Cincinnati), 19 December.
202 ANON. (1989). Les grandes inconnues des océans. *Terre Sauvage*, No. 27 (March): 6.
203 *REYES, J.C., MEAD, J.G., & WAEREBEEK, K. VAN (1991). A new species of beaked whale *Mesoplodon peruvianus* sp. n. (Cetacea: Ziphiidae) from Peru. *Marine Mammal Science*, 7 (January): 1-24.
204 ANON. (1989). Golden-crowned lemur found in Madagascar. *New Scientist*, 121 (25 February): 41; *SIMONS, E. (1988). A new species of *Propithecus* (Primates) from northeast Madagascar. *Folia Primatologica*, 50: 143-51.
205 MEREDITH, D. (1999). Lemur 'Juliet' may be new subspecies; no mate for 'Romeo'. http://www.dukenews.duke.edu/Research/pcnew.htm/ 3 November.
206 SHUKER, K.P.N. (1990). The Kellas cat: reviewing an enigma. *Cryptozoology*, 9: 26-40; ANON. (1992). Mystery of the black cat. *Wild About Animals*, (June): 42.
207 *LORINI, M.L. & PERSSON, V.G. (1990). New species of *Leontopithecus* Lesson 1840 from southern Brazil (Primates: Callithricidae). *Boletin Museo Nacional Rio de Janeiro Zoologia*, No. 338: 1-14.
208 *MITTERMEIER, R.A. et al. (1992). A new species of marmoset ... from the Rio Maués region, State of Amazonas, Central Brazilian Amazonia. *Goeldiana Zoologia*, 14: 1-17.
209 *MENZIES, J.I. (1990). Notes on spiny bandicoots, *Echymipera* spp. ... from New Guinea and description of a new species. *Science in New Guinea*, 16: 86-98.
210 ANON. (1991). 'Extinct' armadillo comes out of its shell. *New Scientist*, 131 (10 August): 15.
211 ANON. (1991). Closet skeleton brought to life. *BBC Wildlife*, 9 (October): 682-3; OLIVER, W. (1997). First catch your hogs. Ibid., 15 (September): 74-8.
212 SUNQUIST, M., et al. (1994). Rediscovery of the Bornean bay cat. *Oryx*, 28 (January): 67-70; Bay cat, Wikipedia entry, accessed 12 July 2011.
213 *GRAY, J.E. (1874). Description of a new species of cat (*Felis badia*) from Sarawak. *P.Z.S.L.*, (19 May): 322-3.
214 JOHNSON, W.E. et al. (1999). Molecular genetic characterization of two insular Asian cat species, Bornean bay cat and Iriomote cat. In: WASSER, S.P. (Ed.), *Evolutionary Theory and Processes: Modern Perspectives*,

215 BILLE, M.A. (1995). *Rumors of Existence*. Hancock House (Blaine); SHUKER, K.P.N. (1996). Wild thing, I think I love you. *Fortean Times*, No. 91 (October): 42-3; BILLE, M.A. (1999). Update: mammals of southeast Asia. *Exotic Zoology*, 6 (July-September): 1-6; COLEMAN, L. & CLARK, J. (1999). *Cryptozoology A To Z*. Fireside Books (New York).

216 ANON. (1993). New evidence confirms new Vietnam species. WWF-UK press release, 30 March; *DUNG, V.V., MACKINNON, J., et al. (1993). A new species of living bovid from Vietnam. *Nature*, 363 (3 June): 443-5; DUNG, V.V. et al. (1994). Discovery and conservation of the Vu Quang ox in Vietnam. *Oryx*, 28 (January): 16-21; ANON. (1995). Boost for conservation of Vu Quang. WWF-UK press release, 12 June.

217 ANON. (1994). Vu Quang ox found—live! WWF-UK press release, 24 June; CONNOR, S. (1994). Lost worlds rich in unique wildlife. *Independent On Sunday* (London), 3 July; ANON. (1994). Oxen of newly-identified species die, months after capture for study. *Post-Dispatch* (St. Louis), 10 October; ANON. (1994-5). New mammals pop up in 'Nam. *Fortean Times*, No. 78 (December-January): 19.

218 ANON. (1996). Rare species captured alive. *Hong Kong Standard* (Hong Kong), 25 January; ETTER, M. & RUGGERI, N. (1996). First adult saola sighted. *Wildlife Conservation*, (March-April): 9.

219 ANON. (1998). Vietnam—sao la released in Thua Thien-Hue. *Saigon Daily Times* (Ho Chi Minh City), 3 June; WHITFIELD, J. (1998). A saola poses for the camera. *Nature*, 396 (3 December): 410; PILLINGER, C. (1998). Camera traps 'spindlehorn' in wild for first time. *Sunday Telegraph* (London), 6 December.

220 ANON. (1994). New mammal found in Vietnam. WWF-UK press release, 21 April; SCOTT, K. (1994). Vietnam explosion—new creatures just keep on coming. *BBC Wildlife*, 12 (June): 12; *TUOC, D. et al. (1994). Introduction of a new large mammal species in Viet Nam. *Science and Technology News* [in Vietnamese].

221 SCOTT, K. (1994). One small step for muntjac. *BBC Wildlife*, 12 (August): 10.

222 TIMMINS, R. et al. (1994). A new species of living muntjac from the Lao P.D.R. Unpublished paper, 11 pp.

223 SCHALLER, G.B. & VRBA, E.S. (1996). Description of the giant muntjac (*Megamuntiacus vuquangensis*) in Laos. *Journal of Mammalogy*, 77(3): 675-83; BAUER, K. (1997). Historic record and range extension for giant muntjac, *Muntiacus vuquangensis* (Cervidae). *Mammalia*, 61(2): 265-7; TIMMINS, R.J. et al. (1998). Status and conservation of the giant muntjac *Megamuntiacus vuquangensis*, and notes on other muntjac species in Laos. *Oryx*, 32 (January): 59-67.

224 MILLET, F. (1930). *Les Grands Animaux Sauvages de l'Annam*. Libraire Plon (Paris).

225 TIMMINS, R. (1996). Another muntjac materialises. *BBC Wildlife*, 14 (March): 22-3; SCHALLER, G.B. (1995). An unfamiliar "bark". *Wildlife Conservation*, (June): 8.

226 AMATO, G. et al. (1999). Rediscovery of Roosevelt's barking deer (*Muntiacus rooseveltorum*). *Journal of Mammalogy*, 80: 639-43; *OSGOOD, W.H. (1932). Mammals of the Kelley-Roosevelts and Delacour Asiatic Expeditions. *Field Museum of Natural History Publication* 312. *Zoological Series* 18, No. 10 (Chicago).

227 NUTTALL, N. (1997). New deer found in a lost world. *The Times* (London), 23 August; ANON. (1997). Tiny deer discovered in Vietnamese jungle. *Guardian* (Manchester), 23 August; *GIAO, P.M. et al. (1998). Description of *Muntiacus truongsonensis*, a new species of muntjac (Artiodactyla: Muntiacidae) from Central Vietnam, and implications for conservation. *Animal Conservation*, 1: 61-8.

228 TIMMINS, R. (1999). Personal communication, 20 November.

229 RABINOWITZ, A. & KHAING, S.T. (1998). Status of selected mammal species in North Myanmar. *Oryx*, 32 (July): 201-8; O'MAHONY, J. (1999). Feast your eyes on deer little thing. *New York Post* (New York Post), 12 July; *AMATO, G., EGAN, M.G., & RABINOWITZ, A. (1999). A new species of muntjac, *Muntiacus putaoensis* (Artiodactyla: Cervidae) from northern Myanmar. *Animal Conservation*, 2: 1-7.

229a BAGLA, P. (2003). Scientists say smallest deer hidden in Arunachal forests. *Indian Express*, 19 February.

230 VENTURA, E. & HUNG, D. (1998). Small wonder. Fourth new species of large mammal discovered in Vietnam. *BBC Wildlife*, 16 (April).

231 HEUDE, P.-M. (1892). *Mémoires d'Histoire Naturelle de l'Empire Chinois*, Vols. 2, 4. Musée du Zikawel (Shanghai).

232 BROWNE, M.W. (1995). In Indochina, tantalizing traces of an elusive pig. *New York Times* (New York), 30 May.

233 GROVES, C.P., SCHALLER, G.B., et al. (1997). Rediscovery of the wild pig *Sus bucculentus*. *Nature*, 386 (27 March): 335.

233a GROVES, C.P. (2010). Personal communication, 17 March 2010.

234 HARPER, F. (1945). *Extinct and Vanishing Mammals of the Old World*. American Committee For International Wild Life Protection (New York).

235 GILES, F.H. (1937). The riddle of *Cervus schomburgki*. *Journal of the Siam Society, Natural History Supplement*, 11(1): 1-34.

236 SCHROERING, G.B. (1995). Swamp deer resurfaces. *Wildlife Conservation*, 98 (December): 22.

237 RICHARDSON, D. (1995). Worldwide work. *Lifewatch*, (spring 1995): 10; RICHARDSON, D. (1996). Personal communication, 30 July.

238 SHUKER, K.P.N. (1997). *From Flying Toads to Snakes with Wings*. Llewellyn Publications (St. Paul, Minnesota).

239 LINDEN, E. (1994). Ancient creatures in a lost world. *Time*, 20 June; SHUKER, K.P.N. (1995). Vietnam—why scientists are stunned. *Wild About Animals*, (March): 32-3; GROVES, C. (1999). Personal communication, 15 December.

240 ANON. (1994). Asian treasure trove. *Wild About Animals*, (September): 10; ADLER, H.J. (1995). Antelope exposé. *BBC Wildlife*, 13 (January): 10; *PETER, P. & FEILER, A. (1994). Horner von einer unbekannten Bovidenart aus Vietnam (Mammalia: Ruminantia). *Faun. Abhandlungen*, 19: 247-53; GROVES, C. (1995). Indo-China's cornucopia of mammals. *Canberra Times* (Canberra), 24 October.

241 DIOLI, M. (1995). A clarification about the morphology of the horns of the female kouprey. A new unknown bovid species from Cambodia. *Mammalia*, 59(4): 663-7; DIOLI, M. (1996). Personal communications, 28 June, 18 July; DIOLI, M. (1997). Notes on the morphology of the horns of a new artiodactyl mammal from Cambodia: *Pseudonovibos spiralis*. *Journal of Zoology*, 241: 527-31.

241a KIMCHHAY, H., et al. (1998). The distribution of tiger, leopard, elephant and wild cattle (gaur, banteng, buffalo, khting vor and kouprey) in Cambodia. *Interim report of the Cambodia National Tiger Survey*, (July), at: http://www.felidae.org/weiler%20prelim%20report.html/ accessed 14 April 2000.

242 WANG CHI & WANG SI YI (1607). SAN CAI TU HUI; MACDONALD, A.A. & YANG, L.N. (1997). Chinese sources suggest early knowledge of the 'unknown' ungulate (*Pseudonovibos spiralis*) from Vietnam and Cambodia. *Journal of Zoology*, 241: 523-6.

243 HAMMER, S.E. et al. (1999). Mitochondrial DNA sequence relationships of the newly described enigmatic Vietnamese bovid, *Pseudonovibos spiralis*. *Naturwissenschaften*, 86: 279-80; TIEDEMANN, R. (1999). Personal communication, 12 November.

244 ANON. (1995). Vietnam says another possible new species found. Reuter press report, 4 January; ANON. (1995). 'Goat' may be a new species. *Express and Star* (Wolverhampton), 4 January; TORODE, G. (1995). Unique species eaten before proof. *South China Morning Post*, 7 January.

244a ANON. (2000). Science declares rare species a bum steer. *The Age* (Melbourne), 18 December; HASSANIN, A. et al. (2001). Evidence from DNA that the mysterious 'linh duong' (*Pseudonovibos spiralis*) is not a new bovid. *C.R.A.S, Série III, Sciences de la Vie*, 324: 71-80; THOMAS, H., et al. (2001). The enigmatic new Indochinese bovid, *Pseudonovibos spiralis*: an extraordinary forgery. Ibid., 324: 81-6; TIMM, R.M. & BRANDT, J.H. (2001). *Pseudonovibos spiralis* (Artiodactyla: Bovidae): new information on this enigmatic Southeast Asian ox. *Journal of Zoology*, 253: 157-66; COPLEY, J. (2001). Load of old bull. *New Scientist*, 169 (17 February): 12; GROVES, C.P. (2001). Personal communication, 13 March.

244b KUZNETSOV, G.V. et al. (2001). The "linh duong" *Pseudonovibos spiralis* (Mammalia, Artiodactyla) is a new buffalo. *Naturwissenschaften*, 88: 123-5.

245 *HERSHKOVITZ, P. (1990). *Titis*, new world monkeys of the genus *Callicebus* (Cebidae, Platyrrhini): a preliminary taxonomic review. *Fieldiana Zoologia*, 55.; *VIVO, M. DE (1991). *Taxonomia de Callithrix Erxleben 1777 (Callithricidae, Primates)*. Fundação Biodiversitas (Belo Horizonte, Brazil).

246 *FERRARI, S.F. & LOPES, M.A. (1992). A new species of marmoset, genus *Callithrix* Erxleben 1777 (Callithricidae, Primates), from western Brazilian Amazonia. *Goeldiana Zoologia*, 12: 1-13.

247 FERRARI, S.F. & QUEIROZ, H.L. (1994). Two new Brazilian primates discovered, endangered. *Oryx*, 28 (January): 31-6.

248 QUEIROZ, H.L. (1992). A new species of capuchin monkey, genus *Cebus* Erxleben 1777 (Cebidae: Primates), from eastern Brazilian Amazonia. *Goeldiana Zoologia*, 15: 1-13.

249 *SILVA JR, J.S. & NORONHA, M. DE A. (1998). On a new species of bare-eared marmoset, genus *Callithrix* Erxleben 1777, from Central Amazonia, Brazil (Primates: Callithricidae). Ibid., 21: 1-28.

250 LINE, L. (1996). New branch of primate family tree. *New York Times* (New York), 18 June; NUTALL, N. (1996). Unknown monkey saved from poachers in Brazilian rainforest. *The Times* (London), 22 June; BARNETT, A. (1996). Primate from Brazil. *BBC Wildlife*, 14 (August): 26.

251 GOERING, L. (1999). Amazon primatologist shakes family tree for new monkeys. *Chicago Tribune* (Chicago), 11 July.

252 ASTOR, M. (1998). Primatologist finds new species. *Hong Kong Standard* (Hong Kong), 6 January.

253 ROOSMALEN, M. VAN (1999). Personal communication, 5 November; *ROOSMALEN, M. VAN et al. (2000). Two new species of marmoset, genus *Callithrix* Erxleben, 1777 (Callitrichidae, Primates), from the Tapajos/Madeira Interfluvium, south central Amazonia, Brazil. *Neotropical Primates*, 8: 2-18; ROOSMALEN, M. VAN (2000). Personal communication, 3 August.

254 PIETTE, C. (1997). New monkey business in hotbed of evolution. *Guardian* (Manchester), 11 August; ANON. (1997). New monkey species found in Brazil. *Washington Post* (Washington D.C.), 19 August; ANON. (1997). New monkey species discovered in Amazon. *Record* (Chacrensack, New Jersey), 19 August; *ROOSMALEN, M. VAN, MITTERMEIER, R.A., & FONSECA, G.A.B. (1998). A new and distinctive species of marmoset (Callitrichidae, Primates) from the Lower Rio Aripuanã, State of Amazonas, Central Brazilian Amazonia. *Goeldiana Zoologia*, 22 (27 April): 1-23; *ROOSMALEN, M. VAN, ROOSMALEN, T. VAN, & MITTERMEIER, R.A. (2002). A taxonomic review of the titi monkeys, genus *Callicebus* Thomas, 1903, with the description of two new species, *Callicebus bernhardi* and *Callicebus stephennashi*, from Brazilian Amazonia. *Neotropical Primates*, 10 (Supplement), (June): 1-53.

254a *VOSS, R.S. & SILVA, M.N.F. DA (2001). Revisionary notes on Neotropical porcupines (Rodentia: Erethizontidae). 2. A review of the *Coendou vestitus* group with descriptions of two new species from Amazonia. *American Museum Novitates*, No. 3351: 24-32.

255 NIGHTINGALE, N. (1992). Land of the big pigeon. *BBC Wildlife*, 10 (March): 44-53.

256 *FLANNERY, T.F., BOEADI, & SZALAY, A.L. (1995). A new tree-kangaroo (Dendrolagus: Marsupialia) from Irian Jaya, Indonesia, with notes on ethnography and the evolution of tree-kangaroos. *Mammalia*, 59(1): 65-84.

257 CHAMPKIN, J. (1994). Out of the trees, the kangaroo family's new branch. *Daily Mail* (London), 21 July; ANON. (1994). New species of kangaroo tempted down from trees. *Daily Telegraph* (London), 21 July; RODGERS, P. (1997). Kangaroo caught descending from the trees. *New Scientist*, (30 July): 8.

257a COOPER, M. (2009). Local explorer returns to Papua New Guinea: John Lane hopes to discover more new species. http://www.newsreview.com/chico/content?oid=1004993 4 June.

258 YOON, C.K. (1995). Woolly flying squirrel, long thought extinct, shows up in Pakistan. *New York Times* (New York), 14 March.

259 ANON. (1994). Potoroo 'back from the dead'. *Daily Telegraph* (London), 8 December; ANON. (1995). Jumping for joy. *Daily Mail* (London), 28 February; ANON. (1995). Kanga's little friend hops into the frame—after 126 years, student finds 'extinct' potoroo. Ibid., 1 March.

260 *FLANNERY, T.F. & BOEADI (1995). Systematic revision within the *Phalanger ornatus* complex, with description of a new species and subspecies. *Australian Mammalogy*, 18: 35-44; *FLANNERY, T.F. (1987). A new species of *Phalanger* (Phalangeridae: Marsupialia) from montane western Papua New Guinea. *Records of the Australian Museum*, 39: 183-93.

261 ANON. (1996). Extinct rat still rocking on. *Daily Telegraph* (Australia), 11 September; ANON. (1996). Rat rediscovered. *Hong Kong Standard* (Hong Kong), 11 September; WINKLER, T. (1997). Rangers round up rare rat. *Age* (Melbourne), 6 November.

262 ANON. (1996). Rare quoll back from extinction. *Courier Mail* (Brisbane), 11 May.

263 *REYES, J.C. et al. (1995). *Mesoplodon bahamondi* sp. n. (Cetacea, Ziphiidae), a new living beaked whale from the Juan Fernández Archipelago, Chile. *Bolitín del Museo Nacional de Historia Natural de Chile*, 45: 31-44; PAPASTAVROU, V. (1997). New South whale. *BBC Wildlife*, 15 (March): 20.

263a HELDEN, A. VAN, et al. (2002). Resurrection of *Mesoplodon traversii* (Gray, 1874), senior synonym of *M. bahamondi* Reyes, van Waerebeek, Cárdenas and Yáñez, 1995 (Cetacea: Ziphiidae). *Marine Mammal Science*, 18 (July): 609-21.

264 *SCHWARTZ, J.H. (1996). *Pseudopotto martini*: A new genus and species of extant lorisiform primate. *Anthropological Papers of the American Museum of Natural History*, No. 78: 1-14; GROVES, C. (1997-98). *Pseudopotto martini*: A new potto? *African Primates*, 31(1-2): 42-3; BEARDER, S.K. (1997-8). *Pseudopotto*: When is a potto not a potto? Ibid., 31(1-2): 43-4.

265 *HONESS, P.E. (1996). *Speciation Among Galagos (Primates, Galagidae) in Tanzanian Forests*. PhD thesis, Oxford Brookes University; BEARDER, S.K. (1999). Physical and social diversity among nocturnal primates: A new view based on long term research. *Primates*, 40(1): 267-82.

266 BURNHAM, O. (1997). Personal communications, March; BEARDER, S.K. (1997). Personal communications, 26 March, 1 April.

267 SHUKER, K.P.N. (1998). A supplement to Dr Bernard Heuvelmans' checklist of cryptozoological animals. *Fortean Studies*, 5: 208-29.

268 EVANS, C. (1996). Squeaking in dialect. *Daily Mail* (London), 6 March; PARK, K.J., ALTRINGHAM, J.D., & JONES, G. (1996). Assortative roosting in the two phonic types of *Pipistrellus pipistrellus* during the mating season. *Proceedings of the Royal Society of London, Series B*, 263 (22 November): 1495-9.

269 BARRATT, E.M. et al. (1997). DNA answers the call of pipistrelle bat species. *Nature*, 387 (8 May): 138-9; BARLOW, K.E., JONES, G., & BARRATT, E.M. (1997). Can skull morphology be used to predict ecological relationships between bat species? *Proceedings of the Royal Academy of London, Series B*, 264: 1695-1700; VAUGHAN, N., JONES, G., & HARRIS, S. (1997). Habitat use by bats (Chiroptera) assessed by means of a broad-band acoustic method. *Journal of Applied Ecology*, 34: 716-730; BARLOW, K.E. & JONES, G. (1999). Roosts, echolocation calls and wing morphology of two phonic types of *Pipistrellus pipistrellus*. *Zeitschrift für Säugetierkunde*, 64: 257-68.

270 JONES, G. & BARRATT, E.M. (1999). *Vespertilio pipistrellus* Schreber, 1774 and *V. pygmaeus* Leach, 1825 (currently *Pipistrellus pipistrellus* and *P. pygmaeus*; Mammalia, Chiroptera): proposed designation of neotypes. *Bulletin of Zoological Nomenclature*, 56 (September): 182-6; *LEACH, W.E. (1825). Description of the *Vespertilio pygmaeus*, a new species, recently discovered in Devonshire by Dr. Leach. *Zoological Journal*, 1(4): 559-61.

271 ANON. (1996). New flying fox. *Oryx*, 30 (October): 242.

272 ANON. (1996). A new brocket deer from Brazil. Ibid., 30 (October): 247; ANON. (1997). Brocket breakthrough. *BBC Wildlife*, 15 (May): 20.

273 SHUKER, K.P.N. (1999). Resurrecting the Barbary lion? *Quest*, 2 (September): 38-9.

274 ANON. (1996). Jah lion roars on. *Fortean Times*, 92 (November): 8; HARRISON, D. (1996). 'Extinct' Barbary lion rescued from circus. *Observer* (London), 22 December; INGHAM, J. (1996). Lion snatched from the jaws of extinction. *Sunday Express* (London), 22 December; DUNKLEY, R. (1997). Circus lion of ancient Rome rises from dead. *Sunday Telegraph* (London), 8 June.

275 NAISH, D. (1997). Lost lion renaissance. *Animals and Men*, No. 12 (January): 14-15.

276 WATSON-SMYTH, K. (1999). 'Barbary lions' to be mated in plan to resurrect lost species. *Independent* (London), 30 June.

276a BURGER, J. & HEMMER, H. (2006). Urgent call for further breeding of the relic zoo population of the critically endangered Barbary lion (*Panthera leo leo* Linnaeus 1758). *European Journal of Wildlife Research*, 52: 54-8

277 *JENKINS, P.D. & BARNETT, A.A. (1997). A new species of water mouse, of the genus *Chibchanomys* (Rodentia, Muridae, Sigmodontinae) from Ecuador. *Bulletin of the Natural History Museum of London (Zoology)*, 63(2): 123-8.

278 ASHTON, P. (1998). High-fishing. *BBC Wildlife*, 16 (August): 62; MORGAN, A. (1998). Found: the mouse that kills fish. *Sunday Telegraph* (London), 16 August.

279 ASHTON, P. (1994). Small wonder. *BBC Wildlife*, 12 (May): 10.

280 KERIN, J. (1998). 'Extinct' mouse in comeback. *Australian*, 5 January.

281 *FLANNERY, T.F. & GROVES, C.P. (1998). A revision of the genus *Zaglossus* (Monotremata, Tachyglossidae), with description of new species and subspecies. *Mammalia*, 62: 367-96.

281a ANON. (2007). New hope over 'extinct' echidna. *BBC News* (London) [http://news.bbc.co.uk/1/hi/sci/tech/6897977.stm], 15 June.

282 BILLE, M.A. (1999). Discovery: a new coati. *Exotic Zoology*, 6 (October-December): 1.

283 *EMMONS, L.H. (1999). A new genus and species of abrocomid rodent from Peru (Rodentia: Abrocomidae). *American Museum Novitates*, No. 3279; HARRISON, D.

(2000). New mammal discovered in South America. *Sunday Telegraph* (London), 28 February.
284 *KOBAYASHI, S. & LANGGUTH, A. (1999). A new species of titi monkey, *Callicebus* Thomas, from northeastern Brazil... *Revista Brasileira de Zoologia*, 16: 531-51.
285 *SOKOLOV, V.E., ROZHNOV, V.V., & PHAM TRONG ANH. (1997). New species of viverrids of the genus *Viverra* (Mammalia, Carnivora) from Vietnam. *Russian Journal of Zoology*, 1(2): 204-7 [translated from *Zoologicheskii Zhurnal*, 76(5): 585-9 (1997)]; ROZHNOV, V.V. & PHAM TRONG ANH. (1999). A note on the Tainguen civet—a new species of viverrid from Vietnam (*Viverra tainguensis* Sokolov, Rozhnov & Pham Trong Anh, 1997). *Small Carnivore Conservation*, No. 20 (April): 11-14; MOURANT, A. & NEALE, G. (1999). Rarest civet pictured for first time. *Sunday Telegraph* (London), 18 April; ANON. (1998). New civet record in Vietnam. *Oryx*, 33 (July).
286 NHAT, P. (1995). [Report re two strange civets from northwestern Vietnam's Lao Cai province—in Vietnamese.] *Lam Nghiep*, 2: 22; WEITZEL, V. (1999). Personal communication, 7 November.
287 SHUKER, K.P.N. (1997). A surfeit of civets? *Fortean Times*, 102 (September): 17.
287a *GROVES, C.P. et al. (2009). The taxonomy of the endemic golden palm civet of Sri Lanka. *Zoological Journal of the Linnean Society*, 155 (January): 238-51.
288 SURRIDGE, A.K. et al. (1999). Striped rabbits in Southeast Asia. *Nature*, 400 (19 August): 726; CONNOR, S. (1999). Striped rabbit found in Laos. *Independent* (London), 19 August; NUTTALL, N. (1999). New species of rabbit found. *The Times* (London), 19 August; COLEMAN, L. (1999). Into the rabbit hole. *Fortean Times*, 128 (November): 48; *AVERIANOV, A.O., ABRAMOV, A.V., & TIKHONOV, A.N. (2000). A new species of *Nesolagus* (Lagomorpha, Leporidae) from Vietnam with osteological description. *Contributions from the Zoological Institute, St. Petersburg*, 3 :1-24.

Section 2: The Birds
1 *OGILVIE-GRANT, W.R. (1896). On a new species of bird from the island of Samar. *B.B.O.C.*, 6: 16-17.
2 *HARTERT, E. (1901). On new birds from the Solomon Islands. Ibid., 12: 24-5.
3 BURTON, J.A. (Ed.) (1992). *Owls of the World* (3rd Edit.). Eurobook/Peter Lowe (London).
4 *ROTHSCHILD, W. (1903). On a new species of *Chalcurus*. *B.B.O.C.*, 13: 41-2.
5 DELACOUR, J. (1977). *The Pheasants of the World* (2nd Edit.). Spur Publications/World Pheasant Association (Hindhead).
6 JACOBSON, E. (1937). The alovot, a bird presumably living in the island of Simalur (Sumatra). *Temminckia*, 2: 159-60; SHUKER, K.P.N. (1990). A selection of mystery birds. *Avicultural Magazine*, 96 (spring): 30-40.
7 *ROTHSCHILD, W. (1903). Description of a new rail from Wake Island. *B.B.O.C.*, 13: 78.
8 GREENWAY, J.C. (1967). *Extinct and Vanishing Birds of the World* (2nd Edit.). Dover (New York); FULLER, E. (1987). *Extinct Birds*. Viking/Rainbird (London).
9 *OGILVIE-GRANT, W.R. (1906). On new species of birds from Formosa. *B.B.O.C.*, 16: 118-23.
10 FISHER, J., SIMON, N., & VINCENT, J. (1969). *The Red Book: Wildlife in Danger*. O.c.
11 *ROTHSCHILD, W. (1909). Description of a new bird from Africa. *Ibis*: 690-1.
12 COLLAR, N.J. & STUART, S.N. (1985). *Threatened Birds of Africa and Related Islands. The ICBP/IUCN Bird Red Data Book, Part 1* (3rd Edit.). ICBP & IUCN (London).
12a Sapayoa, in: http://en.wikipedia.org/wiki/Sapayoa accessed 23 October 2010.
13 *STRESEMANN, E. (1912). *Leucopsar*, gen. n. (Sturnidae). *B.B.O.C.*, 31: 4.
14 KING, W.B. (1981). *Endangered Birds of the World: The ICBP Bird Red Data Book*. Smithsonian Institution Press/ICBP (Washington); Mountfort, G. (1988). *Rare Birds of the World*. Collins (London).
15 COCKER, M. (1990). SOS! Bali starling campaign. *World*, No. 38 (June): 8.
16 *KURODA, N. (1917). On one new genus and three new species of birds from Corea and Tsushima. *Tori*, No. 5 (Supplement): 1-6.
17 MADGE, S. & BURN, H. (1988). *Wildfowl*. Christopher Helm (London).
18 *LOWE, P. (1923). *Atlantisia*, gen. nov. *B.B.O.C.*, 43: 174-6; FRASER, M.W., DEAN, W.R.J., & BEST, I.C. (1992). Observations on the Inaccessible Island rail *Atlantisia rogersi*: the world's smallest flightless bird. Ibid., 112: 12-22.
18a ANON. (1996). 'Extinct pheasants' are rediscovered. *The Times* (London), 5 September.
19 *DELACOUR, J. & JABOUILLE, P. (1924). New races of *Tropicoperdix*, *Hierophasis* ... *Aethopyga*. *B.B.O.C.*, 45: 28-35.
19a HENNACHE, A., et al. (2003). Hybrid origin of the imperial pheasant *Lophura imperialis* (Delacour and Jabouille, 1924) demonstrated by morphology, hybrid experiments, and DNA analyses. *Biological Journal of the Linnean Society*, 80 (No. 4; December): 573-600.
20 *CHAPIN, J. (1929). A new bowerbird of the genus *Xanthomelus*. *American Museum Novitates*, No. 367: 1-3.
21 GILLIARD, E.T. (1969). *Birds of Paradise and Bower Birds*. Weidenfeld & Nicolson (London).
22 *LÖNNBERG, E. (1931). A remarkable gull from the Gobi Desert. *Arkiv foer Zoologi*, 23B: 1-5; KITSON, A. (1980). *Larus relictus*—a review. *B.B.O.C.*, 100: 178-84.
23 *GRISCOM, L. (1929). Studies of the Dwight Collection of Guatemala birds. I. *American Museum Novitates*, No. 379: 1-13.
24 LABASTILLE, A. (1991). *Mama Poc. The Account of a Species Extinction*. Norton (New York).
25 *BOULTON, R. (1932). A new species of tree partridge from Szechuan, China. *P.B.S.W.*, 45: 235-6; KING, B. & LI GUIYUAN. (1988). China's most endangered galliform. *Oryx*, 22 (October): 216-17.
26 HEUVELMANS, B. (1958). *On the Track of Unknown Animals*. O.c.; Wendt, H. (1959). *Out of Noah's Ark*. O.c.
27 CHAPIN, J.P. (1938). The Congo peacock. In: *Compte-Rendu du IX Congrès Ornithologique International* (Rouen). pp. 101-9.

28 *Chapin, J.P. (1936). A new peacock-like bird from the Belgian Congo. *R.Z.B.A.*, 29 (20 November): 1-6.

29 Austin, O.L. & Singer, J. (1961). *Birds of the World*. Hamlyn (London).

30 *Moltoni, E. (1938). Zavattariornis stresemanni novum genus et nova species Corvidarum. *Ornithologische Monatsberichte*, 46: 80-3.

31 *Benson, C.W. (1942). A new species and ten new races from southern Abyssinia. *B.B.O.C.*, 63: 8-19; Benson, C.W. (1946). Notes on the birds of southern Abyssinia. *Ibis*, 88: 444-61.

32 Ripley, S.D. (1955). Anatomical notes on *Zavattariornis*. Ibid., 97: 142-5; Goodwin, D. (1976). *Crows of the World*. British Museum–Natural History (London).

33 Lowe, P.R. (1949). On the position of the genus *Zavattariornis*. *Ibis*, 91: 102-4.

34 Hides, J.G. (1936). *Papuan Wonderland*. Blackie & Son (London).

35 *Stonor, C.R. (1939). A new species of bird of paradise of the genus *Astrapia*. *B.B.O.C.*, 59: 57-61.

36 *Foerster, F. & Rothschild, W. (1906). *Two New Birds of Paradise*. (Tring).

37 *Rothschild, W. & Hartert, E. (1911). Preliminary descriptions of some new birds from Central New Guinea. *Novitates Zoologiae*, 18: 159-60.

38 *Rand, A.L. (1940). Results of the Archbold Expeditions. No. 25. *American Museum Novitates*, No. 1072: 1-14.

39 *Salvadori, T. (1896). Uccelli raccolti da Don Eugenio Dei Principi Ruspoli duvante l'ultimo suo viaggio nelle regioni dei Somali e dei Galla. *Annali del Museo Curico di Storia Naturale di Genova*, 16: 43-6.

40 Murphy, R.C. & Mowbray, L.S. (1951). New light on the cahow, *Pterodroma cahow*. *Auk*, 68: 266-80.

41 *Murphy, R.C. (1949). A new species of petrel from the Pacific. *Festschrift von Erwin Stresemann* (for 1949): 89-91.

42 Mann, R. (1990). Petrel explosion in South Pacific. *BBC Wildlife*, 8 (June): 360.

43 *Owen, R. (1848). [Description of moho.] *Transactions of the Zoological Society of London*, 3: 347, 366.

44 Burton, M. (1948). Unseen for fifty years and now rediscovered in New Zealand: the 'extinct' takahe... *Illustrated London News*, 213 (11 December): 658.

45 *Meyer, A.B. (1883). [Description of takahe.] *Abbildungen von Vogel-Skeletten*, 4-5: 34-7; Trewick, S.A. (1996). Morphology and evolution of two takahe flightless rails of New Zealand. *Journal of Zoology*, 228 (February): 221-37.

46 Phillipps, W.J. (1959). The last (?) occurrence of *Notornis* in the North Island. *Notornis*, 8 (April): 93-4.

47 Orbell, G.B. (1949). In search of the "extinct" takahe. *Illustrated London News*, 214 (1 January): 18-19; Falla, R.A. (1949). *Notornis* rediscovered. *Emu*, 48: 316-22.

48 *Schouteden, H. (1952). Un strigidé nouveau d'Afrique noire: Phodilus prigoginei nov. sp. *R.Z.B.A.*, 46: 423-8; Bille, M. (1996). Rediscoveries: the Congo bay owl. *Exotic Zoology*, 3 (November-December): 1.

49 *Sassi, M. (1914). Einige neue Formen der innerafrikanischen Ornis aus der Kollektic Grauer. *Anzeiger–Akademie der Wissenschaften in Wien* (for 1914): 308-12; *Chapin, J.P. (1932). Fourteen new birds from tropical Africa. *American Museum Novitates*, No. 570: 1-18; *Prigogine, A. (1960). Une nouvelle martinet du Congo. *R.Z.B.A.*, 62: 103-5; *Prigogine, A. (1983). Un nouveau *Glaucidium* de l'Afrique Centrale (Aves, Strigidae). Ibid., 97: 886-95.

50 Ogilvie, M. & Ogilvie, C. (1986). *Flamingos*. Alan Sutton (Gloucester).

51 Walters, M. (1980). *The Complete Birds of the World*. David & Charles (Newton Abbot).

52 Watson, J. (1980). The case of the vanishing owl. *Wildlife*, 22 (April): 38-9.

53 *Benson, C.W. (1960). The birds of the Comoro Islands: Results of the British Ornithologists' Union Centenary Expedition 1958. *Ibis*, 103b (1 March): 5-106.

54 Scott, K. (1992). Sound out at last. *BBC Wildlife*, 10 (October): 12.

55 Davey, K. (1983). *Our Arid Environment*. O.c.

56 Condon, H.T. (1962). [Rediscovery of the Eyrean grasswren.] *Emu*, 62.

57 Hill, R. (1967). *Australian Birds*. Thomas Nelson (London).

58 Anon. (1969). Not extinct after all. *Oryx*, 10 (December): 174.

59 Serventy, D.L. (1962). Die Wiederentdeckung von *Atrichornis clamosus* (Gould) in Westaustralien. *Journal für Ornithologie*, 103: 213-14; Serventy, V. (1975). The noisy scrubbird calls again. In: *Our Magnificent Wildlife*. Reader's Digest (London). pp. 250-1.

60 Moreau, R.E. (1964). The re-discovery of an African owl *Bubo vosseleri*. *B.B.O.C.*, 84: 47-52.

61 *Ripley, S.D. (1966). A notable owlet from Kenya. *Ibis*, 108: 136-7.

62 *Hartert, E. (1907). On a new species of *Larvivora* from the Tsin-Ling Mountains, N. China. *B.B.O.C.*, 19: 50.

63 McClure, H.E. (1963). Is this one of the rarest birds on the world? *Malayan Nature Journal*, 17: 1857; McClure, H.E. (1964). A thrush rediscovered. *Animals*, 4 (2 June): 40.

64 Collar, N.J. & Andrew, P. (1988). *Birds to Watch: The ICBP World Checklist of Threatened Birds*. ICBP (Cambridge).

65 Robson, C. (1989). Pheasants in Vietnam. *Garrulax*, 3: 2-3; Eames, J.C., Robson, C., & Wolstencroft, J.A. (1989). Pheasant surveys in Vietnam. *World Pheasant Association News*, No. 23 (February): 18-22; Anon. (1990). Pheasant rediscoveries. *Oryx*, 24 (October): 194; Anon. (1990). Making a meal of extinction. *New Scientist*, 128 (8 December): 18; Pringle, A. (1999). Personal communication, October.

66 Barloy, J-J. & Civet, P. (1980). *Fabuleux Oiseaux*. Robert Laffont (Paris).

67 *Wetmore, A. (1964). A revision of the American vultures of the genus *Cathartes*. *Smithsonian Miscellaneous Collections*, 146: 1-18; Mayr, E. (1971). New species of birds described from 1956 to 1965. *Journal für Ornithologie*, 112: 302-16.

68 Falla, R.A. (1967). An Auckland Island rail. *Notornis*, 14 (September): 107-113; Whitten, A. (1972). Not quite extinct. *Animals*, 14 (June): 254; Anon. Auckland Islands rail *Lewinia muelleri*. http://www.birdlife.org/datazone/speciesfactsheet.php?id=2875 Accessed 22 November 2010.

69 LOCKLEY, R. (1984). Poor old kakapo. *BBC Wildlife*, 2 (January): 8-12; HELTON, D. (1989). May the kakapo for ever boom. Ibid., 7 (October): 687.

70 *BENSON, C.W. & PENNY, M.J. (1968). A new species of warbler from the Aldabra Atoll. *B.B.O.C.*, 88: 102-8; Aldabran warbler. Wikipedia entry, accessed 22 November 2010.

70a *THONGLONGYA, K. (1968). A new martin of the genus *Pseudochelidon* from Thailand. *Thai National Scientific Papers, Fauna Series*, No. 1; White-eyed river martin, BirdLife International fact sheet, http://www.birdlife.org/datazone/speciesfactsheet.php?id=7077 accessed 13 August 2011.

71 *LOWERY, G.H. & TALLMAN, D.A. (1976). A new genus and species of nine-primaried oscine of uncertain affinities from Peru. *Auk*, 93 (July): 415-28.

72 *CASEY, T.L.C. & JACOBI, J.D. (1974). A new genus and species of bird from the island of Maui, Hawaii (Passeriformes: Drepanididae). *B.P. Bishop Museum Occasional Papers*, 24: 216-226; CASEY, T.L.C. (1975). Po'o-uli. Hawaii's newly discovered honeycreeper. *Wildlife*, 17 (June): 272-3; ANON. (2000). A Hawaiian bird on the edge of extinction. *Oryx*, 34 (January): 17; Po'o-uli. Wikipedia entry, accessed 22 November 2010.

72a *RUMBOLL, M.A.E. (1974). Una nueva especie de macá (Podicipedidae). *Communicaciones del Museo Argentino de Ciencias Naturales Bernardino Rivadavia, Zoologia*, 4(5): 33-5.

73 *VIELLARD, J. (1976). Un nouveau témoin relictuel de la spéciation dans la zone mediterraneanne: *Sitta ledanti* (Aves: Sittidae). *C.R.A.S.*, 283D: 1193-5; BELLATRECHE, M. & CHALABI, B. (1990). Données nouvelles sur l'aire de distribution de la sittelle Kabyle *Sitta ledanti*. *Alauda*, 58: 95-7.

73a COUZENS, D. (2010). *Atlas of Rare Birds*. New Holland Publishers (London).

74 *O'NEILL, J.P. & GRAVES, G.R. (1977). A new genus and species of owl (Aves: Strigidae) from Peru. *Auk*, 94: 409-16; ANON. (1977). New owl in the Andes. *New Scientist*, 76 (3 November): 284.

75 *FLEMING, J.H. (1935). A new genus and species of flightless duck from Campbell Island. *Occasional Papers of the Royal Ontario Museum*, No. 1: 1-3; ANON. (1977). Flightless teal refound. *Oryx*, 14 (January): 22; Campbell Island teal. Wikipedia entry, accessed 4 December 2010.

76 ANON. (1978). Rediscovered—after a century of 'extinction'. *Wildlife*, 20 (April): 151; MACEDO-RUIZ, H. DE (1979). 'Extinct' bird found in Peru. *Oryx*, 15 (January): 33-7.

77 BOURNE, W.R.P. (1964). The relationship between the magenta petrel and the Chatham Island taiko. *Notornis*, 11: 139-44; ANON. (1978). Ornithologists identify new species of bird. *The Times* (London), 9 February.

78 SICK, H. (1979). Die Herkunft von Lear's Ara (*Anodorhynchus leari*) entdeckt! *Gefiederte Welt*, 103: 161-2; SICK, H. & TEIXEIRA, D.M. (1980). Discovery of the home of the indigo macaw in Brazil with notes on field identification of 'blue macaws'. *American Birds*, 34: 118-19, 122; LOW, R. (1984). *Endangered Parrots*. Blandford Press (Poole); ANON. (1995). New Lear's. *Wild About Animals*, 13 (December): 63.

79 JUNIPER, A. (1990). A very singular bird. *BBC Wildlife*, 8 (October): 674-5; ANON. (1995). Lone macaw gets a mate. *Oryx*, 29 (October): 240; JUNIPER, T. (2002). *Spix's Macaw. The Race to Save the World's Rarest Bird*. Fourth Estate (London).

80 *YAMASHINA, Y. & MANQ, T. (1981). A new species of rail from Okinawa island. *Journal of the Yamashina Institute of Ornithology*, 13: 1-6.

81 STOKES, T. (1979). On the possible existence of the New Caledonian wood rail *Tricholimnas lafresnayanus*. *B.B.O.C.*, 99: 47-54; STOKES, T. (1980). Notes on the landbirds of New Caledonia. *Emu*, 80: 81-6.

82 *HUMPHREY, P.S. & THOMPSON, M.C. (1981). A new species of steamer duck (*Tachyeres*) from Argentina. *University of Kansas Museum of Natural History, Occasional Papers*, 95: 1-12.

83 ANON. (1981). Bird feared as extinct doing well in New Guinea. *Patriot-News* (Harrisburg–Pennsylvania), 11 November; DIAMOND, J.M. (1982). Rediscovery of the yellow-fronted gardener bowerbird. *Science*, 216 (23 April): 431-4.

83a SHUKER, K.P.N. (2006). A new Eden in New Guinea. *Fortean Times*, No. 209 (May): 38-40.

84 *WESKE, J.S. & TERBORGH, J.W. (1981). *Otus marshalli*, a new species of screech-owl from Peru. *Auk*, 98: 1-7.

85 ANON. (1982). Owl rediscovered. *Wildlife*, 24 (January): 5; Laughing owl. Wikipedia entry, accessed 6 December 2010.

86 MARTIN, B.P. (1987). *World Birds*. Guinness Books (London).

87 HENSHAW, H.W. (1902). *Birds of the Hawaiian Islands*. Thrum (Honolulu).

88 PYLE, R.L. & RALPH, C.J. (1982). The autumn migration. August 1-November 30, 1981. Hawaiian islands region. *American Birds*, 36: 221-3; PRATT, J.D. et al. (1987). *A Field Guide to the Birds of Hawaii and the Tropical Pacific*. Princeton University Press (Princeton).

89 *ROUX, J-P. et al. (1983). Un nouvel albatross *Diomedea amsterdamensis* n. sp. découvert sur l'Île Amsterdam (37°50´S, 77°35´E). *Oiseau*, 53: 1-11.

90 WATLING, D. & LEWANAVUNUA, R.F. (1985). A note to record the continuing survival of the Fiji (MacGillivray's) petrel *Pseudobulweria macgillivrayi*. *Ibis*, 127: 230-3; GREENWELL, J.R. (1985). Extinct landing. *ISC Newsletter*, 4 (summer): 8; Fiji petrel. http://www.birdlife.org/datazone/speciesfactsheet.php?id=3881 accessed 6 December 2010.

91 *FRY, C. & SMITH, D. (1985). A new swallow from the Red Sea. *Ibis*, 127: 1-6; Red Sea swallow. http://www.birdlife.org/datazone/speciesfactsheet.php?id=7138 accessed 6 December 2010.

92 *RIDGELY, R.S. & ROBBINS, M.B. (1988). *Pyrrhura orcesi*, a new parakeet from southwestern Ecuador... *Wilson Bulletin*, 100 (June): 173-82.

93 *O'NEILL, J.P., MUNN, C.A., & FRANKE J., I. (1991). *Nannopsittaca dachilleae*, a new species of parrotlet from eastern Peru. *Auk*, 108: 225-9; STAP, D. (1990). *A Parrot Without a Name: The Search for the Last Unknown Birds on Earth*. Alfred A. Knopf (New York).

94 BHUSHAN, B. (1986). Rediscovery of the Jerdon's or double-banded courser *Cursorius bitorquatus* (Blyth).

Journal of the Bombay Natural History Society, 83: 1-14.

95 DENNIS, J.V. (1968). Return of the ivory-bill. *Animals*, 10 (March): 492-7; ANON. (1969). An ivory-billed woodpecker. *Pursuit*, 2 (July): 49.

95a HODGES, S. (2000). Could a woodpecker widely believed extinct still exist? *Mobile Express* (Alabama), 10 July; GORMAN, J. (2002). Searchers say rare woodpecker was possibly heard, if not seen. *New York Times* (New York), 21 February; Associated Press (2002). Study: sound was not rare woodpecker, but distant gunfire. http://ap.tbo.com/ap/breaking/MGA6JNEJ92D.html 9 June [no longer online].

95b FITZPATRICK, J.W., et al. (2005). Ivory-billed woodpecker (*Campephilus principalis*) persists in continental North America. *Science*, 308 (3 June): 1460-2; MOSS, S. & FOERSTEL, K. (2005). Long thought extinct, ivory-billed woodpecker rediscovered in Big Woods of Arkansas. http://www.eurekalert.org/pub_releases/2005-04/potn-lte042705.php 28 April; DALTON, R. (2010). Still looking for that woodpecker. *Nature*, 463 (10 February): 718-19.

95c HILL, G.E., et al. (2006). Evidence suggesting that ivory-billed woodpeckers (*Campephilus principalis*) exist in Florida. *Avian Conservation and Ecology*, 1(3): 2; COLLINS, M.D. (2011). Putative audio recordings of the ivory-billed woodpecker (*Campephilus principalis*). *Journal of the Acoustical Society of America*, 129 (March): 1626-30.

96 GREENWELL, J.R. (1986). Ivory-billed woodpecker found alive in Cuba. *ISC Newsletter*, 5 (summer): 3-5; KIRBY, T. (1986). The tick of a lifetime. *BBC Wildlife*, 4 (July): 342; LAMMERTINK, M. (1995). No more hope for the ivory-billed woodpecker. *Cotinga*, 3 (February).

97 GOULD, A.B. (1986). Hot tip to a pitta. Ibid., 4 (October): 502-3; COLLAR, N.J., ROUND, P.D., & WELLS, D.R. (1986). The past and future of Gurney's pitta *Pitta gurneyi*. *Forktail*, 1: 29-51.

97a CORCORAN, P. (2003). Rare forest bird rediscovered after 89 years. http://ens-news.com/jun2003/2003-06-06-01.asp 6 June [no longer online].

98 ANON. (1989). [Rediscovery of Schneider's pitta.] *Oryx*, 23 (July): 167.

99 ANON. (1987). Weaver rediscovery. Ibid., 21 (April): 116.

100 ANON. (1987). Rare woodpecker sighting. Ibid., 21 (October): 256; ANON. (1988). Rediscovery of antwren. Ibid., 22 (April): 119.

101 WOOD, G.L. (1982). *The Guinness Book of Animal Facts and Feats* (3rd Edit.). O.c.; RAXWORTHY, C.J. & COLSTON, P.R. (1992). Conclusive evidence for the continuing existence of the Madagascar serpent-eagle *Eutriorchus astur*. *B.B.O.C.*, 112: 108-11.

102 MCRAE, J. (1989). Flying visits show up birds and scientists. *BBC Wildlife*, 7 (January): 12-13; ANON. (1994). Madagascar serpent eagle captured and released. *Oryx*, 28 (October): 229; MORLAND, H.S. (1994). Eagle photo a first. *Wildlife Conservation*, (November-December): 16.

103 MCKEAN, J. (1985). Birds of the Keep River National Park (Northern Territory), including the night parrot *Geopsittacus occidentalis*. *Australian Bird Watcher*, 11: 114-30; ANON. (1991). [Rediscovery of night parrot.] *New Scientist*, 129 (19 January): 64; HUXLEY, J. (2007). Twitchers cry foul in case of the deceased parrot. *Sydney Morning Herald* (Sydney), 23 June; Night parrot. http://www.birdlife.org/datazone/speciesfactsheet.php?id=1491 accessed 8 December 2010.

104 BURNHAM, O. (1991). Personal communication, 20 August; COLLAR, N.J. (1992). Personal communication, 12 March.

105 ANON. (1991). Cocha antshrike refound. *Oryx*, 25 (July): 133; ANON. (1992). Vanuatu still has its starling. Ibid., 26 (April): 80; ANON. (1992). Rare pochard captured. Ibid., 26 (April): 73; ANON. (1993). Madagascar pochard. http://www.birdlife.org/datazone/speciesfactsheet.php?id=477 accessed 8 December 2010; ANON. (1993). Flowerpecker rediscovery. *Oryx*, 27 (July): 139.

106 ANON. (1990). Seabird capture fuels debate over unknown species. *Daily Telegraph* (London), 9 July; ANON. (1992). Tests uncover bird mystery. Ibid., 14 December.

107 ANON. (1993). Kiwi discovery. *Wild About Animals*, (August): 25; HEWSON, P. (1994). Personal communication, 5 May; NAISH, D. (1999). So how many kiwi species are there? *Animals and Men*, No. 20 (December): 27-30; *TENNYSON, A.J.D. et al. (2003). A new species of kiwi (Aves, Apterygiformes) from Okarito, New Zealand. *Records of the Auckland Museum*, 40: 55-64.

108 *DINESEN, L. et al. (1994). A new genus and species of perdicine bird (Phasianidae, Perdicini) from Tanzania: a relict with Indo-Malayan affinities. *Ibis*, 136: 2-11; ANON. (1994). A rare bird in the wrong place. *New Scientist*, 141 (12 February): 11.

108a *BOWIE, R.C.K. & FJELDSA, J. (2005): Genetic and morphological evidence for two species in the Udzungwa forest partridge. *Journal of East African Natural History*, 94(1): 191-201

109 *SAFFORD, R.J. et al. (1995). A new species of nightjar from Ethiopia. *Ibis*, 137: 301-7; HORNSBY, M. (1995). Wing spotted in torchlight hailed as new nightjar. *The Times* (London), 16 August; LEMOULT, C. (2009). A single wing starts quest for mystery bird. http://www.npr.org/templates/story/story.php?storyId=106749870 19 July.

110 BY, R.A. de (2000). Recently described bird species. http://www.itc.nl/~deby/SM/NewSpecies.html/ updated July 2000.

111 *LENCIONI-NETO, F. (1994). Une nouvelle espèce de Chordeiles (Aves, Caprimulgidae) de Bahia (Brésil). *Alauda*, 62: 241-5.

112 ANON. (1999). Franklin's night jar [sic] sighted. *Gulf News* (Gulf States), 2 May.

112a *SMITH, E.F.G., ARCTANDER, P. et al. (1991). A new species of shrike (Laniidae: Laniarius) from Somalia, verified by DNA sequence data from the only known individual. *Ibis*, 133: 227-35; LECROY, M. & VUILLEUMIER, F. (1992). Guidelines for the description of new species in ornithology. *B.B.O.C.*, 112A: 191-8; COLLAR, N.J. (1999). New species, high standards and the case of *Laniarius liberatus*. *Ibis*, 141 (July): 358-67; NGUEMBOCK, B. et al. (2008). Phylogeny of *Laniarius*: molecular data reveal *L. liberatus* synonymous with *L. erlangeri* and "plumage coloration" as unreliable morphological characters for defining

species and species groups. *Molecular Phylogenetics and Evolution*, 48: 396-407.
113 YOUNG, S. (1995). Back on the rails. *BBC Wildlife*, 13 (December): 24; ANON. (1996). Rail refound. *Oryx*, 30 (January): 15.
114 ANON. (1994). Bird rediscovered in proposed national park. Ibid., 28 (July): 157.
115 BUDEN, D.W. (1996). Rediscovery of the Pohnpei mountain starling (*Aplonis pelzelni*). *Auk*, 113: 229-30.
116 *SHIRIHAI, H., SINCLAIR, I., & COLSTON, P.R. (1995). A new species of *Puffinus* shearwater from the western Indian Ocean. *B.B.O.C.*, 115(2): 75-87.
117 *PACHECO, J.F., WHITNEY, B.M., & GONZAGA, L.A.P. (1996). A new genus and species of furnariid (Aves: Furnariidae) from the cocoa-growing region of southeastern Bahia, Brazil. *Wilson Bulletin*, 108: 397-433; ANON. (1996). Bird world turned upside down. *Express* (London), 16 November; Line, L. (1996). New bird found in Brazil owes its survival to chocolate. *New York Times* (New York), 19 November.
*118 BORNSCHEIN, M.R., REINERT, B.L., & TEIXERA, D.M. (1996). Um nova Formicariidae do sul do Brasil (Aves, Passeriformes). *Publicação Técnico-Cientifica do Instituto Iguaçu*, No. 1: 1-18; ANON. (1996). New Brazilian bird. *Oryx*, 30 (October): 247.
119 ANON. (1997). Scientists find 'mystery bird'. *Sun Journal* (Lewiston, Maine), 31 December; DUTTER, B. (1997). 'Extinct' owl may rescue reputation. *Daily Telegraph* (London), 31 December; DERBYSHIRE, D. (1997). Bird hoax man's owl was not a flight of fancy. *Daily Mail* (London), 31 December.
120 SHUKER, K.P.N. (1993). *The Lost Ark: New and Rediscovered Animals of the 20th Century*. HarperCollins (London).
121 ANON. (1994). Red owl discovered. *Oryx*, 28 (July): 155; ANON. (1996). A remarkable discovery. Conservation International press release.
122 *ROBBINS, M.B. & HOWELL, S.N.G. (1995). A new species of pygmy-owl (Strigidae: *Glaucidium*) from the eastern Andes. *Wilson Bulletin*, 107: 1-6.
123 *ROBBINS, M.B. & STILES, F.G. (1999). A new species of pygmy-owl (Strigidae: *Glaucidium*) from the Pacific slope of the northern Andes. *Auk*, 116: 305-15.
124 *LAMBERT, F.R. & RASMUSSEN, P.C. (1998). A new scops owl from Sangihe Island, Indonesia. *B.B.O.C.*, 118(4): 204-17.
125 *RASMUSSEN, P.C. (1998). A new scops-owl from Great Nicobar Island. Ibid., 118(3): 141-53.
126 *LAFONTAINE, R-M. & MOULAERT, N. (1998). Une nouvelle espèce de petit-duc (*Otus*, Aves) aux Comors: taxonomie et statut de conservation. *Journal of African Ornithology*, 112: 163-9.
127 NIXON, T. (1997). Dog team helps discover new bird species. *Southland Times* (New Zealand), 18 November; ANON. (1997). New bird species on Campbells. *Christchurch Press* (Christchurch), 18 November; MISKELLY, C. (2010). New Zealand's newest bird named after vessel that nearly destroyed it. Department of Conservation media release, 16 February.
128 PACHECO, J.F. (1998). Cherry-throated tanager re-discovered. *Cotinga*, 9 (spring): 41; VENTURINI, A.C. et al. (2005). A new locality and records of Cherry-throated Tanager *Nemosia rourei* in Espirito Santo, south-east Brazil, with fresh natural history data for the species. *Cotinga*, 24: 60-70.
129 *BORNSCHEIN, M.R., REINERT, B.L., & PICHORIM, M. (1998). Descriçao, ecologia e conservação de um novo *Scytalopus* (Rhinocryptidae) do sul do Brasil, com comentários sobre a morfologia da família. *Ararajuba*, 6: 3-36; ANON. (1998). New bird discovered in southern Brazil. http://cnn.com/EARTH/9804/25/brazil.newbird.ap/index.html/ 25 April.
130 ANON. (1998). New rail. *Oryx*, 32 (July): 185; *LAMBERT, F.R. (1998). A new species of *Gymnocrex* from the Talaud Islands, Indonesia. *Forktail*, 13: 1-6.
131 ANON. (1998). New species of robin is no redbreast. *Daily Telegraph* (London), 22 August 1998; ANON. (1998). Scientists discover new robin in Africa. *News Journal* (Lewiston, Maine), 22 August; *BERESFORD, P. & CRACRAFT, J. (1999). Speciation in African forest robins (*Stiphrornis*): species limits, phylogenetic relationships, and molecular biogeography. *American Museum Novitates*, No. 3270: 1-22
132 ANON. (1998). Lost and found. *New Scientist*, 159 (26 September): 21.
133 RASMUSSEN, P.C. & COLLAR, N.J. (1999). On the hybrid status of Rothschild's parakeet *Psittacula intermedia* (Aves, Psittacidae). *Bulletin of the Natural History Museum, London (Zoology)*, 65: 31-50.
134 *KRABBE, N. et al. (1999). A new species of antpitta (Formicariidae: *Grallaria*) from the southern Ecuadorian Andes. *Auk*, 116: 882-90; NUSSBAUM, P. (1998). Bird expert discovers new animal in Andes. *Philadelphia Inquirer* (Philadelphia), 11 June; VARADARAJAN, T. (1998). Winged treasure found in jungle. *The Times* (London), 12 June; AHUJA, A. (1998). New bird of the Andes. Ibid., 6 July.
135 *RASMUSSEN, P.C. (1999). A new species of hawk-owl *Ninox* from North Sulawesi, Indonesia. *Wilson Bulletin*, 111(4): 457-64.
136 SURESH, R., SINGH, K., & SINGH, P. (1999). Discovery of a new monal from Arunachal Pradesh. *Oriental Bird Club Bulletin*, 30: 35-8; SURESH KUMAR R. & SINGH, P. (2004). A new subspecies of Sclater's monal *Lophophorus sclateri* from western Arunachal Pradesh, India. *Bulletin of the British Ornithologists Club*, 124(1): 16-27.

Section 3: The Reptiles and Amphibians

1 *BOULENGER, G.A. (1900). A list of the batrachians and reptiles of the Gaboon (French Congo), with descriptions of new genera and species. *P.Z.S.L.*, (8 May): 433-56.
2 PI, J.S. (1970). The hairy frog. *Animals*, 13 (October): 282-3; DURRELL, G. (1954). *The Bafut Beagles*. Rupert Hart-Davis (London).
3 *MOCQUARD, F. (1900). Diagnoses d'espèces nouvelles ... de batraciens recueillis par M. Alluaud dans le sud de Madagascar. *Bulletin du Muséum d'Histoire Naturelle*: 345-8.
4 GRZIMEK, B. (Ed.) (1972-5). *Grzimek's Animal Life Encyclopedia* (13 vols). O.c.
5 *BOULENGER, G.A. (1903). Report on the batrachians and reptiles. *Fasciculi Malayensis Zoologiae*, 1: 131-76.

6 *SIEBENROCK, F. (1903). Ueber zwei seltene und eine neue Schildkröte des Berliner Museums. *Anzeiger der k-Akademie der Wissenschaften*, 40: 106-8.

7 *BOULENGER, G.A. (1906). Descriptions of new batrachians discovered by Mr. G.L. Bates in South Cameroon. *A.M.N.H.* (Series 7): 17: 317-23.

8 MATTHEWS, P. & MCWHIRTER, N. (Eds.) (1992). *The Guinness Book of Records* 1993. O.c.

9 PI, J.S. (1985). Contribution to the biology of the giant frog (*Conraua goliath*, Boulenger). *Amphibia-Reptilia*, 6: 143-53.

10 *ANDERSSON, L.C. (1908). A remarkable new gecko from South Africa... *Jahrbücher des Nassauischen Vereins für Naturkunde*, 61: 299-306.

11 *BARBOUR, T. (1911). New lizards ... from the Dutch East Indies, with notes on other species. *P.B.S.W.*, 24: 15-21.

12 HEUVELMANS, B. (1958). *On the Track of Unknown Animals*. O.c.

13 WENDT, H. (1959). *Out of Noah's Ark*. O.c.

14 *OUWENS, P.A. (1912). On a large *Varanus* species from the island of Komodo. *Bulletin du Jardin Botanique de Buitenzorg*, 2: 1-3.

15 ATTENBOROUGH, D. (1957). *Zoo Quest for a Dragon*. Lutterworth (London).

16 WOOD, G.L. (1982). *The Guinness Book of Animal Facts and Feats* (3rd Edit.). O.c.

17 BURDEN, W.D. (1928). *The Dragon Lizards of Komodo*. Putnam's (New York).

18 KERN, J.A. (1968). Dragon lizards of Komodo. *National Geographic Magazine*, 134 (December): 872-80.

19 DIAMOND, J.M. (1987). Did Komodo dragons evolve to eat pygmy elephants? *Nature*, 326 (30 April): 832; MITCHELL, P.B. (1987). Here be Komodo dragons. Ibid., 329 (10 September): 111.

20 AUFFENBERG, W. (1981). *The Behavioural Ecology of the Komodo Monitor*. University of Florida Press (Gainesville).

20a WATTS, P.C. et al. (2006). Parthenogenesis in Komodo dragons. *Nature*, 444 (21 December): 1021-2; FRY, B.G. et al. (2009). A central role for venom in predation by *Varanus komodoensis* (Komodo Dragon) and the extinct giant *Varanus* (*Megalania*) *priscus*. *Proceedings of the National Academy of Sciences*, 106 (2 June): 8964-74.

21 CROOK, I. & CROOK, G. (1972). New Zealand's rarest frog. *Animals*, 14 (April): 188-90.

22 *MCCULLOUGH, A.R. (1919). A new discoglossid frog from New Zealand. *Transactions of the New Zealand Institute*, 51: 447-9; *BELL, B.D.; et al. (1998). *Leiopelma pakeka*, n. sp. (Anura: Leiopelmatidae), a cryptic species of frog from Maud Island, New Zealand, and a reassessment of the conservation status of *L. hamiltoni* from Stephens Island. *Journal of the Royal Society of New Zealand*, 28: 39-54.

23 *TURBOTT, E.G. (1942). The distribution of the genus *Leiopelma* in New Zealand with a description of a new species. *Transactions and Proceedings of the Royal Society of New Zealand*, 71: 247-53.

24 *AMARAL, A. DO (1921). Contribuiçao para o conhecimento dos ofidios do Brazil: Parte i Quatro noves espécies de serpentes brasileires. *Anexos das Memórias do Instituto de Butantan, Secção de Ofiologia*, 1: 1-88; Ditmars, R.L. (1931). *Snakes of the World*. Macmillan (London).

25 MAY, J. & MARTEN, M. (1982). *The Book of Beasts*. Hamlyn (London).

26 HOGE, A.R. et al. (1960). Sexual abnormalities in *Bothrops insularis* (Amaral) 1921 Serpentes. *Memórias do Instituto de Butantan*, 29: 17-87; EVELEIGH, R. (1999). Viper isle. *BBC Wildlife*, 17 (November): 34-5, 37, 39-40; O'SHEA, M. (1999). Personal communication, 9 December.

27 OLIVER, J.A. (1958). The taipan, Australia's deadliest snake. *Animal Kingdom*, 61 (February): 23-6; *COVACEVICH, J. & WOMBEY, J. (1976). Recognition of *Parademansia microlepidotus* (McCoy) (Elapidae), a dangerous Australian snake. *Proceedings of the Royal Society of Queensland*, 87: 29; *DOUGHTY, P. et al. (2007). A new species of taipan (Elapidae: *Oxyuranus*) from central Australia. *Zootaxa*, No. 1422: 45-58.

28 *AHL, E. et al. (1930). Beiträge zur Lurch- und Kriechtierfauna Kwangsi's. *S.G.N.F.B.*: 310-32.

29 *SCHMIDT, K.P. (1928). A new crocodile from New Guinea. *Field Museum Publications, Chicago* (*Zoology Series 12*), 247: 177-81.

30 *SCHMIDT, K.P. (1935). A new crocodile from the Philippine islands. *Chicago Field Museum of Natural History* (*Zoological Series*), 20: 67-70.

31 GUGGISBERG, C.A.W. (1972). *Crocodiles*. David & Charles (Newton Abbot).

32 *LAFRENTZ, K. et al. (1930). Beiträge zur Herpetologie Mexikos. *Abhandlungen–Berlin Museum Naturwissenschaften*, 6: 91-161.

33 *DUNN, E.R. (1933). Amphibians and reptiles from El Valle de Anton, Panama. *Occasional Papers of the Boston Natural History Society*, 8: 65-79.

34 FISHER, J.; SIMON, N.; & VINCENT, J. (1969). *The Red Book: Wildlife in Danger*. O.c.

35 *MYERS, G.S. (1942). The black toad of Deep Springs Valley, Inyo County, California. *Occasional Papers of the Museum of Zoology, University of Michigan*, No. 460 (16 September): 1-13.

35a *DUNN, E.R. & CONANT, R. (1936). Notes on anacondas, with descriptions of two new species. *Proceedings of the Academy of Natural Sciences of Philadelphia*, 88: 503-6; PETZOLD, H-G. (1995). *Die Anakondas* (New Edit.). Spektrum (Germany); STRIMPLE, P.D. et al. (1997). On the status of the anaconda *Eunectes barbouri* Dunn and Conant. *Journal of Herpetology*, 31(4): 607-9; O'SHEA, M. (1999). Anaconda. *Pet Reptile*, (March): 24-8; *DIRKSEN, L. (2002). *Anakondas*. Natur und Tier Verlag (Münster).

36 *CARR, A.F. (1939). *Haideotriton wallacei*, a new subterranean salamander from Georgia. *Occasional Papers of the Boston Natural History Society*, 8: 333-6.

37 *MCCRADY, E. (1954). A new species of *Gyrinophilus* (Plethodontidae) from Tennessee caves. *Copeia* (for 1954): 200-6.

38 *MENDELSSOHN, H. & STEINITZ, H. (1943). A new frog from Palestine. Ibid., (for 1943): 231-3.

39 DAY, D. (1989). *The Encyclopedia of Vanished Species*. O.c.

40 *STUART, L.C. (1941). A new species of *Xenosaurus* from Guatemala. *P.B.S.W.*, 54: 47-8; *TAYLOR, E.H. (1949). A preliminary account of the herpetology of

the state of San Luis Potosi, Mexico. *Kansas University Science Bulletin*, 33: 169-215; *KING, W. & THOMPSON, F.G. (1968). A review of the American lizards of the genus *Xenosaurus* Peters. *Bulletin of the Florida State Museum (Biological Science)*, 12: 93-123; *SMITH, H.M. & IVERSON, J.B. (1993). A new species of knobscale lizard (Reptilia: Xenosauridae) from Mexico. *Bulletin of the Maryland Herpetological Society*, 29 (30 June): 51-66

41 *MYERS, G.S. & FUNKHAUSER, J.W. (1951). A new toad from southwestern Colombia. *Zoologica*, 36: 279-82; SHUKER, K.P.N. (2011). Giant toads and South American sasquatches? ShukerNature, http://karlshuker.blogspot.com/2011/01/giant-toads-and-south-american.html 30 January.

42 *SIEBENROCK, F. (1901). Beschreibung einer neuen Schildkrätengattung aus der Familie Chelydidae von Australien: *Pseudemydora*. *Anzeiger der k-Akademie der Wissenschaften*, No. 22; WILLIAMS, E.E. (1958). Rediscovery of the Australian chelid genus *Pseudemydora* Siebenrock (Chelidae, Testudines). *Breviora*, No. 84: 1-8.

43 DUGES, A. (1888). La tortuga polifemo. *La Naturaleza*, 1: 146-7; *LEGLER, J.M. (1959). A new tortoise, genus *Gopherus* from Northcentral Mexico. *University of Kansas Publications, Museum of Natural History*, 11: 335-43.

44 GROOMBRIDGE, B. & WRIGHT, L. (1982). *The IUCN Amphibia-Reptilia Red Data Book*. Part 1: Testudines. Crocodylia. Rhynchocephalia. IUCN (Gland–Switzerland).

45 *TYLER, M.J. (1963). An account of collections of frogs from Central New Guinea. *Records of the Australian Museum*, 26: 113-30.

46 *NEILL, W.T. (1964). A new species of salamander, genus *Amphiuma*, from Florida. *Herpetologica*, 20: 62-6.

46a *TAYLOR, E.H. (1968). *The Caecilians of the World*. University of Kansas Press (Lawrence); NUSSBAUM, R.A. & WILKINSON, M. (1995). A new genus of lungless tetrapod: a radically divergent caecilian (Amphibia: Gymnophiona). *Proceedings of the Royal Society, London, Series B*, 261 (22 September): 331-5; WILKINSON, M. et al. (1998). The largest lungless tetrapod: report on a second specimen of *Atretochoaa eiselti* (Amphibia: Gymnophiona: Typhlonectidae) from Brazil. *Journal of Natural History*, 32(4): 617-27.

46b *WAKE, M.H. & DONNELLY, M.A. (2009). A new lungless caecilian (Amphibia: Gymnophiona) from Guyana. *Proceedings of the Royal Society, London, Series B*, doi: 10.1098/rspb.2009.1662 18 November.

47 REYNOLDS, R.P. & MARLOW, R.W. (1982). Lonesome George, the Pinta Island tortoise: a case of limited alternatives. *Noticias de Galapagos*, No. 37: 14-17; PRITCHARD, P.C.H. (1984). Further thoughts on "Lonesome George". Ibid., No. 39: 20-3; WIGMORE, B. (1994). The tragedy of Lonesome George. *Today* (London), 10 May: 19-20; SULLOWAY, F.J. (2006). Is Lonesome George really lonesome? *Skeptic*, http://www.skeptic.com/eskeptic/06-07-28/ 28 July; NICHOLLS, H. (2006). *Lonesome George: The Life and Loves of a Conservation Icon*. Palgrave Macmillan (London).

48 *MYERS, C.W. et al. (1978). A dangerously toxic new frog (*Phyllobates*) used by Embera Indians of western Colombia with discussion of blowgun fabrication and dart poisoning. *Bulletin of the American Museum of Natural History*, 161: 309-65; Wikipedia entry, accessed 4 February 2011.

49 *LIEM, D.S. (1973). A new genus of frog of the family Leptodactylidae from SE. Queensland, Australia. *Memoirs of the Queensland Museum*, 16: 459-70; CORBEN, C.J., INGRAM, G.J., & TYLER, M.J. (1974). Gastric brooding: Unique form of parental care in an Australian frog. *Science*, 186: (6 December): 946-7; TYLER, M.J. (Ed.) (1983). *The Gastric Brooding Frog*. Croom Helm (Beckenham).

50 *MAHONEY, M., TYLER, M.J., & DAVIES, M. (1984). A new species of the genus *Rheobatrachus* ... from Queensland. *Transactions of the Royal Society of Australia*, 108: 155-62; MCDONALD, K.R. & TYLER, M.J. (1984). Evidence of gastric brooding in the Australian leptodactylid frog *Rheobatrachus vitellinus*. Ibid., 108: 226; CHERFAS, J. (1984). Ulcer studies rescued by reincarnated frog. *BBC Wildlife*, 2 (April): 172-3.

51 INGRAM, G. (1991). The earliest record of the extinct platypus frog. *Memoirs of the Queensland Museum*, 30: 454.

52 AUFFENBERG, W. (1979). A monitor lizard in the Philippines. *Oryx*, 15 (January): 38-46.

53 ANON. (1979). Legless lizard rediscovered. Ibid., 15 (April): 126; DAVEY, K. (1983). *Our Arid Environment*. O.c.

54 ANON. (1979). New banded iguana. *Oryx*, 15 (January): 24; *GIBBONS, J.R.H. (1981). The biogeography of *Brachylophus* (Iguanidae) including the description of a new species, *B. vitiensis*, from Fiji. *Journal of Herpetology*, 15: 255-73; ANON. (1981). Fijians protect the crested iguana. *World Wildlife News*, (summer): 26-7.

54a *KEOGH, J. et al. (2008)). Molecular and morphological analysis of the critically endangered Fijian iguanas reveals cryptic diversity and a complex biogeographic history. *Philosophical Transactions of the Royal Society, London, Series B, Biological Sciences*, 363 (No. 1508): 3413-26; COOPER, D. (2008). Hello, it's a new species of Pacific iguana. http://www.abc.net.au/science/articles/2008/09/16/2365110.htm 16 September.

55 *STORR, G.M. (1980). A new *Brachyaspis* (Serpentes: Elapidae) from Western Australia. *Record of the West Australian Museum*, 8: 397-9.

56 *SANCHIZ, F.B. & ADROVER, R. (1977). Anfibios fosiles del Pleistoceno de Mallorca. *Donana Acta Vertebratica*, 4: 5-25; MAYOL, J. & ALCOVER, J.A. (1981). Survival of *Baleaphryne* Sanchiz and Adrover, 1979 (Amphibia: Anura; Discoglossidae) on Mallorca. *Amphibia-Reptilia*, 1: 343-5.

57 *HENDERSON, J.R. (1912). Preliminary note on a new tortoise from south India. *Records of the Indian Museum*, 7: 217; GROOMBRIDGE, B., MOLL, E.O., & VIJAYA, J. (1982). Rediscovery of a rare Indian turtle. *Oryx*, 17 (July): 130-4.

58 CORKE, D. (1984). Maria Islands—home of the world's rarest snake. *World Wildlife News*, (spring): 8-9; ANON. (1986). [Couresse.] *Times Higher Education Supplement* (London), 17 August.

59 ANON. (1987). [Ensaf's giant gecko.] *New Scientist*, 113 (19 March): 19; *BALOUTCH, M. & THIREAU, M. (1986). A new species of gecko, *Eublepharis ensafi* (Sauria, Gekkonidae, Eublepharinae), from Khouzistan (southwestern Iran). *Bulletin Mensuel de la Société Linnéene de Lyon*, 55 (August): 281-8

59a *ANDERSON, S.C. & LEVITON, A.E. (1966). A new species of *Eublepharis* from Southwestern Iran (Reptilia: Gekkonidae). *Occasional Papers of the California Academy of Sciences*, No. 53: 1-5.

60 GRISMER, L.L. (1989). *Eublepharis ensafi* Baloutch and Thireau, 1986: A junior synonym of *E. angramainyu* Anderson and Leviton, 1966. *Journal of Herpetology*, 23: 94-5.

61 *BAUER, A. & RUSSELL, A.P. (1986). *Hoplodactylus delcourti* n. sp. (Reptilia: Gekkonidae), the largest known gecko. *New Zealand Journal of Zoology*, 13: 141-8; GREENWELL, J.R. (1988). World's largest gecko discovered. *ISC Newsletter*, 7 (spring): 1-4.

62 BAUER, & RUSSELL, A.P. (1987). *Hoplodactylus delcourti* (Reptilia: Gekkonidae) and the kawekaweau of Maori folklore. *Journal of Ethnobiology*, 7 (summer): 83-91; MAIR, W.G. (1873). Notes on Rurima rocks. *Transactions of the New Zealand Institute*, 5: 151-3; RAYNAL, M. & DETHIER, M. (1990). Lézards géants des Maoris ... La vérité derrière la légende. *Bulletin Mensuel de la Société Linnéene de Lyon*, 59 (March): 85-91.

63 HELLABY, D. (1984). Giant geckos 'sighted' 20 years ago. *Dominion* (Wellington), 11 September; GRANT, P. (1990). Lizards live on radio. *BBC Wildlife*, 8 (June): 360; WHITAKER, A.H. & THOMAS, B.W. (1990). Large Lizard Sightings in the Gisborne Region: Report on a National Museum Investigation. *National Museum of New Zealand* (Wellington).

64 BURTON, J.A. (1986). Golden find on holy hill. *BBC Wildlife*, 4 (June): 262.

65 BÖHME, W. et al. (1987). Neuentdeckung einer Grossechse (Sauria: *Varanus*) aus der Arabischen Republik Jemen. *Herpetofauna*, 9 (February): 13-20; GREENWELL, J.R. (1989). T.V. show leads to reptile discovery. *ISC Newsletter*, 8 (winter): 5-6; *BÖHME, W. et al. (1989). A new monitor lizard (Reptilia: Varanidae) from Yemen, with notes on ecology, phylogeny and zoogeography. *Fauna of Saudi Arabia*, 10: 433-48.

66 ANON. (1989). Snake rediscovered on St. Vincent. *Oryx*, 23 (October): 226; HENDERSON, R.W. & HAAS, G.T. (1993). Status of the West Indian snake *Chironius vincenti*. Ibid., 27 (July): 181-4.

67 *BULLER, W.L. (1877). [(*Hatteria*) *Sphenodon guntheri*, sp. n., from Brothers Island, near Cook Strait, New Zealand...] *Transactions and Proceedings of the New Zealand Institute* (for 1876), 9: 317-325; BARTON, M. (1989). Brother's cousins. *BBC Wildlife*, 7 (December): 790.

68 DAUGHERTY, C.H. et al. (1900). Neglected taxonomy and continuing extinctions of tuatara (*Sphenodon*). *Nature*, 347 (13 September): 177-9.

68a HAY, J.M. et al. (2010). Genetic diversity and taxonomy: a reassessment of species designation in tuatara (*Sphenodon*: Reptilia). *Conservation Genetics*, 11(3): 1063-81.

68b ANON. (1995). Lizard [sic] discovery baffles scientists. *The Times* (London), 27 May; ANON. (2009). Our first baby tuatara! http://www.visitzealandia.com/Site/Zealandia_Home/Inside/News/Media_Releases_2009/Tuatara_baby.aspx 18 March.

69 GLEDHILL, R. (1990). Explorer may have found new species. *The Times* (London), 8 May; RAXWORTHY, C. (1992). Personal communication, 19 July.

70 ANON. (1991). Jamaican iguana rediscovered. *Oryx*, 25 (July): 133.

71 *INGRAM, G. & CORBEN, C. (1990). *Litoria electrica*: a new treefrog from western Queensland. *Memoirs of the Queensland Museum*, 28: 475-8.

72 *MYERS, C.W., PAOLILLO O, A., & DALY, J.W. (1991). Discovery of a defensively malodorous and nocturnal frog in the family Dendrobatidae... *American Museum Novitates*, No. 3002 (7 March): 1-33; HARDING, K. (1992). The secrete weapon. *BBC Wildlife*, 10 (June): 12.

73 ANON. (1992). Road kill reveals lizard not extinct. *Portland Press Herald* (Portland), 28 December.

74 *WALLACH, V. & JONES, G.S. (1992). *Cryptophidion annamense*, a new genus and species of cryptozoic snake from Vietnam (Reptilia: Serpentes). *Cryptozoology*, 11: 1-37.

75 PAUWELS, O. & MEIRTE, D. (1993-96). The status of *Cryptophidion annamense*. Ibid., 12: 95-100.

76 LAZELL, J.D. (1993-96). *Cryptophidion* is not *Xenopeltis*. Ibid., 12: 101-2; WALLACH, V. & JONES, G.S. (1993-96). *Cryptophidion* is a valid taxon. Ibid., 12: 102-13.

77 *BOSCH, H.A.J. IN DEN & INEICH, I. (1994). The Typhlopidae of Sulawesi (Indonesia): a review with description of a new genus and a new species (Serpentes: Typhlopidae). *Journal of Herpetology*, 28(2): 206-217.

78 *SEÑARIS, J.C. (1998). A new species of *Typhlophis* (Serpentes: Anomalepididae) from Bolivar State, Venezuela. *Amphibia-Reptilia*, 19(3): 303-10.

79 *PLATZ, J.E. (1993). *Rana subaquavocalis*, a remarkable new species of leopard frog (*Rana pipiens* complex) from southeastern Arizona that calls under water. *Journal of Herpetology*, 27(2): 154-62.

80 *RIVA, I. DE LA & LYNCH, J. (1997). New species of *Eleutherodactylus* from Bolivia (Amphibia: Leptodactylidae). *Copeia* (for 1997)(1): 151-7.

81 *WHITE, A. & ARCHER, M. (1994). *Emydura lavarackorum*, a new Pleistocene turtle (Pleurodira: Chelidae) from fluviatile deposits at Riversleigh, Northwestern Queensland. *Records of the South Australian Museum*, 27: 159-67.

82 THOMSON, S., WHITE, A., & GEORGES, A. (1997). Re-evaluation of *Emydura lavarackorum*: Identification of a living fossil. *Memoirs of the Queensland Museum*, 42: 327-36; KENNEDY, F. (1997). 'Living fossil' turtle in cook pot. *The Australian*, 15 May.

83 *CANN, J. & LEGLER, J.M. (1994). The Mary River tortoise: a new genus and species of short-necked chelid from Queensland, Australia (Testudines; Pleurodira). *Chelonian Conservation and Biology*, 1 (July): 81-96; BARNETT, A. (1994). Turtle dreams. *BBC Wildlife*, 12 (November): 12; BAUMANN, J. (1995). Whale of a tail and a tale. *Deseret News* (Deseret, Utah), 16 October.

84 HARRISON, D. (1998). 'Extinct' snake saved from rat pack. *Observer* (London), 29 March.

85 *GÜNTHER, R. & MANTHEY, U. (1995). *Xenophidion*, a new genus with two new species of snakes from Malaysia (Serpentes, Colubridae). *Amphibia-Reptilia*, 16: 229-40; WALLACH, V. & GÜNTHER, R. (1998). Visceral anatomy of the Malaysian snake genus *Xenophidion*, including a cladistic analysis and allocation to a new family (Serpentes: Xenophidiidae). Ibid., 19(4): 385-404.

86 *BOUR, R. (1982). Contribution à la connaissance des tortues terrestres des Seychelles ... et description d'une espèce nouvelle probablement originaire des Îles granitiques et au bord de l'extinction. *C.R.A.S.*, 295 (20 September): 117-22; ANON. (1983). An extinct tortoise rediscovered? *Oryx*, 17 (April): 61.

87 ARNOLD, N. (1992). Personal communication, May; Swingland, I. (1992). Personal communication, 11 February.

88 GERLACH, J. & CANNING, L. (1995). Seychelles giant tortoise rediscovered? *Oryx*, 29 (April): 74-5; GERLACH, J. (1995). Big shell shock. *BBC Wildlife*, 13 (May): 12.

89 MORGAN, A. (1997). Emerging from its shell ... 120 years after dying out. *Sunday Telegraph* (London), 6 April; GERLACH, J. (1998). The rediscovery and conservation of Seychelles giant tortoises. *International Zoo News*, 45 (January-February): 4-10.

90 OLDFIELD, S. (1998). How Darwin evolved from extinction. *Daily Mail* (London), 11 March; ANON. (1999). A trio of tortoise tales. *Fortean Times*, No. 121 (April): 16.

91 PRINGLE, A. (1999). Personal communication, October; GERLACH, J. (1999). The Nature Protection Trust of Seychelles. http://www.bogo.co.uk/gerlach/ accessed 13 November.

92 *LIPS, K.R. & SAVAGE, J.M. (1996). A new species of rainfrog, *Eleutherodactylus phasma* (Anura: Leptodactylidae), from montane Costa Rica. *Proceedings of the Biological Society of Washington*, 109(4): 744-8.

93 PRINCE WILLIAM OF SWEDEN (1923). *Among Pygmies and Gorillas*. Gyldendal (London).

94 ANON. (1997). Monsters of the deep. *Fortean Times*, No. 100 (July): 6; ANON. (1998). Giant turtle sightings set Vietnam capital abuzz. http://cnn.com/EARTH/9804/13/vietnam.turtles.ap/ 13 April; MCDONALD, M. (1998). Legendary turtle creating ripples across Vietnam. *San Jose Mercury News* (San Jose), 2 November; THUY, T.L. (2000). Mock turtles—or a whole new genus? *Vietnam Investment Review* (Hanoi), 10 January.

94a FARKAS, B. & WEBB, R.G. (2003). *Rafetus leloii* Hà Dinh Dúc, 2000—an invalid species of softshell turtle from Hoan Liem Lake, Hanoi, Vietnam (Reptilia, Testudines, Trionychidae). *Zoologische Abhandlungen*, 53: 107-12.

95 *ZIEGLER, T., BÖHME, W., & PHILIPP, K. (1999). *Varanus caerulivirens* sp. n., a new monitor of the *Varanus indicus* group from Halmahera, Moluccas, Indonesia. *Herpetozoa*, 12(1-2): 45-56.

96 BAYLESS, M.K. (1999). Monitoring new discoveries, Parts I-III. Unpublished article, copy emailed to Dr Karl P.N. Shuker on 17, 18, and 19 November.

97 *SPRACKLAND, R.G. (1991). Taxonomic review of the *Varanus prasinus* group with descriptions of two new species. *Memoirs of the Queensland Museum*, 30(3): 561-76.

98 *YANG, D. & LIU, W. (1994). Relationships among species groups of *Varanus* from southern southeastern Asia with description of a new species from Vietnam. *Zoological Research*, 15(1): 11-15.

99 *BÖHME, W. & ZIEGLER, T. (1997). *Varanus melinus* sp. n., ein neuer Waran aus der *V. indicus*-gruppe von den Molukken, Indonesien. *Herpetofauna*, 19: 26-34.

100 *HARVEY, M.B. & BARKER, D. (1998). A new species of blue-tailed monitor lizard (genus *Varanus*) from Halmahera island, Indonesia. *Herpetologica*, 54(1): 34-44.

101 LEMM, J. (1998). Year of the monitor: a look at some recently discovered varanids. *Reptiles*, 6 (September): 70-80.

102 *SPRACKLAND, R.G. (1999). A new species of monitor (Squamata: Varanidae) from Indonesia. *Reptile Hobbyist*, 4 (June): 20-7.

103 ANON. (1994). Isla de El Hierro. *El Semanal*, (29 May): 52-3; *HUTTERER, R. (1985). Neue Funde von Rieseneidechsen (Lacertidae) auf der Insel Gomera. *Bonner Zoologische Beiträge*, 36: 365-94; VALIDO, A. et al. (2000). 'Fossil' lizard found alive in the Canary Islands. *Oryx*, 34 (January): 71-2; RANDO, J.C. et al. (2004). Discovery of a new population of the spotted lizard of the Canary Islands. Ibid., 38 (April): 134.

Section 4: The Fishes

1 *JORDAN, D.S. (1898). Description of a species of fish (*Mitsukurina owstoni*) from Japan, the type of a distinct family of lamnoid sharks. *Proceedings of the Californian Academy* (Series 3), 1: 199-201.

2 ELLIS, R. (1982). *The Book of Sharks* (Rev. Edit.). Robert Hale (London); STEEL, R. (1985). *Sharks of the World*. Blandford (Poole).

3 *REGAN, C.T. (1903). On a collection of fishes from the Azores. *A.M.N.H.* (Series 7), 12: 344-8.

4 GRZIMEK, B. (Ed.) (1972-5). *Grzimek's Animal Life Encyclopedia* (13 vols). O.c.

5 MARSHALL, N.B. (1961). A young Macristium and the ctenothrissid fishes. *B.B.M.N.H.Z.*, 7: 355-70; *BERRY, F.H. & ROBINS, C.R. (1967). *Macristiella perlucens*, a new clupeiform fish from the Gulf of Mexico. *Copeia* (for 1967): 46-50; ROSEN, D.E. (1971). The Macristidae, a ctenothrissiform family based on juvenile and larval scolepomorph fishes. *American Museum Novitates*, No. 2452: 1-22.

6 *PARR, A.E. (1933). Deep sea Berycomorphi and Percomorphi from the waters around the Bahama and Bermuda Islands. *Bulletin of the Bingham Oceanographic Collection*, 3: 1-51; JOHNSTON, R.K. (1974). A second record of *Korsogaster nanus* Parr. *Copeia* (for 1970): 758-60.

7 *TUCKER, D.W. (1954). Report on the fishes collected by S.Y. "Rosaura" in the North and Central Atlantic, 1937-38. Part I. Families Carcharhinidae, Torpedinidae, Rosauridae (Nov.)... *B.B.M.N.H.Z.*, 2: 163-214.

8 *REGAN, C.T. (1909). The Asiatic fishes of the family Anabantidae. *P.Z.S.L.*: 767-87.

9 *REGAN, C.T. (1903). Descriptions de poissons nouveaux faisant partie de la collection du Musée d'Histoire Naturelle de Genève. *Revue Suisse de Zoologie*, 11: 413-18.

10 *SMITH, H.M. (1912). Two squaloid sharks of the Philippine Archipelago, with descriptions of new genera and species. *Smithsonian Institution National Museum Proceedings*, 42: 677-85.

11 NELSON, J.S. (1984). *Fishes of the World* (2nd Edit.). John Wiley (New York); WHEELER, A. (1985). *The World Encyclopedia of Fishes*. Macdonald (London).

12 *FOWLER, H.W. & BALL, S.E. (1924). Descriptions of new fishes obtained by the Tanager Expedition of 1923 in the Pacific Islands west of Hawaii. *Proceedings of the Academy of Natural Science, Philadelphia*, 76: 269-74; NORMAN, J.R. (1957). *Draft Synopsis of the Orders, Families and Genera of Recent Fishes and Fish-Like Vertebrates*. British Museum (Natural History) (London).

13 *PRASHAD, B. & MUCKERJI, D.D. (1929). The fish of the Indawgyi Lake and the streams of the Myitkyina District (Upper Burma). *Records–Indian Museum* (Calcutta), 31: 161-223; BANISTER, K.E. (1970). The anatomy and taxonomy of *Indostomus paradoxus*... *B.B.M.N.H.Z.*, 19: 179-209.

13a *BRITZ, R. & KOTTELAT, M. (1999). Two new species of gasterosteiform fishes of the genus *Indostomus* (Teleostei: Indostomidae). *Ichthyological Exploration of Freshwaters*, 10(4): 327-36.

14 *CHEVEY, P. (1930). Sur un nouveau silure géant du basin du Mékong *Pangasianodon gigas* nov. g., nov. sp. *Bulletin Société Zoologique de France*, 55: 536-42; ANON. (1998). Giant catfish demise. *Oryx*, 32 (July): 184.

15 WOOD, G.L. (1982). *The Guinness Book of Animal Facts and Feats* (3rd Edit.). O.c.

16 *BEEBE, W. (1932). A new deep-sea fish. *Bulletin of the New York Zoological Survey*, 35: 175-7; BEEBE, W. (1934). *Half-Mile Down*. Harcourt, Brace (New York).

17 DEMBECK, H. (1965). *Animals and Men*. Natural History Press (Garden City).

18 *MYERS, G.S. (1936). A new characid fish of the genus *Hyphessobrycon* from the Peruvian Amazon. *P.B.S.W.*, 49: 115-16.

19 *MYERS, G.S. & WEITZMAN, S.H. (1956). Two new Brazilian fresh water fishes. *Stanford Ichthyological Bulletin*, 7: 1-4.

20 *HUBBS, C.L. & INNES, W.T. (1936). The first known blind fish of the family Characidae: a new genus from Mexico. *Occasional Papers of the Museum of Zoology, University of Michigan*, No. 342: 1-7.

21 HERISSE, J. (1965). Blind fish. *Animal Life*, No. 29 (January): 8-9; TEYKE, T. (1990). Morphological differences in neuromasts of the blind cave fish *Astyanax hubbsi* and the sighted river fish *Astyanax mexicanus*. *Brain, Behavior and Evolution*, 35: 23-30.

22 *CLARK, H.W. (1937). New fishes from the Templeton Crocker Expedition of 1934-5. *Copeia* (for 1937): 88-91.

23 COURTENAY-LATIMER, M. (1989). Reminiscences of the discovery of the coelacanth. *Cryptozoology*, 8: 1-11.

24 SMITH, J.L.B. (1956). *Old Fourlegs. The Story of the Coelacanth*. Longmans (London).

25 ANON. (1939). A coelacanth fish. *The Times* (London), 17 March.

26 *SMITH, J.L.B. (1939). A living fish of Mesozoic type. *Nature*, 143 (18 March): 455-6; SMITH, J.L.B. (1939). The living coelacanthid fish from South Africa. Ibid., 143 (6 May): 748-50.

27 SMITH, J.L.B. (1941). A living coelacanthid fish from South Africa. *Transactions of the Royal Society of South Africa*, 28: 1-106.

28 *HUBBS, C.L. & BAILEY, R.M. (1947). Blind catfishes from Artesian waters of Texas. *Occasional Papers of the Museum of Zoology, University of Michigan*, No. 499 (28 April): 1-15.

29 SMITH, A. (1990). *Blind White Fish in Persia* (Rev. Edit.). Penguin (London).

30 *BRUUN, A.F. & KAISER, E.W. (1950). *Iranocypris typhlops* n. g., n. sp., the first true cave-fish from Asia. *Dan. Scient. Invest. Iran*, 4: 1-8; *TREWAS, E. (1955). A blind fish from Iraq, related to *Garra*. *A.M.N.H.* (Series 12), 8: 551-5.

31 *GREENWOOD, P.H. (1976). A new and eyeless cobitid fish (Pisces, Cypriniformes) from the Zagros Mountains. *Journal of Zoology*, 180: 129-37.

32 *CHU SINLUO & CHEN YINRUI. (1979). A new blind cobitid fish ... from subterranean waters in Yunnan, China. *Acta Zoologica Sinica*, 25: 285-7.

33 *REGAN, C.T. (1925). New ceratioid fishes from the North Atlantic, the Caribbean Sea, and the Gulf of Panama collected by the 'Dana'. *A.M.N.H.* (Series 9), 15: 561-67.

34 GÜNTHER, K. & DECKERT, K. (1956). *Creatures of the Deep Sea*. George Allen & Unwin (London).

34a *REGAN, C.T. (1930). A ceratioid fish (*Caulophryne polynema*, sp. n.), female with male, from off Madeira. *Journal of the Linnean Society of London, Zoology Series*, 37 (August): 191-5; BLUNDELL, N. & KENDALL, P. (2001). Devils of the deep. *Daily Mail* (London), 29 August.

35 IDYLL, C.P. (1964). *Abyss: The Deep Sea and the Creatures that Live in It*. Constable (London).

36 *BRUUN, A. (1953). *Galatheas Jordomsejling*. Schultz Forl. (Copenhagen); BERTELSEN, E. & STRUHSAKER, P.J. (1977). The ceratioid fish of the genus *Thaumatichthys*: Osteology, relationships, distribution and biology. *Galathea Report*, 14: 7-40.

37 SMITH, J.L.B. (1953). The second coelacanth. *Nature*, 171 (17 January): 99-101; FOREY, P.L. (1988). Golden jubilee for the coelacanth *Latimeria chalumnae*. Ibid., 336 (22/29 December): 727-32; GORR, T. et al. (1991). Close tetrapod relationships of the coelacanth *Latimeria* indicated by haemoglobin sequences. Ibid., 351 (30 May): 394-7.

38 JACKMAN, B. (1972). The five figure fish. *Sunday Times* (London), 2 April.

39 FRICKE, H. (1988). Coelacanths. The fish that time forgot. *National Geographic Magazine*, 173 (June): 824-38.

39a BRUTON, M.N., CABRAL, A.J.P., & FRICKE, H. (1992). First capture of a coelacanth, *Latimeria chalumnae* (Pisces, Latimeriidae), off Mozambique. *South African Journal of Science*, 88 (April): 225-7; SCHLIEWEN, U., FRICKE, H., et al. (1993). Which home for coelacanth? *Nature*, 363 (3 June): 405.

39b HEEMSTRA, P.C. et al. (1996). First authentic capture of a coelacanth, *Latimeria chalumnae* (Pisces: Latimeriidae), off Madagascar. *South African Journal of Science*, 92 (March): 150-1; VICENTE, N. (1997). Un

coelacanthe à Madagascar. *Océanorama*, No. 27 (January): 11-15.
39c STODDARD, E. (2000). "Living fossils" discovered off South Africa coast. Reuters News Service, 1 December; ANON. (2000). Divers find six live coelacanths at St Lucia. SAPA [South African Press Association] Domestic News Wire, 1 December.
39d ANON. (2001). Tourism set to cash in on coelacanth discovery. Africa News Service, 10 October.
40 *BERTELSEN, E. & MARSHALL, N.B. (1956). The Mirapinnati, a new order of teleost fishes. *Dana Report*, No. 42: 1-34; Johnson, G.D. et al. (2009). Deep-sea mystery solved: astonishing larval transformations and extreme sexual dimorphism unite three fish families. *Biology Letters*, 5(2): 235-9.
41 *COHEN, D.M. (1958). *Bathylychnops exilis*, a new genus of argentinid fish from the North Pacific. *Stanford Ichthyological Bulletin*, 7: 47-52; PEARCY, W. et al. (1965). A 'four-eyed' fish from the deep-sea: *Bathylychnops exilis* Cohen, 1958. *Nature*, 207 (18 September): 1260-2; STEIN, D.L. & BOND, C.E. (1985). Observations on the morphology, ecology, and behaviour of *Bathylychnops exilis* Cohen. *Journal of Fish Biology*, 27: 215-28.
42 *CLAUSEN, H.S. (1959). Denticipitidae, a new family of primitive isospondylous teleosts from West African freshwater. *Videnskabelige Meddelelser fra Dansk Naturhistorisk Forening*, 121: 141-56; GREENWOOD, P.H. (1968). The osteology and relationships of the Denticipitidae, a family of clupeomorph fishes. *B.B.M.N.H.Z.*, 16: 213-73.
43 *MEES, G.F. (1961). Description of a new fish of the family Galaxiidae from Western Australia. *Journal of the Royal Society of Western Australia*, 44: 33-8; BANISTER, K.E. & CAMPBELL, A. (Eds.) (1985). *The Encyclopedia of Underwater Life*. George Allen & Unwin (London).
44 *GÉRY, J. (1964). Une nouvelle famille de poissons dulcaquicales africains: les Grasseichthyidae. *C.R.A.S.*, 259: 4805-7.
45 *ROBINS, C.R. & SYLVA, D.P. DE (1965). The Kasidoroidae, a new family of mirapinniform fishes from the Western Atlantic Ocean. *Bulletin of Marine Science*, 15: 189-201.
46 *THORP, C.H. (1969). A new species of mirapinniform fish (family Kasidoroidae) from the Western Indian Ocean. *Journal of Natural History*, 3: 61-70.
47 SYLVA, D.P. DE & ESCHMEYER, W.N. (1977). Systematics and biology of the deep-sea fish family Gibberichthyidae, a senior synonym of the family Kasidoroidae. *Proceedings of the California Academy of Sciences*, 41: 215-31.
48 *PARR, A.R. (1933). Deep sea Berycomorphi and Percomorphi from the waters around the Bahama and Bermuda Islands. *Bulletin of the Bingham Oceanographic Collection*, 3: 1-51.
49 HEUVELMANS, B. (1968). *In the Wake of the Sea-Serpents*. Rupert Hart-Davis (London).
50 NIELSEN, J.G. & LARSEN, V. (1970). Remarks on the identity of the giant *Dana* eel-larva. *Videnskabelige Meddelelser fra Dansk Naturhistorisk Forening*, 133: 149-57.
51 *CASTLE, P.H.J. (1959). A large leptocephalid (Teleostei, *Apodes*) from off South Westland, New Zealand. *Transactions of the Royal Society of New Zealand*, 87: 179-84.
52 SMITH, D.G. (1970). Notacanthiform leptocephali on the Western North Atlantic. *Copeia* (2 March): 1-8; Colocongridae. Wikipedia entry, accessed 18 February 2011.
53 DUNFORD, B. (1976). Huge shark may be new species. *Star-Bulletin* (Honolulu), 17 November; TAYLOR, L.R. (1977). Megamouth, a new family of sharks. *Oceans*, 10: 46-7.
54 SOULE, G. (1981). *Mystery Monsters of the Deep*. Franklin Watts (New York).
55 *TAYLOR, L.R., COMPAGNO, L.J.V., & STRUHSAKER, P.J. (1983). Megamouth—a new species, genus, and family of lamnoid shark (*Megachasma pelagios*, family Megachasmidae) from the Hawaiian Islands. *Proceedings of the California Academy of Sciences*, 43 (6 July): 87-110; YANO, K. et al. (Eds.) (1997). *Biology of the Megamouth Shark*. Tokai University Press (Tokyo).
56 GREENWELL, J.R. (1985). Second megamouth shark found. *ISC Newsletter*, 4 (spring): 5.
57 GREENWELL, J.R. (1991). Megamouth VI caught alive and studied. Ibid., 10 (summer): 1-3.
58 GREENWELL, J.R. (1988). Third megamouth found. Ibid., 7 (winter): 4; ANON. (1988). Megamouth leviathan. *New Scientist*, 119 (1 September): 30.
59 NAKAYA, K. (1989). Discovery of a megamouth shark from Japan. *Japanese Journal of Ichthyology*, 36: 144-6.
60 TAKADA, K. (1994). Stranding of a megamouth shark in Hakata-Bay. *Report of the Japanese Society of Elasmobranch Studies*, 31: 13-16.
61 SÉRET, B. (1995). Première capture d'un réquin grande gueule (Chondrichthyes, Megachasmidae) dans l'Atlantique, au large du Sénégal. *Cybium*, 19(4): 425-7.
62 COLEMAN, L. & CLARK, J. (1999). *Cryptozoology A to Z*. O.c.
63 ELIZAGA, E.T. (1998). Megamouth shark found in Cagayan de Oro. http://cdo.weblinq.com/~econews/Megamouth.html/ 9 April.
64 PETERSEN, D. (1999). Megamouth shark caught off California. ELASMO-L@artemis.it.luc.edu 5 November.
65 PECCHIONI, P. (1998). Megamouth. Ibid., 4 September; PECCHIONI, P. & BENOLDI, C. (1999). Sperm whales spotted attacking megamouth shark. http://www.flmnh.ufl.edu/fish/Sharks/Megamouth/Mega13.htm
65a PETERSEN, D. (1999). Megamouth shark #14 caught off California. http://www.flmnh.ufl.edu/fish/sharks/megamouth/Mega14.htm 5 November; PETERSEN, D. (2001). Megamouth shark #15 caught off California. Ibid., 19 October.
65b WHITE, W.T. et al. (2004). A juvenile megamouth shark *Megachasma pelagios* (Lamniformes: Megachasmidae) from Northern Sumatra, Indonesia. *Raffles Bulletin of Zoology*, 52(2): 603-7.
65c ANON. (2004). Rare shark dies on shore of Iloilo town; bewilders, awes townfolk. *Manila Bulletin* (Manila), 15 November.
65d Wikipedia (2011). List of megamouth shark specimens and sightings. http://en.wikipedia.org/wiki/List_of_megamouth_shark_specimens_and_sightings accessed 23 August.
66 *DAILEY, M.D. & VOGELBEIN, W. (1982). Mixodigmatidae, a new family of cestode ... from a deep sea,

planktivorous shark. *Journal of Parasitology*, 68: 145-9.
67 *HEEMSTRA, P.C. & SMITH, M.M. (1980). Hexatrygonidae, a new family of stingrays (Myliobatiformes: Batoidea) from South Africa... *Ichthyological Bulletin of the J.L.B. Smith Institute of Ichthyology*, No. 43: 1-17; WHEELER, A. (1981). A new stingray from South Africa. *Nature*, 289 (22 January): 221; SMITH, J.L.B et al. (2003). *Smith's Sea Fishes*. Random House Struik (Cape Town).
68 STERRY, P. (1988). Up the Amazon without a puddle. *BBC Wildlife*, 6 (August): 422-5; HENDERSON, P.A. & WALKER, I. (1990). Spatial organization and population density of the fish community of the litter banks within a central Amazonian blackwater stream. *Journal of Fish Biology*, 37: 401-11; HENDERSON, P.A. (1991). Personal communication, 29 May; HENDERSON, P.A. (1999). Personal communication, 24 November.
69 GREENWELL, J.R. (1986). Giant fish reported in China. *ISC Newsletter*, 5 (autumn): 7-8; SHUKER, K.P.N. (1997). *From Flying Toads to Snakes with Wings*. O.c.
69a SHUKER, K.P.N. (2010). *Karl Shuker's Alien Zoo: From the Pages of Fortean Times*. CFZ Press (Bideford).
70 *CHAROUSSET, F. (1986). Un nouveau poisson trouvé en Mediterranée. *Clin d'Oeil*, 13 (February): 10-17.
71 *MONKOLPRASIT, S. & ROBERTS, T.R. (1990). *Himantura chaophraya*, a new giant freshwater stingray from Thailand. *Japanese Journal of Ichthyology*, 37: 203-8.
72 ANON. (1991). Once-extinct fish found living in Mexico in disgusting pond. *Belleville News-Democrat* (Belleville–Illinois), 26 February.
73 TURNER, M. (1992). New fish species found in outback. *Courier-Mail* (Brisbane), 11 January.
74 HOMEWOOD, B. (1994). Vampire fish show their teeth. *New Scientist*, 144 (3 December): 7.
75 ANON. (1996). Vietnam nature reserve reveals another "new" species. WWF News press release, 7 August; HIGHFIELD, R. (1996). New fish found in Vietnam. *Daily Telegraph* (London), 28 September; *NGUYEN, T.T. et al. (1999). Natural fish resources and aquaculture in Vu Quang natural conservation area. In: NGUYEN, T.T. (Ed.), *Selected paper of seminar on North Truongson Biodiversity* (the second). Publishing House of Hanoi National University (Hanoi): 24-9.
76 *NGUYEN T.T. (1995). *Parazacco vuquangensis*, a new species of Cyprinidae from Vietnam. *Ichthyol. Explor. Freshwater*, 6(1): 77-80; *KOTTELAT, M. (2001). *Freshwater Fishes of Northern Vietnam. A Preliminary Check-List of the Fishes Known or Expected to Occur in Northern Vietnam with Comments on Systematics and Nomenclature*. Environment and Social Development Unit, East Asia and Pacific Region. The World Bank. Freshwater Fish (Vietnam); *NGUYEN, T.T. (1987). Genus *Opsarichthys* [sic] Bleeker, 1863 Leuciscini-Cyprinidae of the Lam River Basin (Prov. Nghe-Tinh). *Tap Chi Sinh Hoc* (*J. Biol.*), 9(2): 32-6.
77 YOON, C.K. (1997). Amazon's depths yield strange new world of unknown fish. *New York Times* (New York), 18 February: B9-B10.
78 *LUNDBERG, J.G. et al. (1996). *Magosternarchus*, a new genus with two new species of electric fishes (Gymnotiformes: Apteronotidae) from the Amazon River basin, South America. *Copeia* (for 1996): 657-70.

79 *FRIEL, J.P. & LUNDBERG, J.G. (1996). *Micromyzon akamai*, gen. et sp. nov., a small and eyeless banjo catfish (Siluriformes: Aspredinidae) from the river channels of the lower Amazon basin. Ibid., 641-8.
80 ANON. (1997). A shark once feared extinct is rediscovered. IUCN press release, April; ANON. (1997). Shark riddle solved. *The Times* (London), 28 August; HIGHFIELD, R. (1997). Rare shark from Borneo surfaces in Birmingham. *Daily Telegraph* (London), 28 August; *COMPAGNO, L.J.V. et al. (2010). *Glyphis fowlerae* sp. nov., a new species of river shark (Carcharhiniformes; Carcharhinidae) from northeastern Borneo. In: LAST, P.R. et al. (Eds.), Descriptions of New Sharks and Rays from Borneo. *CSIRO Marine and Atmospheric Research Paper*, No. 32: 29-44.
81 *DIDIER, D.A. (1998). The leopard *Chimaera*, a new species of chimaeroid fish from New Zealand (Holocephali, Chimaeriformes, Chimaeridae). *Ichthyological Research*, 45(3): 281-9; ANON. (1999). Ichthyologist discovers a new fish species in the waters of the South Pacific. Academy of Natural Sciences news release, http://www.acnatsci.org/press/newfish.html/ 30 March; JAFFE, M. (1999). In New Zealand, new fish species is discovered by Phila. researcher. *Philadelphia Inquirer* (Philadelphia), 30 March.
81a *LUCHETTI, E.A. et al. (2011). *Chimaera opalescens* n. sp., a new chimaeroid (Chondrichthyes: Holocephali) from the north-eastern Atlantic Ocean. *Journal of Fish Biology*, 79(2): 399-417.
82 ANON. (1999). Lost fish found—85 years later. *ScienceDaily Magazine*, http://www.sciencedaily.com/releases/1999/04/990426212332.htm/ 27 April.
83 ANON. (1999). [four new nototheniids.] National Science Foundation press release, http://www.nsf.gov/od/lpa/news/press/99/pr9914.htm/ 4 March.
84 SANDERS, R. (1998). UC Berkeley biologist finds second population of primitive coelacanths among islands off the coast of Sulawesi. University of California, Berkeley, news release, 23 September; BROWNE, M.W. (1998). Second home of fish from dinosaur age is found. *New York Times* (New York), 24 September.
85 ERDMANN, M., CALDWELL, R.L., & MOOSA, M.K. (1998). Indonesian 'king of the sea' discovered. *Nature*, 395 (24 September): 335.
86 FRICKE, H. et al. (2000). Biogeography of the Indonesian coelacanths. Ibid., 403 (6 January): 38.
87 SHUKER, K.P.N. (1993). *The Lost Ark: New and Rediscovered Animals of the 20th Century*. O.c.; SHUKER, K.P.N. (1995). *In Search of Prehistoric Survivors*. O.c.
88 *POUYAUD, L. et al. (1999). Une nouvelle espèce de coelacanthe. Preuves génétiques et morphologiques. *C.R.A.S.* (Series 3), 322: 261-7.
89 HOLDER, M.T., ERDMANN, M.V., et al. (1999). Two living species of coelacanths? *Evolution*, 96 (26 October): 12616-20.
90 RAYNAL, M. (1998). Personal communications, 1 and 3 October.
91 MCCABE, H. & WRIGHT, J. (2000). Tangled tale of a lost, stolen and disputed coelacanth. *Nature*, 406 (13 July): 114.

Section 5: The Invertebrates

1. CHUN, C. (1903). Ueber Leuchtorgane und Augen von Tiefsee-Cephalopoden. *Verhandlungen. Deutsche Zoologische Gesellschaft*, 13: 67-91.
2. DARWIN, C. (1877). *The Various Contrivances by Which Orchids are Fertilized by Insects* (2nd Edit.). John Murray (London); *ROTHSCHILD, W. & JORDAN, K. (1903). A revision of the lepidopterous family Sphingidae. *Novitates Zoologiae*, 9 (Suppl.): 1-972.
3. BURTON, M. & BURTON, R. (Eds.) (1968-70). *Purnell's Encyclopedia of Animal Life* (6 vols). O.c.
4. ANGIER, N. (1992). It may be elusive but moth with 15-inch tongue should be out there. *New York Times* (New York), 14 January.
5. *SINETY, R. DE (1901). *Recherches sur la Biologie et l'Anatomie des Phasmes*. Joseph Van In & Cie (Lierre); BRUNNER VON WATTENWYL, K. & REDTENBACHER, J. (1908). *Die Insektenfamilie der Phasmiden*. (Leipzig).
6. *ROTHSCHILD, W. (1907). *Troides alexandrae* sp. n. *Novitates Zoologiae*, 14: 96.
7. WELLS, S.M., PYLE, R.M., & COLLINS, N.M. (1983). *The IUCN Invertebrate Red Data Book*. IUCN (Gland–Switzerland).
8. *AGASSIZ, A. & CLARK, H.L. (1907). Preliminary report on the Echini collected in 1906 ... by the U.S. Fish Commission Steamer "Albatross"... *Bulletin of the Museum of Comparative Zoology, Harvard College*, 51: 107-39.
9. *SILVESTRI, F. (1907). Descrizione di un nuovo genere di Insetti Apterigoti rappresentante di un novo ordine. *Bollettino ... Scuola Superiore di Agricoltura...*, 1: 296-311; HEALEY, I.N. (1978). Proturans. In: BEER, G. DE et al. (Eds.), *Encyclopedia of the Animal World* (3 vols). Bay Books (London). pp. 1486-7; Protura. Wikipedia entry, accessed 5 March 2011.
10. REITTER, E. (1961). *Beetles*. Hamlyn (London); KLAUSNITZER, B. (1983). *Beetles*. Exeter Books (New York); ZAHL, P.A. (1959). Giant insects of the Amazon. *National Geographic Magazine*, 115 (May): 632-69; *Titanus giganteus*. Wikipedia entry, accessed 10 March 2011.
11. *HICKMAN, S.J. (1911). On *Ceratopora*, the type of a new family of Alcyonaria. *Proceedings of the Royal Society of London* (B), 84: 195-200.
12. *HARTMAN, W.D. & GOREAU, T.F. (1970). Jamaican coralline sponges. *Symposia of the Zoological Society of London*, 25: 205-43.
13. *LISTER, J.J. (1900). *Astrosclera willeyana*, the type of a new family of sponges. In: WILLEY, A., *Zoological Results...* , 4: 459-82.
14. *KIRKPATRICK, R. (1908). On two new genera of recent pharetronid sponges. *A.M.N.H.* (Series 8), 2: 503-15.
15. *HARTMAN, W.D. & GOREAU, T.F. (1975). A Pacific tabulate sponge, living representative of a new order of sclerosponges. *Postilla*, No. 167: 1-14; VACELET, J. (1985). Coralline sponges and the evolution of the Porifera. *Systematics Association Special*, 28: 1-13.
16. *SILVESTRI, F. (1913). Descrizione di un nuovo ordine di Insetti. *Bollettino ... Scuola Superiore di Agricoltura...*, 7: 193-209.
17. LINDENMAIER, W. (1972). *Insects of the World*. McGraw-Hill (New York).
18. GRZIMEK, B. (Ed.) (1972-5). *Grzimek's Animal Life Encyclopedia* (13 vols). O.c.
18a. *CHOE, J.C. (1989). *Zorotypus gurneyi*, new species, from Panama and redescription of *Zorotypus barberi* Gurney (Zoraptera, Zorotypidae). *Annals of the Entomological Society of America*, 82(2): 149-55; CHOE, J.C. (1994). Sexual selection and mating system in *Zorotypus gurneyi* Choe (Insecta, Zoraptera). 1. Dominance and mating success. *Behavioral Ecology and Sociobiology*, 34(2): 87-93; CHOE, J.C. (1994). Sexual selection and mating system in *Zorotypus gurneyi* Choe (Insecta, Zoraptera). 2. Determinants and dynamics of dominance. Ibid., 34(4): 233-7.
19. *WALKER, E.M. (1914). A new species of Orthoptera, forming a new genus and family. *Canadian Entomology*, 46: 93-8.
20. *GURNEY, A.B. (1961). Further advances in the taxonomy and distribution of the Grylloblattidae (Orthoptera). *P.B.S.W.*, 74: 67-76.
21. *CHUN, C. (1914). Cephalopoda from the "Michael Sars" North Atlantic Deep-Sea Expedition 1910. *Report of Scientific Research, "Michael Sars" North Atlantic Deep Sea Expedition*, 3: 1-28; NESIS, K.N. (1987). *Cephalopods of the World*. T.F.H. Publications (Neptune City).
22. *CHAPIN, E.A. (1921). Remarks on the genus *Hystrichopsylla* Tasch., with description of a new species. *Proceedings of the Entomological Society of Washington*, 23: 25-7; SCHEFFER, V.B. (1969). Super flea meets mountain beaver. *Natural History*, 78 (May): 54-5.
23. *KOEHLER, R. (1922). Ophiurans of the Philippine Seas and adjacent waters. *Bulletin of the U.S. National Museum*, 5: 1-486.
24. *MONOD, T. (1924). Sur un type nouveau de Malacostracé: *Thermosbaena mirabilis*. *Bulletin–Société Zoologique de France*, 49: 58-68.
25. SCHMIDT, W.L. (1973). *Crustaceans*. David & Charles (Newton Abbot); SCHRAM, F.R. (1986). *Crustacea*. Oxford University Press (Oxford).
25a. OLIVER, P. (2003). Colossal squid a formidable customer. *New Zealand Herald* (Auckland), 3 April; OWEN, J. (2003). "Colossal squid" revives legends of sea monsters. *National Geographic News*, 23 April.
25b. *ROBSON, G.C. (1925). On *Mesonychoteuthis*, a new genus of oegopsid, Cephalopoda. *A.M.N.H.*, Series 9, 16: 272-7.
25c. List of colossal squid specimens and sightings. Wikipedia entry, accessed 10 August 2011.
25d. LILLEY, R. (2007). New Zealand fishermen catch rare squid. Associated Press report, 22 February.
26. *HAMPSON, G.F. (1926). *Descriptions of New Genera and Species of Lepidoptera ... (Noctuidae) in the British Museum (Natural History)*. BMNH (London).
27. BÄNZIGER, H. (1969). The extraordinary case of the blood-sucking moth. *Animals*, 12 (July): 135-7.
28. BÄNZIGER, H. (1976). In search of the blood-sucker. *Wildlife*, 18 (August): 366-9.
29. *SPENGEL, J.W. (1932). *Planctosphaera pelagica*. *Scientific Results "Michael Sars" North Atlantic Deep-Sea Expedition*, 5: 1-27; HORST, C.J. VAN DER. (1936). Planctosphaera and tornaria. *Quarterly Journal of Microscopical Science*, 78: 605-13; HYMAN, L.H. (1959). *The Invertebrates: Smaller Coelomate Groups*. Vol. 5. McGraw-Hill (New York); DAMAS, D. & STIASNY, G. (1961). Les larves planctoniques d'entéropneustes.

Mémoires de l'Académie Royale de Belgique Cl. Science, 15: 1-68.

30 HADFIELD, M.G. & YOUNG, R.E. (1983). Planctosphaera (Hemichordata: Enteropneusta) in the Pacific Ocean. *Marine Biology*, 73: 151-3.

31 *CALMAN, W.T. (1933). A dodecapodous pycnogonid. *Proceedings of the Royal Society of London*, 113B: 107-15.

32 SAVORY, T. (1964). Ricinuleids. *Animals*, 4 (18 August): 376-8; Ricinulei. Wikipedia entry, accessed 12 March 2011.

33 FINNEGAN, S. (1935). Rarity of the archaic arachnids Podogona (Ricinulei). *Nature*, 136 (3 August): 186; SANDERSON, I.T. (1937). *Animal Treasure*. Macmillan (London).

34 *CLARK, J. (1934). Notes on Australian ants, with descriptions of new species and a new genus. *Memoirs of the National Museum of Victoria*, 8: 5-20; TAYLOR, R.W. (1978). Nothomyrmecia macrops: A living-fossil ant rediscovered. *Science*, 201 (15 September): 979-85.

35 *PENNAK, R. & ZINN, D.J. (1943). Mystacocarida, a new order of Crustacea from intertidal beaches in Massachusetts and Connecticut. *Smithsonian Miscellaneous Collections*, 103: 1-11.

36 IVANOV, A.V. (1963). *Pogonophora*. Academic Press (New York).

37 *CAULLERY, M. (1914). Sur les Siboglinidae, type nouveau d'invertebrés, recueilli par l'expédition du Siboga. *C.R.A.S.*, 158: 2014; CAULLERY, M. (1948). Le genre Siboglinum Caullery 1914. In: GRASSÉ, P.P. (Ed.), *Traité de Zoologie*, 11: 494-9.

38 *USCHAKOW, P. (1933). Eine neue Form der Familie Sabellidae. *Zoologischer Anzeiger*, 104: 705.

39 JOHANSSON, K.E. (1939). Lamellisabella zachsi Uschakow, ein Vertreter einer neuen Tierklasse Pogonophora. *Zoologiska Bidrag fran Uppsala*, 18: 253-68.

40 BEKLEMISHEV, V.N. (1944). [*Foundations of a Comparative Anatomy of Vertebrates*.] (Moscow).

41 IVANOV, A.V. (1951). On the affiliation of the genus Siboglinum Caullery with the class Pogonophora. *Doklady Akademii Nauk USSR*, 76: 739-42.

42 BRATTSTRÖM, H. & FAUCHAULD, K. (1961). Pogonophora in Norwegian inshore waters. *Sarsia*, 2: 51-2.

43 SOUTHWARD, E. (1961). Pogonophora. *Siboga Expedition*, 25(3): 1-22.

44 *WEBB, M. (1966). Lamellibrachia barhami, gen. nov., sp. nov. (Pogonophora), from the north-east Pacific. *Bulletin of Marine Science*, 19: 18-47; *LAND, J. VAN DER. & NØRREVANG, A. (1975). The systematic position of Lamellibrachia (Annelida, Vestimentifera). *Zeitschrift für Zoologische Systematik und Evolutionforschung*, 1: 86-101; WASOWICZ, L. (2000). Huge worms live 250 years under the sea. UPI news report, http://www.vny.com/cf/News/upidetail.cfm?QID=62364/ 4 February.

45 *CHUN, C. (1903). *Aus den Tiefen des Weltmeeres* (2nd Edit.). G. Fischer Verlag (Jena).

46 PICKFORD, G.E. (1946). Vampyroteuthis infernalis Chun: an archaic dibranchiate cephalopod. I. Natural history and distribution. *Dana Report*, No. 29: 1-40.

47 *LEMCHE, H. (1957). A new living deep-sea mollusc of the Cambro-Devonian class Monoplacophora. *Nature*, 179 (23 February): 413-16; BARNES, R.D. (1987). *Invertebrate Zoology* (5th Edit.). W.B. Saunders (New York).

48 *CLARKE, A.M. & MENZIES, R.J. (1959). Neopilina (Vema) ewingi, a second living species of the Paleozoic class Monoplacophora. *Science*, 129 (17 April): 1026-7; HYMAN, L.H. (1967). *The Invertebrates: Vol. VI, Mollusca I*. McGraw-Hill (New York).

49 LOWENSTAM, H.A. (1978). Recovery, behaviour and evolutionary implications of live Monoplacophora. *Nature*, 272 (18 May): 231-2.

50 *SANDERS, H.L. (1955). The Cephalocarida, a subclass of Crustacea from Long Island Sound. *Proceedings of the National Academy of Science*, 41: 61-6; SANDERS, H.L. (1957). The Cephalocarida and crustacean phylogeny. *Systematic Zoology*, 6: 112-29; *HESSLER, R.R. & WAKABARA, Y. (2000). Hampsonellus brasiliensis n. gen., n. sp., a cephalocarid from Brazil. *Journal of Crustacean Biology*, 20(3): 550-8.

51 *GORDON, I. (1957). On Spelaeogriphus, a new cavernicolous crustacean from South Africa. *B.B.M.N.H.Z.*, 5: 31-47.

52 *SOUTHCOTT, R.V. (1956). Studies on Australian Cubomedusae, including a new genus and species apparently harmful to man. *Australian Journal of Marine and Freshwater Research*, 7: 254-80.

53 BARNES, J.H. (1966). Studies on three venomous Cubomedusae. *Symposia of the Zoological Society of London*, 16: 307-22.

54 WOOD, G.L. (1982). *The Guinness Book of Animal Facts and Feats* (3rd Edit.). O.c.

54a *LEWIS, C. & BENTLAGE, B. (2009). Clarifying the identity of the Japanese Habu-kurage, Chironex yamaguchii, sp nov (Cnidaria: Cubozoa: Chirodropida). *Zootaxa*, No. 2030: 59-65

55 MARGULIS, L. & SCHWARTZ, K. (1988). *Five Kingdoms* (2nd Edit.). W.B. Saunders (New York).

56 AX, P. (1956). Die Gnathostomulida, eine rätselhafte Wurmgruppe aus dem Meeresand. Akademie der Wissenschaften und der Litteratur in Mainz. *Abhandlungen der Mathematischen-Naturwissenschaftlichen Klasse*, 8: 1-32.

57 AX, P. (1960). *Die Entdeckung neue Organisationstypen im Tierreich*. Die Neue Brehm Bücherei, Band 258 (Wittenberg).

58 RIEDL, R. (1969). Gnathostomulida from America. First record of the new phylum from North America. *Science*, 163: 445-52.

59 DANCE, S.P. (1966). *Shell Collecting*. Faber & Faber (London).

60 DANCE, S.P. (1969). *Rare Shells*. Faber & Faber (London).

61 *RUSSELL, F.S. (1959). A viviparous deep-sea jellyfish. *Nature*, 184 (14 November): 1527-9; RUSSELL, F.S. & REES, W.J. (1960). The viviparous scyphomedusa Stygiomedusa fabulosa Russell. *Journal of the Marine Biological Association U.K.*, 39: 303-17.

62 REPELIN, R. (1967). Stygiomedusa fabulosa n. sp. Scyphomedusé géante des profondeurs. Cahiers. *Office de la Recherche Scientifique et Technique d'Outre-Mer (Oceanographie)*, 5: 23-8.

63 ULRICH, W. (1972). Ein drittes Exemplar der grossen, viviparen Tiefseemeduse Stygiomedusa Russ...

S.G.N.F.B., 12: 48-60; CORNELIUS, P.F.S. (1972). Second occurrence of *Stygiomedusa fabulosa* (Scyphozoa). *Journal of the Marine Biological Association U.K.*, 52: 487-8.

64 HARBISON, G.R. et al. (1973). *Stygiomedusa fabulosa* from the North Atlantic: Its taxonomy, with a note on its natural history. Ibid., 53: 615-17; *Stygiomedusa gigantea*, Wikipedia page, accessed 20 April 2011; *BROWNE, E.T., In: BELL, J. (Ed.) (1910). [Description of *Diplulmaris gigantea*]. *National Antarcic Expedition 1901-1904, Natural History, Vol II, Zoology*. BMNH: London; BOURTON, J. (2010). Giant deep sea jellyfish filmed in Gulf of Mexico. http://news.bbc.co.uk/earth/hi/earth_news/newsid_8638000/8638527.stm 23 April; http://www.youtube.com/watch?v=sLGkBnw6X6U&feature=player_embedded accessed 20 April 2011.

65 CLENCH, W.J. (1960). *Cypraea leucodon* Broderip, 1828. *Journal of the Malacological Society of Australia*, No. 4: 14-15.

66 *SOWERBY, B.B. (1903). Mollusca of South Africa. *Marine Investigations in South Africa*, 2: 213-32.

67 ANON. (1961). New "living fossil" discovered. *The Times* (London), 21 December; FELL, H.B. (1962). A surviving somasteroid from the eastern Pacific Ocean. *Science*, 136 (18 May): 633-6; FELL, H.B. (1962). A living somasteroid *Platasterias latiradiata* Gray. *University of Kansas Palaeontological Contributions; Echinodermata*, 6: 1-16; FELL, H.B. (1962). A living somasteroid. *Zoologicheskii Zhurnal*, 41: 1353-66; MADSEN, F.J. (1966). The recent sea-star *Platasterias* and the fossil Somasteroidea. *Nature*, 209 (26 March): 1367; BLAKE, D.B. (1967). Skeletal elements in asteroids. *Progr. Ann. Mtgs. of the Geological Society of America* (for 1967): 15-16; BLAKE, D.B. (1972). Sea star *Platasterias*: ossicle morphology and taxonomic position. *Science*, 176 (3 February): 306-7; BLAKE, D.B. (1972). Ossicle morphology of some recent asteroids and description of some west American fossil asteroids. *University of California Publications in Geological Science*, 104: 1-59.

68 BARNES, J.H. (1964). Cause and effect in irukandji stingings. *Medical Journal of Australia*: 897-904.

69 *SOUTHCOTT, R.V. (1967). Revision of some Carybdeidae (Scyphozoa: Cubomedusae), including a description of the jellyfish responsible for the "irukandji syndrome". *Australian Journal of Zoology*, 15: 651-71; Irukandji, Wikipedia page, accessed 20 April 2011.

69a ANON. (2002). Tourist dies after tiny jellyfish sting. *Express and Star* (Wolverhampton), 1 February; SHEARS, R. & IRWIN, J. (2002). Briton killed by sting of an 'invisible' jellyfish. *Daily Mail* (London), 2 February.

69b *GERSHWIN, L-A. (2007). *Malo kingi*: A new species of irukandji jellyfish (Cnidaria: Cubozoa: Carybdeida), possibly lethal to humans, from Queensland, Australia. *Zootaxa*, No. 1659: 55-68; *GERSHWIN, L-A. 2005 Two new species of jellyfishes (Cnidaria: Cubozoa: Carybdeida) from tropical Western Australia, presumed to cause Irukandji syndrome. Ibid., No. 1084: 1-30

70 LOVERIDGE, A. (1971). Giant earwig. *Animals*, 13 (March): 507.

71 ZEUNER, F.E. (1962). A subfossil giant dermapteran from St. Helena. *P.S.Z.L.*, 138: 651-3.

72 PEARCE-KELLY, P. (1988). Project Hercules update. *Zoo News*, (summer): 3; ANON. (1993). Search for shy earwig. *Express and Star* (Wolverhampton), 22 September.

73 *DOWNEY, M.E. (1972). *Midgardia xandaros* new genus, new species, a large brisingid starfish from the Gulf of Mexico. *P.B.S.W.*, 84 (29 February): 421-6.

74 MILLER, R.L. (1971). Observations on *Trichoplax adhaerens* Schulze 1883. *American Zoologist*, 11: 698-9; MILLER, R.L. (1971). *Trichoplax adhaerens* Schulze, 1883: return of an enigma. *Biological Bulletin*, 41: 374; IVANOV, A.V. (1973). *Trichoplax adhaerens*, a phagocytella-like animal. *Zoologischeskii Zhurnal*, 52: 1117-31; GRELL, K.G. & RUTHMANN, A. (1991). Chapter 2: Placozoa. In: HARRISON, F.W. (Ed.), *Microscopic Anatomy of Invertebrates, Volume 2: Placozoa, Porifera, Cnidaria, and Ctenophora*. Wiley-Liss, Inc (New York). pp. 13-27.

75 *GERTSCH, W.J. (1973). The cavernicolous fauna of Hawaiian lava tubes, 3. Araneae (Spiders). *Pacific Insects*, 15: 163-80.

76 ANON. (1991). Blind spiders are a sight for sore eyes. *Daily Telegraph* (London), 6 August.

77 *ROUX, M. (1976). Découverte dans le Golfe de Gascogne de deux espèces actuelles du genre cenozoïque *Conocrinus* (Echinodermes, Crinoïdes pedonculés). *C.R.A.S.*, 282D: 757-60; ANON. (1973). Des fossiles vivants dans le golfe de Gascogne. *La Recherche*, 8 (February): 161.

78 *FOREST, J. & SAINT LAURENT, M. DE. (1975). Présence dans la faune actuelle d'un représentant du groupe mésozoïque des glyphéides: *Neoglyphea inopinata* gen. nov., sp. nov. (Crustacea Decapoda Glypheidae). *C.R.A.S.*, 281D: 155-8; SAINT LAURENT, M. DE. & CHACE, F.A. (1976). *Neoglyphea inopinata*: a crustacean "living fossil" from the Philippines. *Science*, 192 (28 May): 884.

79 FOREST, J. & SAINT LAURENT, M. DE. (1976). Capture aux Philippines de nouveaux exemplaires de *Neoglyphea inopinata* (Crustacea Decapoda Glypheidae). *C.R.A.S.*, 283D: 935-8; ANON. (1976). Fossil crustacean lives! *New Scientist*, 70 (27 May): 466.

80 *CRESSEY, R.F. (1976). *Nicothoe tumulosa*, a new siphonostome copepod parasite on the unique decapod *Neoglyphea inopinata*... *P.B.S.W.*, 89: 119-26.

80a *RICHER DE FORGES, B. (2006). Découverte en mer du Corail d'une deuxième espèce de glyphéide (Crustacea, Decapoda, Glypheoidea). *Zoosystema*, 28(1): 17-28; FOREST, J. (2006). *Laurentaeglyphea*, un nouveau genre pour la seconde espèce de Glyphéide récemment découverte (Crustacea Decapoda Glypheidae). *C. R. Biologies*, 329(10): 841-6.

81 ANON. (1978). Curious Australian cricket founds new family. *New Scientist*, 79 (7 September): 686; *RENTZ, D.C.F. (1980). A new family of ensiferous orthoptera from the coastal sands of southeast Queensland. *Memoirs of the Queensland Museum*, 20: 49-63.

81a *RENTZ, D.C. (1986). The Orthoptera Family Cooloolidae, including description of two new species and observations on biology and food preferences. *Systematic Entomology*, 11: 231-46; *RENTZ, D.C. (1999). Pearson's Monster, a new species of Cooloola Rentz from Queensland (Orthoptera: Cooloolidae). *Journal of Orthoptera Research*, 8: 25-32; RENTZ, D.C.F.

(1994). Personal communication, 7 February; C.S.I.R.O. (1993). A new 'Cooloola monster'. *Australian National Insect Collection News*, 2: 13.

82 CORLISS, J.B. & BALLARD, R.D. (1977). Oases of life in the cold abyss. *National Geographic Magazine*, 152 (October): 440-53; GROSVENOR, G.M. (1979). Strange world without sun. Ibid., 156 (November): 680-8; BALLARD, R.D. & GRASSLE, J.F. (1979). Return to oases of the deep. Ibid., 156 (November): 689-705; LIPSCOMB, J. (Producer) (1980). *Dive to the Edge of Creation* [a National Geographic TV documentary first screened (in U.S.A.) on 7 January 1980]. Stylus Video (London); HESSLER, R. (1981). Oasis under the sea—where sulphur is the staff of life. *New Scientist*, 81 (10 December): 741-7; BRIGHT, M. (1984). *Unlocking Nature's Secrets*. BBC (London); HESSLER, R. et al. (1988). Patterns on the ocean floor. *New Scientist*, 117 (24 March): 47-51; KAHARL, V.A. (1990). *Water Baby: The Story of Alvin*. Oxford University Press (Oxford); DOVER, C.L. VAN (1996). *The Octopus's Garden: Hydrothermal Vents and Other Mysteries of the Deep Sea*. Helix Books (New York).

83 *WILLIAMS, A.B. (1980). A new crab family from the vicinity of submarine thermal vents on the Galapagos Rift... *P.B.S.W.*, 93: 943-72.

84 *BOSS, K.J. & TURNER, R.D. (1980). The giant white clam from the Galapagos Rift, *Calyptogena magnifica* species novum. *Malacologia*, 20: 161-94.

85 *PUGH, P.R. (1983). Benthic siphonophores: a review of the family Rhodaliidae... *Philosophical Transactions of the Royal Society (B)*, 301: 165-300.

86 *WOODWICK, K.H. & SENSENBAUGH, T. (1985). *Saxipendium coronatum*, new genus, new species (Hemichordata: Enteropneusta): The unusual spaghetti worms of the Galapagos rift hydrothermal vents. *P.B.S.W.*, 98: 351-65.

87 *JONES, M.L. (1980). *Riftia pachyptila*, new genus, new species, the vestimentiferan worm from the Galapagos Rift geothermal vents (Pogonophora). Ibid., 93: 1295-1313; JONES, M.L. (1981). *Riftia pachyptila* Jones: Observations on the vestimentiferan worm from the Galapagos rift. *Science*, 213 (17 July): 333-6; JONES, M.L. (1984). The giant tube worms. *Oceanus*, 27: 47-52.

88 JONES, M.L. (1985). On the Vestimentifera, new phylum: six new species, and other taxa, from hydrothermal vents and elsewhere. *Bulletin of the Biological Society of Washington*, No. 6: 117-58.

89 *NEWMAN, W.A. & HESSLER, R. (1989). A new abyssal hydrothermal verrucomorphan (Cirripedia; Sessilia): The most primitive living sessile barnacle. *Transactions of the San Diego Society of Natural History*, 21 (15 February): 259-73; *SCHEIN-FATTON, E. (1985). Découverte sur la ride du Pacifique oriental à 13°N d'un Pectinidae (Bivalvia, Pteromorphia) d'affinités paléozoïques. *C.R.A.S.*, 301: 491-6; *DESBRUYRES, D. & LAUBIER, L. (1980). *Alvinella pompejana* gen. sp. nov., aberrant Ampharetidae from the East Pacific Rise hydrothermal vents. *Oceanologica Acta*, 3: 267-74; CARY, S.C. et al. (1998). Worms bask in extreme temperatures. *Nature*, 391 (5 February): 545-6.

90 *ROSENBLATT, R.H. & COHEN, D.M. (1986). Fishes living in deepsea thermal vents in the tropical eastern Pacific, with descriptions of a new genus and two new species of eelpouts (Zoarcidae). *Transactions of the San Diego Society of Natural History*, 21 (24 February): 71-9; *COHEN, D.M. et al. (1990). Biology and description of a bythitid fish from deep-sea thermal vents in the tropical eastern Pacific. *Deep-Sea Research*, 37 (February): 267-83.

91 SMITH, C.R. et al. (1989). Vent fauna on whale remains. *Nature*, 341 (7 September): 27-8.

92 ANON. (1993). Gas guzzlers. *Fortean Times*, No. 68 (April-May): 19.

93 ANON. (1993). Boffins find huge springs underwater. *Express and Star* (Wolverhampton), 11 March; ANON. (1994). In ocean volcanoes, giant worms are turning. Reuter (London), 20 October; MACQUITTY, M. (1988). Sulphur on the menu cuisine for the hairy snail. *New Scientist*, 117 (24 March): 50; *OKUTANI, T. & OHTA, S. (1988). A new gastropod mollusk associated with hydrothermal vents in the Mariana Back-Arc Basin, Western Pacific. *Venus [Japan Journal of Malacology]*, 47: 1-9.

93a *VAN DOVER, C.L. et al. (2001). Biogeography and ecological setting of Indian ocean hydrothermal vents. *Science*, 294: 818-23.

94 DAYTON, S. (1988). The underwater light fantastic. *New Scientist*, 119 (25 August): 32; FLANAGAN, R. (1997). The light at the bottom of the sea. Ibid., 156 (13 December): 42-6.

95 FRANKLIN, C. (1994). Starstruck. *BBC Wildlife*, 12 (June): 11.

96 WOESE, C.R. et al. (1990). Towards a natural system of organisms: Proposal for the domains Archaea, Bacteria, and Eucarya. *Proceedings of the National Academy of Science USA*, 87: 4576-9; WINKLER, S. & WOESE, C.R. (1991). A definition of the domains Archaea, Bacteria and Eucarya in terms of small subunit ribosomal RNA characteristics. *System. Appl. Microbiology*, 15: 513-21; WADE, N. (1996). Scientists discover a clue to the origin of life in a microbe rather than a myth. *New York Times* (New York), 23 August.

97 *VACELET, J. (1977). Une nouvelle relique du secondaire: un répresentant actuel des éponges fossiles sphinctozoaires. *C.R.A.S.*, 285D: 509-11; ANON. (1978). Une fossile vivant résout un problème de systématique. *La Recherche*, No. 85 (January): 46.

98 PICKETT, J. (1982). *Vaceletia progenitor*, the first Tertiary sphinctozoan (Porifera). *Alcheringa*, 6: 241-7; BASILE, L.L. et al. (1984). Sclerosponges, pharetronids, and sphinctozoans (relict cryptic hard-bodied Porifera) in the modern reefs of Enewetak Atoll. *Journal of Paleontology*, 58: 636-56.

99 *VACELET, J. (1979). Une éponge tétractinellide nouvelle des grottes sous marines de la Jamaïque, associées à des membranes étrangères. *Bulletin. Museum Nationale d'Histoire Naturelle (Zool. Biol. Ecol. Anim.)*, No. 1: 33-9.

100 *YAGER, J. (1981). Remipedia, a new class of crustacean from a marine cave in the Bahamas. *Journal of Crustacean Biology*, 1: 328-33; GORDON, G. (1985). A dive into the Ice Age. *Daily Mail* (London), 24 October.

101 *VALDECASAS, A.G. (1984). Morlockiidae new family of Remipedia (Crustacea) from Lanzarote (Canary Islands). *Eos*, 60: 329-33; PALMER, R. (1986). Ecology beneath the Bahama Banks. *New Scientist*, 110 (8

May): 44-8; PALMER, R. (1987). In the land of the lusca. *Natural History*, 96 (January): 42-7; Remipedia, Wikipedia page, accessed 22 April 2011.

102 KNIGHT, J. (1997). Subterranean blues. *New Scientist*, 155 (13 September): 26; DERBYSHIRE, D. (1997). The cave dwellers. *Daily Mail* (London), 10 September.

103 MITCHELL, A.W. (1981). *Operation Drake—Voyage of Discovery*. Severn House (London).

104 *RUSSELL, A.B. (1981). A spectacular new *Idea* from Celebes (Lepidoptera, Danaidae). *Systematic Entomology*, 6: 225-8.

105 MESSER, A.C. (1984). *Chalicodoma pluto*: the world's largest bee rediscovered living communally in termite nests (Hymenoptera: Megachilidae). *Journal of the Kansas Entomological Society*, 57: 165-8; SHUKER, K.P.N. (1991). *Extraordinary Animals Worldwide*. O.c.

106 *BECKER, K-H. (1975). *Basipodella harpacticola* n. gen., n. sp. *Helgoländer Wissenschaftliche Meeresunters*, 27: 96-100; *BRADFORD, J.M. & HEWITT, G.C. (1980). A new maxillopodan crustacean parasitic on a myodocopid ostracod. *Crustaceana*, 38: 67-72; BOXSHALL, G.A. & LINCOLN, R.J. (1983). Tantulocarida, a new class of Crustacea ectoparasitic on other crustaceans. *Journal of Crustacean Biology*, 3: 1-16; Tantulocarida, Wikipedia page, accessed 22 April 2011.

106a *MOHRBECK, I. et al. (2010). Tantulocarida Crustacea) from the Southern Ocean deep sea, and the description of three new species of *Tantulacus* Huys, Andersen & Kristensen, 1992. *Systematic Parasitology*, 77(2): 131-51.

107 LEWIN, R. (1983). New phylum discovered, named. *Science*, 222 (14 October): 149; *KRISTENSEN, R.M. (1983). Loricifera, a new phylum with Aschelminthes characters from the meiobenthos. *Zeitschrift für Zoologische Systematik und Evolutionsforschung*, 21: 163-80; KRISTENSEN, R.M. (1991). Chapter 9: Loricifera. In: HARRISON, F.W. & RUPPERT, E.E. (Eds.), *Microscopic Anatomy of Invertebrates, Volume 4: Aschelminthes*. Wiley-Liss, Inc (New York). pp. 351-75.

108 *HIGGINS, R.P. & KRISTENSEN, R.M. (1986). New Loricifera from southeastern United States coastal waters. *Smithsonian Contributions to Zoology*, No. 438: 1-70.

109 *KRISTENSEN, R.M. & SHIRAYAMA, Y. (1988). *Pliciloricus hadalis* (Pliciloricidae), a new loriciferan species collected from the Izu-Ogasawara Trench, Western Pacific. *Zoological Science*, 5: 875-81.

110 KRISTENSEN, R.M. (1992). Personal communication, 11 February; KRISTENSEN, R.M. (1999). Personal communications, 14 November, 8 December.

110a *HEINER, I. & KRISTENSEN, R.M. (2005). Two new species of the genus *Pliciloricus* (Loricifera, Pliciloricidae) from the Faroe Bank, North Atlantic. *Zoologischer Anzeiger*, 243(3): 121-38.

111 *BAKER, A.N., ROWE, F.W.E., & CLARK, H.E.S. (1986). A new class of Echinodermata from New Zealand. *Nature*, 321 (26 June): 862-4; ANON. (1986). Sea daisy, a star in a class of its own. *New Scientist*, 111 (3 July): 30.

112 *ROWE, F.W.E., BAKER, A.N., & CLARK, H.E.S. (1988). The morphology, development and taxonomic status of *Xyloplax* ..., with the description of a new species. *Proceedings of the Royal Society of London (B)*, 233: 431-59.

112a *MAH, C.L. (2006). A new species of *Xyloplax* (Echinodermata: Asteroidea: Concentricycloidea) from the northeast Pacific: comparative morphology and a reassessment of phylogeny. *Invertebrate Biology*, 125 (No. 2; June): 136-53; VOIGHT, J.R. (2005). First report of the enigmatic echinoderm *Xyloplax* from the North Pacific. *Biological Bulletin*, 208: 77-80.

113 *ROUX, M. (1985). Découverte d'un représentant actuel des crinoïdes pédonculés paléozoïques Inadunata (Echinodermes) dans l'étage bathyal de l'Île de la Reunion (Océan Indien). *C.R.A.S.*, 301: 503-6; *BOURSEAU, J-P. et al. (1991). Echinodermata: les crinoïdes pédonculés de Nouvelle-Calédonie. pp. 229-333. In: Crosnier A, (Ed.) *Résultats des Campagnes Musorstom, Vol. 8*. Mémoires de la Muséum Nationale d'Histoire Naturelle (Paris).

114 ANON. (1988). A singular fly. *New Scientist*, 117 (14 January): 30; *HUTTON, F.W. (1901). Synopsis of the Diptera Brachycera of New Zealand. *Transactions and Proceedings of the New Zealand Institute*, 31: 1-95; PATRICK, B.H. (1996). The status of the bat-winged fly, *Exsul singularis* Hutton (Diptera: Muscidae: Coenosiinae). *New Zealand Entomologist*, 19: 31-3.

115 ANON. (1990). She's got ants in her plants. *New York Post* (New York), 16 October; *WILSON, E.O. (2003). *Pheidole in the New World. A Dominant, Hyperdiverse Ant Genus*. Harvard University Press (Cambridge, MA).

116 ANON. (1991). Bulldozers millipeded. *Express & Star* (Wolverhampton), 29 May; BURROWS, F.J. et al. (1994). Aquatic millipedes in Australia: a biological enigma and a conservation saga. *Australian Zoologist*, 29 (Nos. 3-4): 213-16; BLACK, D.G. (1997). Diversity and biogeography of Australian millipedes (Diplopoda). *Memoirs of Museum Victoria*, 56(2): 557-61; GOLOVATCH, S.I. & KIME, R.D. (2009). Millipede (Diplopoda) distributions: A review. *Soil Organisms*, 81(3): 565-97.

117 DOUBLE, T. & DOUBLE, A. (1992). Here be giants. *BBC Wildlife*, 10 (May): 34-40; COLLARD, M. & YONOW, N. (1992). Looks familiar? Ibid., 10 (November): 89; YONOW, N. (2000). Personal communication, 29 February.

118 *SILVESTRI, F. (1902). Contribuzione alla conoscenza dei Meliponidid del Bacino del Rio de la Plata. *Rivista de Patologia Vegetale*, Portici, 10: 121-74; ANON. (1982). No taste of honey for flesh-eating bees. *New Scientist*, 95 (19 August): 486; ROUBIK, D.W. (1982). Obligate necrophagy in a social bee. *Science*, 217 (10 September): 1059-60.

119 *CAMARGO, J.M.F. & ROUBIK, D.W. (1991). Systematics and bionomics of the apoid obligate necrophages: the *Trigona hypogea* group (Hymenoptera: Apidae; Meliponinae). *Biological Journal of the Linnean Society*, 44: 13-39; SERRÃO, J.E. (2000). Personal communications, 28 and 31 January.

119a *WARD, P.S. (1994). *Adetomyrma*, an enigmatic new ant genus from Madagascar (Hymenoptera: Formicidae), and its implications for ant phylogeny. *Systematic Entomology*, 19: 159-75; ANON. (2001). Evolutionary biologist discovers missing pieces of the evolutionary

puzzle in ant evolution. *California Academy of Sciences*, press release, 8 January; ANON. (2001). 'Dracula ants' may be key evolutionary link. http://www.cnn.com/2001/NATURE/01/10/science.ants.reut/index.html 10 January.
120 *DILLY, P.N. (1993). *Cephalodiscus graptolitoides* sp. nov. a probable extant graptolite. *Journal of Zoology*, 229: 69-78.
121 ZALASIEWICZ, J. & RIGBY, S. (1993). The creatures that time forgot. *New Scientist*, 140 (18 December): 38-40; URBANEK, A. (1994). Living non-graptolite. *Lethaia*, 27: 18; DILLY, P.N. (1994). When is a graptolite not a graptolite? Ibid., 27: 34.
122 NORMAN, M. (1999). Personal communications, 15 and 29 November.
123 KENNEDY, J. (1999). The great pretender. *BBC Wildlife*, 17 (August): 28-33.
123a NORMAN, M.D. et al. (2001). Dynamic mimicry in an Indo-Malayan octopus. *Proceedings of the Royal Society (Series B)*, 268 (7 September): 1755-8; *NORMAN, M.D. & HOCHBERG, F.G. (2005). The "mimic octopus" (*Thaumoctopus mimicus* n. gen. et sp.), a new octopus from the tropical Indo-West Pacific (Cephalopoda: Octopodidae). *Molluscan Research*, 25 (No 2; 31 August): 57-70.
123b *HOCHBERG, F.G., NORMAN, M.D. & FINN, J. (2006). *Wunderpus photogenicus* n. gen. and sp., a new octopus from the shallow waters of the Indo-Malayan Archipelago (Cephalopoda: Octopodidae). Ibid., 26 (No 3; December): 128-40.
124 VACELET, J. & BOURY-ESNAULT, N. (1995). Carnivorous sponges. *Nature*, 373 (26 January): 333-5; ANON. (1995). 'Horror star' of ocean depths. *Express and Star* (Wolverhampton), 26 January; *VACELET, J. & BOURY-ESNAULT, N. (1996). A new species of carnivorous sponge (Demospongiae: Cladorhizidae) from a Mediterranean cave. pp 109-15. In: WILLENZ, P. (Ed.), *Recent Advances in Sponge Biodiversity Inventory and Documentation*. Bulletin de l'Institut Royal des Sciences Naturelles de Belgique. Biologie, 66.
125 DUBRANA, D. (1992). La caverne des fossiles vivants. *Science et Vie*, No. 903 (December): 74-9; LASCU, C., POPA, R., SARBU, S., et al. (1993). La grotte de Movilé: une faune hors du temps. *La Recherche*, 24 (October): 1092-8; *DÉCU, V., GRUIA, M., KEFFER, S.L., & SARBU, S.M. (1994). A stygobiotic waterscorpion, *Nepa anophthalma* n. sp. (Hemiptera, Nepidae), from Movile Cave, Romania. *Annals of the Entomological Society of America*, 87: 755-61; HAWKES, N. (1996). Explorers discover a lost world the apes left behind. *The Times* (London), 14 February; WILKIE, T. (1996). 30 new species emerge to see the light of day. *Independent* (London), 14 February; SARBU, S.M. et al. (1996). A chemoautotrophically based cave ecosystem. *Science*, 272 (28 June): 1953-5.
126 ANDERSEN, P.F. [=DR. PETER FUNCH] (1993). En hidtil ukendt gruppe af hvirvelløse dyr. *Dansk Naturhistorisk Forening*, 5: 58.
127 *FUNCH, P. & KRISTENSEN, R.M. (1995). Cycliophora is a new phylum with affinities to Entoprocta and Ectoprocta. *Nature*, 378 (14 December): 711-14; MORRIS, S.C. (1995). A new phylum from the lobster's lips. Ibid., 378 (14 December): 661-2; FUNCH, P. & KRISTENSEN, R.M. (1997). Chapter 6: Cycliophora. In: HARRISON, F.W. & WOOLLACOTT, R.M. (Eds.), *Microscopic Anatomy of Invertebrates, Volume 13: Lophophorates, Entoprocta, and Cycliophora*. Wiley-Liss (New York). pp. 409-74.
128 WINNEPENNINCKX, B.M.H., BACKELJAU, T., & KRISTENSEN, R.M. (1998). Relations of the new phylum Cycliophora. *Nature*, 393 (18 June): 636-8.
128a *OBST, M.; FUNCH, P. & KRISTENSEN, R.M. (2006). A new species of Cycliophora from the mouthparts of the American lobster, *Homarus americanus* (Nephropidae, Decapoda). *Organisms Diversity & Evolution*, 6: 83-97.
129 *DUFFY, J.E. (1996). *Synalpheus regalis*, new species, a sponge-dwelling shrimp from the Belize Barrier Reef, with comments on host specificity in *Synalpheus*. *Journal of Crustacean Biology*, 16: 564-73; DUFFY, J.E. (1996). Eusociality in a coral-reef shrimp. *Nature*, 381: 512-14; ADLER, T. (1996). A shrimpy find: communal crustaceans. *Science News Online*, 8 June.
130 KENNEDY, B. (1997). Scientists discover methane ice worms on Gulf of Mexico sea floor. Pennsylvania State University press release, http://www.bio.psu.edu/fisher/main.html/ 30 July; RADFORD, T. (1997). Methane-mad worms may hold key to new world underwater. *Guardian* (Manchester), 12 August; *DESBRUYÈRES, D. & TOULMOND, A. (1998). A new species of hesionid worm, *Hesiocaeca methanicola* sp. nov. (Polychaeta: Hesionidae), living in ice-like methane hydrates in the deep Gulf of Mexico. *Cahiers de Biologie Marine*, 39(1): 93-8.
131 CROWDER, W. (1926). The life of the moon-jelly. *National Geographic Magazine*, 50: 187-202; HALSTEAD, B.W. (1965). *Poisonous and Venomous Marine Animals of the World, Vol. 1. Invertebrates*. U.S. Government Printing Office (Washington DC).
132 MARTIN, J.W. & KUCK, H.G. (1991). Faunal associates of an undescribed species of *Chrysaora* (Cnidaria, Scyphozoa) in the Southern California Bight, with notes on unusual occurrences of other warm water species in the area. *Bulletin of the South California Academy of Sciences*, 90(3): 89-101.
133 *MARTIN, J.W. et al. (1997). *Chrysaora achlyos*, a remarkable new species of scyphozoan from the eastern Pacific. *Biological Bulletin*, 193 (August): 8-13.
134 ANON. (1997). Monsters lurk beneath the ice. *The Times* (London), 9 September; OULTON, C. (1997). Monsters trapped in icy wastes. *Express* (London), 9 September.
135 DERBYSHIRE, D. (1998). The mosquito line. *Daily Mail* (London), 26 August; BLACKMAN, S. (1998). Mind the gnat. *BBC Wildlife*, 16 (September); BURDICK, A. (2001). Insect from the Underground. *Natural History*, (February); London Underground mosquito, Wikipedia page, accessed 29 April 2011.
136 *AHYONG, S.T. & MANNING, S.B. (1998). Two new species of *Erugosquilla* from the Indo-West Pacific (Crustacea: Stomatopoda: Squillidae). *Proceedings of the Biological Society of Washington*, 111(3): 653-62; SHEARS, R. (1999). Put another Jurassic prawn on the barbie. *Daily Mail* (London), 7 April; ROSENBERG, Y. (1999). [Article re *Erugosquilla grahami*]. Newsweek.com Periscope, http://www.newsweek.com/

nw-srv/tnw/today/ps/ps03th_1.htm/ 9 April; SHUKER, K.P.N. (1999). Jurassic prawn. *Fortean Times*, No. 124 (July): 13; AHYONG, S.T. (2000). Personal communication, 14 February.

137 *HORITA, T. (2000). An undescribed lobate ctenophore, *Lobatolampea tetragona* gen. nov. & spec. nov., representing a new family, from Japan. *Zoologische Mededelingen*, 73 (March): 457-64.

138 KRISTENSEN, R.M. & FUNCH, P. (1995). En ny aschelminth med gnathostomulid-lignende kæber fra en kold kilde ved isunngua. In: Ehrhardt, C. (Ed.), *Arktisk Biologisk Feltkursus Qeqertarsuaq/Godhavn 1994*. Botanisk Institut (Copenhagen). pp. 73-82; KRISTENSEN, R.M. (1995). Are Aschelminthes pseudo-coelomate or acoelomate? In: LANZAVECCHIA, R. et al. (Eds.), *Body Cavities: Function and Phylogeny*. U.Z.I. (Modena). pp. 41-3; KRISTENSEN, R.M. (1999). Personal communications, 14 and 15 November, 13 December.

139 KRISTENSEN, R.M. (2000). Personal communications, 3 and 7 August; *KRISTENSEN, R.M. & FUNCH, P. (2000). Micrognathozoa: A new class with complicated jaws like those of Rotifera and Gnathostomulida. *Journal of Morphology*, 246: 1-49.

PART 2:
21ST-CENTURY NEW AND
REDISCOVERED ANIMALS

Section 1: The Mammals

1 BROOK, S. (2000). Scientists pull new phascogale out of the hat. *The Australian* (Sydney), 1 June.

2 ANON. (2000). DNA research reveals a new whale species. Wildlife Conservation Society press release, 15 November.

3 *THALMANN, U. & GEISSMANN, T. (2000). Distribution and geographic variation in the western woolly lemur (*Avahi occidentalis*) with description of a new species (*A. unicolor*). *International Journal of Primatology*, 21 (December): 915-41; *GROVES, C.P. (2000). The genus *Cheirogaleus*: unrecognized biodiversity in dwarf lemurs. Ibid., 21 (December): 943-62; *RASOLOARISON, R.M. et al. (2000). Taxonomic revision of mouse lemurs (*Microcebus*) in the western portions of Madagascar. Ibid., 21 (December): 963-1019.

4 ANON. (2001). Rare salt-water camel may be separate species. AOL News/Reuters, 12 February; SMITH, J. (2001). Distinctive camel. *BBC Wildlife*, 19 (April).

5 *ANDERSON, R & HANDLEY, C. (2001). A new species of three-toed sloth (Mammalia: Xenarthra) from Panama, with a review of the genus *Bradypus*. *Proceedings of the Biological Society of Washington*, 114(1): 1-33.

6 *LE SOUEF, A.S. (1924). [Description of *Petrogale purpureicollis*.] *Australian Journal of Zoology*, 3: 274; ELDRIDGE, M.D.B. et al. (2001). Taxonomy of rock-wallabies, *Petrogale* (Marsupialia: Macropodidae). III. Molecular data confirms the species status of the purple-necked rock-wallaby (*Petrogale purpureicollis* Le Souef). Ibid., 49(4): 323-43.

7 ANON. (2002). Rare African predator photographed for first time. *National Geographic News*, 20 June.

8 *DALEBOUT, M. et al. (2002): A new species of beaked whale, *Mesoplodon perrini* sp. n. (Cetacea: Ziphiidae), discovered through phylogenic analysis of mitochondrial DNA sequences. *Marine Mammal Science*, 18(3): 577-608.

8a *SHORTRIDGE, G.C. & CARTER, T.D. (1938). New genus and new species and subspecies of mammals from Little Namaqualand and the north-west Cape Province... *Annals of the South African Museum*, 32 (July): 281-91; HELGEN, K.M. & WILSON, D.E. (2003). Additional material of the enigmatic golden mole *Cryptochloris zyli*, with notes on the genus *Cryptochloris* (Mammalia: Chrysochloridae). *African Zoology*, 36(1): 110-12; *BROOM, R. (1950). Some further advances in our knowledge of the Cape golden moles. *Annals of the Transvaal Museum*, 21 (9 May): 234-41; *SIMONETTA, A.M. (1968). A new golden mole from Somalia with an appendix on the taxonomy of the family Chrysochloridae (Mammalia, Insectivora). *Monitore Zoologico Italiano, New Series*, 2 (Supplemento): 27-55; BLAKE, M. (2010). Cryptic super-moles. *Animals and Men*, No. 48: 78-80.

9 *WADA, S. et al. (2003). A newly discovered species of living baleen whale. *Nature*, 426: 278-81.

10 *LUNA, L. & PATTERSON, B.D. (2003). A remarkable new mouse (Muridae: Sigmodontinae) from southeastern Peru: with comments on the affinities of *Rhagomys rufescens* (Thomas, 1886). *Fieldiana Zoology, (New Series)*, 101: 1-24.

11 *ROOSMALEN, M. et al. (2007). A new species of living peccary (Mammalia: Tayassuidae) from the Brazilian Amazon. *Bonner Zoologischen Beiträge*, 55(2): 105-12; ROOSMALEN, M. VAN (2007). Several personal communications, May; GONGORA, J. et al. (2007). Re-examining the evidence for a 'new' peccary species, 'Pecari maximus', from the Brazilian Amazon. *Suiform Soundings*, 7 (No. 2; December): 19-26.

12 *KÖNIG, C. (1962). Eine neue Wühlmaus aus der Umgebung Garmisch-Partenkirchen (Oberbayern): *Pitymys bavaricus*. *Senckenbergiana Biologica*, 43: 1-10; ANON. (2001). Bavarian vole still alive. *Oryx*, 35 (July): 185.

13 *SINHA, A. et al. (2005). *Macaca munzala*: a new species from western Arunachal Pradesh, northeastern India. *International Journal of Primatology*, 26(4): 977-89; *New York Times* (New York), BBC News (London), and http://www.newkerala.com/news-daily/news/features.php?action=fullnews&id=48281 all 16 December 2004.

13a *MEIJAARD, E. & GROVES, C.P. (2004). A taxonomic revision of the *Tragulus* mouse-deer (Artiodactyla). *Zoological Journal of the Linnean Society*, 140(1): 63-102.

13b *COTTERILL, F.P.D. (2005). The Upemba lechwe: an antelope new to science emphasizes the conservation importance of Katanga, Democratic Republic of Congo. *Journal of Zoology*, 265: 113-32

14 *BEASLEY, I. et al. (2005): Description of a new dolphin, the Australian snubfin dolphin *Orcaella heinsohni* sp. n. (Cetacea, Delphinidae). *Marine Mammal Science*, 21(3): 365-400.

15 *THALMANN, U. & GEISSMANN, T. (2005). New species of woolly lemur *Avahi* (Primates: Lemuriformes) in Bemaraha (Central Western Madagascar). *American Journal of Primatology*, 67: 371-6.

16 *Louis, E.E. et al. (2006). Molecular and morphological analyses of the sportive lemurs (family Megaladapidae: genus *Lepilemur*) reveals 11 previously unrecognized species. *Special Publications, Museum of Texas Tech University*, 49: 1-47; Tattersall, I. (2007). Madagascar's lemurs: cryptic diversity or taxonomic inflation? *Evolutionary Anthropology: Issues, News, and Reviews*, 16: 12-23.

17 Ammann, K., The Bondo mystery apes? What exactly has Karl found in the DRC? http://www.karlammann.com/bondo.html accessed 28 May 2004; Roach, J. (2003). Elusive African apes? Giant chimps or new species? *National Geographic News*, 14 April; Walton, M. (2003). Seeking answers to big 'mystery ape'. CNN News report, 9 August; Bedlan, B. (2003). Omaha Zoo testing DNA of mystery apes. Associated Press report, 22 September; Hanlon, M. (2004). King Congo: Scientists claim they have discovered a new species of giant killer ape which stands 7 ft tall and eats lions. *Daily Mail* (London), 14 October; Young, E. & Barnett, A. (2006). DNA tests solve mystery of giant apes. *New Scientist*, No. 2558 (30 June); Randerson, J. (2007). Found: the giant lion-eating chimps of the magic forest. *Guardian* (London), 14 July; Barone, J. (2007). Bondo mystery ape proves to be a chimpanzee with unusual habits. *Discover* (online), 15 March; Hanlon, M. (2007). Chimps who eat lions. *Daily Mail* (London), 16 July.

18 *Jenkins, P.D. et al. (2004): Morphological and molecular investigations of a new family, genus and species of rodent (Mammalia: Rodentia: Hystricognatha) from Lao PDR. *Systematics and Biodiversity*, 2(4): 419-54; Dawson, M.R. et al. (2006). *Laonastes* and the "Lazarus effect" in Recent mammals. *Science*, 311: 1456-8.

19 *Jones, T. et al. (2005). The highland mangabey *Lopocebus kipunji*: A new species of African monkey. *Science*, 308 (No. 5725): 1161-4; Davenport, T.R.B. et al. (2006). A new genus of African monkey, *Rungwecebus*: Morphology, ecology, and molecular phylogenetics. Ibid., 312 (No. 5778): 1378-81.

19a *Wallace, R.B. et al. (2006). On a new species of titi monkey, genus *Callicebus* Thomas (Primates, Pitheciidae), from western Bolivia with preliminary notes on distribution and abundance. *Primate Conservation*, 20: 29-39; The Official Website of the Golden Palace Monkey, http://www.goldenpalacemonkey.com/ accessed 18 August 2011.

20 *Cucchi, T. et al. (2006). A new endemic species of the subgenus *Mus* (Rodentia, Mammalia) on the island of Cyprus. *Zootaxa*, No. 1241: 1-36; Naish, D. (2006). The first new European mammal in 100 years? You must be joking. http://darrennaish.blogspot.com/2006/10/first-new-european-mammal-in-100-years.html 18 October.

21 *Muchhala, N. et al. (2005). A new species of *Anoura* (Chiroptera: Phyllostomidae) from the Ecuadorian Andes. *Journal of Mammalogy*, 86(3): 457-61.

22 *Goodman, S.M. et al. (2007). The description of a new species of *Myzopoda* (Myzopodidae: Chiroptera) from western Madagascar. *Mammalian Biology*, 72: 65-81.

23 *Buckley-Beason, V.A. et al. (2006). Molecular evidence for species-level distinctions in clouded leopards. *Current Biology*, 16 (5 December): 2371-6; Kitchener, A.C. et al. (2006). Geographical variation in the clouded leopard, *Neofelis nebulosa*, reveals two species. *Current Biology*, 16 (5 December): 2377-83.

24 *Patterson, B.D. & Velazco, P.M. (2006). A distinctive new cloud-forest rodent (Hystricognathi: Echimyidae) from the Manu Biosphere Reserve, Peru. *Mastozoología Neotropical*, 13(2): 175-91.

24a Marcgrave, G. (1648). *Liber Sextus: De Quadrupedibus, et Sepentibus, Historiae Rerum Naturalium Brasiliae*. Franciscus Hackius (Lugdunum Batavorum); *Schreber, J.C.D. von (1774), *Die Säugethiere in Abbildungen nach der Natur, mit Beschreibungen*. Siegfried Leberecht Crusius (Leipzig); Mendes Pontes, A. R. et al. (2006), A new species of capuchin monkey, genus *Cebus* Erxleben (Cebidae, Primates): Found at the very brink of extinction in the Pernambuco Endemism Centre. *Zootaxa*, No. 1200: 1-12; Oliveira, M.M. de & Langguth, A. (2006), Rediscovery of Marcgrave's capuchin monkey and designation of a neotype for *Simia flavia* Schreber, 1774 (Primates, Cebidae). *Boletim do Museu Nacional: Nova Série: Zoologia*, 523: 1-16.

25 Amazonian Association for the Preservation of Nature. http://www.marcvanroosmalen.org/; Shuker, K.P.N. (2007). Amazing Amazonia. *Fortean Times*, No. 226 (August): 42-5; Naish, D. (2007). Multiple new species of large, living mammal (Parts I-IV). http://scienceblogs.com/tetrapodzoology 1 June.

26 *Roosmalen, M.G.M. van (2008). A new species of living lowland tapir (Mammalia: Tapiridae) from the Brazilian Amazon. http://marcvanroosmalen.org/dwarftapir.htm

27 *Roosmalen, M.G.M. van (2008). A new living species of manatee from the Amazon. http://www.marcvanroosmalen.org/dwarfmanatee.htm; Hammer, J. (2008). Trials of a primatologist. http://www.smithsonianmag.com/science-nature/roosmalen-200802.html February.

28 *Roosmalen, M.G.M. van & Hooft, P. van (2008). A new species of living brocket deer (Mammalia: Cervidae) from the Brazilian Amazon. http://www.marcvanroosmalen.org/fairbrocketdeer.htm

29 Anon. (2007). New species from PNG "lost world". http://www.mongabay.com 16 December; Anon. (2009). New giant rat species discovered. http://articles.cnn.com/2009-09-07/world/giant.rat.papua_1_papua-new-guinea-new-species-rat?_s=PM:WORLD 7 September.

30 *Rovero, F. et al. (2008). A new species of giant sengi or elephant-shrew (genus *Rhynchocyon*) highlights the exceptional biodiversity of the Udzungwa Mountains of Tanzania. *Journal of Zoology*, 274: 126-33; Barley, S. (2010). Scientists trumpet new elephant shrew species. *Guardian* (London), http://www.guardian.co.uk/environment/2010/sep/16/new-elephant-shrew-species?CMP=twt_gu 16 September.

31 *Boubli, J.-P. et al. (2008). A taxonomic reassessment of black uakari monkeys, *Cacajao melanocephalus* group, Humboldt (1811), with the description of two new species. *International Journal of Primatology*, 29: 723-49; *Defler, T.R. et al. (2010). *Callicebus caquetensis*: a new and critically endangered titi monkey from southern Caquetá, Colombia. *Primate*

Conservation, No. 25: 1-9; ANON. (2011). Scientists discover new monkey species in Amazon. http://wwf.panda.org/wwf_news/?201430/Scientists-discover-new-monkey-species-in-Amazon 24 August.

32 *LEI, R. et al. (2008). Nocturnal lemur diversity at Masoala National Park. *Special Publications of the Museum of Texas Tech University*, 53: 1-48.

33 ANON. (2008). "Lost" deer rediscovered in Indochina. http://www.reuters.com/article/2008/10/10/us-deer-idUSTRE4995GN20081010 10 October; *ROBINSON, H. C. & KLOSS, C.B. (1918). Results of an expedition to Korinchi Peak, Sumatra. I. Mammals. *Journal of the Federation of Malay States Museum*, 8(2): 1-81.

34 MACEY, R, (2008). DNA discovery shows dolphin one of a new breed. *Sydney Morning Herald* (Sydney), 20 November; MÖLLER, L.A. et al. (2008). Multi-gene evidence for a new bottlenose dolphin species in southern Australia. *Molecular Phylogenetics and Evolution*, 49: 674-81; Charlton-Robb, K. et al. (2011) A new dolphin species, the burrunan dolphin *Tursiops australis* sp. nov., endemic to southern Australian coastal waters. *PLoS ONE* 6(9): 1-17.

35 *DURBIN, J. et al. (2010). Investigations into the status of a new taxon of *Salanoia* (Mammalia: Carnivora: Eupleridae) from the marshes of Lac Alaotra, Madagascar. *Systematics and Biodiversity*, 8(3): 341-55.

36 *THINH, V.N. et al. (2010). A new species of crested gibbon, from the central Annamite mountain range. *Vietnamese Journal of Primatology*, 1(4): 1-12; ANON. (2010). New gibbon species discovered in Indochina. http://www.innovations-report.com/html/reports/life_sciences/gibbon_species_discovered_indochina_161988.html 29 September.

37 HANCE, J. (2011). Egyptian jackal is actually ancient wolf. http://news.mongabay.com/2011/0126-hance_africanwolf.html 26 January.

Section 2: The Birds

1 STOLZENBURG, W. (2000). Scientists pluck a new sage grouse from the old. *Nature Conservancy Magazine*, (November/December); *YOUNG, J.R. et al. (2000). A new species of sage-grouse (Phasianidae: Centrocercus) from southwestern Colorado. *Wilson Bulletin*, 112: 445-53.

2 ANON. (2001). Long-lost bird raises its head. http://www.academicpress.com/inscight/03142001/grapha.htm 14 March; BONTHRONE, P.J. (2001). 'Extinct' turkey is gobbled up. *Daily Telegraph* (London), 6 August; Bruijn's brush-turkey *Aepypodius bruijnii* BirdLife International species fact sheet. http://www.birdlife.org/datazone/speciesfactsheet.php?id=112, accessed 9 June 2011.

3 OLMOS, F. & PACHECO, J. F. (2003). Rediscovery of golden-crowned manakin *Lepidothrix vilasboasi*. *Cotinga*, 20: 48-50.

4 Indigo-winged parrot *Hapalopsittaca fuertesi* BirdLife International species fact sheet. http://www.birdlife.org/datazone/speciesfactsheet.php?id=112, accessed 9 June 2011.

5 BROWN, P. (2002). This parrot's alive—and bald. *Guardian* (London), 20 May; ASTOR, M. (2002). New parrot discovered. Associated Press report, 31 May; *GABAN-LIMA, R. et al. (2002). Description of a new species of *Pionopsitta* (Aves: Psittacidae) endemic to Brazil. *Auk*, 119: 815-19.

5a *SILVEIRA, L.F. et al. (2005). A new species of *Aratinga* parakeet (Psittaciformes:Psittacidae) from Brazil, with taxonomic remarks on the *Aratinga solstitialis* complex. *Auk*, 122: 292-305; Sulphur-breasted parakeet. Wikipedia page, accessed 13 July 2011.

6 *OLSEN, J. et al. (2002). A new *Ninox* owl from Sumba, Indonesia. *Emu*, 102: 223-31.

7 *WHITTAKER, A. (2003). A new species of forest-falcon (Falconidae: *Micrastur*) from southeastern Amazonia and the Atlantic rainforests of Brazil. *Wilson Bulletin*, 114: 421-45.

8 *CARDOSA DA SILVA, J.M. et al. (2003). Discovered on the brink of extinction: a new species of pygmy-owl (Strigidae: *Glaucidium*) from Atlantic forest of northeastern Brazil. *Ararajuba*, 10: 123-30.

9 KIRBY, A. (2003). Fiji's 'extinct' bird flies anew. http://thewe.cc/contents/more/archive2003/november/fijis_thought_extinct_bird_still_flies.htm 28 November.

10 FLOOD, R. (2003). The New Zealand storm-petrel is not extinct. *Birding World*, 16: 479-83; New Zealand storm-petrel *Oceanites maorianus* Birdlife International species fact sheet. http://www.birdlife.org/datazone/speciesfactsheet.php?id=30105 accessed 11 June 2011; SZIMULY, G. (2011). Another new storm-petrel discovered. http://birdingblogs.com/2011/szimi/another-new-storm-petrel-discovered 25 February.

11 BLUNT, D. & FRITH, C.B. (2004). Rawnsley's bowerbird (satin x regent). http://www.gondwanaguides.com.au/Rawnsley%27s%20Bowerbird.htm

12 *ALLEN, D. et al. (2004). A new species of *Gallirallus* from Calayan island, Philippines. *Forktail*, 20: 1-7; *GOODMAN, S. et al. (2011). Patterns of morphological and genetic variation in the *Mentocrex kioloides* complex (Aves: Gruiformes: Rallidae) from Madagascar, with the description of a new species. *Zootaxa*, No. 2776: 49-60

13 BANKS, R.C. et al. (2004). Forty-fifth supplement to the American Ornithologists' Union Check-list of North American Birds. *Auk*, 121: 985-95.

14 KIRBY, T. (2006). Paradise found! *Independent* (London), 7 February; *BEEHLER, B. et al. (2007). A new species of smoky honeyeater (Meliphagidae: *Melipotes*) from western New Guinea. *Auk*, 124(3): 1000-9.

15 ROUND, P.D. et al. (2007) Lost and found: the enigmatic large-billed reed warbler *Acrocephalus orinus* rediscovered after 139 years. *Journal of Avian Biology*, 38(2): 133-8; TIMMINS, R.J. et al. (2009). The discovery of large-billed reed warblers *Acrocephalus orinus* in north-eastern Afghanistan. *BirdingASIA*, 12: 42-5.

16 *LECROY, M. & BARKER, F.K. (2006). A new species of bush-warbler from Bougainville Island and a monophyletic origin of southwest Pacific *Cettia*. *American Museum Novitates*, No. 3511: 1-20; BEEHLER, B. (1983). Thoughts on an ornithological mystery from Bougainville Island, Papua New Guinea. *Emu*, 83: 114-15; NAISH, D. (2010). *Tetrapod Zoology Book One*. CFZ Press (Bideford).

17 CLEERE, N. et al. (2007). A new genus of frogmouth (Podargidae) from the Solomon Islands—results from

a taxonomic review of *Podargus ocellatus inexpectatus* Hartert 1901. *Ibis*, 149: 271-86.

18 ANON. (2009)). Long feared extinct, rare bird rediscovered. http://www.eurekalert.org/pub_releases/2009-10/msu-lfe100909.php 9 October; *ROTHSCHILD, W. & HARTERT, E. [Description of *Corvus unicolor*]. *B.B.O.C.*, 11: 29.

19 *MURPHY, R.C. & MATTHEWS, G.M. (1928). Birds collected during the Whitney South Sea Expedition, 5. *American Museum Novitates*, No. 337: 1-18; SHIRIHAI, H. (2008). Rediscovery of Beck's Petrel *Pseudobulweria becki*, and other observations of tubenoses from the Bismarck archipelago, Papua New Guinea. *B.B.O.C.*, 128: 3-16.

20 *WOXVOLD, I.A. et al. (2009). An unusual new bulbul (Passeriformes Pycnonotidae) from the limestone karst of Lao. *Forktail*, 25 (July): 1-12; ANON. (2009). Bizarre bald songbird discovered in Asia. *Science Daily*, http://www.sciencedaily.com/releases/2009/07/090729203655.htm 31 July.

21 *PYLE, P. et al. (2011). A new species of shearwater (*Puffinus*) recorded from Midway Atoll, northwestern Hawaiian Islands. *Condor*, 113 (August): 518-27.

Section 3: The Reptiles and Amphibians

1 *HARVEY, M.B. et al. (2000). Systematics of pythons of the *Morelia amethistina* complex (Serpentes: Boidae) with the description of three new species. *Herpetological Monographs*, 14: 139-85.

2 *WÜSTER, W. & BROADLEY, D.G. (2003). A new species of spitting cobra from northeastern Africa (Serpentes: Elapidae: *Naja*). *Journal of Zoology*, 259: 345-59.

3 *BIJU, S.D. & BOSSUYT, F. (2003). New frog family from India reveals an ancient biogeographical link with the Seychelles. *Nature*, 425: 711-14.

4 ANON. (2003). Rare lizard sighted for first time since 1917. *Independent Online* [*IOL*], 29 September; *Barkudia insularis*, Wikipedia page, accessed 12 July 2011; *ANNANDALE, N. (1917). A new genus of limbless skinks from an island in the Chilika lake. *Records of the Indian Museum*, 13: 17-21.

5 Terror skink, Wikipedia page, accessed 20 June 2011; http://www.iucnredlist.org/apps/redlist/details/17008/0 accessed 20 June 2011.

6 *MURPHY, J.C. et al. (2005). A new species of *Enhydris* (Serpentes: Colubridae: Homalopsinae) from the Kapuas river system, West Kalimantan, Indonesia. *Raffles Bulletin of Zoology*, 53 (No. 2; 31 December): 271-5.

7 *WÜSTER, W. & BROADLEY, D.G. (2007). Get an eyeful of this: a new species of giant spitting cobra from eastern and north-eastern Africa (Squamata: Serpentes: Elapidae: *Naja*). *Zootaxa*, No. 1532: 51-68.

8 *GRISMER, L.L. et al. (2007). A new species of *Chiromantis* Peters 1854 (Anura: Rhacophoridae) from Phnom Samkos in the Northwestern Cardamom Mountains, Cambodia. *Herpetologica*, 63(3): 392-400; SMITH, L. (2008). Green-blooded frog makes first appearance for scientists. *Times* (London), 18 December.

9 *GENTILE, G. & SNELL, H.L. (2009). *Conolophus marthae* sp.nov. (Squamata, Iguanidae), a new species of land iguana from the Galápagos archipelago. *Zootaxa*, No. 2201: 1-10; GENTILE, G. et al. (2009). An overlooked pink species of land iguana in the Galapagos. *Proceedings of the National Academy of Sciences of the United States of America*, 106 (13 January): 507-11; MADRIGAL, A. (2009). Pink iguana that Darwin missed holds evolutionary surprise. *Wired Science*, http://www.wired.com/wiredscience/2009/01/pinkiguana/ 5 January.

10 *WELTON, L.J. et al. (2010). A spectacular new Philippine monitor lizard reveals a hidden biogeographic boundary and a novel flagship species for conservation. *Biology Letters*, 6(5): 654-8; ZABARENKO, D. (2010). Philippines dragon-sized lizard is a new species. Reuters http://uk.reuters.com/article/2010/04/06/us-science-lizard-idUSTRE6355L920100406 7 April.

11 *WEIJOLA, V.S-Å. & SWEET, S.S. (2010). A new melanistic species of monitor lizard (Reptilia: Squamata: Varanidae) from Sanana Island, Indonesia. *Zootaxa*, No. 2434: 17-32; ANON. (2010). New monitor lizard discovered in Indonesia. *Science Daily*, 27 April.

12 *ROWLEY, J.J.L. et al. (2010). A new tree frog of the genus *Rhacophorus* (Anura: Rhacophoridae) from southern Vietnam. *Zootaxa*, No. 2727: 45-55; CHOI, C. (2011). "Vampire flying frog" found; tadpoles have black fangs. *National Geographic Daily News*, 7 January.

Section 4: The Fishes

1 ANON. (2003). Amazon 'mystery fish' is a new species. http://www.rense.com/general38/newfish.htm 4 July; ANON. (2003). http://www.sun-sentinel.com/news/custom/fringe/sfl-73mysteryfish,0,1152499.story?coll=sfla-news-fringe/ 3 July; *JOHNSON, G.D. et al. (2011). A 'living fossil' eel (Anguilliformes: Protoanguillidae, fam. nov.) from an undersea cave in Palau. *Proceedings of the Royal Society, Series B*, http://rspb.royalsocietypublishing.org/content/early/2011/08/16/rspb.2011.1289.full.pdf 16 August.

2 MACEY, R. (2004). What lies beneath. *Sydney Morning Herald* (Sydney) [http://www.smh.com.au/articles/2004/03/21/1079823237850.html], 22 March.

3 *SUTTON, T.T. & HARTEL, K.E. (2004). New species of *Eustomias* (Teleostei: Stomiidae) from the Western North Atlantic, with a review of the subgenus *Neostomias*. *Copeia*, (9 February): 116-21.

4 *WATSON, W. & WALKER, H.J. (2004). The world's smallest vertebrate, *Schindleria brevipinguis*, a new paedomorphic species in the family Schindleriidae (Perciformes: Gobioidei). *Records of the Australian Museum*, 56: 139-42.

5 *KOTTELAT, M. et al. (2006). *Paedocypris*, a new genus of Southeast Asian cyprinid fish with a remarkable sexual dimorphism, comprises the world's smallest vertebrate. *Proceedings of the Royal Society of Edinburgh, Section B (Biology)*, 273: 895-9.

6 *REGAN, C.T. (1925). Dwarfed males parasitic on the females in oceanic angler-fishes (Pediculati Ceratioidea). *Proceedings of the Royal Society of London, Series B*, 97 (No. 684): 386-400

7 LEIDIG, M. (2004). Cuddles the red 'shaggy shark' makes her debut. *Daily Telegraph* (London) [http://www.telegraph.co.uk/news/worldnews/europe/germany/1470495/Cuddles-the-red-shaggy-shark-

makes-her-debut.html], 29 August; ANON. (2004). Cuddles the shark is some fin new. *Sydney Morning Herald* (Sydney) [http://www.smh.com.au/articles/2004/08/29/1093717842562.html], 30 August; HEMMLER, M. (2011). Personal communications, 2 July.

8 *RODILES-HERNÁNDEZ, R. et al. (2005). *Lacantunia enigmatica* (Teleostei: Siluriformes), a new and phylogenetically puzzling freshwater fish from Mesoamerica. *Zootaxa*, No. 1000: 1-24.

9 *MARSHALL, A.D. et al. (2009). Redescription of genus *Manta* with resurrection of *Manta alfredi* (Krefft, 1868) (Chondrichthyes; Myliobatoidei; Mobulidae). *Zootaxa*, No. 2301: 1-28; WALKER, M. (2009). Manta ray's secret life revealed. http://news.bbc.co.uk/earth/hi/earth_news/newsid_8347000/8347024.stm 9 November; Manta ray, Wikipedia page, accessed 5 July 2011; Manta ray research, http://marinemegafauna.org/mantarays/ accessed 5 July 2011.

10 *BRITZ, R. et al. (2009). Spectacular morphological novelty in a miniature cyprinid fish, *Danionella dracula* n. sp. *Proceedings of the Royal Society, Series B*, 276: 2179-86.

11 *PIETSCH, T. et al. (2009). A bizarre new frogfish of the genus *Histiophryne* (Lophiiformes: Antennariidae) from Ambon and Bali, Indonesia. *Copeia*, (for 2009): 37-45.

12 *DE CARVALHO, M.R. & LOVEJOY, N.R. (2011). Morphology and phylogenetic relationships of a remarkable new genus and two new species of Neotropical freshwater stingrays from the Amazon basin (Chondrichthyes: Potamotrygonidae). *Zootaxa*, No. 2776: 13-48.

Section 5: The Invertebrates

1 ANON. (2000). Millennium bug found by CSIRO entomologist. CSIRO media release, 1 January; HEINSELMAN, C. (2000). Veliidae millennium bug. *Crypto*, 3 (January): 4; *ANDERSON, N.M. & Weir, T. (2001). New genera of Veliidae (Hemiptera: Heteroptera) from Australia, with notes on the generic classification of the subfamily Microveliinae. *Invertebrate Taxonomy*, 15: 217-58.

2 BARKHAM, P. (2001). The walking sausage lives again. *Guardian* (London), 14 February; *Dryococelus australis*, Wikipedia page, accessed 6 July 2011.

3 SIMPSON, S. (2001). Colony of giant sea sponges discovered off coast of B.C. *Vancouver Sun* (Vancouver), 9 March; DEPARTMENT OF FISHERIES AND OCEANS. (2000). Hexactinellid sponge reefs on the British Columbia continental shelf: geological and biological structure. *DFO Pacific Region Habitat Status Report 2000/02*. Department of Fisheries and Oceans (Vancouver); WHITNEY, F. et al. (2005). Oceanographic habitat of sponge reefs on the western Canadian continental shelf. *Continental Shelf Research*, 25: 211-26; The Sponge Reef Project website, http://www.porifera.org/a/cif1.htm accessed 6 July 2011.

4 VECCHIONE, M. et al. (2001). Worldwide observations of remarkable deep-sea squids. *Science*, 294 (21 December): 2505-6; YOUNG, R.E. (1991). Chiroteuthid and related paralarvae from Hawaiian waters. *Bulletin of Marine Science*, 49: 162-85; *VECCHIONE, M. & YOUNG, R.E. (1998): The Magnapinnidae, a newly discovered family of oceanic squids (Cephalopoda; Oegopsida). *South African Journal of Marine Science*, 20: 429-37; Bigfin squid, Wikipedia page, accessed 6 July 2011.

5 KLASS, K.-D. et al. (2002). Mantophasmatodea: a new insect order with extant members in the Afrotropics. *Science*, 296 (No. 5572): 1456-9; ADIS, J. et al. (2002). Gladiators: a new order of insect. *Scientific American*, 287 (November): 42-7; *ZOMPRO, O. et al. (2002). A new genus and species of Mantophasmatidae (Insecta: Mantophasmatodea) from the Brandberg Massif, Namibia, with notes on behaviour. *Cimbebasia*, 19: 13-24; ARILLO, A. & ENGEL, M.S. (2006). Rock crawlers in Baltic amber (Notoptera: Mantophasmatodea). *American Museum Novitates*, No. 3539 (7 December): 1-10.

6 *MATSUMOTO, G. I. et al. (2003). *Tiburonia granrojo* n. sp., a mesopelagic scyphomedusa from the Pacific Ocean representing the type of a new subfamily (class Scyphozoa: order Semaeostomeae: family Ulmaridae: subfamily Tiburoniinae subfam. nov.). *Marine Biology*, 143(1): 73-7; PERLMAN, D. (2003). New jellyfish: Big Red has clusters of arms, not tentacles. *San Francisco Chronicle* (San Francisco), 7 May.

7 *RUNYON, J.B. & HURLEY, R.L. (2004) A new genus of long-legged flies displaying remarkable wing directional asymmetry. *Proceedings of the Royal Society of London Series B (Supplement)*, 271: S114-16.

8 *RASKOFF, K.A. & MATSUMOTO, G.I. (2004). *Stellamedusa ventana*, a new mesopelagic scyphomedusa from the eastern Pacific representing a new subfamily, the Stellamedusinae. *Journal of the Marine Biological Association U.K.*, 84: 37-42.

9 *SANTOS-FLORES, C.J. et al. (2004). *Dumontia oregonensis* n. fam., n. gen., n. sp., a cladoceran representing a new family of 'Water-fleas' (Crustacea, Anomopoda) from U.S.A., with notes on the classification of the Order Anomopoda. *Hydrobiologia*, 500(1-3): 145-55.

10 LEAHY, S. (2004). Mysteries of the ocean deepen. *Wired* [http://www.wired.com/news/technology/0,1282,64483,00.html?tw=wn_tophead_3], 6 August.

11 *ROUSE, G.W. et al. (2004). *Osedax*: bone-eating marine worms with dwarf males. *Science*, 305 (No. 5684): 668-71; *GLOVER, A.G. et al. (2005). Worldwide whale worms? A new species of *Osedax* from the shallow north Atlantic. *Proceedings of the Royal Society B: Biological Sciences*, 272 (22 December): 2587-92; *DALY, M. & GUSMAO, L. (2007). The first sea anemone (Cnidaria: Anthozoa: Actinaria) from a whale fall. *Journal of Natural History*, 41: 1-11.

12 MULFORD, K. (2004). New leech discovered. *Courier Post* (Camden, NJ), 4 October; MULFORD, K. (2004). 'C-P' reader helps Rutgers study by catching leeches. *Ibid.*, 16 October; ANON. (2009). Camden research team names new species of leech for South Jersey family. http://news.rutgers.edu/medrel/news-releases/2009/09/rutgers-camden-resea-20090929 29 September; *WIRCHANSKY, B.A. & SHAIN, D.H. (2010). A new species of *Haemopis* (Annelida: Hirudinea): evolution of terrestrial leeches. *Molecular Phylogenetics and Evolution*, 54 (No. 1; January): 226-34.

13 HADDOCK, S.H. et al. (2005). Bioluminescent and red-fluorescent lures in a deep-sea siphonophore. *Science*,

309 (8 July): 263; PERLMAN, D. (2005). Bright lights lure prey in deep sea. *San Francisco Chronicle* (San Francisco), 8 July.

14 *MACPHERSON, E. et al. (2006). A new squat lobster family of Galatheoidea (Crustacea, Decapoda, Anomura) from the hydrothermal vents of the Pacific-Antarctic Ridge. *Zoosystema*, 27(4): 709-23.

15 ANON. (2006). Unique underground ecosystem revealed by Hebrew University researchers uncovers eight previously unknown species. Hebrew University press release [http://www.huji.ac.il/cgi-bin/dovrut/dovrut_search_eng.pl?mesge114907691205976587], 31 May; *The Independent* (London), *Guardian* (London), *Calgary Herald* (Calgary), all 1 June 2006.

16 *COPPARD, S.E. & SCHULTZ, H.A.G. (2006). A new species of *Coelopleurus* (Echinodermata: Echinoidea: Arbaciidae) from New Caledonia. *Zootaxa*, No. 1281: 1-19; O'BRIEN, H. (2006). New urchin leaves eBayers all at sea. *Guardian* (London), 17 August.

17 *WESTBLAD, E. (1949). *Xenoturbella bocki* n.g., n.sp., a peculiar, primitive turbellarian type. *Arkiv för Zoologi*, 1: 3-29; *ISRAELSSON, O. (1999). New light on the enigmatic Xenoturbella (phylum uncertain): ontogeny and phylogeny. *Proceedings of the Royal Society of London B*, 266: 835-41; BOURLAT, S.J. et al. (2006). Deuterostome phylogeny reveals monophyletic chordates and the new phylum Xenoturbellida. *Nature*, 444 (2 November): 85-8.

18 *BRAGG, P., In: HENNEMANN, F.H. & CONLE, O.V. (2008). Revision of Oriental Phasmatodea: The tribe Pharnaciini Günther, 1953, including the description of the world's longest insect, and a survey of the family Phasmatidae Gray, 1835 with keys to the subfamilies and tribes (Phasmatodea: "Anareolatae": Phasmatidae). *Zootaxa*, No. 1906: 1-138; ANON. (2008). World's longest insect named after KK [Kota Kinabalu] naturalist. *The Star* (Petaling Jaya, Selangor, Malaysia) [http://www.thestar.com.my/news/story.asp?file=/2008/10/18/nation/2316389&sec=nation] 18 October; Beccaloni, G. (2010). *Big Bugs Life-Size*. Natural History Museum (London).

19 *ROWSON, B. & SYMONDSON, O.C. (2008). *Selenochlamys ysbryda* sp. nov. from Wales, UK: a *Testacella*-like slug new to western Europe (Stylommatophora: Trigonochlamydidae). *Journal of Conchology*, 39(5): 537-52.

20 *KUNTNER, M. & CODDINGTON. J.A. (2009). Discovery of the largest orbweaving spider species: The evolution of gigantism in *Nephila*. *PLoS ONE*, 4 (20 October): e7516. doi:10.1371/journal.pone.0007516.

21 *OSBORN, K.J. et al. (2009). Deep-sea, swimming worms with luminescent "bombs". *Science*, 325 (21 August): 964.

22 *OSBORN, K.J. et al. (2011). The remarkable squidworm is an example of discoveries that await in deep-pelagic habitats. *Biology Letters*, 7 (23 June): 449-53; WATSON, T. (2010). "Flamboyant" new squid worm surprises, delights experts. *National Geographic Daily News*, 24 November.

23 *BORGONIE, G. et al. (2011). Nematoda from the terrestrial deep subsurface of South Africa. *Nature*, 474 (2 June): 79-82.

Rare picture postcard from the early 20th century depicting a possible mountain nyala exhibited at Egypt's Cairo Zoo *before* this species' formal scientific description in 1910 (John Edwards).

INDEX OF NEW AND REDISCOVERED ANIMALS

Acanthochaetetes wellsi, 226
Acerentomon doderoi, 224
'Acinonyx rex', 43
Acouchy, agouti-furred, 285
Acrobat bird, 157
Acrobatornis fonsecai, 157
Acrocephalus orinus, 293
Adelocosa anops, 248
Adetomyrma venatrix, 262
Aepypodius bruijnii, 289-90
Afropavo congensis, 114-5, 123-5
Agak, 173
Agouti, grey, 285
Ailuropoda melanoleuca, 22, 46-9
Albatross, Amsterdam, 147
Allenopithecus nigroviridis, 79, 280, 362
Allocebus trichotis, 69
Allotoca, opal, 216
Allotoca maculata, 216
Alovot, 117
Alsophis antiguae, 185-6
Alvinella pompejana, 252
Alviniconcha hessleri, 253
Alytes muletensis, 178
Amblyornis flavifrons, 144-5
Amphiuma, one-toed, 173-4
Amphiuma pholeter, 173-4
Amytornis barbatus, 133
  goyderi, 133
  housei, 133
Anaconda, Barbour's, 170
  Bolivian, 171
  de Schauensee's, 170-1
Anas aucklandica nesiotis, 141
Angel wings, 226
Anglerfish, hairy, 202
  paisley, 300
  Prince Axel's, 203
  three-starred, 196
Anoa, mountain, 37
Anoa quarlesi, 37
Anodorhynchus leari, 142
Anoptichthys jordani, 198
Anoura fistulata, 281
Ant, dinosaur, 233
  Dracula, 262

Ant *(cont.)*, Fuller's, 260-1
  primitive Australian, 233
Antarctic giant fauna, 268-9
Anteater, arboreal giant, 284
Anthosactis pearseae, 306
Antpitta, Jocotoco, 160
Antshrike, Cocha, 153
Antwren, black-hooded, 152
  Paraná, 158-9
Aplonis pelzelni, 157
  santovestris, 153
Aproteles bulmerae, 76-7
Apteryx rowi, 154
Aratinga pintoi, 290
Arborophila rufipectus, 123
Archaea, 254
Archboldia papuensis, 127
Arctocephalus doriferus, 34
  galapagoensis, 34
Armadillo, Brazilian three-banded, 88
Aromobates nocturnus, 183
Artedidraco n.sp., 219
Asbestopluma hypogea, 264-5
Astrapia, Huon, 126
Astrapia mayeri, 126
  rothschildi, 126
Astrosclera willeyana, 226
Astyanax jordani, 198
'*Ateles aripuanensis*', 103
  '*kamayurensis*', 103
Atelopus zeteki, 169-70
Athene blewitti, 158
Atlantisia rogersi, 120-1
Atretochoana eiselti, 174
Atrichornis clamosus, 133-4
Avahi cleesei, 277
  mooreorum, 287
  unicolor, 274
Ayalon Cave invertebrates, 307
Aythya innotata, 153

Baiji, 41-2
Balaenoptera omrai, 276
Bandicoot, Fly River spiny, 88
Barb, Vu Quang, 217
Barkudia insularis, 295-6
Barnacle, verrucomorphan vent, 252

Barycypraea fultoni, 243
Basipodella harpacticola, 257
Bat, banana, 65-6
  Bulmer's fruit, 76-7
  Kitti's hog-nosed (bumblebee), 72-3, 279
  Mount Iglit fruit, 109
  Ridley's leaf-nosed, 73
  Salim Ali's fruit, 73
  tube-lipped nectar, 281
  western sucker-footed, 281
Batfish, unicorn, 298
Bathyceratias trilychnus, 196
Bathyembrix istiophasma, 196
Bathylychnops exilis, 208
Bathypecten vulcani, 252
Bathysidus pentagrammus, 196
Bathysphaera intacta, 196
Beaked whale, Andrews', 35, 67
  Bahamonde's, 108
  Hubbs's, 67
  Japanese, 61-2
  lesser (Peruvian), 86
  Longman's, 44
  Perrin's, 275
  Shepherd's, 51-2
  True's, 40
Bear, Shennongjia white, 49
Bee, king, 256-7
  vulture, 261-2
Beetle, titan longhorn, 225
Betta fasciata, 193
  splendens, 193
  taeniata, 193
Bigfin, 302-3
Bird of paradise, Berlepsch's six-wired, 193
  Huon six-wired, 127
  ribbon-tailed, 125-6
Birdwing, Queen Alexandra's, 223-4
Blindcat, wide-mouthed, 200
Blue-eye, red-finned, 216
Bondegezou, 105
Bonobo, 45-6, 279
Boophis spp., 183
Bororo, 110
Bos sauveli, 53-5, 97

Bostrychia bocagei, 152
Bothrops insularis, 166-7
Bowerbird, Archbold's, 127
   Baker's regent, 122
   Rawnsley's, 291-2
   yellow-fronted gardener, 144-5, 292
Brachyaspis [*Echiopsis*] *atriceps*, 178
Brachylophus bulabula, 178
   vitiensis, 177-8
Bradypus pygmaeus, 274
Branta hutchinsii, 292
Bristlebird, western, 134-5
Brittle star, chopsticks, 254
   oegophiuridan, 228-9
Broadbill, African green, 118-9
Brocket, dwarf, 109
   small red, 109-10
   van Tienhoven's fair, 285-6
Bronzeback, 177
Bubo vosseleri, 135
Budorcas taxicolor bedfordi, 65
Bufo blombergi, 172-3
   exsul, 170
Bug, millennium, 301
Bulbul, bare-faced, 294-5
Bumpy, 305
Burramys parvus, 70
Bushbaby, mystery giant, 109
Bush-crow, Ethiopian, 125
Bythites hollisi, 252
Bythograea thermydron, 250

Cacajao ayresi, 287
   hosomi, 287
Caecilian, Lafrentz, 169
   lungless, 174
Caecilita iwokramae, 174
Cahow, 127-8
Calcochloris tytonis, 275
Callibella humilis, 102-4
Callicebus 'aripuanensis', 103
   aureipalatii, 280-1
   bernhardi, 103
   caquetensis, 287
   coimbrai, 112
   hoffmannsi, 100
   stephennashi, 103
Callimico goeldii, 33, 362
Calodactylodes aureus, 181
Caloprymnus campestris, 50-1
Caluromysiops irrupta, 59
Calyptogena magnifica, 250
Calyptra eustrigata, 231
Camel, wild Bactrian, 274
Camelus ferus, 274
Campephilus imperialis, 364
   principalis, 149-51
Canirallus beankaensis, 292
Caprimulgus affinis, 156
   noctitherus, 133
   solala, 155-6
Caprolagus hispidus, 72

Capromys angelcabrerai, 57
   garridoi, 57
   nana, 56
   sanfelipensis, 57
Capuchin, blond, 283
   Ka'apor, 101
Carausius morosus, 223
Carn-pnag, 173
Carp, Iranian blind, 200-1
   Vu Quang river, 217
Carukia barnesi, 244-5
   shinju, 245
Cat, Bornean bay, 88-90
   Chinese desert, 44
   Iriomote, 68-9
   Kellas, 87
   Tsushima, 69
Catagonus wagneri, 73-4
Catfish, land, 214-5
   Mekong giant, 195-6
Cathartes melambrotus, 136
Catopuma badia, 88-90
Caulophryne polynema, 202
Cebus kaapori, 101
   queirozi, 283
Centrocercus minimus, 289
Cephalocarids, 238-9
Cephalodiscus graptolitoides, 262-3
Cephalophus jentinki, 57
Ceratoporella nicholsoni, 225-6
Ceratotherium simum cottoni, 23
Cercopithecus dryas, 79
   salongo, 79
   solatus, 79
Cervus schomburgki, 95-6
Cettia haddeni, 293-4
Chaetomys subspinosus, 84
Charmosyna diadema, 144
Cheetah, king, 43
Cheirodon axelrodi, 197
Cheirogaleus adipicaudatus, 274
   crossleyi, 274
   minusculus, 274
   ravus, 274
   sibreei, 274, 363
Chelonoidis nigra abingdoni, 174-6
   nigra phantastica, 176
Chevrotain, silver-backed, 277
Chibchanomys orcesi, 110
Chiltoniella elongata, 239
Chimaera, leopard, 218
Chimaera opalescens, 218
   panthera, 218
Chimpanzee, pygmy, 45-6
Chiromantis samkosensis, 296
Chironex fleckeri, 239-40
   yamaguchii, 240
Chironius vincenti, 182
Choeropsis liberiensis, 6, 38-40
Chordeiles vielliardi, 156
Chrotogale owstoni, 40
Chrysaora achlyos, 268
Chrysochloris visagiei, 275

Cirrothauma murrayi, 227-8
Civet, giant palm, 78
   golden dry-zone palm, 113
   golden wet-zone palm, 113
   Owston's banded, 40
   Seram mystery, 113
   Sri Lankan brown palm, 113
   Taingu, 112-3
   water, 41
Clam, giant vent, 250
Cliona patera, 365
Co, 217
Coati, Apurimac mountain, 112
   orange, 285
Cobitis yeni, 217
Cobra, Ashe's spitting, 296
   Nubian spitting, 295
Cochito, 62
Coelacanth, Comoros, 191, 198-200, 203-7, 280, 365
   Indonesian, 219-20, 361
Coelopleurus exquisitus, 308
Coendou koopmani, 84
Coenocorypha aucklandica perserverance, 159
Cologonger giganteus, 209-10
Conch, bull, 242
Cone, glory of the sea, 241
Coneheads, 223-4
Conocrinus cabiochi, 248
   cherbonnieri, 248
Conolophus marthae, 296-7
Conraua goliath, 162-3
Conus gloriamaris, 241
Cooloola monster, 221, 249
   South Percy Island, 249
Cooloola dingo, 249
   pearsoni, 249
   propator, 249
   ziljan, 249
Corvus unicolor, 294
Couresse (=St. Lucia racer), 179
Courser, Jerdon's, 148-9, 158
Cowry, Fulton's, 243
   white-toothed, 242-3
Crab, blind vent, 250
Craseonycteris thonglongyai, 72-3
Craugastor phasma, 187-8
Crocias, grey-crowned, 157
Crocias langbianis, 157
Crocodile, Morelet's, 169
   New Guinea, 168
   Philippine, 169
   sacred, 364
Crocodylus mindorensis, 169
   moreletii, 169
   novaeguineae, 168
   suchus, 364
Crossocheilus vuha, 217
Crow, Banggai, 294
Cryptochloris zyli, 275
Cryptophidion annamense, 183-4
Cryptosyringa membranophila, 255

Crysomallon squamiferum, 253
Ctenothrissids, 192-3
Cuddles, 299
Culex (pipiens) molestus, 269
Cuscomys ashaninka, 112
  oblativus, 112
Cusk-eel, Hollis's, 252
Cuscus, Gebe Island, 107
  Telefomin, 107
Cyanolimnas cerverai, 56
Cyanopsitta spixii, 142
Cycliophorans, 266-7, 308
Cyclotyphlops deharvengi, 184
Cyclura collei, 183
Cypraea leucodon, 242-3

Dancer, giant Spanish, 261
Danionella dracula, 300
Dasogale fontoynonti, 44-5
Dasycyon hagenbecki, 58-9
Dasyornis longirostris, 134-5
Dasyurus spartacus, 85-6
  viverrinus, 107-8
Deer, black, 97
  Schomburgk's, 95-6
  slow-running, 96-7
Dendrolagus mbaiso, 104-6
  scottae, 86
Denticeps clupeoides, 208
Deoterthron dentatum, 257
Dermophis oaxacae, 169
Derocheilocaris typicus, 233
Dibbler, 71
Dicaeum quadricolor, 153-4
Dicerorhinus sumatrensis, 84-5
Dingiso, 104-6
Dingo monster, 249
Dinomys branickii, 30-1, 49
Diomedea amsterdamensis, 147
Dipsochelys arnoldi, 186
  hololissa, 186-7
Discoglossus nigriventer, 171-2, 365
Disko Island invertebrate, 269-70
Djoongari, 111
Dodecolopoda mawsoni, 232
Dolphin, Australian snubfin, 277
  burrunan, 272, 288
  Chinese river, 41-2
  Fraser's, 61
  helmet, 78-9
  Rio Aripuanã river, 285
Dormouse, Selevin's, 55-6
Dragon, Komodo, 2, 161, 163-5,
  2
Dragonfish, Jim Craddock's, 298
Drepanovelia millennium, 301
Dryococelus australis, 301-2
Dryocopus galeatus, 152
Duck, white-headed steamer, 144
Duiker, Jentink's, 57
  Walter's, 363
Dumontia oregonensis, 305
Dunnart, long-tailed, 76
  sandhill, 76

Eagle, Madagascan serpent-, 152
  monkey-eating, 115-6
Earwig, St. Helena giant, 245-6
Echidna, Attenborough's long-nosed, 112
Echiopsis atriceps, 178
Echymipera echinista, 88
Eel, *Dana* giant, 209-10
Eelpout, Anderson's, 252
  Cerberus, 252
Eleutherodactylus pluvicanorus, 184-5
Elseya irwini, 2
  lavarackorum, 185
Elusor macrurus, 185
Enhydris gyii, 296
Erebomyia exalloptera, 304-5
Erenna sp., 306-7
Erugosquilla grahami, 269
  serenei, 269
Eublepharis angramainyu, 180
  ensafi, 179-80
Eunectes barbouri, 170
  beniensis, 171
  deschauenseei, 170-1
Euoticus pallidus, 109
Eupetaurus cinereus, 106
Eustomias jimcraddocki, 298
Eutrichomyias rowleyi, 157
Eutriorchis astur, 152
Exsul singularis, 260

Falco fascinucha, 127
Falcon, cryptic forest-, 291
  Teita, 127
Felis bieti, 44
Feresa attenuata, 59
Ferminia cerverai, 56
Ferret-badger, Vietnam, 363-4
Ficedula bonthaina, 157
Fighting fish, banded, 193
  Bornean, 193
  Siamese, 193-4
Fish, Dracula, 300
  five-lined constellation, 196
  gibber, 209
  Grassé's, 208-9
  hairy, 207-8
  Lake Hanas/Kanasi giant, 215
  Lake Indawgyi, 195
  Manaus eel-like mystery, 297-8
  siphonophore, 209
  six-eyed spook, 208
  untouchable bathysphere, 196
  vampire, 300
  Widdowson's blind, 201
Flamingo, James's, 20, 131-2
Flea, Scheffer's giant, 228
Flowerpecker, four-coloured, 153-4
Fly, bat-winged, 260
Flycatcher, cerulean paradise, 157
  Chapin's, 131
  Lompobattang, 157
Fregetta grallaria titan, 128

Frog, Archey's, 166
  buzzing tree, 183
  Colombian arrow-poison, 176, 313
  goliath, 162-3
  hairy, 161-2
  Hamilton's, 166
  Madagascan golden, 162
  Madagascan tree, 183
  Maud Island, 166
  northern platypus (gastric-brooding), 176-7
  Palestinian painted, 171-2, 365
  Panamanian golden, 169-70
  phantom, 187-8
  purple, 295
  Ramsey Canyon leopard, 184
  Rwandan blind white, 188
  Samkos bush, 296
  southern platypus (gastric-brooding), 176-7
  vampire flying (gliding), 297
  Venezuelan skunk, 183
Frogfish, psychedelic, 300-1
Frogmouth, Solomon Island, 294

Galago gabonensis, 109
Galagoides granti, 109
  orinus, 109
  rondoensis, 109
  udzungwensis, 109
Galidictis grandidieri, 83-4
Gallirallus calayensis, 292
  lafresnayanus, 144
  okinawae, 144
  wakensis, 117
Gallotia bravoana, 190
  intermedia, 190
  simonyi, 190
  simonyi gomerana, 190
Gar, abyssal rainbow, 196
Gardener, yellow-fronted, 144-5
Gazella bilkis, 79-80
Gazelle, Bilkis, 79-80
Gecko, Delcourt's giant, 179-81
  Ensaf's giant, 179-80
  golden, 181
  Iraqi eyelid, 180
  web-footed, 163
Genet, giant, 27
  Lowe's servaline, 275
Genetta lowei, 275
  victoriae, 27
Ghostshark, opal, 218
Gibberichthyes pumilus, 209
Gibbon, Mentawi, 30
  northern buff-cheeked, 288
Giraffa camelopardalis rothschildi, 21, 26-7
Giraffe, five-horned, 21, 26-7
Gladiators (=rock crawlers), 303-4
Glaucidium albertinum, 131
  mooreorum, 291

Glaucidium (cont.), nubicola, 159
  parkeri, 159
Glyphis fowlerae, 218
  gangeticus, 218
Gnathostomulids, 240-1
Goanna, canopy, 189
Goat, holy, 97-100
Godzilliognomus frondosus, 256
Godzillus robustus, 256
Golden mole, Somali, 275
  Van Zyl's, 275
  Visagie's, 275
Goose, cackling, 292
Gopherus flavomarginatus, 173
Goral, brown, 64-5
  red, 64-5
Goreauiella auriculata, 226
Gorilla, Bwindi, 29
  Cross River, 29
  eastern lowland, 29
  mountain, 20, 28-9
  pygmy, 29
Gorilla gorilla beringei, 20, 28-9
  gorilla graueri, 29
Grallaria ridgelyi, 160
Graptolites, 262-3
Grasseichthys gabonensis, 208-9
Grasswren, black, 133
  Eyrean, 133
  grey, 133
Graveteiro, 157-8
Grebe, giant pied-billed (Atitlan), 122-3
  hooded, 140
Gregoryina gygis, 194-5
Grosbeak, São Tomé, 153
Grouse, Gunnison sage, 289
Grylloblatta campodeiformis, 226-7
  chirurgica, 227
Grylloblattids, 226-7
Guan, white-winged, 141
Guenon, dryas, 79
  salongo, 79
  sun-tailed, 79
Guillecrinus neocaledonicus, 260
  reunionensis, 260
Gull, relict, 122
Gymnobelideus leadbeateri, 66-7
Gymnocrex talaudensis, 160
Gymnotid, anteater, 217-8
Gyrinophilus palleucus, 171

Habroptila wallacii, 157
Habu-kurage, 240
Haematopinus oliveri, 72
Haemopis ottorum, 306
Haideotriton wallacei, 171
Halicephalobus mephisto, 309
Halosbaena spp., 229
Hampsonellus brasiliensis, 239
Hamster, golden, 49-50, 196, 314
Hapalemur aureus, 83
  simus, 83
Hapalopsittaca fuertesi, 290

Hare, hispid, 72
Heliotrygon gomesi, 301
  rosai, 301
Heosemys [*Vijayachelys*] *silvatica*, 178-9
Herring, denticle, 208
Hesiocaeca methanicola, 267-8
Hexabranchus sanguineus, 261
Hexatrygon bickelli, 214
Hieremys annandalii, 162
Himantura chaophraya, 216
Hippopotamus, pygmy, 6, 38-40
Hipposideros ridleyi, 73
Hirundo megaensis, 127
  perdita, 148
Hispidopetra miniana, 226
Histiophryne psychedelica, 300-1
Hog, giant forest, 31-3
  pygmy, 69, 71-2
Hololissa, 186-7
Hoplodactylus delcourti, 179-81
Hoplostethus gigas, 218-9
Horseshoe shrimps, 238-9
Hutchinsoniella macracantha, 238-9
Hutia, Cabrera's, 57
  Cuban dwarf, 56
  Cuvier's, 56
  Dominican, 56
  Garrido's, 57
  little earth, 57
  Samaná, 56
Hydrosaurus weberi, 163
Hylobates klossii, 30
Hylochoerus meinertzhageni, 31-3
Hystrichopsylla schefferi, 228

Ibis, São Tomé dwarf, 152
Ice crawlers (=Ice bugs), 226-7
Ichthyaetus relictus, 122
Idea tambusisiana, 256
Iguana, central Fijian banded, 178
  Fijian crested, 177-8
  Jamaican, 183
  pink, 296-7
Ikan fomar, 220
Indopacetus pacificus, 44
Indostomus crocodilus, 195
  paradoxus, 195
  spinosus, 195
Infantfish, stout, 298
Iranocypris typhlops, 200-1
Irukandji, 244-5
Isothrix barbarabrownae, 282

Jack Dempsey, 194
Jaguar, white-throated black, 286
Jellyfish, Big Red, 304-5
  black, 268
  Bumpy, 305
  fabulous Stygian, 242
Jetete, 132

Kakapo, 137

Kangaroo, desert rat, 50-1
  Torricelli tree, 86
Kasidoron edom, 209
  latifrons, 209
Kaweau, 181
Kawekaweau, 181
Kha-nyou (=Laotian rock rat), 279-80, 363
Killer whale, dwarf, 59, 366
  Prudes Bay, 59
  pygmy, 59
Kipunji, 280
Kiwa hirsuta, 307
  puravida, 366
Kiwi, Haast tokoeka, 154
  Okarito brown, 154
  southern tokoeka, 154
Kobus anselli, 277
Korsogaster nanus, 193
Kouprey, 53-5, 97-8
  pygmy, 55
Kting voar, 97-8

Labidura herculeana, 245-6
Lagenodelphis hosei, 61
Lamellibrachia barhami, 235
  luymesi, 235
Lamellisabella zachsi, 234
Lancehead, golden, 166-7
Langur, golden, 60-1
  white-headed, 61
Laniarius liberatus, 156
Laonastes aenigmamus, 279-80
Lascona cristiani, 265
Lasiognathus saccostoma, 202
Latidens salimalii, 73
Latimeria chalumnae, 191, 198-200, 203-7, 280, 365
  menadoensis, 219-20, 361
Laurentaeglyphea neocaledoniae, 249
Leaproach, 365
Lechwe, Upemba, 277
Leiopelma archeyi, 166
  hamiltoni, 166
  pakeka, 166
Lemur, Cleese's woolly, 277
  golden bamboo, 83
  greater bamboo, 83
  hairy-eared dwarf, 69
  Moore's woolly, 287
  pygmy (western rufous) mouse, 69
  Scott's sportive, 287-8
  Sibree's dwarf, 274, 363
Lemurs, new dwarf, 274
  new mouse, 274
  new sportive, 278
  new woolly, 274
Leopard, Bornean (Sunda) clouded, 271, 282
Lepidogalaxias salamandroides, 208
Lepidothrix vilasboasi, 290
Lepilemur scottorum, 287-8
Leptocephalus giganteus, 209-10

Leucopsar rothschildi, 12, 119
Lewinia muelleri, 137
Liberiictis kuhni, 62-3
Lightiella serendipita, 239
Limnognathia maerski, 269-70
Limnosbaena spp., 229
Ling, 98
Linh duong, 97
Linophryne arborifera, 201-2
Lion, Barbary, 110
  Cape, 110
Liophis ornatus, 179
Lipophrys heuvelmansi, 215
Lipotes vexillifer, 41-2
Litoria electrica, 183
Lizard, Chinese crocodile, 168, 172
  El Hierro giant, 190
  La Gomera giant, 190
  Tenerife speckled, 190
  Weber's sailing, 163
Loach, Gejiu cave, 201
  Smith's cave, 201
  Vietnamese, 217
Lobatolampea tetragona, 269
Lobster, yeti, 307, 366
Lophophorus sclateri arunachalensis, 160
Lophura edwardsi, 121
  hatinhensis, 121, 136
  imperialis, 121
Loriciferan, Shuker's, 259
Loriciferans, 257-9
Lorikeet, New Caledonian, 144
Loris, intermediate, 34
  pygmy slow, 34
  tailed, 34-5
  Vietnamese giant slow, 96
Louse, pygmy hog sucking, 72
Luscinia ruficeps, 135-6
Lycoteuthis diadema, 222

Macaca munzala, 277
Macaque, Arunachal, 277
  giant, 80
Macaw, Lear's (indigo), 142-3
  Spix's, 142
Macristiella perlucens, 193
Macristium chavesi, 192-3
Macrogalidia musschenbroeki, 78
Macropus parma, 69-70
Magnapinna spp., 302-3
Magosternarchus duccis, 217
  raptor, 217
Malacochersus tornieri, 162
'Malania anjouanae', 205
Mallomys sp., 286-7
Malo kingi, 245
  maxima, 245
Manakin, golden-crowned, 290
Manatee, dwarf, 284
Mangabey, highland, 280
Mangden, 97
Mangkurtu kutjarra, 239

Mangkurtu (cont.), *mityula*, 239
Manta, Caribbean mystery, 300
  reef, 300
Manta alfredi, 300
Mantella aurantiaca, 162
Mantophasma zephyrum, 303-4
Mantophasmatodeans, 303-4
Marmoset, Acarí, 103-4
  black-headed, 101
  Hershkovitz's, 100
  manicore, 102
  Maués, 87
  Rio Acari (black-capped dwarf), 102-4
  Satéré, 101
  Upper Xingú Amazonian, 285
Martin, white-eyed river, 138
Mazama bororo, 109-10
  chunyi, 109
  tienhoveni, 285-6
Megachasma pelagios, 210-4
Megachile pluto, 256-7
Megamouth, 19, 191, 210-4
Megamuntiacus vuquangensis, 92-3
Megapode, Bruijn's, 289-90
Megastick, Chan's, 308
Melamprosops phaeosoma, 138-9
Melogale cucphuongensis, 363-4
Merlia normani, 226
Mesocapromys angelcabrerai, 57
  nanus, 56
  sanfelipensis, 57
Mesocricetus auratus, 49-50, 314
Mesoplodon bahamondi, 108
  bowdoini, 35
  carlhubbsi, 67
  ginkgodens, 61-2
  mirus, 40
  perrini, 275
  peruvianus, 86
Mico acariensis, 103
  intermedia, 100
  manicorensis, 102
  mauesi, 87-8
  nigriceps, 101
  saterei, 101
Micrastur mintoni, 291
Microcebus berthae, 274
  myoxinus, 69
  sambiranensis, 274
  tavaratra, 274
Micromyzon akamai, 217
Micropotamogale lamottei, 60
  ruwenzorii, 60
Microtus bavaricus, 276-7
Midgardia xandaros, 246
Midget (blenny), 215
Millipede, Australian aquatic, 261
Mink, sea, 29-30
Mirapinna esau, 207-8
Mitsukurina owstoni, 192
Mixodigma leptaleum, 214
Moho, 128-30

Moho bishopi, 147
  braccatus, 147
Monachus schauinslandi, 33-4
Monal, Arunachal Pradesh, 160
Mongoose, giant striped Malagasy, 83-4
  Liberian, 62-3
Monitor, Gray's, 177
  Moluccan blue, 190
  Moluccan yellow, 190
  northern Sierra Madre forest, 297
  peacock, 190
  pygmy mangrove, 190
  Rossel Island, 189
  torch, 297
  tricoloured (black-backed), 190
  turquoise, 189
  Vietnam, 189-90
  Yemen, 181-2
Monkey, Allen's swamp, 280, 362
  black woolly, 285
  Goeldi's, 33, 362
  long-limbed black spider, 285
  orange woolly, 285
  Rio Aripuanã green-backed squirrel, 285
  silver-bellied spider, 285
  Vanzolini's (black) squirrel, 80
  yellow-tailed woolly, 75
Monoplacophorans, 237-8
Morelia clastolepis, 295
  nauta, 295
  tracyae, 295
Morlockia ondinae, 256
Mosquito, London Underground, 269
Moth, predicted hawk, 223
  vampire, 231
Mouse, Alice Springs, 111
  Cypriot, 281
  Orcés's fishing, 110
  Shark Bay, 111
  woodpecker (long-tongued arboreal), 276
Movile Cave fauna, 265-6
Mudminnow, Shannon, 208
Muntiacus atherodes, 78
  feae, 77
  gongshanensis, 77-8
  montanus, 288
  putaoensis, 94
  rooseveltorum, 93-4
  truongsonensis, 94
  vuquangensis, 92-3
Muntjac, Bornean, 78
  Fea's, 77
  giant (large-antlered), 92-3
  Gongshan, 77-8
  leaf, 94
  Pu Hoat, 94-5
  Roosevelts', 93-4
  Sumatran, 288
  Truong Son (Annamite), 94
Mus cypriacus, 281
Muscicapa lendu, 131

Musonycteris harrisoni, 65-6
Mustela felipei, 59
   macrodon, 29-30
Mynah, Rothschild's, 12, 119
Myrientomatans, 223-4
Myrmotherula erythronotos, 152
Mysateles garridoi, 57
Mystacocarids, 233-4
Myzopoda schliemanni, 281

Naja ashei, 296
   nubiae, 295
Nanaloricus mysticus, 258, 266
Nannopsittacus dachilleae, 148
Nasikabatrachus sahyadrensis, 295
Nectarinia rockefelleri, 131
Nemacheilus gejiuensis, 201
Nemertean, Antarctic giant, 268-9
Nemorhaedus baileyi, 65
   cranbrooki, 65
Nemosia rourei, 159
Neocoelia crypta, 254-5
Neofelis diardi, 271, 282
Neoglyphea inopinata, 248-9
Neopilina bacescui, 238
   bruuni, 238
   ewingi, 238
   galatheae, 237-8
   valeronis, 238
Neospiza concolor, 153
Neoverruca brachylepadoformis, 252
Neovison macrodon, 29-30
Nepa anophthalma, 265
Nephelornis oneilli, 138
Nephila komaci, 272, 309
Nesasio solomonensis, 115-6
Nesillas aldabrana, 138
Nesoclopeus poecilopterus, 137
Nesolagus netscheri, 113
   timminsi, 113
Nesophontids, 53
Nicothoe tumulosa, 249
Nighthawk, Bahia, 156
Nightjar, Franklin's, 156
   Nechisar, 155-6
Ningaui, Pilbara, 75
   southern, 76
   Wongai, 75
Ningaui ridei, 75
   timealeyi, 75
   yvonneae, 76
Ninox ios, 160
   sumbaensis, 290-1
Nomascus annamensis, 288
Nothomyrmecia macrops, 233
Notopterans, 304
Notornis mantelli, 115, 128-30
Nototheniids, new, 219
Nuthatch, Kabylian, 140
Nyala, mountain, 35-7, 352
Nycticebus caudatus, 34-5
   intermedius, 34
   pygmaeus, 34

Oceanites maorianus, 291
Octopus, blind, 227-8
   mimic, 263-4
Odedi, 293-4
Okapi, 23-6, 314
Okapia johnstoni, 24-6, 314
Onça-canguçú, 286
Onza, 23, 80-3
O-o, Bishop's, 147
   Kauai, 147
O-o-aa, 147
Ophidiocephalus taeniatus, 177
Opossum, black-shouldered, 59
Opsariichthys bea, 217
   hieni, 217
Orcaella heinsohni, 277
Orcinus glacialis, 59
   nanus, 59
Oreonax flavicauda, 75
Ornithoptera alexandrae, 223
Orthosternarchus tamandua, 217-8
Oryzomys swarthi, 69
Osbornictis piscivora, 41
Osedax frankpressi, 306
   mucofloris, 306
   rubiplumus, 306
Otter, black giant, 284
Otus alius, 159
   capnodes, 132
   collari, 159
   insularis, 132
   ireneae, 135
   marshalli, 145
   moheliensis, 159
   pauliani, 132
Owl, African bay, 131
   Anjouan scops, 132
   cinnabar hawk, 160
   cloudforest pygmy, 159
   cloudforest screech, 145
   fearful, 115-6
   Grand Comoro, 132
   Little Sumba hawk, 290-1
   Madagascan red, 158-9
   Mohéli scops, 159
   Nduk eagle, 135
   Nicobar scops, 159
   Pernambuco pygmy, 291
   Sangihe scops, 159
   Seychelles scops, 132
   Sokoke scops, 135
   subtropical pygmy, 159
Owlet, Albertine, 131
   forest spotted, 158
   long-whiskered, 140
Ox, Cambodian wild, 53-5
   Vu Quang, 90-2, 96, 99
Oxyuranus microlepidotus, 168
   scutellatus, 167-8
   temporalis, 168

Pa beuk, 195-6
Paca, giant, 284

Pacarana, 30-1, 49
Paedocypris progenetica, 298
Palmatogecko rangei, 163
Pan paniscus, 45-6
Panda, giant, 22, 46-9
Pangasianodon gigas, 195-6
Panthera leo leo, 110
   leo melanochaita, 110
Paracheirodon innesi, 12, 197
Paracobitis smithi, 201
Paradigalla, short-tailed, 127
Paradigalla brevicauda, 127
Paradox fish, 195
   Mekong River, 195
   Thailand, 195
Paradoxurus aureus, 113
   montanus, 113
   stenocephalus, 113
Parakeet, El Oro, 148
   Rothschild's, 160
   sulphur-breasted, 290
Parantechinus apicalis, 71
Parazacco vuquangensis, 217
Pardusco, 138
Parotia berlepschi, 193
   wahnesi, 127
Parrot, bald, 290
   Fuertes's (=indigo-winged), 290
   night, 152-3
   owl, 137
Parrotlet, Amazonian, 148
Partridge, Rubeho forest, 155
   Sichuan hill, 123
   Udzungwa forest, 154-5
Peacock, Congo, 114-5, 123-5
Pearson's monster, 249
Pecari maximus, 276
Peccary, Chacoan, 73-4
   giant, 276
   orange, 285
   white-hoofed, 286
Penelope albipennis, 141
Petrel, Beck's, 294
   Bermuda, 127-8
   MacGillivray's, 147-8
   magenta, 141-2
   Murphy's, 128
   Swinhoe's, 154
   Tyne, 154
Petrogale persephone, 76
   purpureicollis, 274-5
Pezoporus occidentalis, 152-3
Phacochoerus aethiopicus, 88
Phalanger alexandrae, 107
   matanim, 107
Phascogale, eastern, 273
   northern, 273
   western, 273
Pheasant, Edwards's, 121
   imperial, 121
   mikado, 117-8
   Rothschild's peacock, 116-7
   Vo Quy's, 121, 136

Pheidole fullerae, 261
Philantomba walteri, 363
Phobaeticus chani, 308
Phoboscincus bocourti, 296
Phocoena dioptrica, 40
   sinus, 62
Phodilus prigoginei, 131
Phoenicoparrus jamesi, 20, 131-2
Photocorynus spiniceps, 299
'Phreatobius walkeri', 214-5
Phyllobates terribilis, 176, 313
Picathartes oreas, 11
Pig, Ryukyu dwarf, 69
   Vietnamese warty (Heude's), 95
Pipistrelle, soprano, 109
Pipistrellus pygmaeus, 109
Pithecophaga jefferyi, 115-6
Pitta, Gurney's, 151-2
   Schneider's, 152
Pitta gurneyi, 151-2
   schneideri, 152
Placozoan, 246-7
Plagiodontia aedium, 56
   hylaeum, 56
   ipnaeum (=*velozi*), 56
Planctosphaera pelagica, 231-2
Platasterias latiradiata, 243
Pliciloricus dubius, 258
   enigmaticus, 258
   gracilis, 258
   hadalis, 258
   orphanus, 258
   profundus, 258
   shukeri, 259
Ploceus aureinuchia, 152
Poc, 122
Pochard, Madagascan, 153
Podiceps gallardoi, 140
Podilymbus gigas, 122
Pogonophorans, 234-5
Polyplectron inopinatum, 116-7
Po'o-uli, 138-9
Porcula salvania, 71-2
Porcupine, black-tailed dwarf, 103
   Koopman's tree, 84
   streaked dwarf, 103
Porphyrio mantelli, 128
Porpoise, Gulf of California, 62
   spectacled, 40
Possum, Leadbeater's, 66-7
   mountain pygmy, 70
   scaly-tailed, 42-3
Potamotrygon tigrina, 365
Potiicoara brasiliensis, 239
Potoroo, Gilbert's 106-7
   long-footed, 78
Potorous gilberti, 106-7
   longipes, 78
Potto, false, 108
Presbytis geei, 60-1
   leucocephalus, 61
Prionailurus iriomotensis, 68-9
Prolemur simus, 83

Propithecus tattersalli, 86-7
Protoanguilla palau, 297-8
Proturans, 223-4
Pseudemydura umbrina, 173
Pseudobulweria becki, 294
   macgillivrayi, 147-8
Pseudocalyptomena graueri, 118-9
Pseudochelidon [=*Eurochelidon*]
   sirintarae, 138
Pseudoginglymostoma nigropunctatum, 299
Pseudogorilla mayema, 29
Pseudois schaeferi, 67-8
Pseudomys fieldi, 111
   praeconis, 111
Pseudonovibos spiralis, 97-100
Pseudopotto martini, 108
Pseudoryx nghetinhensis, 90-2
Psittacula intermedia, 160
Pterodroma cahow, 127-8
   magentae, 141-2
   ultima, 128
Ptilonorhynchus rawnsleyi, 291-2
Puffinus atrodorsalis, 157
   bryani, 295
Pycnonotus hualon, 294-5
Pyrilia aurantiocephala, 290
Pyrrhura orcesi, 148
Python, Halmahera, 295
   Moluccan, 295
   Tanimbar, 295

Quang khem, 96-7
Quemi, 49, 56
Quemisia gravis, 49
Quoll, bronze, 85-6
   eastern, 107-8

Rabbit, Annamite striped, 113
   Sumatran striped, 113
Racer, Antiguan, 185-6
   St. Lucia (see Couresse), 179
Rafetus hoankiemensis, 188
Rail, Auckland Islands, 137
   Calayan, 292
   Fijian barred-wing, 137
   Inaccessible Island, 120-1
   invisible, 157
   Okinawa, 144
   Talaud, 160
   Tsingy wood, 292
   Wake Island, 117
   Zapata, 56
Raja laut, 219
Rana jimiensis, 173
   subaquavocalis, 184
Rat, Asháninka arboreal chinchilla, 112
   Barbara Brown's brush-tailed, 282
   Foja giant woolly, 286-7
   Inca tomb, 112
   James Island rice, 69
   Laotian rock (see Kha-nyou), 279-80, 363

Rat *(cont.)*, Mount Bosavi giant woolly, 287
   thin-spined porcupine, 84
Ray, Gomes's round, 301
   Rosa's round, 301
   six-gilled, 214
Remipedes, 255-6
Rhacophorus vampyrus, 297
Rhagomys longilingua, 276
Rheobatrachus silus, 176-7
   vitellinus, 176-7
Rhinoceros, Cotton's white, 23
   Javan, 85, 363
   Sumatran, 84-5
Rhinoceros sondaicus, 85, 363
Rhinoptilus bitorquatus, 148-9, 158
Rhynchocyon udzungwensis, 287
Ricinoides sjostedti, 232-3
Ricinuleids, 232-3
Riftia pachyptila, 251
Rigidipenna inexpectata, 294
Rimicaris exoculata, 254
Robin, rufous-headed, 135-6
   Sangha forest, 160
Rocio octofasciata, 194
Rock-rat, central, 107
Rosaura rotunda, 193
Roughy, giant, 218-9
Rowi, 154
Rugiloricus carolinensis, 258
   cauliculus, 258
   ornatus, 258
Rungwecebus kipunji, 280

Saguinus imperator, 6, 34
Sailfin, pallid, 196
Saimiri vanzolinii, 80
Saki, grey, 285
   southbank Rio Negro, 285
Salamander, Georgia blind, 171
   Tennessee cave, 171
Salamanderfish, 208
Salanoia durrelli, 288
Saltoblatella montistabularis, 365
Sandersiella acuminata, 239
Saola, 90-2
Satan eurystomus, 200
Sawbelly, giant, 218-9
Saxipendium coronatum, 250
Scallop, vent, 252
Scaturiginichthys vermeilipinnis, 216
Sceloglaux albifacies, 145
Schindleria brevipinguis, 298
Schoutedenapus schoutedeni, 131
Sclerosponges, 225-6
Scrub-bird, noisy, 133-4
Sculpin, mote, 198
Scutisorex congicus, 14, 137-8
   somereni, 137-8
Scytalopus iraiensis, 159-60
Sea anemone, zombie, 306
Sea daisies, 259-60
Sea lilies, 248, 260

Sea slater, Antarctic giant, 268
Sea spider, Antarctic giant, 268
    twelve-legged, 232
Sea urchin, giant, 223
Sea-nettle, black, 268
Sea-wasp, Flecker's, 239-40
Seal, Australian fur, 34
    Galapagos fur, 34
    Hawaiian monk, 33-4
Selenochlamys ysbryda, 309
Selevinia betpakdalaensis, 55-6
Semisalsa dobrogica, 265
Sengi, Boni-Dodori giant, 287
    grey-faced giant, 287
Sericulus bakeri, 122
Serpent-eagle, Madagascan, 152
Shark, Borneo river, 218
    dotted nurse, 299
    furry mystery (Cuddles), 299
    Ganges river, 218
    goblin, 191-2
    megamouth, 191, 210-4
    spined pygmy, 194
Shearwater, Bryan's, 295
    Mascarene, 157
Sheep, dwarf blue, 67-8
Shelduck, crested, 119-20
Shinisaurus crocodilurus, 168
Shrew, Congo hero, 14, 37-8
    Congo pygmy otter, 60
    Guinea pygmy otter, 60
    Ugandan hero, 37-8
Shrike, Bulo Burti boubou, 156
Shrimp, social, 267
    Sydney Harbour mantis, 269
    thermal vent eyeless, 254
Siamang, dwarf, 30
Siboglinum weberi, 234
Sifaka, golden-crowned, 86-7
    mystery diademed, 87
Siphonophore, dandelion, 250
    red-bioluminescing, 306-7
Sitta ledanti, 140
Skink, Barkudia (aka Badakuda)
    Island, 295-6
    Bocourt's giant (terror), 296
    Madagascan, 183
    pygmy bluetongue, 183
Sloth, pygmy three-toed, 274
Slug, ghost, 309
Sminthopsis longicaudata, 76
    psammophila, 76
Snail, hairy hydrothermal vent, 253
    Movile Cave water, 265
    scaly-footed, 253
Snake, Deharveng's blind, 184
    Kapuas mud, 296
    Lake Cronin, 178
    pink Madagascan, 183
    Schaefer's spinejaw, 186
    small-scaled (fierce), 168
    spinejaw, 186
    Vietnamese sharp-nosed, 183-4

Snipe, Campbell Island, 159
Solenodon, Cuban, 52-3
    Hispaniolan, 10, 52-3
    Marcano's, 53
Solenodon cubanus, 52-3
    marcanoi, 53
    paradoxus, 10, 52-3
Somasteroids, 243
Spelaeogriphus lepidops, 239
Speleonectes lucayensis, 255
Sperosoma giganteum, 223
Sphenodon guntheri, 182
Sphiggurus ichillus, 103
    roosmalenorum, 103
Sphinctozoans, 254-5
Spider, Komac's golden orb-web, 272, 309
    Movile Cave blind, 265
    no-eyed big-eyed, 247-8
    Nullarbar Plain eyeless, 248
Sponge, giant glass, 302
    Neptune's cup, 365
Sponge reefs, 302
Sponges, Antarctic giant, 269
    coralline, 226
    flesh-eating, 264-5
Squaliolus laticaudus, 194
Squid, bigfin, 302-3
    jewelled, 222
    long-arm, 302-3
    vampire, 10, 235-6
Squirrel, woolly flying, 106
Starfish, Midgard, 246
Starling, Pohnpei mountain, 157
    Santo Martin, 153
Stellamedusa ventana, 305
Stenella clymene, 78-9
Stick insect, Indian, 223
    Lord Howe Island, 301-2
Stingray, giant freshwater, 216
    sixgill, 214
    tiger, 365
Stiphrornis sanghensis, 160
Storm petrel, New Zealand, 291
    Puerto Montt mystery, 291-2
    Rapa, 128
Strigops habroptilus, 137
Stromatospongia norae, 226
    vermicola, 226
Strombus taurus, 242
Stygiomedusa gigantea, 242
Stymphalornis acutirostris, 158
Sugarcane monster, 249
Sunbird, Rockefeller's, 131
Sus bucculentus, 95
    salvanius, 71-2
    scrofa riukiuanus, 69
Swallow, Red Sea cliff, 148
    white-tailed, 127
Swift, Schouteden's, 131
Swima bombiviridis, 309
Symbion americanus, 267
    pandora, 266-7, 308

Synalpheus regalis, 267
Syrmaticus mikado, 117-8

Tachyeres leucocephalus, 144
Tadorna cristata, 119-20
Tagua, 74
Taiko, Chatham Island, 141-2
Taipan, coastal, 167-8
    Central Ranges, 168
    inland, 168
Takahe, 115, 128-30
Takin, golden, 65
Tamarin, black-faced lion, 87
    Cruz Lima's saddleback, 285
    eastern saddleback, 285
    emperor, 6, 34
Tanager, Cabanis's, 125
    cherry-throated, 159
Tangara cabanisi, 125
Tantulocarids, 257
Tanzaniophasma subsolanum, 303-4
Tapaculo, lowland, 159-60
Tapeworm, megamouth, 214
Tapir, black dwarf lowland, 284
Tapirus pygmaeus, 284
Tarsier, Diane, 35
    pygmy, 35
Tarsius dianae, 35
    pumilus, 35
Tasmacetus shepherdi, 51-2
Tatu bola, 88
Tauraco ruspolii, 127
Tayra, orange, 285
Teal, Campbell Island flightless, 140-1
Tenrec, Fontoynont's, 44-5
Tethysbaena texana, 229
Tetra, cardinal, 197
    Mexican blind cave, 198
    neon, 12, 197
Teuthidodrilus samae, 309
Thamnophilus praecox, 153
Thaumatichthys axeli, 203
Thaumoctopus mimicus, 263-4
Thermarces andersoni, 252
    cerberus, 252
Thermopalia taraxaca, 250
Thermosbaena mirabilis, 229
Thermosbaenaceans, 229
Thrush, forest ground, 131
Thylogale parma, 69-70
Tiburonia granrojo, 304-5
Tiliqua adelaidensis, 183
Titanus giganteus, 225
Titi, Caquetá (red-bearded), 287
    Coimbra's, 112
    Golden Palace, 280-1
    Guariba-Roosevelt Extractive Reserve, 287
    Hoffmanns's, 100
    Prince Bernhard's, 103
    Rio Mamurú, 285
    Rio Purús collared, 285
    Stephen Nash's, 103

Titi *(cont.)*, Upper Rio Xingú, 285
Toad, black, 170
　Blomberg's giant, 172-3
　Mallorcan midwife, 178
Tokoeka, 154
Tolypeutes tricinctus, 88
Tortoise, Abingdon Island giant, 174-6
　Arnold's giant, 186
　Bolson, 173
　Mary River, 185
　Narborough Island giant, 176
　pancake, 162
　Western snake-necked, 173
Touraco, Prince Ruspoli's, 127
Tragelaphus buxtoni, 35-7, 352
Tragulus versicolor, 277
Tree kangaroo, New Britain mystery, 106
Tree-nymph, Tambusisi, 256
Trichechus pygmaeus, 284
Trichobatrachus robustus, 161-2
Trichocichla rufa, 291
Trichoplax adhaerens, 246-7
Trigona hypogea, 61
　necrophaga, 62
Triplophysa geijuensis, 201
Troglodiplura lowryi, 248
Tuatara, Gunther's, 182
Tuoa, 99
Tursiops australis, 272, 288
Turtle, Gulf snapping, 185
　Hoan Kiem Lake, 188-9
　Indian cave, 178-9
　Irwin's, 2
　yellow-headed temple, 162
Typhlogarra widdowsoni, 201
Typhlophis ayarzaguenai, 184
Tyto soumagnei, 158-9

Uakari, Ayres's, 287
　Neblina, 287
　Rio Pauiní white bald-headed, 285
Ufiti, 63-4

Vaceletia crypta, 254-5
Vampyroteuthis infernalis, 10, 235-6

Varanus auffenbergi, 190
　bitatawa, 297
　caerulivirens, 189
　komodoensis, 2, 161, 163-5
　melinus, 190
　obor, 297
　olivaceus (=*grayi*), 177
　telenestes, 189
　teriae, 189
　vietnamensis, 189-90
　yemenensis, 181-2
　yuwonoi, 190
Vestimentiferans, 235
Vijayachelys silvatica, 178-9
Viper, Queimada Grande, 166-7
Viverra tainguensis, 112-3
Vole, Bavarian pine, 276-7
Vontsira, Durrell's, 288
Vulture, greater yellow-headed, 136

Wallaby, Proserpine rock, 76
　purple-throated rock, 274-5
　white-throated (parma), 69-70
Warbler, Aldabra brush-, 138
　Fijian long-legged, 291
　large-billed reed, 293
Warthog, desert, 88
Water flea, hairy, 305
Water scorpion, Movile Cave blind, 265
Weasel, Colombian (aquatic), 59
Weaver, golden-naped, 152
Whale, beaked (see Beaked whale), 35, 40, 44, 51-2, 61-2, 67, 86, 108, 275
　Eden's, 276
　killer (see Killer whale), 59, 366
　North Pacific right, 273
　Omura's, 276
Whip-poor-will, Puerto Rican, 133
Whipsnake, St. Vincent, 182
Wolf, African, 289
　Andean, 58-9
Wombat, Deniliquin, 80
Woodpecker, helmeted, 152
　imperial, 364

Woodpecker *(cont.)*, ivory-billed, 149-51
Woodrail, New Caledonian, 144
Worm, devil, of hell, 309
　giant tubicolous vent, 250
　green bomber, 309
　methane ice, 267-8
　Pompeii, 252
　spaghetti, 250
　squid, 309
Worms, beard, 234-5
　zombie, 305-6
Wren, Zapata, 56
Wunderpus, 264
Wunderpus photogenicus, 264
Wyulda squamicaudata, 42-3

Xanthopan morgani praedicta, 223
Xenoglaux loweryi, 140
Xenonetta nesiotis, 140-1, 159
Xenoperdix obscurata, 155
　udzungwensis, 154-5
Xenophidion acanthognathus, 186
　schaeferi, 186
Xenosaur, flat-headed, 172
　Newman's, 172
　pallid, 172
　Rackham's, 172
Xenosaurus newmanorum, 172
　platyceps, 172
　rackhami, 172
　rectocollaris, 172
Xenoturbella bocki, 308
　westbladi, 308
Xyloplax janetae, 260
　medusiformis, 259-60
　turnerae, 260

Zaglossus attenboroughi, 112
　bartoni, 111-2
Zavattariornis stresemanni, 125
Zoothera oberlaenderi, 131
Zorapterans, 226
Zorotypus gurneyi, 226
Zyzomys pedunculatus, 107

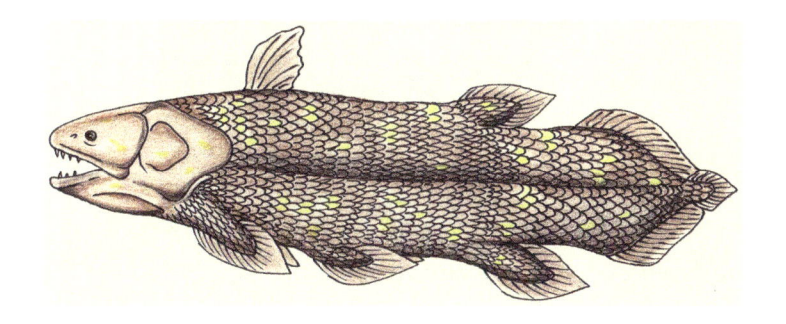

In 1998, scientists were stunned to learn that a second species of coelacanth existed in the waters around Sulawesi, Indonesia, over 6000 miles east of the Comoros and southern Africa (the only previously known coelacanth provenance), and instead of being blue with white blotches like the African species was brown with golden spangles (Tim Morris).

Goeldi's monkey. (Tony Hisgett)

Allen's swamp monkey. (Michael Nordine)

STOP PRESS

SMALL BUT SIGNIFICANT—WALTER'S DUIKER
Until 2010, only two species of duiker within the genus *Philantomba* were recognised, but now a third species has been described and named—*P. walteri*, Walter's duiker. This very small African antelope (standing less than 16 in at the shoulder) is native to Togo, Benin, and Nigeria, and was distinguished from Maxwell's duiker *P. maxwelli* by morphological and DNA studies. It is named after the late Prof. Walter Verheyen from Antwerp University, who was the first scientist to collect a specimen when he encountered it in a bushmeat market at Badou, Togo, in 1968.

*Colyn, M. et al. (2010). Discovery of a new duiker species (Bovidae: Cephalophinae) from the Dahomey Gap, West Africa. *Zootaxa*, no. 2637: 1-30.

SIBREE'S DWARF LEMUR REDISCOVERED
In 1896, a specimen of dwarf lemur found in an eastern Madagascan rainforest and sent to London's Natural History Museum was deemed to represent a hitherto-undescribed species, which was dubbed Sibree's dwarf lemur *Cheirogaleus sibreei*, honouring English missionary-naturalist James Sibree. No further specimen, or even a confirmed sighting, emerged for over a century, and the rainforest where the lone specimen had been found was destroyed, inciting fears that this primate was extinct. In 2001, however, unusual dwarf lemurs were observed by McGill University zoologist Dr. Mitchell Irwin at Tsinjoarivo, and in 2006 specimens were captured at several different sites there. Tests and comparisons with the Natural History Museum's type specimen confirmed that these were Sibree's dwarf lemurs and also that it was indeed a valid, very distinct species—as formally documented by Irwin and co-workers in 2010.

Groeneveld, L.F. et al. (2010). MtDNA and nDNA corroborate existence of sympatric dwarf lemur species at Tsinjoarivo, eastern Madagascar. *Molecular Phylogenetics and Evolution*, 55(3): 833-45.

LAOTIAN LIVING FOSSIL FOUND IN VIETNAM TOO
In September 2011, Fauna and Flora International scientists announced that the kha-nyou or Laotian rock rat *Laonastes aenigmamus*—an extraordinary mammalian 'living fossil' whose discovery in Laos a decade earlier had resurrected an entire taxonomic family of rodents from 11 million years of presumed extinction—also existed in Vietnam. A specimen had been captured in Quang Binh Province's Minh Hoa District, where the scientists learnt that the local Ruc people trap these animals for their meat.

Anon. (2011). Rodent thought to be extinct found in Vietnam. http://en.www.info.vn/science-and-education/nature-and-animals/35142-rodent-thought-to-be-extinct-found-in-vietnam.html 7 September.

JAVAN RHINOCEROS NOW EXTINCT IN MAINLAND ASIA?
The last-known Javan rhinoceros *Rhinoceros sondaicus* living in Vietnam was killed by poachers for its horn in April 2011, following a sustained onslaught by poachers on this country's tiny population of Javan rhinos ever since the species' rediscovery here in 1988. Now, the only surviving specimens, thought to number no more than 60 individuals, are confined to the Ujung Kulon National Park in Java, Indonesia, with the species seemingly extinct in mainland Asia.

Ives, M. (2011). Last Javan rhino in Vietnam killed for horn. Associated Press report, 25 October.

FERRETING OUT A NEW FERRET-BADGER
Although related to badgers, as indicated by their name the ferret-badgers of southeastern Asia are more slender, with much more elongate, ferret-like faces. Belonging to the genus *Melogale*, until very recently only four species were recognised, but in late November 2011 a fifth was formally described and named. Known as the Vietnam ferret-badger *M. cucphuongensis*, the first scientifically-documented specimen was encountered back in March 2005, within central Vietnam's Cuc Phuong National Park. It was discovered in an injured state, and after it died its remains were not retained, but when a second, newly-deceased specimen was found here in January 2006, studies on it were duly conducted. This new ferret-badger species is readily distinguished externally from the previously-recorded quartet by virtue

of its brown head and body plus the distinctive black and white stripe running from its neck to its shoulders. It also exhibits a number of notable cranial differences. Despite being hitherto unknown to science, the Vietnam ferret-badger is a familiar creature to local people, who refer to it as the *chon bac ma* ('silver-cheeked fox').

*Nadler, T. et al. (2011). A new species of ferret-badger, genus *Melogale*, from Vietnam. *Zoologische Garten*, 80(5): 271-86.

IMPERIAL WOODPECKER REDISCOVERED . . . ON FILM

The size of a raven, Mexico's imperial woodpecker *Campephilus imperialis*—a close relative of the ivory-billed woodpecker *C. principalis* (see p. 149)—is the world's largest woodpecker species. Despite a number of unverified reports during the past decade, however, the last confirmed record of this spectacular species came in spring 1956, with the unequivocal observation of a female specimen in southern Durango by dentist and amateur ornithologist Dr. William L. Rhein. Since then, it has popularly been deemed extinct, although it is just possible that a few birds linger undetected in relict patches of Mexican woodland, and searches for it continue. In October 2011, however, the next best thing to a sighting of a living imperial woodpecker occurred—via the public release by Cornell University's Cornell Lab of Ornithology of the only known film footage of this species. The film was a home movie shot in 16-mm colour format by Dr. Rhein during his 1956 observation of the above-mentioned female bird, and was tracked down in the 1990s by woodpecker expert Dr. Martjan Lammertink from the Cornell Lab. Dr. Rhein died in 1999, but in 2005 one of his nephews, Ronald Thorpe, donated the film footage to Cornell, after which it was digitally restored before finally being released for public viewing in 2011. A year earlier, Dr. Lammertink had visited the exact site where Rhein had shot his film 54 years previously, in the hope of locating imperial woodpeckers there, but none was found. Just 1 minute 35 seconds long, Rheim's historic footage can be viewed on YouTube at:

http://www.youtube.com/watch?v=Q0OCd6b1aXU

Lammertink, M. et al. (2011). Film documentation of the probably extinct imperial woodpecker (*Campephilus imperialis*). *Auk*, 128(4): 671-7.

NILE CROCODILE—WHEN ONE BECOMES TWO

In 2011, Fordham University zoologist Dr. Evon Hekkala and her team of co-researchers revealed they had discovered that the world's best-known crocodile, the Nile crocodile *Crocodylus niloticus*, was not one species but two, and that these two newly-delineated species were actually only distantly related to each other. The team had sequenced the genes of 123 living Nile crocodiles

Imperial woodpecker. (John L. Ridgway)

and 57 museum specimens, including several 2,000-year-old Egyptian crocodile mummies, and the results confirmed that in addition to the larger, more savage, familiar Nile crocodile, a second, smaller, tamer species could be readily discerned. Moreover, all of the crocodile mummies examined proved to belong to this latter form, thereby corroborating writings of the Greek historian Herodotus, who had claimed that in their ceremonies the ancient Egyptians had selectively utilised a smaller, tamer crocodile than the Nile crocodile and had considered it sacred. Reviving a name first coined in relation to that intriguing crocodile form by French naturalist Geoffroy Saint-Hilaire in 1807 but subsequently abandoned, Hekkala and her team plan to christen this species *C. suchus*, the sacred crocodile. Probably due to out-competition by the bigger, more aggressive *C. niloticus*, *C. suchus* no longer exists in the Nile region, being confined now to parts of western Africa.

Yong, K. (2011). Nile crocodile is two species. http://www.nature.com/news/2011/110914/full/news.2011.535.html 14 September.

PALESTINIAN PAINTED FROG REAPPEARS

After more than five decades since the last confirmed record of its existence, the Palestinian painted frog *Discoglossus nigroventer* (see p. 171) was rediscovered on Tuesday 15 November 2011 when a mysterious frog was encountered by nature reserve warden Yoram Malka during a routine patrol of Hula Lake. Photographs of the living specimen, a female, were taken, and when amphibian expert Dr. Sarig Gafni from Ruppin Academic Center's School of Marine Sciences arrived to examine the frog in the field, he confirmed that it belonged to this 'lost' species. Its lone survivor has now been taken into captivity for safe-keeping at least for the present time, but scientists plan to search the region in case other specimens exist here.

Rinat, Z. (2011). Long thought extinct, Hula painted frog found once again in Israeli nature reserve. http://www.haaretz.com/print-edition/news/long-thought-extinct-hula-painted-frog-found-once-again-in-israeli-nature-reserve-1.396000 17 November 2011.

SPOTTING A NEW FRESHWATER STINGRAY

Despite being a familiar species in the international pet trade for many years (and especially popular in Japan and China), it was only as recently as 2011 that the tiger stingray, native to the upper Amazon River Basin of Peru, finally received a formal scientific description and name—*Potamotrygon tigrina*. Yet despite its monickers, it is boldly marked not in stripes but in spots, because 'tiger' is the name given in much of Latin America to the jaguar, whose rosette-dappled coat is reminiscent of the patterning exhibited by *P. tigrina*. However, the reason for this particular river stingray's dramatic markings is currently unknown.

*de Carvalho, M.R. et al. (2011) *Potamotrygon tigrina*, a new species of freshwater stingray from the upper Amazon basin, closely related to *Potamotrygon schroederi* Fernandez-Yépez, 1958 (Chondrichthyes: Potamotrygonidae). *Zootaxa*, no. 2827: 1–30.

SECOND POPULATION OF AFRICAN COELACANTHS DISCOVERED

Until now, all specimens of the coelacanth *Latimeria chalumnae* discovered off the south and eastern coasts of Africa have belonged to the same genetic population. In November 2011, however, a team of researchers from Tokyo Institute of Technology led by Prof. Norihiro Okada announced that genetic analyses of more than 20 specimens caught off northern Tanzania and other sites revealed that a second breeding population existed, so genetically distinct from the only previously known one that the researchers believe the two populations diverged 200,000 to 2 million years ago.

Anon. (2011) 2nd coelacanth population found off Tanzania coast. *Daily Yomiuri Online*. http://www.yomiuri.co.jp/dy/features/science/T111108004421.htm 10 November 2011.

LEAPING COCKROACHES!
IT'S THE HIGH-JUMPING LEAPROACH

Jumping cockroaches were thought to have died out during the late Jurassic Period, over 140 million years ago—until 2010, that is, when the first-known living species was formally described. Scientifically christened *Saltoblatella montistabularis* and belonging to a new genus, it is commonly known as the leaproach, and was discovered in the Silvermine Nature Reserve within South Africa's Table Mountain National Park ('*montistabularis*' translates as 'Table Mountain'). Its legs are highly modified for jumping, so much so that its jumping abilities compare favourably with those of grasshoppers.

*Bohn, H. et al. (2010). A jumping cockroach from South Africa, *Saltoblattella montistabularis*, gen. nov., spec. nov. (Blattodea: Blattellidae). *Arthropod Systematics and Phylogeny*, 68(1):53-9.

NEPTUNE'S CUP SPONGE
REDISCOVERED IN SINGAPORE WATERS

Discovered by science in 1822, and originally found off the coasts of Singapore and elsewhere in southeast Asia, the aptly-named Neptune's cup sponge *Cliona patera* grew so large—often exceeding 3 ft in height and also in diameter—that specimens were sometimes used as bathtubs for babies! Tragically, however, because of its usefulness this spectacular demosponge was overharvested; by the 1870s, it had vanished off Singapore, and the last-known living specimens anywhere were recorded off western Java in 1908—since when this remarkable species has been deemed extinct. During the 1990s, however, a few dead specimens were discovered in Australian waters, and now, finally, a couple of small living individuals have been found, both of them near southern Singapore's St. John's Island. The first one was spotted in March 2011 by biologists working with the environmental engineering firm DHI Water & Environment (S) Pte Ltd, and the second was found close by later. Pale yellow to white and resembling shallow bowls standing on sturdy stalks, they were confirmed as immature Neptune's cup sponges by poriferan expert Lim Swee Cheng, from the Tropical Marine Science Institute of the National University of Singapore. The biologists now hope to find more specimens in this area.

Platt, John R. (2011) Amazing Neptune's cup sponge rediscovered in Singapore. Extinction Countdown blog. http://blogs.scientificamerican.com/extinction-countdown/2011/11/17/amazing-neptunes-cup-sponge-rediscovered-singapore/ 17 November 2011.

SECOND SPECIES OF YETI LOBSTER REVEALED

A new, second species of yeti lobster, *Kiwa puravida*, has been described, based upon specimens first discovered and collected in June 2006 (as well as on several later occasions) around a Costa Rican methane seep at a hydrothermal vent on the ocean floor. Of especial interest is that apparently this crustacean actively harvests or farms for food the bacteria growing upon the dense hair-like setae clothing its pincers, by waving its pincers in fluid escaping from the methane seep. Its research team has conjectured that this activity increases the productivity of the bacteria, and, in so doing, increases the yeti lobster's food source.

*Thurber, A.R. et al. (2011). (2011) Dancing for food in the deep sea: bacterial farming by a new species of yeti crab. *PLoS ONE*, 6 (no. 11; 30 November): e26243. doi:10.1371/journal.pone.0026243

Kiwa puravida. (Thurber et al, *PLoS* 2011)

This killer whale calf, accompanied by NOAA marine ecologist Lisa Ballance, is part of a study to determine if there are new species of killer whales in the Antarctic. (NOAA Photo Library)

AUTHOR BIOGRAPHY

Dr. Karl P.N. Shuker BSc PhD FRES FZS is a zoologist who is internationally recognised as a world expert in cryptozoology (the scientific investigation of mystery animals whose existence or identity has yet to be formally ascertained), as well as in animal mythology and allied subjects relating to wildlife anomalies and inexplicabilia. He obtained a BSc (Honours) degree in pure zoology at the University of Leeds (U.K.), and a PhD in zoology and comparative physiology at the University of Birmingham (U.K.). He is now a freelance zoological consultant and writer, living in the West Midlands, England. His website can be accessed at: KarlShuker.com and his award-winning ShukerNature blog can be accessed at http://www.karlshuker.blogspot.com. There is also a detailed entry for him online in Wikipedia (at: http://en.wikipedia.org/wiki/Karl_Shuker).

Dr Shuker is the author of numerous cryptozoological articles and books, and is also a published poet. His books are: *Mystery Cats of the World* (Robert Hale: London, 1989), *Extraordinary Animals Worldwide* (Robert Hale: London, 1991), *The Lost Ark: New and Rediscovered Animals of the 20th Century* (HarperCollins: London, 1993), *Dragons: A Natural History* (Aurum Press: London/Simon & Schuster: New York, 1995), *In Search of Prehistoric Survivors* (Blandford Press: London, 1995), *The Unexplained: An Illustrated Guide to the World's Natural and Paranormal Mysteries* (Carlton Books: London/JG Press: North Dighton, 1996), *From Flying Toads To Snakes With Wings* (Llewellyn Publications: St Paul, Minnesota, 1997), *Mysteries of Planet Earth: An Encyclopedia of the Inexplicable* (Carlton Books: London, 1999), *The Hidden Powers of Animals: Uncovering the Secrets of Nature* (Reader's Digest: Pleasantville/Marshall Editions: London, 2001), *The New Zoo: New and Rediscovered Animals of the Twentieth Century* (House of Stratus: Thirsk, 2002), *The Beasts That Hide From Man: The Search For the World's Last Undiscovered Animals* (Paraview: New York, 2003), *Extraordinary Animals Revisited* (CFZ Press: Bideford, 2007), *Dr Shuker's Casebook: In Pursuit of Marvels and Mysteries* (CFZ Press: Bideford, 2008), *Dinosaurs and Other Prehistoric Animals on Stamps: A Worldwide Catalogue* (CFZ Press: Bideford, 2008), *Star Steeds and Other Dreams: The Collected Poems* (CFZ Press: Bideford, 2009), and *Karl Shuker's Alien Zoo: From the Pages of Fortean Times* (CFZ Press: Bideford, 2010).

Dr. Shuker has acted jointly as consultant and major contributor to four multi-author volumes dealing with cryptozoology and other mysterious phenomena. These are: *Man and Beast* (Reader's Digest: Pleasantville, New York, 1993), *Secrets of the Natural World* (Reader's Digest: Pleasantville, New York, 1993), *Almanac of the Uncanny* (Reader's Digest: Surry Hills, Australia, 1995), and *Chambers Dictionary of the Unexplained* (Chambers: London, 2007). He also acts Senior Life Sciences Consultant to *The Guinness Book of Records/Guinness World Records* (Guinness: London, 1997-present day), and was the consultant to *Monsters* (Lorenz Books: London, 2001), as well as a contributor to *Mysteries of the Deep* (Llewellyn: St Paul, 1998), *Guinness Amazing Future* (Guinness: London, 1999), and *The Earth* (Channel 4 Books: London, 2000). In addition, Dr Shuker has contributed to, and features in, *Of Monsters and Miracles* (Croydon Museum Services & Interactive Designs Ltd: Oxton, 1995), a multi-contributor CD ROM providing an interactive voyage through the fascinating realm of mysteries and the unexplained.

Aside from books, Dr. Shuker has acted as a Contributing Editor and cryptozoological columnist for *Strange Magazine*, and is a longstanding contributor to *FATE Magazine*, both published in the USA In the UK, he is a Special Correspondent for, and also has two cryptozoological columns in, *Fortean Times*; and regularly contributes to many other publications.

Dr. Shuker appears regularly on television and radio, has acted as a consultant for the Discovery TV series *Into the Unknown* and also for the Sony Pictures 'faction' documentary *The Last Dragon* (released in 2005), as a question setter for the BBC's cerebral quiz show *Mastermind*, and has travelled throughout the world during the course of his researches, most recently visiting Easter Island to research the famous moai or giant stone statues for future writings. He is a Scientific Fellow of the prestigious Zoological Society of London, a Fellow of the Royal Entomological Society, an affiliate of several other wildlife-related organisations, the Cryptozoological Consultant and West Midlands Representative for the Centre for Fortean Zoology, and is also a Member of the Society of Authors.

Outside zoology, Dr. Shuker's interests include motorbikes, vintage rock 'n' roll music, the life and career of James Dean, the history of animation, collecting wildlife postage stamps, collecting Venetian masquerade masks and other masks from around the world, global travel to unusual localities, and quizzes. In 2001, Dr. Shuker won £250,000 on the original British edition of the television quiz show *Who Wants To Be A Millionaire*; and in 2005 he was honoured by the naming of a new species of loriciferan invertebrate after him–*Pliciloricus shukeri*–in recognition of the very considerable contribution of *The Lost Ark: New and Rediscovered Animals of the 20th Century*–the first edition of this present book–to the zoological literature.

The author with a life-sized model of a Komodo dragon at Chester Zoo, England. (Dr. Karl P.N. Shuker)

Coachwhip Publications
CoachwhipBooks.com

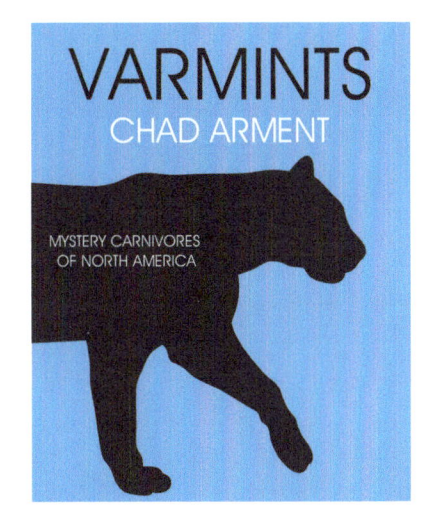

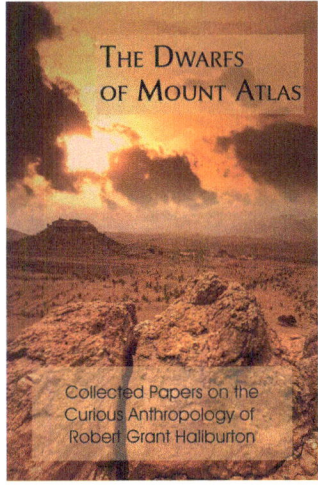

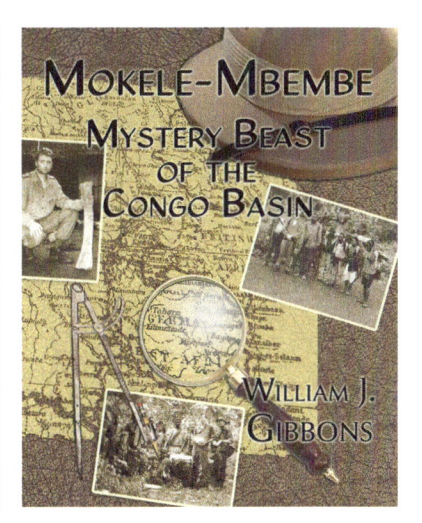

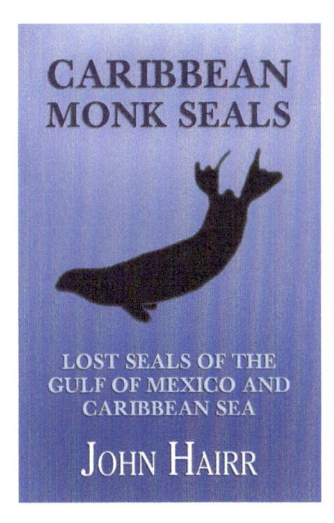

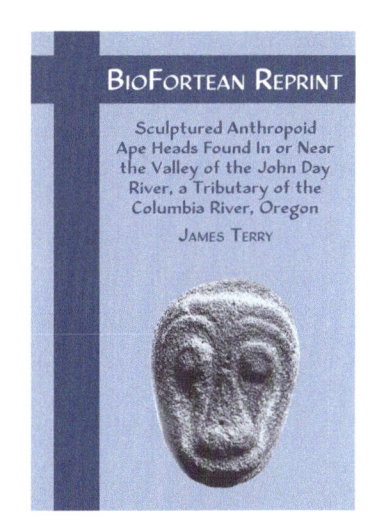

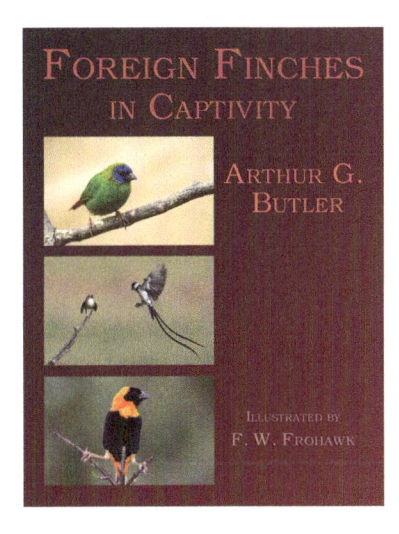

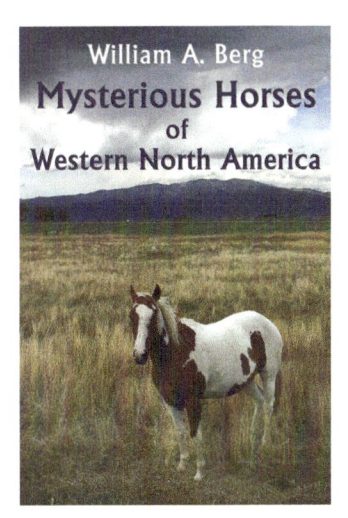

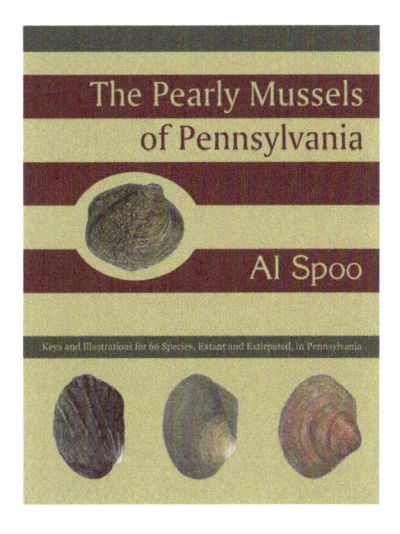